MEMBRANE TECHNOLOGY

AND INDUSTRIAL SEPARATION TECHNIQUES

MEMBRANE TECHNOLOGY AND INDUSTRIAL SEPARATION TECHNIQUES

Peter R. Keller

NOYES DATA CORPORATION
Park Ridge, New Jersey London, England
1976

Library of Congress Catalog Card Number: 76-2325
ISBN: 0-8155-0621-X
Printed in the United States

Published in the United States of America by
Noyes Data Corporation
Noyes Building, Park Ridge, New Jersey 07656

FOREWORD

The detailed, descriptive information in this book is based on U.S. patents issued since 1960 and relating to the technology of membranes and industrial separation techniques. Included also are one published patent application and one reissue of a patent published earlier.

This book serves a double purpose in that it supplies detailed technical information and can be used as a guide to the U.S. patent literature in this field. By indicating all the information that is significant, and eliminating legal jargon and juristic phraseology, this book presents an advanced, technically oriented review of membrane technology and industrial separation techniques.

The U.S. patent literature is the largest and most comprehensive collection of technical information in the world. There is more practical, commercial, timely process information assembled here than is available from any other source. The technical information obtained from a patent is extremely reliable and comprehensive; sufficient information must be included to avoid rejection for "insufficient disclosure." These patents include practically all of those issued on the subject in the United States during the period under review; there has been no bias in the selection of patents for inclusion.

The patent literature covers a substantial amount of information not available in the journal literature. The patent literature is a prime source of basic commercially useful information. This information is overlooked by those who rely primarily on the periodical journal literature. It is realized that there is a lag between a patent application on a new process development and the granting of a patent, but it is felt that this may roughly parallel or even anticipate the lag in putting that development into commercial practice.

Many of these patents are being utilized commercially. Whether used or not, they offer opportunities for technological transfer. Also, a major purpose of this book is to describe the number of technical possibilities available, which may open up profitable areas to research and development. The information contained in this book will allow you to establish a sound background before launching into research in the field.

Advanced composition and production methods developed by Noyes Data are employed to bring our new durably bound books to you in a minimum of time. Special techniques are used to close the gap between "manuscript" and "completed book." Industrial technology is progressing so rapidly that time-honored, conventional typesetting, binding and shipping methods are no longer suitable. We have bypassed the delays in the conventional book publishing cycle and provide the user with an effective and convenient means of reviewing up-to-date information in depth.

The Table of Contents is organized in such a way as to serve as a subject index. Other indexes by company, inventor and patent number help in providing easy access to the information contained in this book.

15 Reasons Why the U.S. Patent Office Literature Is Important to You –

1. The U.S. patent literature is the largest and most comprehensive collection of technical information in the world. There is more practical commercial process information assembled here than is available from any other source.

2. The technical information obtained from the patent literature is extremely comprehensive; sufficient information must be included to avoid rejection for "insufficient disclosure."

3. The patent literature is a prime source of basic commercially utilizable information. This information is overlooked by those who rely primarily on the periodical journal literature.

4. An important feature of the patent literature is that it can serve to avoid duplication of research and development.

5. Patents, unlike periodical literature, are bound by definition to contain new information, data and ideas.

6. It can serve as a source of new ideas in a different but related field, and may be outside the patent protection offered the original invention.

7. Since claims are narrowly defined, much valuable information is included that may be outside the legal protection afforded by the claims.

8. Patents discuss the difficulties associated with previous research, development or production techniques, and offer a specific method of overcoming problems. This gives clues to current process information that has not been published in periodicals or books.

9. Can aid in process design by providing a selection of alternate techniques. A powerful research and engineering tool.

10. Obtain licenses — many U.S. chemical patents have not been developed commercially.

11. Patents provide an excellent starting point for the next investigator.

12. Frequently, innovations derived from research are first disclosed in the patent literature, prior to coverage in the periodical literature.

13. Patents offer a most valuable method of keeping abreast of latest technologies, serving an individual's own "current awareness" program.

14. Copies of U.S. patents are easily obtained from the U.S. Patent Office at 50¢ a copy.

15. It is a creative source of ideas for those with imagination.

CONTENTS AND SUBJECT INDEX

INTRODUCTION

Semipermeable membranes have recently gained commercial importance as a result of their employment in the treatment of aqueous solutions in order that water having greatly reduced amounts of impurities is caused to flow through the semipermeable membrane. These treatments are carried out using high pressure, above the osmotic pressure of the solution being treated, which causes water flux through the membrane to occur, and has been commonly referred to by the term "reverse osmosis."

Semipermeable membranes are now being used in the reverse osmosis treatment of aqueous solutions either for production of relatively pure water or for concentration of the solution being treated or both. Such semipermeable membranes have been produced from various polymeric materials, such as esters of cellulose. More recently, semipermeable membranes of this type have been fabricated from polyamides, polyimides, polyphenyl esters, polysulfonamides, polybenzimidazoles, polyarlyene oxides, polyvinyl methyl ether, and other polymeric organic materials.

In general, cellulose derivatives, such as cellulose acetate, cellulose acetate butyrate, cellulose propionate, ethylcellulose and other cellulose esters and mixed esters, have found ready use in fabricating so-called asymmetric membranes where the semipermeability of the membrane results from a thin, dense skin located at one surface of the membrane, with the remaining major part of the membrane providing a more porous supporting layer formed of the same polymer.

Somewhat more recently, a great deal of work has been done in developing what are now known as "composite" membranes, where semipermeability results from an ultrathin polymeric film of one material which is formed separately and is supported on an underlying porous substrate that may be made from a different polymer than the ultrathin film.

Membranes for carrying out hyperfiltrations have very diversified applications, including for instance:

(1) The recovery of water from saline solutions, e.g.,

the desalination of seawater;
the demineralization of fresh water;
the purification of industrial effluents; and
the purification of rivers and waterways.

(2) The recovery of small quantities of dissolved or colloidally dispersed sub-

stances from solutions containing the same, e.g.,

the recovery of metals from industrial effluents; and
the concentration of hormones, vitamins, inoculants and vaccines, etc.

(3) The concentrations of solutions or dispersions of available products that are thermally and/or chemically unstable or fugitive, e.g.,

the concentration of fruit and vegetable juices;
the concentration of sugar solutions;
the preparation of beverage extracts (milk, coffee, tea, etc.); and
the concentration of sera.

(4) The separation and purification of macromolecular or colloidal substances from solutions containing low molecular weight impurities, e.g.,

the purification of blood;
the use in artificial kidney machines; and
the separation of alkaloids, amino acids, glycosides, etc.

This book presents over 280 processes as described in the U.S. patent literature during the past fifteen years. Overall, hundreds of formulations, casting procedures, fabrication techniques and end use applications are described for every major industrial phase of membrane technology.

The first four chapters are devoted to the preparation of membranes, separators and modular units employing cellulosics and a wide variety of other polymeric materials. While largely described for use in desalination processes by reverse osmosis, it is obvious that many of these systems will find broad commercial utility as this rapidly growing area of technology matures over the next decade.

Ion exchange membranes are described in the fifth chapter and some specific industrial membrane separation processes relating to gas and liquid separation, catalyst and heavy metal removal and a very active area, hemodialysis and other medical applications, are described in the final three chapters.

CELLULOSIC MEMBRANES–REVERSE OSMOSIS

SOLVENTS AND ADDITIVES FOR CASTING BATHS

Basic Process

A process described by *S. Loeb and S. Sourirajan; U.S. Patent 3,133,132; May 12, 1964; assigned to The Regents of the University of California* is directed to a porous membrane capable of desalinizing brine solutions at much higher flow rates of the desalinized water than has been found possible by use of porous membranes previously known. The membranes of this process have been found capable of reducing the concentration of a 5.25% sodium chloride solution to 500 parts per million in a single pass at a flow rate on the order of 8 gallons per square feet of membrane surface per 24-hour day under a pressure differential of 1,500 psig, the membranes having been approximately 1½" and 4¼" in diameter, respectively, and approximately 0.004" in thickness. The membranes receive a permanent set or strain such that their final thickness may be 0.0025".

The process involves the preparation of a film-casting solution from which a film is cast and then treated and prepared as described. To provide an understanding generally of the method of preparing the films or membranes, it may be pointed out at the outset that the film-casting solution from which the films are cast contains a pore-producing agent (i.e., this agent produces a structure which allows appreciable rate of passage of fresh water under suitable condition) of the nature of aqueous perchlorate salt solution in a suitable ratio to a film material such as cellulose acetate.

Magnesium perchlorate is preferred but other inorganic perchlorate salts may be utilized, such as sodium perchlorate salts. A solvent, preferably acetone, is added to prevent the solution from becoming too viscous. Methyl ethyl ketone, ethyl alcohol and/or methyl alcohol may also be used but not as efficaciously, as the solvent or solvents. The film material may also be other cellulosic esters and cellulosic derivatives.

Next, film casting and subsequent solvent evaporation into air is accomplished most conveniently and efficiently, but not essentially, at reduced temperatures to reduce the solvent evaporation rate and permit effective initiation of the desired organization of the water-cellulose acetate structure; then, after a predetermined time such that the solvent is not completely evaporated, the film is immersed in water, preferentially but not essentially ice water, to prevent complete air-drying of the film. Complete air-drying has been found to be harmful in that it reduces the desalinizing capacity of the film and is believed to damage the water-cellulose acetate structure which has been initiated.

Finally, the film is preferably heated prior to use to complete the organization of the film for high-flow desalinization of seawater. For certain applications, where the ions to be separated are large relative to those predominant in seawater it would not be necessary to heat the film. The following is a detailed description of a specific preferred example of the formation of the solution and casting, treatment, and preparation of the film in accordance with this process. A solution is prepared containing components as follows.

(A) Aqueous magnesium perchlorate. The concentration of magnesium perchlorate in this aqueous solution may be varied from 5% by weight up to saturation concentration, approximately 50% by weight of $Mg(ClO_4)_2$. The 10% concentration is preferred. As the aqueous concentration of magnesium perchlorate is increased above this figure, the flow rate of the water through the completely treated membrane is progressively reduced. If the water is omitted completely from the casting solution, the flow rate from the subsequently made film is very low, and, in the two tests made, did not appear to be influenced by the amount of magnesium perchlorate present. At aqueous concentrations below 10%, there is inadequate magnesium perchlorate to maintain the desired relation between water and cellulose acetate such that in the film the desired organization can be obtained. The aqueous magnesium perchlorate solution is referred to as MP.

(B) Cellulose acetate containing preferably 54 to 56% by weight of combined acetic acid (CA).

(C) Acetone (A).

The proportions of these materials in the foregoing solution is subject to some variation. However, the region wherein the best results are obtainable is found to be that where the ratio of A to CA is to be between 2:1 and 4:1 by weight; and the ratio of MP to CA is to be between 1:1 and 1:3 by weight. A preferred and representative solution is MP, 11%; CA, 22% and A, 67%. Another representative solution is a mixture of cellulose acetate, acetone, water and magnesium perchlorates in the percentages 22.2, 66.7, 10.0 and 1.1%, respectively. An example of the process follows.

Example: The casting was done on glass plates with runners on the sides, of a thickness, i.e., height, above an associated glass plate to maintain the desired thickness of casting. Preferred runner thicknesses are approximately 0.010" for a film of approximately corresponding thickness. Both solution and the plates on which the film was cast were cooled in a cold box to a temperature of approximately -7.5° to -16°C. The film was cast by means of a doctor blade or knife which rested on the raised runners at the edges of the glass plate. The doctor blade guided the film on, and by the suitably spaced runners, was pulled across the plate in predetermined spaced relationship thereto at a rate such that the entire width of the plate, which was 8" (representatively) was accomplished in 0.17 minute although this time range has been varied from 0.13 to 0.25 minute without adversely affecting the results.

Too slow a period of time permits excessive evaporation. Too fast a sweep time results in imperfect surface formation or even rupture of the film. The doctor knife also served as the back wall of an open-bottomed box, ½" high, filled with 40 cc of casting solution, prior to casting. During casting, 35% of the solution was spread on the plate, leaving enough solution in the box so that the static head available for flow under the knife did not decrease excessively as the knife was pulled from one end of the plate to the other.

After casting the film on the plate, solvent evaporation was allowed to proceed in a cold box at -7.5° to -16°C (optimum -11°C). It was allowed to remain in the cold box at the same temperature for a period of from 2 to 4 minutes to permit evaporation of a major portion of solvent and initial organization of the film after which it was immersed in water at approximately 1° to 5°C. This completed removal of the solvent and probably some of the magnesium perchlorate by diffusion into the water. The time of 2 to 4 minutes between casting and immersion was critical. Too short a time prevented the formation of a firm film. Too long a time prevented the film from having good desalinization qualities.

Therefore, it is clear that the primary function of the immersion is not to extract the materials, but to obtain the film structure necessary for optimum results. The optimum time appeared to be 3 minutes. The optimum permeability structure in the finished membrane apparently need not necessarily be the same for all desalinizing membranes; it can vary depending upon the chemical nature of the membrane and the saline water solution. The thickness of the finished films is not critical in view of the apparent surface nature of the barrier effect. A film of membrane of the order of 0.004" is appropriate.

A preferred technique for heating the films consists in setting them or placing them under water on glass plates which are separated by washers, as of brass, 1/32" thick, that is, with one on top of another. The brass washers merely keep the top glass from bearing directly on the film. The assembly is then heated up on a hot plate. The film after heating in the water bath is left there for a period of time before being taken out to use. The water bath is between 77° and 83°C and the optimum temperature is 82°C. It is preferable to heat the water up to the desired temperature for one-half hour gradually. The water may be put in at 50°C and then heated up to 82°C and held for at least approximately an hour but can be left overnight or longer and then allowed to cool.

The foregoing results in what may be described as an opalescent film in that it is transparent but has a certain characteristic opalescent appearance. It has a porosity, i.e., ratio of open volume to total volume, in the order of 25 to 40%. A significant aspect of the film is that one side is useful whereas the other side is not. The side which is away from the glass plate during casting is the side which must be used against the salt solution. The other side is assembled against a porous plate.

Tests have been carried out in both a laboratory-scale and a bench-scale test cell providing: (1) a constant high pressure on the saline water in contact with one side of the porous medium; (2) a continuous source of flowing or turbulent food saline water to the high pressure side of the porous medium; (3) a valve for the continuous or intermittent removal of concentrated solution to prevent the accumulation of salt concentration in the cell; and (4) a means of exit for the demineralized water to the outside of the cell at atmospheric pressure.

Best results are obtained when the pressure on the concentrated brine is increased to the operating pressure in one or more stages rather than all at once. For example, if the operating pressure will finally be 1,500 psig, it is desirable first to maintain 1,000 psig for ½ hour; then after ½ hour, increase it to 1,500 psig. If desired, it may then be increased to 2,000 psig, at least 1 hour later.

Figure 1.1a shows diagrammatically such a form of apparatus or test cell in which tests were run. The apparatus is shown exploded in Figure 1.1b. In these figures, the apparatus comprises an upper plate **10** and a lower plate **11**. The upper plate has an annular groove **12** in which is an O-ring **13** which seals the plates together. The lower plate has a flange or shoulder **15** which interfits with a corresponding shoulder on the upper plate.

The upper plate has a central opening or bore **16** and a somewhat enlarged counterbore **17** of relatively shallow depth in which is received the membrane or film member **20**. The lower plate has a circular recess or depression **21** of the same diameter as the counterbore so that these spaces when matched form a cylindrical opening as shown. In this space in addition to the membrane **20** is a paper filter or spacer member **22** which is between the membrane and a porous plate or disc **24** which may be of suitable material such as stainless steel.

The paper filter facilitates easy transmission of demineralized water to the porous plate. The porous plate fits against spacer member **25** which is in the bottom of the depression or space **21**. The upper plate has an annular groove **27** in which is disposed a sealing O-ring **28** which seals the peripheral edges of the membrane **20**. Numeral **31** designates an opening in the upper plate for the inlet of the feed water or solution which passes through channel **32** which has a branch **33** leading to the solution reservoir **16**.

FIGURE 1.1: HIGH-FLOW POROUS MEMBRANES

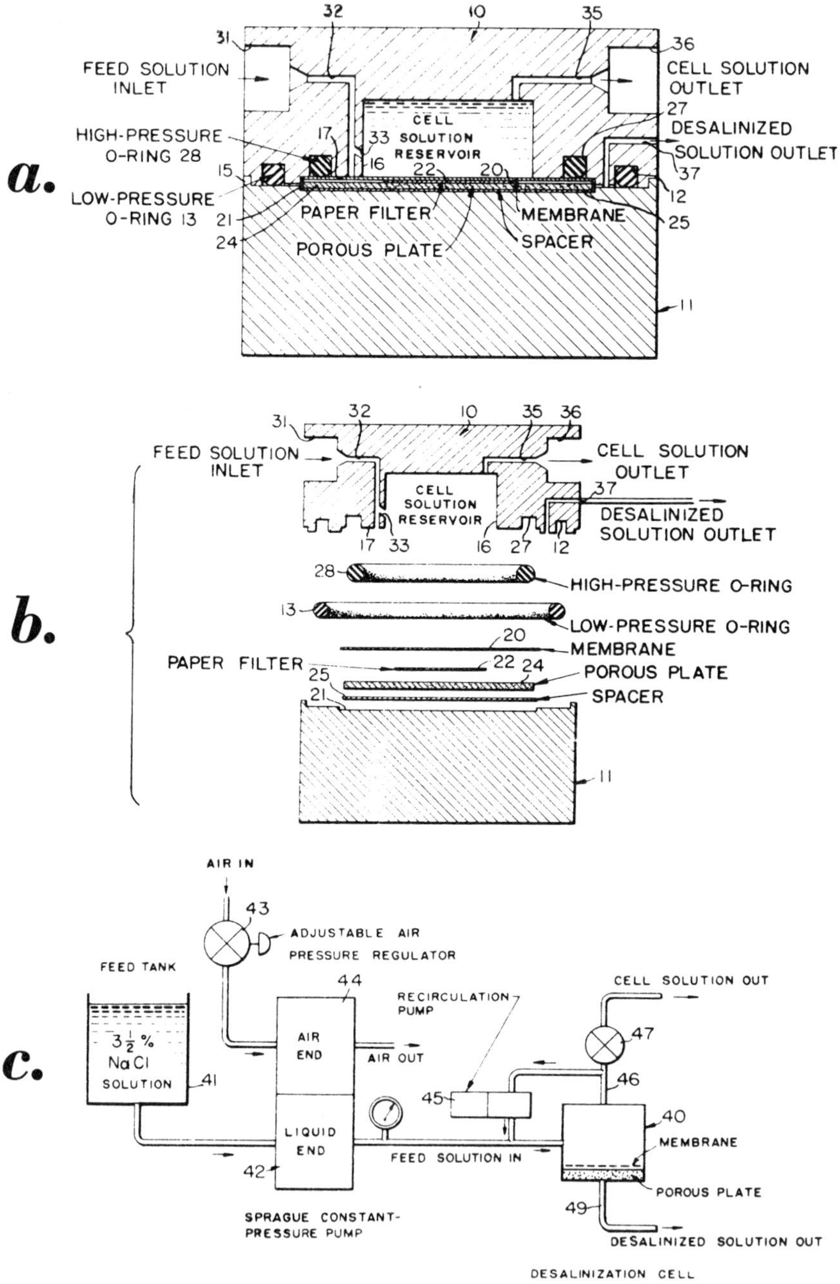

(a)(b) Detail of Filtration Cell
(c) Flow Diagram, Desalinization Assembly (continued)

FIGURE 1.1: (continued)

d.

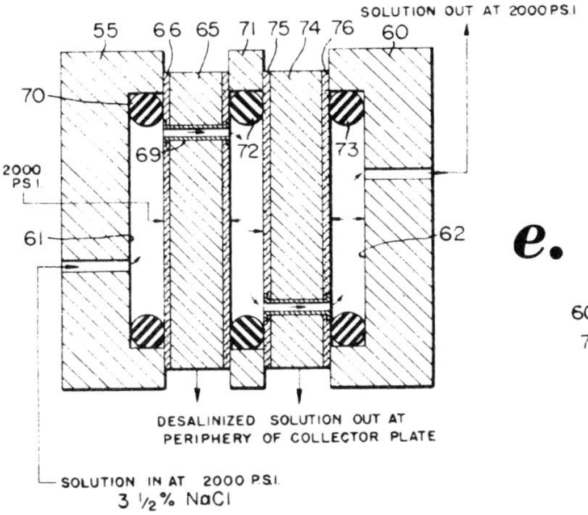

e.

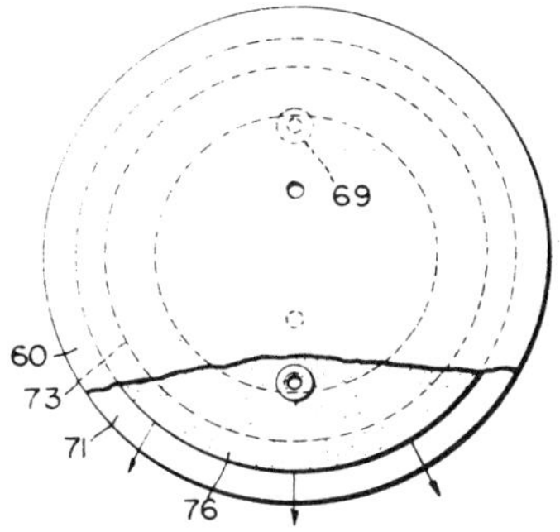

f.

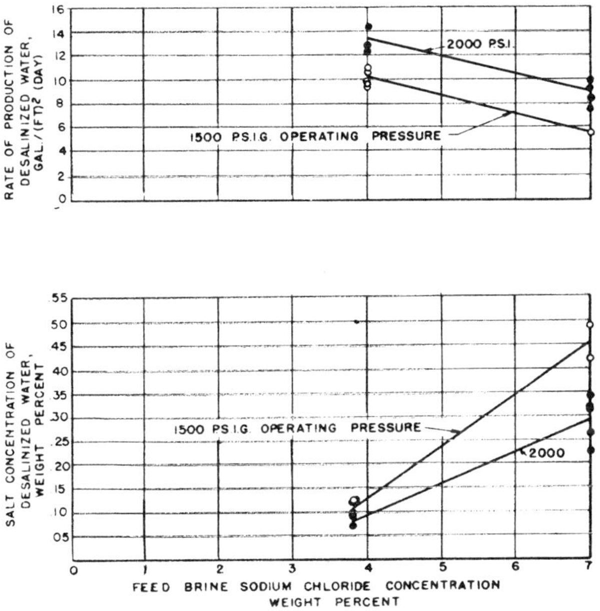

(d) Commercial-Type Cell
(e) End View of Figure 1.1d
(f) Effect of Feed Brine Salt Concentration and Cell Operating Pressure on Salt Content and Flow Rate of Product

(continued)

FIGURE 1.1: (continued)

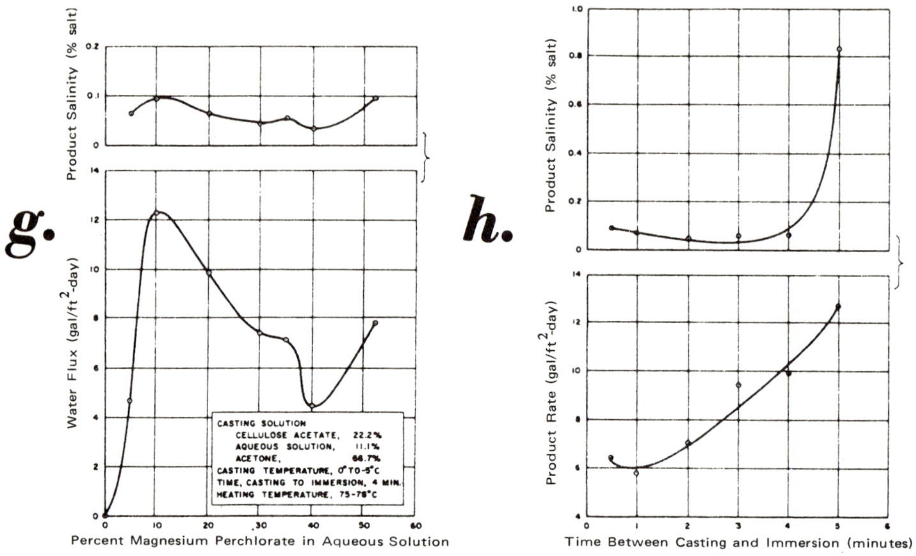

(g) Effect of Magnesium Perchlorate
(h) Effect of Time Intervals Between Casting and Immersion

Source: S. Loeb and S. Sourirajan; U.S. Patent 3,133,132; May 12, 1964

Passage **32** also communicates directly with the membrane **20**. Numeral **35** designates an outlet passage from the reservoir **16** for withdrawing solution from the cell at the outlet **36**. The demineralized water as described filters through the membrane of the cell into the central circular space and to the porous plate **24**. It is drained out of porous plate **24** to the desalinized solution outlet **37**.

Figure 1.1c shows the system utilized with the cell of Figure 1.1a and in which the membranes have been tested. The test cell is shown generally at **40** and a source of 3½% salt solution is shown generally at **41**. The salt solution is pressurized in the test cell by pressure pump means **42** which may be a reciprocating pump **44** driven by air pressure applied through a pressure regulator **43** as shown. Preferably, the solution in the test cell is recirculated by a pump as shown at **45**.

More concentrated solution is withdrawn from the cell through conduit **46** and valve **47**. The desalinized solution is withdrawn from the cell through the outlet **49**. The results summarized in the table on the following page were obtained in a laboratory-scale test cell as shown in Figure 1.1a. Approximately 3.5% solution of sodium chloride in water was used as the saline water in all experiments.

The graphs, Figure 1.1f, illustrate characteristic results of operation of the cell such as shown in Figure 1.1a. Figure 1.1g illustrates the result of varying the percentage content of perchlorate salts, i.e., magnesium perchlorate in aqueous solution. As will be noted above 10% of the rate of fresh water production decreases. Figure 1.1h illustrates the effect of varying the time interval between casting and immersion. The films fabricated in accordance with the method and techniques as described may be used in a commercial or plant type of assembly as shown in Figure 1.1d.

Film No.	Casting Solution Composition, %: $Mg(ClO_4)_2$	H_2O	Cellulose Acetate (398-3)	Acetone	Thickness of Film as Cast, inches	Thermal Treatment Temp., °C	Time Between Casting and Immersion in Water, minutes	Flow Rate*	Desalinized Water Salt Content, percent
3C	4.5	4.5	18.2	72.8	0.015	74	6	20	0.420
3A	4.5	4.5	18.2	72.8	0.015	83	6	10	0.124
9B	4.5	4.5	18.2	72.8	0.015	83	10	12	0.42
A2	4.5	4.5	18.2	72.8	0.010	74	6	17	0.1025
A1	4.5	4.5	18.2	72.8	0.010	83	6	9	0.080
2B	4.5	4.5	18.2	72.8	0.015	74	6	8.9	0.185
2A	4.5	4.5	18.2	72.8	0.015	83	11	4.3	0.070
1A	5.6	5.6	22.2	66.6	0.015	83	6	2.0	0.043
1A	4.5	4.5	18.2	72.8	0.007	74	1	11	0.246
1A	5.6	5.6	22.2	66.6	0.007	83	1	6.8	0.159
2A	5.6	5.6	22.2	66.6	0.007	83	3	4	0.216

*(gal/ft^2-day)

Conditions: casting temperature, 11° to 15°C; immersion water temperature, 30° to 31.5°C; time at thermal treating temperature, 1 hr; operating pressure, 1,500 psig; and feed solution, 3.5% NaCl.

A cell may consist of a stainless steel porous plate upon which the film rests with the salt water over the top of the film itself. The seals are O-rings and the arrangement is that the saline solution circulates through the cells over the top of the film. Circulation is important both in improving the rate and in improving the desalinization of the water. The cell is enclosed by plates **55** and **60** having circular depressions as shown at **61** and **62**. The plate **55** has adjacent to it a plate **65** which is a collector plate made of stainless steel preferably ⅛" thick and between this plate and the opening **61** is the semipermeable or porous membrane **66**, 0.005" thick sealed by O-ring **70**.

The incoming liquid to be desalinized may enter at a pressure of 2,000 psi and may, for example, have in it 3½% of sodium chloride. Numeral **69** represents a channel through the collector plate so that the water to be desalinized passes into an intermediate cell **71**. This cell is also sealed by an O-ring **72** around the periphery of an opening in plate **71**. Numeral **60** represents another end plate like the one **55** having a depression **62** in it sealed by O-ring **73**.

Separating the plates **60** and **71** is another stainless steel collector plate **74** which is porous and there being semipermeable membranes **75** and **76** between this collector plate and the plates **60** and **71**. The pure water which is filtered out of the salinized water is collected in the pores in the collector plate and drains out at the bottom. It has been found that one important requirement for readying the films for service is to increase the pressure in stages rather than all at once to the final pressure, for example, a preferred staging for a final pressure of 2,000 psi is 1,000 pounds for ½ hour, 1,500 pounds for 1½ hours and 2,000 pounds from then on.

Results include a production rate of 8 gallons per square foot of film per day. The feed solution was 5¼% sodium chloride. It may be said that seawater is 3½% total salts. Accordingly, the feed of 5¼% would correspond to a ratio of rejected seawater brine, of 5¼% concentration, to water produced of 2 to 1. These results were obtained at a total pressure of 1,500 psi and the circulation rate on the apparatus used of 6 gallons per hour.

The salt concentration of the water produced was 500 ppm. Referring again to the characteristics of the cast film, it is to be pointed out that once the film is made, that is, once it is immersed in water, it should always be kept in a completely wet condition. If the film is allowed to dry at any time, it will physically shrink on the order of 20 or 30% of its former size and lose its desalinizing effectiveness.

In a continuation of this work, *S. Loeb and G.R. Nagaraj; U.S. Patent 3,170,867; Feb. 23, 1965; assigned to The Regents of the University of California* describe the addition of aluminum ion to the brine. Typical experiments were set up to determine the results of the improved process with aluminum ion added to the brine to be desalinized. A distilled water brine containing 5.25% of sodium chloride was desalinized. The desalinized product contained 0.18% salt. The same desalinizing film was used for desalinizing the same brine except that 2 ppm of aluminum ion had been added to the brine. The desalinized product then contained only 0.09% salt. It was found that it was necessary that the treatment of the brine to be desalinized with aluminum ion added must be accomplished under pressure.

This was demonstrated by the fact that the membranes used in obtaining the results just stated were stored in local Los Angeles tap water which does contain aluminum ions for a period of 3 months. They did not produce potable water from 5.25% distilled water brine. This indicates that though there are minute quantities of aluminum ion content in local Los Angeles tap water, in contact with membranes at atmospheric pressure, it does not improve the desalinization features of the high-flow membrane. From an analysis of this process therefore, it is concluded that the mere presence of the aluminum ion in water does not improve the characteristics of the membrane, but that the application of pressure is necessary.

High-flow porous membranes are also described by *S. Loeb, S. Sourirajan and D.E. Weaver; U.S. Patent 3,133,137; May 12, 1964; assigned to The Regents of the University of Calif.*

Formamide and Acetone

A process described by *M.A. Justice and J.W. McCutchan; U.S. Patent 3,666,508; May 30, 1972; assigned to The Regents of the University of California* is directed to dry semipermeable membranes which can advantageously be dry-stored and used for the selective separation of various components of a solution such as salt water. The membranes are prepared by casting a solution of a cellulosic derivative to form a membrane. The membrane is then immersed in an aqueous medium for a time sufficient for the formation of a membrane possessing a tough outer skin on the side exposed to air during casting. The membrane is then dried at room temperature. A preferable cellulosic solution is composed of cellulose acetate dissolved in acetone containing formamide as an additive.

The relative proportions of the cellulosic derivative in the casting solution may range from 18 to 30% by weight while the remaining weight percentage is constituted by solvents such as dimethyl sulfoxide, dimethylformamide, acetic acid, acetaldoxime and diacetone alcohol or a solvent additive mixture composed of acetone and formamide or a mixture of acetone and glyoxal. A cellulosic solution composed of from about 18 to 30% by weight of cellulose acetate, from about 33 to 52% by weight acetone and from about 18 to 48% of formamide has yielded dry membranes which upon use in the reverse osmotic systems have exhibited an excellent balance between the yield of treated water and the desalination or separation effected thereby.

Dimethylformamide and Acetone

According to a process described by *V.J. Shayler and G.A.L. Wells; U.S. Patent 3,657,401; April 18, 1972; assigned to Pressed Steel Fisher Limited, England* a reverse osmosis membrane is prepared by casting a solution of cellulose acetate, dimethylformamide and acetone where the dimethylformamide is present in the proportion of from 50 to 70 parts by weight to 25 parts by weight of cellulose acetate, and the cast membrane is then washed in cold water and annealed in hot water as in known methods.

The use of dimethylformamide in the above proportions allows membranes to be prepared which give improved flux rates as compared with membranes cast from known ternary casting solutions that comprise cellulose acetate, acetone and known swelling or flux-inducing additives.

Blends of Cellulose Acetates in Acetone, Dioxane and Formamide

D. Walmsley; U.S. Patent 3,807,571; April 30, 1974; assigned to United Kingdom Atomic Energy Authority, England describes a cellulose acetate reverse osmosis membrane, a method of making such a membrane by casting and a solution suitable for use in casting the membrane. The particular features of the process lie in the selection of a narrow range of acetyl values for the cellulose acetate, obtained by blending cellulose acetates having acetyl values falling outside the selected range, and in the composition of the casting solution which includes acetone, dioxane and formamide.

According to this process, a reverse osmosis membrane is thus formed from a solution of cellulose acetate comprising a blend of at least one cellulose acetate having an acetyl value of less than 40.9% and at least one cellulose acetate having an acetyl value of greater than 41.4% and having an acetyl content of from 40.9 to 41.4% in a mixture of acetone, dioxane and formamide. The solution has the following composition with percentages being by weight on the weight of the solution.

	Percent by Weight
Cellulose acetate	15.0 - 30.0
Acetone	15.0 - 50.0
Dioxane	10.0 - 45.0
Formamide	5.0 - 30.0

Glycol Ethers, Acetone and Dioxane

According to a process described by *D.L. Hoernschemeyer; U.S. Patent 3,878,276; Apr. 15, 1975; assigned to the U.S. Secretary of the Interior* cellulose acetate reverse osmosis membranes which have an improved product flux and are suitable for use in the desalination of water can be prepared from casting solutions which consist of a cellulose ester having an acetyl content of 41.5 to 42.7% by weight of the cellulose ester, a solvent, and an ether alcohol or more specifically an alkyl half-ether of ethylene glycol or diethylene glycol. The casting solutions do not contain a swelling agent or nonsolvent.

Example 1: Casting formulations were prepared using 10 grams of E-398-3 and 10 grams of A-432-130B cellulose acetates (Eastman Chemical Corp., having 39.8 and 43.2 weight percent acetyl content respectively) to achieve an average acetyl content of 41.5 weight percent. To this solution acetone, dioxane and 2-methoxyethanol were added in the amounts given in the following table. Flat membranes were cast and after exposure for 5 seconds were immersed in ice water for 5 minutes. These membranes were then annealed in hot water at 65°C for 5 minutes. The resulting membranes were tested at 800 psig with a 1% sodium chloride solution, the standard test conditions used in these examples. The osmotic properties of the membranes are given in the following table.

				Reverse Osmosis Performance	
Test	Dioxane (grams)	Acetone (grams)	2-Methoxyethanol (grams)	Flux (gfd)	Selectivity (percent)
1	60	30	6	23	97
2	60	30	9	34	96
3	60	30	12	51	93
4	60	30	15	53	92
5	50	40	18	28	94
6	50	30	22	71	84

Example 2: A polymer solution was prepared exactly as in Example 1 with the exception that 2-ethoxyethanol was used in place of 2-methoxyethanol.

The membranes were then prepared and tested as in the previous example and the results are shown in the table on the following page.

Test	Dioxane (grams)	Acetone (grams)	2-Ethoxyethanol (grams)	Reverse Osmosis Performance Flux (gfd)	Selectivity (percent)
1	60	30	6	15	97
2	60	30	12	23	97

Acetone, Water and Polar Organic Compounds

T. Asada, F. Komatsu, H. Tsuge, C. Yanagi and T. Kuwahara; U.S. Patent 3,917,777; November 4, 1975; assigned to Kobe Steel, Ltd., Japan describe a process for the preparation of improved semipermeable membranes useful under dry conditions which overcome the disadvantages involved in the storage, transportation and handling of conventional, Loeb-type semipermeable membranes because of their hydrous characteristics.

Briefly, this is achieved by a process for the preparation of semipermeable membranes which comprises casting a film from a solution which comprises 9 to 13% by weight, of a cellulose derivative having a viscosity of 20 to 100 seconds as measured by the ASTM D-1343-56 and D-871-63 methods, water or a substantially water-soluble alcohol inclusive of polyhydric alcohols where the weight ratio of water or the alcohol to the cellulose derivative ranges from 1:10 to 2:1, and a substantially water-soluble additive capable of hydrogen bond formation with the balance of the casting solution being an organic solvent. After evaporating a portion of the organic solvent from the cast film, the film is dipped into water to extract the remaining organic solvent and other soluble residues. Finally, the film is dried directly or by a heat treatment by dipping the film into hot water.

The second component is selected from the group consisting of water and substantially water-soluble alcohols inclusive of polyhydric alcohols. The third component includes organic solvents such as acetone, dioxane and methyl acetate. According to various experiments, acetone has been confirmed to be a particularly good solvent.

A polar organic compound is used as the fourth component, and it is a substantially water-soluble additive. It is essential that such compounds should possess active hydrogen atoms capable of hydrogen bond formation. Suitable examples of these compounds include hydroxy acids, carboxylic acids, alcohols, aldehydes, acid imides, phosphoric acid esters, carbamides, cyclic acid amides, acid amides, and the like. In addition to these organic compounds, inorganic compounds such as hydrogen peroxide, ammonium sulfate, phosphoric acid, and the like may also be used effectively.

Example: A casting solution containing 11% by weight cellulose acetate, 68% by weight acetone, 10% by weight water and 11% by weight lactic acid was prepared, and was cast onto a glass sheet in a thickness of 0.2 mm. The preparation of the casting solution and the casting thereof were conducted at room temperature. For a period of about 45 seconds, as measured from the initiation of the casting operation, acetone was evaporated at room temperature. Then, the cast film together with the glass sheet was dipped in ice water maintained at a temperature of from 1° to 2°C for about 1 hour.

The resulting film was heated at 80°C for about 5 minutes, and then placed on a porous supporting plate. The film was subjected to a permeation test wherein an aqueous solution containing 0.5% sodium chloride was placed on the film and passed therethrough under a pressure of 50 kg/cm^2. The amount of water which permeated was 2.07 cc/cm^2 of the membrane surface area per hour. The salt concentration in the permeated water was 148 ppm. The viscosity of the cellulose acetate derivatives used in this example was 40 to 50 seconds as measured by the ASTM D-1343-56 and D-871-63 methods.

Formic Acid

B.M. Brown and E.L. Ray; U.S. Patent 3,789,993; February 5, 1974; assigned to Eastman Kodak Company describe cellulose ester dope compositions in which the solvent portion contains a high proportion of formic acid.

These membranes have high flux and salt rejection properties, an unexpected resistance to compaction at high pressures over prolonged periods of time and salt rejection properties that are practically independent of applied pressure. For optimum results, cellulose acetates containing from about 37.5 to 40.5% acetyl and having an intrinsic viscosity of from about 0.8 to 1.2 should be utilized in the manufacture of the membranes. Also, it is preferred that formic acid represent from about 60 to 75 weight percent of the volatile (at 105°C under atmospheric pressure) fraction of the dope compositions, and acetone constitute from about 25 to 40 weight percent.

Preferred pore-producing materials are those effective amine salts of inorganic strong acids disclosed in U.S. Patent 3,522,335, such as, for example, dipyridine sulfate, ditriethyl ammonium sulfate, ditriethanol ammonium sulfate, and the like.

Ethyl Acetate

According to a process described by *R.R. Stana; U.S. Patent 3,646,179; February 29, 1972; assigned to Westinghouse Electric Corporation* a high-strength semipermeable membrane, consisting of a supportive layer and an osmotic skin layer of dense polymeric cellulosic material, is made by mixing polymeric film-forming cellulosic material, ethyl acetate and additives such as swelling additives to form a casting solution; casting a film from the solution; drying the film; leaching the film to form a semipermeable membrane; and optionally curing the membrane in hot water.

Many solvents have been used in reverse osmosis membrane-casting solutions but acetone has become the standard solvent. The use of ethyl acetate in membrane-casting solutions, however, gives the improved result of enabling the production of membranes having tensile strengths in the order of 850 psi. A typical casting solution would have the following composition: cellulose acetate, 4.5 oz (20.8 weight percent); ethyl acetate, 10.0 oz (46.3 weight percent); formamide (methanamide), 7.0 oz (32.5 weight percent); and triacetin (triacetate glycerol), 2.0 ml (0.4 weight percent).

Alcohol Nonsolvents

R.E. Kesting; U.S. Patent 3,884,801; May 20, 1975; assigned to the U.S. Secretary of the Interior describes a method for preparing asymmetric reverse osmosis membranes by preparing a dilute solution of cellulose acetate in a solvent system consisting of a solvent and at least one pore-forming material selected from the nonsolvents for cellulose acetate, casting the solution to form a thin film, and allowing the solvent system to evaporate completely. The dry membranes so formed can be stored in the absence of water and rewet when necessary for use in reverse osmosis separations. The membranes can be repeatedly dried and rewet without any special treatment steps and without harm to the membrane.

Solvents useful in the process include any complete solvent for the particular cellulose ester or ether used. Among these solvents particularly useful in preparing dry membranes of cellulose acetate are methylene chloride, acetone, and dioxane. In addition, the solution should contain at least one material which aids in the formation of pores in the substratum of the asymmetric membranes. Some confusion in semantics is involved in denominating such materials as either swelling agents, weak swelling agents, or nonsolvents.

Part of this uncertainty stems from the fact that the properties of a given material which would put it into a particular classification depend on the concentration and temperature at which this material is employed. Thus, there may be some confusion as to whether a particular material to be used in the process can be termed a weak swelling agent or a nonsolvent. Although some of the materials suitable as pore-producing agents might be classified as weak swelling agents, unlike the pore-forming materials used in the prior art processes, all pore-forming constituents used in this process have a very low affinity for the polymer, i.e., they are all essentially nonsolvents for the polymer. Because of this common property, this term is used to describe them, although it may not be the only one which might accurately be applied.

In addition, pore-forming materials should not react with the polymer or solvent, should be volatile, and should have a boiling point at least about 30° to 40°C higher than that of the solvent. Suitable nonsolvents include isopropanol, tert-butyl alcohol, methanol, ethanol, and water. Isomeric butanols and pentanols, in general, meet these requirements.

While the process is described with particular regard to the desalination of water, the improved membranes prepared by the process may also be used for the separation of water from electrolyte solutions such as seawater, brackish water, acid mine water, and industrial brines and bitterns; the separation of organic liquids; the purification and concentration of liquid feeds such as citrus juices, beer, and syrups; the purification of liquid wastes such as urine; and the separation of gaseous mixtures.

Example 1: A casting solution was prepared using the following formulation.

	Grams
Cellulose acetate (Eastman E-394-60)	5
Cellulose acetate (Eastman E-432-130B)	1
Methylene chloride	69
Ethanol	20
tert-Butyl alcohol	13
Water	2.5

The solution was cast onto a glass surface at room temperature and allowed to dry until the membrane appeared opaque. The membrane was then placed in a test cell with the skin side facing the (about 0.4%) aqueous sodium chloride feed solution. Test conditions were 800 psi and approximately 25°C. After each test cycle the membranes were dried completely at 25°C overnight and rewet before testing. This cycle was repeated seven times. The results are shown in the following table.

Cycle (dry-rewet)	Flux (gfd)	Salt Rejection (percent)
1	13.3	94.2
2	12.17	94.0
3	11.2	92.0
4	13.3	90.0
5	12.17	89.0
6	9.3	91.5
7	10.6	85.5

Example 2: Casting solutions were prepared varying the pore-forming materials and the quantity of these materials included in the solution. The following solutions produced dry membranes by the simple process described in the previous example.

Solution A

	Grams
Cellulose acetate (Eastman E-394-60)	5
Cellulose acetate (Eastman E-432-130B)	1
Methylene chloride	69
Ethanol	20
tert-Butyl alcohol	13
Water	2.5

Solution B

Cellulose acetate (Eastman E-394-60)	12
Acetone	65

(continued)

	Grams
tert-Butyl alcohol	65
Water	1

Solution C

Cellulose acetate (Eastman E-400-25)	15
Acetone	65
tert-Butyl alcohol	65

Solution D

Cellulose acetate (Eastman E-394-60)	9
Dioxane	67
1-Butanol	35

The membranes were tested in the same manner and conditions as in Example 1 and the results are given below in the table.

Solution	Salt Rejection (percent)	Flux (gfd)
A	94.2	13.3
B	93	2.5
C	96	2.2
D	86	1.0

The data in the above table confirms the requirement that a difference of about 30° to 40°C in the boiling points of the solvent and nonsolvent is desirable. When the desolvation of solvent and nonsolvent is concurrent rather than sequential, the quality of the membrane is reduced. Solutions B, C and D of Example 2 were prepared from reagents having a boiling point difference of less than 30°C. A marked decrease in flux is found in the membranes produced from these solutions from those of the process as illustrated by Solution A. Where the nonsolvent possesses a volatility close to the solvent, then both species will evaporate from the membrane surface almost simultaneously, resulting in the formation of a dense almost impermeable membrane of low flux.

Pyridine

S. Manjikian; U.S. Patent 3,483,282; December 9, 1969; assigned to Universal Water Corp. describes the production of a semipermeable membrane by casting the membrane from a solution comprising a cellulosic derivative, for example, cellulose acetate; a flux promoter; a water-miscible organic solvent; and from about 2 to 50% pyridine by weight. The following example illustrates the process.

Example: A casting solution was prepared consisting of the following ingredients in amounts to provide the indicated percentages by weight.

	Percent
Cellulose acetate E-398-3	25
Formamide (practical grade)	25
Acetone (reagent grade)	30
Pyridine (reagent grade)	20

The ingredients were mixed in a rotating jar mixer for 24 hours. Then the casting solution was spread over the surface of a flat glass plate using a straight edge doctor blade and side spacers 0.008" thick. After the film of casting solution had been formed the glass plate and film were immersed in water at 1°C for a period of 90 minutes. The glass plate and film were then removed from the water bath and the film removed from the glass plate.

The so-formed film was subsequently immersed in water at a temperature of 74°C for a period of 10 minutes to accomplish heat treatment. The heat-treated membrane, 0.004" thick, was clear and transparent. A specimen of the membrane was cut off for tensile testing. Ultimate tensile strength was 940 psi determined at a constant load increase of 16 psi per minute. Yield strength was 530 psi, and total elongation to break was 33%.

A section of the membrane was tested for initial flux and selectivity in a laboratory test cell which comprised a porous stainless steel plate supporting the membrane, means for supplying impure feedwater under pressure to the exposed surface of the membrane and means for collecting purified water passing through the membrane. Using brackish water feed of 4,200 ppm sodium chloride content and an operating pressure of 600 psi, flux through the membrane was determined to be 22 gfd and selectivity, that is percentage of the feedwater salt content rejected by the membrane was 96.7%. A conductivity meter was used to determine salt content of both feed and product water. Another membrane was produced by a procedure similar to that described above, but was cast from a solution containing no pyridine and composed of the following ingredients.

	Percent
Cellulose acetate (E-398-3)	25
Formamide (practical grade)	30
Acetone (reagent grade)	45

After casting and immersion in cold water it was heat-treated at 78°C. The membrane appeared translucent. Strength test of this membrane showed ultimate tensile strength of 800 psi and yield strength of 530 psi. Total elongation to break was 22%. Operating tests in the laboratory cell showed flux of 26 gfd and selectivity of 94%.

It is seen that the membrane cast from the pyridine-containing solution was stronger, that it had higher ultimate strength and also greater elongation before breaking. The flux was slightly less and the selectivity of the pyridine membrane was greater, although these could be considered substantially equivalent since a decrease in selectivity is often accompanied by an increase in flux. The results are tabulated below, for ready comparison.

Membrane Casting Solution Composition	Ultimate Tensile Strength (psi)	Yield Strength (psi)	Total Elongated (percent)	Flux (gfd)	Selectivity (percent)
CA, Ac, formamide, pyridine	940	530	33	22	96.7
CA, Ac, formamide	800	530	22	26	94

C.R. Cannon; U.S. Patent 3,285,765; November 15, 1966; assigned to the U.S. Secretary of the Interior describes the use of nonaqueous casting solvents such as methanol.

Amines and Amine Salts

M.E. Rowley and N.G. Baumer; U.S. Patent 3,527,853; September 8, 1970; assigned to Eastman Kodak Company describe the preparation of semipermeable membranes directly from conventional so-called hydrolysis baths in which cellulose triacetate has been partially hydrolyzed. It has been found that cellulose acetate hydrolysis baths can be used practically directly, provided that (a) the sulfuric acid hydrolysis catalyst is converted into a new type of pore-producing material (an effective organic amine sulfate); and (b) the evaporation step is encouraged by forcing passage of a gas such as air across the surface of the cast film of the concentrated dope (to thereby cause the formation of the desired specially selective skin on the surface of the film).

The conversion of the sulfuric acid in the hydrolysis bath into the effective pore-producing organic amine sulfate can readily be accomplished by simply blending into the bath an effective amount (preferably from about 0.5 to 2.5 mols) of a useful organic amine per mol of sulfuric acid in the bath. For optimum results, about 2 mols of the amine per mol of sulfuric acid should be used. Typical examples of amines that can be utilized successfully include: pyridine, triethylamine, triethanolamine, diethanolamine, α-picoline, β-picoline, lutidine, N,N-dimethylaniline, 2-aminoethanol, monoisopropanolmonoethylamine and diisopropanolamine (all of these amines having molecular weights of at most about 400).

Example: Thirty pounds of cellulose containing 7% moisture are loaded into a conventional acetylation mixture along with 95 pounds of acetic acid. The resulting mixture is blended for 30 minutes at a temperature of 110°F. Into this mixture is then added 124.3 grams of sulfuric acid dissolved in 200 grams of acetic acid. The resulting blend is mixed at 110°F for 5 minutes. Its temperature is then quickly lowered to 75°F and 85 pounds of acetic anhydride are then blended into it. The resulting anhydrous mixture is subsequently cooled to 60°F before 790 grams of sulfuric acid dissolved in 900 grams acetic acid is intermixed with it. The acetylation bath (containing sulfuric acid catalyst) is then warmed to a temperature of 140°F and maintained at this temperature for approximately 30 minutes until an intrinsic viscosity of 1.2 is obtained.

1,362 grams of the mixture (containing cellulose in the form of triacetate) is then hydrolyzed by first blending into it a mixture of 136 grams of water and 250 grams of acetic acid and subsequently maintaining the temperature of the resulting hydrolyzing reaction mixture at a temperature of 100°F for about 24 hours until a partially hydrolyzed cellulose acetate containing 40% acetyl results. The hydrolysis reaction is then stopped by blending into it 2 mols of pyridine per mol of sulfuric acid in the hydrolysis bath.

The concentrated dope is then cast on a glass plate at room temperature to thereby form a film having a thickness of 10 mils. The film is permitted to stand for 10 seconds under ambient conditions and then is subjected to a stream of moving air at room temperature. The volume of the stream of moving air is about 1 liter per square inch of film per minute. The plate is then submerged into water maintained at a temperature of 34°F for 5 minutes and subsequently washed for an additional 10 minutes with water at room temperature and stripped from the glass plate.

The membrane is then tempered for 4 minutes in warm (170°F) water and tested for its ability to function well as a reverse osmosis membrane by subjecting the skin side of it to brackish water containing 5,500 ppm of sodium chloride under 600 psi pressure. Results of this test indicate that a flux of 33.5 gfd can be obtained with a salt rejection rate of 82.5%.

In related work, *M.E. Rowley; U.S. Patent 3,522,335; July 28, 1970; assigned to Eastman Kodak Company* have found that certain members of the class of materials known as organic amine salts (of strong inorganic acids) and certain quaternary ammonium compounds have the peculiar ability to function as effective pore-producing materials. Typical examples of such effective organic amine salts include pyridine sulfate, triethylamine sulfate, triethanolamine sulfate, diethanolamine sulfate, the picoline sulfates, the lutidine sulfates, N,N-dimethylaniline sulfate, 2-aminoethanol sulfate and triethanolamine phosphate.

S. Manjikian, S. Loeb and J.W. McCutchan; U.S. Patent 3,344,214; Sept. 26, 1967; assigned to the Regents of the University of California describe the preparation of cellulosic membranes in the presence of water-soluble liquid amides.

Lipids

R.R. Stana; U.S. Patent 3,593,855; July 20, 1971; and U.S. Patent 3,780,147; December 18, 1973; both assigned to Westinghouse Electric Corporation describes a semipermeable membrane containing a supportive layer and a lipid-containing osmotic skin layer of polymeric film-forming cellulosic material.

The membrane is made by mixing the film-forming cellulosic material, solvent, swelling additive and lipid to provide a casting solution; casting a film from the solution; drying the film; leaching the film to form a semipermeable osmotic skin membrane containing a residue of lipid; and optionally curing the membrane. The semipermeable lipid-containing membranes can be used for applications on tap water for use in water softening and taste improvement, seawater and brackish water for desalination applications, waste effluents for waste impurity disposal and urea water for biological applications. The lipid-containing membranes exhibit up to two times the flux, at a given solute rejection and pressure, as standard unmodified cellulose acetate membranes. They also can give an increase in solute rejection by the initial uncured membrane. The lipids most useful are esters of carboxylic acids. These esters are derived from glycerol ($HOCH_2CHOHCH_2OH$) and therefore are known as glycerides. The basic structure of these glycerol ester lipids is as follows.

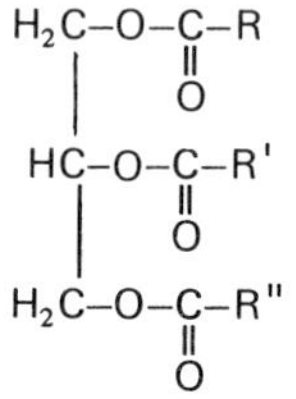

In this structure R, R', R" are each alkyl groups containing 1 to 20 carbons and preferably 1 to 8 carbons. These glycerol ester lipids are mixed glyceryl esters of various saturated and unsaturated, straight chain, even-numbered carbon acids. The most abundant saturated acids are the straight chain even-numbered carbon acids: caprylic [$CH_3(CH_2)_6COOH$]; capric [$CH_3(CH_2)_8COOH$]; lauric [$CH_3(CH_2)_{10}COOH$]; myristic [$CH_3(CH_2)_{12}COOH$]; palmitic [$CH_3(CH_2)_{14}COOH$]; and stearic [$CH_3(CH_2)_{16}COOH$]. The common unsaturated acids are oleic [$CH_3(CH_2)_7CH{=}CH(CH_2)_7COOH$], linoleic [$CH_3(CH_2)_4CH{=}CHCH_2CH{=}CH(CH_2)_7COOH$], etc. The major constituents of the glycerol ester lipids are the acids, which represent up to 90% of their weight.

Pore-Forming Metal Salts

K. Kamide, S. Manabe, E. Osafune and T. Sakamoto; U.S. Patent 3,883,626; May 13, 1975; assigned to Asahi Kasei Kogyo KK, Japan describe a process for the manufacture of a cellulose acetate porous membrane which comprises dissolving in a main organic solvent, cellulose acetate having an acetyl content of 20 to 62.5%, and adding to the solution a diluent solvent having a boiling point higher than the main solvent.

To the solution is added a metal salt, the metal portion of the salt being selected from Groups I through III of the Periodic Table to provide a homogeneous solution. The resulting solution is cast onto a polished plane surface to form a thin film and the solvent is removed from the film by evaporation at elevated temperatures, and subsequently, the metal salt is removed from the film by dissolving the salt to produce a porous membrane.

Pore-Forming Bromide Salts

S. Loeb and G.R. Nagaraj; U.S. Patent 3,283,042; November 1, 1966; and S. Loeb; U.S. Patent 3,364,288; January 16, 1968; both assigned to The Regents of the University of California describe a method of preparing a semipermeable membrane adapted to permit substantially selective diffusion of a component of a solution. The method involves, first, dissolving in an organic solvent a cellulosic ether or ester. Also dissolved in such organic solvent is a pore-producing salt having an anion from the class consisting of iodide, bromide, tetraphenyl boron, salicylate, chlorate, tetraiodomercurate and fluosilicate.

In addition, preferably, a different pore-regulating inorganic compound is dissolved in the organic solvent which has an anion from the class consisting of halides and oxygenated halides. Such solution is then cast to form a membrane of substantially uniform thickness and a major portion of the organic solvent is then evaporated for a predetermined period of time. The cast membrane may be then immersed in water and heated, with such immersion and heating being done either in sequence or simultaneously. Typical results from using cellulose acetate and such pore-producing salts in the method are set forth in the table on the following page.

Replacements for Magnesium Perchlorate as Primary Additive in Casting Solution

Pore-Producing Salt	Composition of Casting Solution (wt %) Pore-Producing Salt	Water	Cellulose Acetate	Acetone	Time Between Casting and Immersion, minutes	Heating Temp., °C	Desalinization Characteristics - Product Flux, gfd	Product Salt Content, percent
$CdBr_2$ (hydrated)	2.2	8.8	22.2	66.7	3	82.3	4.1	0.05
	4.4	6.6	22.2	66.7	6	70.0	6.9	0.30
$CdBr_2$ (anhydrous)	1.1	10.0	22.2	66.7	10	81.0	15.3	0.60
	2.2	8.9	22.2	66.7	3	69.5	9.3	0.19
KI	5.5	5.5	22.2	66.7	6	82.5	7.4	0.08
	5.5	5.5	22.2	66.7	3	84.5	6.7	0.10
NaI	2.2	3.9	22.2	66.7	10	90.0	11.0	0.80
	3.3	7.8	22.2	66.7	6	81.6	15.7	0.60
$NaB(C_6H_5)_4$	1.1	10.0	22.2	66.7	-	84.5	5.6	0.16
	2.2	8.9	22.2	66.7	-	84.0	4.8	0.13
Sodium salicylate	1.1	10.0	22.2	66.7	10	82.6	8.6	0.70
$NaClO_3$	3.3	7.8	22.2	66.7	12	72.5	17.4	0.85
K_2HgI_4	17.5	15.8	16.7	50.0	1.5	72.0	13.8	0.78
	17.5	15.8	16.7	50.0	3	72.0	6.1	0.14

Di- and Tribasic Aliphatic Acids as Swelling Agents

W.M. King and P.A. Cantor; U.S. Patent 3,673,084; June 27, 1972; assigned to Aerojet-General Corporation describe a casting solution for the preparation of a cellulose membrane containing cellulose acetate, an organic solvent and a water-soluble organic swelling agent, which is capable of hydrogen bonding with ketone and hydroxyl groups of the cellulose acetate. The organic swelling agent is present in an amount between about 1.0 to 10 parts by weight of each 10 parts of the cellulose acetate. The organic swelling agent is of a different composition than the organic solvent and may be, for example, a di- or a tribasic aliphatic acid having from 2 to 13 carbon atoms. To illustrate the process several membranes were prepared and tested with the data being presented in the following table.

Swelling Agent	Grams	Additional Amount of H_2O (grams)	Flux	Salt Permeation
Malonic acid	3	2.0	13.0	315
Oxalic acid	3.28	0.92	13.0	480
Succinic acid	3.0	2.0	15.0	420
Glutaric acid	3.0	2.0	14.0	315
Formic acid	4.0	4.0	11.2	350
Acetic acid	3.0	3.0	12.8	595
Glycolic acid	3.0	0.5	14.6	445
Lactic acid	3.0	1.5	14.2	605
Maleic acid	3.28	2.80	12.8	380
Tartaric acid	2.5	–	17.0	535
Citric acid	2.5	0.2	15.5	510
Tert-butyl alcohol	8.0	1.0	12.0	700
Methanol	6.0	–	5.4	810
Ethanol	7.0	–	6.5	615
Isopropanol	5.0	–	6.0	800
Tert-butyl-amine	6.0	–	14.8	700
n-Butylamine	6.0	–	18.0	2,350
Isobutylamine	6.0	–	14.0	5,900
Formamide	6.0	–	15.0	1,350
Propionamide	6.0	3.0	16.0	480
Octanoic acid	7	5	4.4	24,000

(continued)

Swelling Agent	Grams	Additional Amount of H_2O (grams)	Flux	Salt Permeation
Phthalic (ortho) acid	4	8.0	10.4	420
Propylamine	5	2.0	14.0	720
Diethylamine	6	5.0	17.0	5,500
Acetamide	6	5.0	3.0	26,400

A basic cellulose acetate casting formulation was employed which contained 10 grams of E-398-3 (Eastman Chemical Company) cellulose acetate (having an acetyl group concentration of 39.8 weight percent and an ASTM falling ball viscosity of 3 seconds) in a mixture with 30 grams of reagent grade acetone and 5 grams of distilled water. This basic formulation was employed in forming each of the films identified in the table.

In the first column of the table is set forth the swelling agent which was added to 45 grams of the basic formulation. In the second column is set forth the additional amount of water (in grams) which was added to the basic formulation. In column three is set forth the flux through the finished membrane (gallons of water per square foot of film surface per 24-hour day) using a typical test cell. In the fourth column is set forth the salt concentration in the product stream issuing from the test cell in parts per million of sodium chloride.

The figures set forth in columns three and four were obtained by subjecting a circular membrane portion having a diameter of 3 inches to a salt stream containing approximately 35,000 parts per million of sodium chloride at a pressure of 1,500 psig. As shown in the table, the process produces a superior cellulose acetate membrane for desalination when utilizing a wide variety of organic swelling agents.

Mineral Acid Swelling Agents

W.M. King and P.A. Cantor; U.S. Patent 3,444,286; May 13, 1969; assigned to Aerojet-General Corporation describe a casting solution for the preparation of a reverse osmosis cellulose ester membrane. The solution comprises a cellulose ester, an organic polar solvent for the cellulose ester, a mineral acid swelling agent, and a solvating agent. The mineral acid swelling agent is present on an anhydrous basis in an amount within the range of about 0.1 to 6 parts by weight for each 10 parts of the cellulose ester.

It has been found that mineral acids are superior swelling agents for the preparation of reverse osmosis cellulose ester membranes. It is believed that swelling agents act by associating large quantities of the solvating agent, usually water, with the functional groups of the cellulose ester, that is, in the swelling phenomenon sizeable spheres of water hydrogen bond with the hydroxyl and the ketone groups of the cellulose ester.

Various mineral acids may be employed as swelling agents in the casting solution. Perchloric acid is a preferred material. Hydriodic acid, nitric acid, phosphoric acid are all desirable materials and can be used in casting solutions for producing superior membranes. Hydrolysis of the cellulose ester by perchloric, hydriodic and nitric acid will occur if the casting solutions are stored at room temperature for more than a day. Low storage temperature slows the hydrolysis sufficiently to permit storing of these solutions for a week or more without detrimental effects. Phosphoric acid does not cause hydrolysis of the casting solution in the instance of cellulose acetate where held in storage periods up to one month at room temperature. The acid may be supplied to the casting solution in conventional concentrations.

Acid Pretreatment

F.E. Littman; U.S. Patent 3,642,773; February 15, 1972; and U.S. Patent 3,560,232; February 2, 1971; both assigned to McDonnell Douglas Corporation describes cellulose ester, especially cellulose acetate, semipermeable membranes having superior permeation properties of flux and solute rejection which are particularly suited for desalination of aqueous salt solutions by reverse osmosis. These membranes are produced by a method which includes pretreatment of the cellulose ester material in the form of an aqueous slurry with catalytic amounts of a strong acid prior to casting the cellulose ester material into a film.

Example 1: A reaction vessel equipped with a stirrer and a thermometer was charged with 1,500 ml of 0.4 N HCl and 300 grams of E-398-10 (Eastman Kodak) cellulose acetate understood to have a normal acetyl content of 39.8% corresponding to about 55.6% combined acetic acid. The overall concentration of acid in the reaction mixture was about 0.3 N. Agitation of the flask contents at ambient temperature was begun and continued for 10 minutes after which the cellulose acetate was recovered by filtration, washed several times with water to remove any residual acid, and allowed to dry at room temperature. The modified cellulose acetate material was then used to prepare a cellulose acetate membrane as described below.

A casting solution was prepared by charging a vessel with 450 grams of acetone and 300 grams of formamide and 250 grams of the treated cellulose acetate material. The resulting solution was mixed by rolling overnight. A film was then cast from the viscous, almost colorless solution using an 8-inch-square plate glass base and a film casting knife positioned above the plate at a distance calculated to give a 0.10-inch-thick film. After casting, the film was air-dried for 1½ minutes after which the plate carrying the film was immersed in 4°C ice water for 1 hour, followed by immersion in hot 76°C water for 5 minutes. The film was then stripped from the glass plate and stored in distilled water until used for desalination of salt water as described below.

A disc having a 2-inch diameter was cut from the film and mounted in a reverse osmosis pressure cell with the air-dried surface of the membrane in contact with the salt solution, with the backup disc of filter paper placed between the side of the membrane exposed to the glass plate and the porous stainless steel backup plate.

Figure 1.2 shows the concentration of salt in the product against flux rate for the modified cellulose acetate membrane prepared as described above in comparison with an unmodified cellulose acetate membrane. The plot of Figure 1.2 shows that for flux rates in excess of about 10 gallons per square feet of membrane area per day, the product effluent for the modified membrane has a substantially smaller salt concentration than for the unmodified membrane, and that this difference increases with increased flux.

Thus, for a flux of 20 gallons per square feet per day, the effluent product using the modified membrane has a salt concentration of about 1,300 parts per million whereas for the unmodified membrane the salt concentration in the product is about 2,300 parts per million, thus indicating the improved salt rejection properties of the membrane.

Example 2: A reaction vessel equipped with a stirrer and a thermometer is charged with 300 grams of cellulose acetate and sufficient water to give a total volume of 5.95 liters. Agitation is begun in the reaction vessel and 34.2 milliliters of concentrated sulfuric acid (96%, specific gravity 1.84) is added dropwise over a period of about 17 minutes.

Agitation is continued at ambient temperature for an additional 40 minutes. At the end of this time, the treated cellulose acetate is recovered by filtration, washed several times with water, and allowed to air-dry at room temperature.

The modified cellulose acetate material is then used to prepare a cellulose acetate membrane according to the procedure outlined in Example 1.

When used in a reverse osmosis cell, as described in Example 1, the membrane shows improved flux rate and high salt rejection as well as a high resistance to compaction.

FIGURE 1.2: ACID-MODIFIED CELLULOSE ESTER SEMIPERMEABLE MEMBRANE

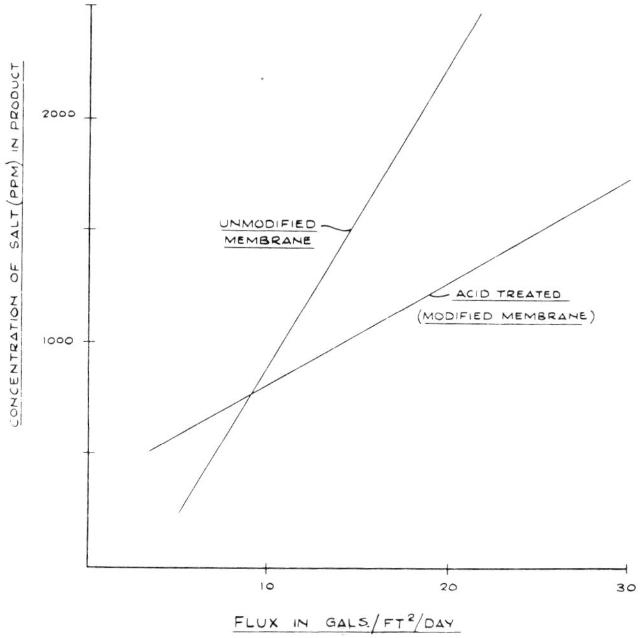

Source: F.E. Littman; U.S. Patent 3,642,773; February 15, 1972

MEMBRANE TREATMENTS

Sodium Phosphate

R.R. Stana; U.S. Patent 3,721,623; March 20, 1973; assigned to Westinghouse Electric Corporation describes a method of improving the salt rejection of supported semipermeable membranes which comprises contacting the membrane with sodium phosphate and alkyl or alkylaryl sulfonates.

Example: A casting solution was prepared containing about 24 weight percent cellulose acetate, 47 weight percent acetone and 29 weight percent formamide (methanamide). The solid cellulose acetate was added to the other liquid ingredients to give a viscous liquid casting solution. This viscous casting solution was poured into a film casting apparatus containing a blade with an opening of about 16 mils. The casting apparatus was 1-foot long and it was hand-drawn across a 1' x 1' glass plate in about 5 seconds to give a casting about 7¼ inches square. The liquid solution poured through the opening in the casting apparatus during this time to form a film on the glass plate.

The casting was allowed to air dry for about 5 seconds to give a total air dry time of about 10 seconds. The glass plate and casting apparatus were at room temperature. The glass plate with the casting was then immersed in a leaching bath at 32°F, prepared from a mixture of water and crushed ice. After 20 minutes, the semipermeable membrane casting, about 5 mils thick and having an osmotic skin layer, was separated from the glass plate. A small sample about 1.75 inches in diameter was cut out of the uncured semipermeable membrane. This sample was mounted in a standard stainless reverse osmosis apparatus with the osmotic skin side of the membrane facing the feedwater entrance. Tap water containing 100 ppm dissolved solids was fed into the reverse osmosis cell against and contacting the membrane at 200 psi and 25°C. The uncured semipermeable membrane operated at 57% rejection of undissolved solids with a purified water flux of 45 gfd.

Then 80 ppm of a detergent additive material was added to the tap water, containing 100 parts per million dissolved solids, and fed into the reverse osmosis apparatus described above at 200 psi and 25°C. The additive contained about 50 weight percent sodium tripolyphosphate, 40 weight percent sodium lauryl sulfonate and 10 weight percent bulk as carboxymethylcellulose and sodium sulfate. The uncured semipermeable membrane operated at 69% rejection of undissolved solids with a purified water flux of 41.4 gfd. As can be seen a dramatic improvement in percent rejection resulted from the additive with only moderate decrease in pure water flux.

A variety of other impurities were added to the tap water, containing 100 ppm dissolved solids, fed against the same supported membrane using the same test cell and operating conditions, and the following fluxes and rejections were obtained.

Feed	Flux (gfd)	Percent Rejection
Tap water (about 100 ppm dissolved solids)	45	57
Tap water + 80 ppm additive*	41.4	69
Tap water + 80 ppm additive* + 100 ppm $MgSO_4$	38	70
Tap water + 80 ppm additive* + 100 ppm $MgSO_4$ + 100 ppm NaCl	37.7	57
Tap water + 160 ppm additive* + 100 ppm $MgSO_4$ + 100 ppm NaCl	37	62
Tap water through treated membrane after 7 days	42.5	68

*Approximately 50 weight percent phosphate, 40 weight percent sulfonate, 10 weight percent bulk

Each time the additive was dissolved in the feedwater contacting the membrane, the rejection increased with the only decrease in flux being due to the increase in osmotic pressure of the feed from the addition of the detergent additive. When returning to a feed of tap water containing only 100 ppm dissolved solids using the same membrane which had been treated by additive contact after 7 days of running, the flux was almost back to the initial level, but the rejection stayed at the increased level.

Polyvinyl Methyl Ether

F.E. Martin; U.S. Patent 3,373,056; March 12, 1968; assigned to Aerojet-General Corp. describes a method for improving the salt retention characteristics of a cellulose reverse osmosis membrane by treatment with various modifiers such as polyvinyl methyl ether, guartec gum, activated charcoal and copper ferrocyanide.

Method No. 1: A membrane prepared by the conventional methods is treated in a solution of polyvinyl methyl ether (PVM). The membrane is soaked in a PVM solution, ranging from an effective amount of PVM to about a 10% weight solution. Preferably, the PVM solution is about 1 to 5% by weight. The membrane is soaked in the PVM solution for a time ranging from 30 minutes to 24 hours. Generally, at least about 4 hours is required for optimum results. When treated in this manner, the membrane may be dried, stored in air for an indefinite period, rewetted, and then replaced in a cell, and used for desalination with the same resulting flux and purity obtained prior to being dried.

Example 1: Six membranes were prepared in the following manner. A solution of 100

parts of cellulose acetate, 300 parts acetone, 4.5 parts magnesium perchlorate, 4.5 parts magnesium bromide, 4.5 parts zinc bromide, and 45 parts of water was cooled to -11°C, then cast 10 mils thick on glass at -11°C. After drying for 3 minutes, the film was set by immersing in ice water. The film was soaked loose from the glass plate, heated at 89°C for 5 minutes, then stored in fresh water.

Two of the membranes were left untreated, two membranes were soaked 4 hours in aqueous 1% PVM, and two membranes were soaked 4 hours in aqueous 5% PVM. These six membranes were then placed between layers of absorbent paper and allowed to dry overnight, then removed and stored exposed to laboratory air for 2 weeks. These membranes were then examined as osmotic membranes for the desalination of a 3.5 weight percent NaCl solution of 1,500 psig by reverse osmosis. The data collected are shown in Table 1.

TABLE 1

Membrane	Treatment	Run	Product Flux (gfd)	Product Salt (ppm)
1	Untreated	1	Nil	Salty
		2	Nil	Salty
2	Untreated	1	Nil	Salty
		2	Nil	Salty
3	1% PVM,	1	4.4	960
	4 hours	2	5.2	960
4	1% PVM,	1	8.8	790
	4 hours	2	12.4	700
5	5% PVM,	1	10.8	790
	4 hours	2	13.6	900
6	5% PVM	1	10.0	555
	4 hours	2	12.8	580

The runs designated 1 and 2 are short runs taken over a period of 2 hours. The membrane used in this example shows the results for desalination of a 3.5 weight percent NaCl solution of 1,500 psi, when used directly as prepared without being dried, in Table 2.

TABLE 2

Run	Product Flux (gfd)	Product Salt (ppm)
1	10.5	750
2	10.5	725
3	10.5	580

Method No. 2: The salt permeation characteristics of the above-discussed osmotic membranes are improved by the addition of a trace of polyvinyl methyl ether to the feed solution. It appears that the PVM effectively plugs small holes which may exist in the osmotic membrane. As the feed solution containing PVM passes over the membrane, the PVM selectively plugs small holes and more open permeation sites. When the desired degree of improvement in the semipermeability of the membrane has been obtained, the PVM modifier may be removed from the feed solution.

Materials other than PVM which are effective as membrane modifiers in plugging microscopic as well as macroscopic holes include powders, such as microscopic plastic balls; colloidal clays; gels, such as aluminum hydroxide, ferric hydroxide, and aluminum sulfate; organic colloids, such as maleic acid-vinyl alkyl ethers; and colloidal inorganic precipitates formed in situ, such as copper ferrocyanide. Generally, an effective amount to about 1.0% by weight of the modifier may be employed in the feed solution; preferably an amount ranging from 100 to 500 ppm is employed.

Example 2: Six osmotic membranes were examined for the desalination of salt water by reverse osmosis, according to the following schedule: (A) with a feed solution of 3.5 weight percent NaCl at a linear velocity of 1 fps and a pressure of 1,500 psig; (B) then with a

3.5 weight percent NaCl solution containing 100 ppm PVM at the same linear velocity and pressure; (C) after flushing the system with fresh water, again with a 3.5 weight percent NaCl solution at the same linear velocity and osmotic pressure. The product flux and salt content were determined and the data shown in Table 3 were collected.

TABLE 3

Membrane	Product A (from 3.5 weight percent NaCl)		Product B (from 3.5 weight percent NaCl and 100 ppm PVM)		Product C (from 3.5 weight percent NaCl)	
	Flux (gfd)	NaCl (ppm)	Flux (gfd)	NaCl (ppm)	Flux (gfd)	NaCl (ppm)
1	9.4	300	6.8	195	8.0	150
2	9.4	275	8.0	195	8.2	140
3	25.0	5,100	16.5	440	16.5	430
4	22.0	3,800	14.8	348	15.5	320
5	80.0	27,000	9.6	620	12.8	740
6	76.0	27,000	10.0	525	14.0	635

Membranes 1 through 6 were prepared from a solution of 100 grams of cellulose acetate, 300 grams acetone, 45 grams water, 4.5 grams magnesium perchlorate, 4.5 grams magnesium bromide hexahydrate, 4.5 grams zinc bromide, cast at -11°C, dried 3 minutes, set at 0°C.

Acetylation of Dried Membrane

R.L. Riley and H.K. Lonsdale; U.S. Patent 3,806,564; April 23, 1974; assigned to the U.S. Secretary of the Interior have found that the properties of an asymmetric membrane can be altered by chemically treating the membrane with the appropriate reagent after the membrane has been dried. In particular, it has been found that asymmetric cellulose acetate membranes suitable for use in desalination by reverse osmosis can be dried and subjected to acetylation to enhance their selectivity.

Example 1: An asymmetric cellulose acetate membrane was prepared from the following casting solution.

Component	Parts/100
Acetone	66
Cellulose acetate (39.9% acetyl)	22
Water	11
Magnesium perchlorate (anhydrous)	1

This solution was cast on a glass surface at a temperature between 0° and 5°C. After exposing the membrane to air for 30 seconds to allow for evaporation of the solvent, the membrane was placed in ice water for a time sufficient to allow the membrane to gel (1 to 10 minutes). The membrane was annealed for 30 minutes at 85°C and dried by the method using a surface active agent described in the article by Vos and Burris [*Industrial and Engineering Chemistry,* volume 8, pages 84 to 89 (1969)]. This membrane was used as a control for comparison with the membranes subsequently subjected to acetylation.

A membrane (3 square inches) identically prepared was placed in 200 ml of an inert liquid medium, trifluorotrichloroethane, to which 0.3 ml of acetyl chloride was added. After two days at room temperature the cellulose acetate membrane was removed and found to be fully acetylated as evidenced by its insolubility in acetone and its solubility in chloroform.

The control membrane and the acetylated membrane were then tested in reverse osmosis desalination of a 1.0% aqueous sodium chloride solution at a temperature of 19°C and an applied pressure of 1,500 psi. The average results of tests on two membranes are given in the table. The data clearly show a significant improvement in selectivity and a decrease in the water flux caused by the acetylation.

Reverse Osmosis Results for Chemically Modified Asymmetric Membranes*

Example	Treating Agent	Membrane Constant After 1 Hour**	Water Flux After 1 Hour (gfd)	Salt Rejection (percent)
Control	None	0.65	12.9	97.8
1	Acetyl chloride	0.56	11.2	99.15
2	Methacrylyl chloride	0.73	14.6	98.4
3	Succinyl chloride	0.50	9.95	98.5
4	Sebacoyl chloride	0.14	2.6	97.4

*All tests were conducted on 1.0% weight NaCl solution of 19°C and applied pressure of 1,500 psi. The data given for each example represents the average results of two membranes tested.

**10^{-5} grams per square centimeter per second per atmosphere.

Example 2: A membrane was prepared in the same manner as the control membrane in Example 1. This membrane was then esterified by being placed in 200 ml of trifluorotrichloroethane containing 0.6 ml of methacrylyl chloride. After fifty hours at room temperature (25°C) the membrane was removed and the osmotic properties were then tested at the same conditions as in the previous example. These properties are listed in the above table.

Example 3: Membranes were prepared and tested as in Example 2 except that esterification was accomplished using 0.6 ml of succinyl chloride in 200 ml of trifluorotrichloroethane. Membrane properties are given in the above table.

Example 4: Membranes were prepared and tested as in Example 2 except that esterification was accomplished using 0.6 ml of sebacoyl chloride in 200 ml of trifluorotrichloroethane. Membrane properties are given in the table above.

Example 5: Again a membrane was prepared in the same manner as the control membrane in Example 1. In this case acetylation was accomplished by contacting the membrane with the acetylation agent in gaseous form. The membrane was suspended in a closed container which also housed an open flask of liquid acetyl chloride. The membrane was acetylated by reacting with the acetyl chloride vapors for 15 hours at 25°C. Following acetylation the osmotic properties of the membrane were measured on a 2.0 weight percent sodium chloride solution at 19°C using an applied pressure of 1,500 psi. The membrane had a salt rejection of 97.5%, a flux of 5.7 gfd, and a membrane constant of 0.31×10^{-5} grams per square centimeter per second per atmosphere.

High-Humidity Pretreatment of Film Prior to Surface Acetylation

According to a process described by *J.J. Hillman and R.H. Horowitz; U.S. Patent 3,671,516; June 20, 1972; assigned to the U.S. Secretary of the Interior* reverse osmosis membranes are prepared by surface acetylation of cellulose film. A pretreatment consisting of exposure of the film to high relative humidity, prior to acetylation, is employed to provide superior salt rejection.

Examples 1 through 9: The base material in each of these examples consisted of uncoated cellophane film having dimensions of 9 x 16 inches. These films were initially stored at room temperature and 45% relative humidity. The films were then acetylated, either directly or after activation for a period of 3 to 4 days at the relative humidity given in the table. Acetylation was accomplished by reaction of one side of the film with a solution of acetyl chloride in benzene (1:2 ratio) at 25°C for the time given in the following table.

The films were then washed with excess water to remove any acetyl chloride and the resulting membranes were tested at 10,000 ppm NaCl at 1,500 psi in a reverse osmosis test cell of the flat sheet type for a period of time given in the table. Results, i.e., percent rejection and flux, are also given in the table.

Example	Activation (percent RH)	Acetylation Time (minutes)	Duration of Test (hours)	Rejection (percent)	Flux (gfd)
1	None	30	23	37.2	2.33
2	None	60	5	34.5	1.80
3	None	90	23	32.0	2.06
4	79	15	4	51.9	1.77
5	79	22.5	22	77.3	0.90
6	79	30	24	71.4	1.10
7	100	3.75	23	37.4	1.34
8	100	7.5	25	91.9	0.62
9	100	15	2	72.5	0.69

The marked improvement in rejection obtained by activation, particularly at 100% relative humidity, is evident from the data in the table.

Sterilization

According to a process described by *C.E. Lyall; U.S. Patent 3,850,797; November 26, 1974; assigned to Culligan International Company* the semipermeable membrane of a reverse osmosis water treatment unit is treated during periods of down time with a solution having a pH preferably in the range of 4.5 and 6.0 and tending both to sterilize the membrane as well as to retard bacterial and fungal growth. The method comprises: passing an aqueous solution of the desired pH and containing formaldehyde and an acid selected from the group consisting of acetic, formic, citric, fumaric and boric, into the water treatment unit and through the membrane thereof and maintaining the aqueous solution in the unit and in contact with the membrane during extended periods of unit down time to both sterilize the unit and extend the life of its membrane.

Nonaqueous Organic Leach Solvents

In a process described by *A. Sharples and W. Banks; U.S. Patent 3,412,184; November 19, 1968; assigned to Arthur D. Little Research Institute, Scotland and the U.S. Secretary of the Interior* a method for preparing reverse osmosis membranes is presented. The procedure involves casting a thin film from a mixture comprising cellulosic ester and casting solvent, removing a portion of the casting solvent from the film, and contacting the film with a liquid comprising an organic leaching solvent which is miscible with the casting solvent to remove leachable components in the film. The casting solvent may be acetone and the leaching solvent may be methanol. The process can be illustrated by the following diagram.

Casting mix comprising cellulosic ester and casting solvent
↓
Casting into thin film
↓
Evaporating a portion of casting solvent
↓
Contacting with organic leaching solvent
↓
Immersion in water bath (optional)
↓
Product membrane

Control Membrane: A reverse osmosis membrane was prepared using the method described in U.S. Patent 3,133,132. The casting dope was prepared by mixing 22.2 grams cellulose acetate (Eastman 39.8% acetyl content), 66.8 grams acetone, 10.0 grams water and 1.0 gram anhydrous magnesium perchlorate.

A film 0.010 inch in thickness is cast on a glass plate at -8°C, and the solvent is permitted to evaporate for a period of 4 minutes at that temperature. The film is then immersed in cold water at about 0° to 2°C for at least 2 hours. The final desalination properties are effected by heating the membrane in water for 15 minutes at 80°C. A membrane prepared in this manner was tested in a standard reverse osmosis cell using a feed solution of aqueous 3.5 weight percent NaCl (synthetic seawater) at an input pressure of 1,500 psi.

A water permeation rate of 98 ml/hour was obtained and the salt rejection was 98% for the control membrane.

Extraction Annealing Step

C.R. Cannon, P.A. Cantor and W.M. King; U.S. Patent 3,432,584; March 11, 1969; assigned to Aerojet-General Corporation have found that a cellulosic membrane suitable for reverse osmosis may be produced by a process employing a nonheating annealing procedure and more specifically an annealing procedure involving solvent extraction of the excess water from the swollen gel structure. In the process the heating of the membrane is eliminated and replaced with a solvent treatment or annealing step.

In forming the membrane, a solution of cellulose acetate or other cellulose ester is prepared in an organic solvent, for example, acetone, or other known solvents for the cellulose ester. A typical casting solution will comprise 22.2 parts by weight of cellulose acetate and 66.7 parts by weight of the acetone. The solution of cellulose acetate is then mixed with a swelling agent such as an aqueous solution of a perchlorate salt, for example, magnesium perchlorate, or with an organic swelling agent of the type described in U.S. Patent 3,444,286.

A particularly desirable organic swelling agent is tartaric acid. In accordance with known practice, the casting solution is then chilled to about -13°C (in the instance of an acetone solvent) and cast as a membrane at that low temperature. The casting may be accomplished by feeding the chilled solution through a hollow doctor blade with the blade resting on raised brackets at the edges of a glass plate. Typically, the blade is pulled across the plate at a rate such that the film which is formed has a thickness between 10 and 20 mils.

The conventional casting rate is about 18 inches of film per second. Following casting, the film in the instance of an acetone solution is allowed to set for about 3 minutes at -13°C to allow for the evaporation of a portion of the solvent. During the casting operation, the doctor blade and casting surface which may be a glass plate, are all maintained at about -13°C.

Following the casting of the film and partial evaporation of the solvent, the casting plate and the film are dipped into cold water which ranges in temperature from about 1° to 10°C. For an acetone casting solution, the temperature of the water bath is preferably maintained at about 1.5° to 2°C. The casting plate is thrust into the ice water bath in one continuous motion with the plate making an angle of between 30° to 60° with the surface of the water. The film will soon float off the glass casting plate, at which time it will be strong enough to manipulate. The membrane film is rolled up in a damp state. If the film is permitted to dry, it will lose its desirable properties and is unsuitable for desalination.

In this process, syneresis water is extracted from the primary gel structure during annealing by a solvent treatment which comprises immersing the swollen primary gel structure in a water-miscible, organic compound containing polar liquid which is capable of extracting water from the swollen gel structure and thus substantially reducing the water content of the membrane. The water-miscible extracting liquid has little or substantially no solubility for the cellulosic ester.

Among the solvents which may be employed in the annealing process are the lower alkanols exemplified by methanol, ethanol, n-propanol, isopropanol, tert-butanol, isobutanol, and sec-butanol. Methanol and ethanol may be used without water dilution since they have no appreciable solubility for cellulose acetate.

Other solvents which may be used are the aliphatic ketones exemplified by acetone and methyl ethyl ketone. Acetone is a particularly desirable solvent; however, because of its high solubility for the cellulose esters it is necessary to use acetone in solution with water. Among the heterocyclic solvents which may be employed are 1,3- and 1,4-dioxane, tetrahydrofuran, pyrrolidine and pyrazine. Various ones of the water-miscible ethers may be utilized including 2-ethoxyethanol and 2-methoxyethanol. Each of the membranes in the table below was made by forming a casting solution of 22.2 grams of cellulose acetate in 66.7 grams of acetone. This solution was then mixed with 13 grams of a solution containing 10 grams of water and 3 grams of magnesium perchlorate. Following the mixing the solutions were placed in a stoppered bottle which was then sealed, placed in a bottle roller, mixed on the roller for 250 minutes and chilled to about -13°C. They were then fed through a doctor blade which was moved across a glass plate to form a cellulose acetate film.

During casting the doctor blade and glass plate were maintained at -13°C. Following casting, the films were allowed to set for 3 minutes at -13°C, after which they were removed from the glass plate by dipping the plate into ice water at 1.5° to 2°C, the plate being placed into the ice water in one continuous motion making an angle of 30° to 60° with the water level. The films floated off the glass and were then rolled up and kept damp until they were treated by solvent-annealing.

As shown in the table, all of the membranes were capable of removing salt from water. The membranes had, in general, a high flux and a correspondingly high salt permeation. Potable water has a salt content not in excess of 500 ppm. Although most of the films were not capable of producing potable water in one pass through the test cell from a feed stream containing 35,000 ppm of sodium chloride, certain of the films did produce potable water in the one pass from a feed stream containing 5,000 ppm of sodium chloride.

		1,500 psig, 35,000 ppm NaCl				750 psig, 5,000 ppm NaCl					
Solvent	Immersion Time (min)	Flux (gfd)		Salt Permeation (ppm)		Flux (gfd)			Salt Permeation (ppm)		
20% acetone	2	57	-------	20,000	---------	-------	----	----	------	------	------
	4	51	-------	16,000	---------	-------	----	----	------	------	------
	6	46	-------	13,800	---------	-------	----	----	------	------	------
	8	42	-------	14,500	---------	-------	----	----	------	------	------
	10	40	-------	11,900	---------	-------	----	----	------	------	------
	12	40	-------	14,000	---------	-------	----	----	------	------	------
22.5% acetone	4	54	61	10,500	9,200	-------	----	----	------	------	------
	6	72	63	7,800	9,300	-------	----	----	------	------	------
	8	74	69	7,600	8,000	-------	----	----	------	------	------
25% acetone	4	31.5	21	6,900	3,200	41	48	----	950	900	------
	6	31.5	19.5	5,000	3,500	40	30	----	800	800	------
	8	24	19.5	3,500	3,000	40	37	----	950	900	------
	10	28.5	-------	4,800	---------	-------	----	----	------	------	------
	12	25.5	-------	3,400	---------	-------	----	----	------	------	------
	14	26	-------	4,000	---------	-------	----	----	------	------	------
26% acetone	4	-------	-------	-----------	---------	34	31	----	610	500	------
	6	-------	-------	-----------	---------	30.5	28	----	550	500	------
	8	-------	-------	-----------	---------	27	28	----	480	500	------
27% acetone	4	-------	-------	-----------	---------	24	17	----	445	380	------
	6	-------	-------	-----------	---------	20.5	15	----	410	400	------
	8	-------	-------	-----------	---------	22.5	17	----	390	410	------
30% dioxane	4	18.4	-------	3,200	---------	24	----	----	610	------	------
	6	18.4	-------	2,700	---------	21.5	----	----	530	------	------
	8	16.4	-------	2,700	---------	25	----	----	610	------	------
	10	17.6	-------	2,700	---------	31.5	----	----	690	------	------
	12	17.0	-------	2,700	---------	23.5	----	----	580	------	------
	14	14.2	-------	1,950	---------	28	----	----	660	------	------
15% tetra-hydrofuran	2	41	-------	10,500	---------	-------	----	----	------	------	------
	4	38	-------	7,400	---------	-------	----	----	------	------	------
	6	36	-------	8,400	---------	-------	----	----	------	------	------
	8	35	-------	7,400	---------	-------	----	----	------	------	------
	10	33	-------	7,400	---------	-------	----	----	------	------	------
	12	35	-------	6,900	---------	-------	----	----	------	------	------
70% diethylene-glycol ethyl ether (methyl Cellosolve)	4	80	-------	30,000	---------	-------	----	----	------	------	------
	6	66	-------	22,000	---------	-------	----	----	------	------	------
	8	58	-------	25,500	---------	-------	----	----	------	------	------
100% methanol	8	8.4	-------	740	---------	28	39	26	530	540	580
						24.5	31	28	610	690	640

Aqueous Quench

A process described by *B.M Brown and E.L. Ray; U.S. Patent 3,792,135; February 12, 1974; assigned to Eastman Kodak Company* involves manufacturing cellulosic reverse osmosis membranes. The method comprises the utilization of a very hot initial quench bath, in combination with dope compositions that contain acetic acid, formic acid or mixtures thereof, either with or without other organic solvents.

The process comprises the following steps in the order given: (1) casting onto a casting web an appropriate concentrated dope of a film-forming cellulose ester in the form of a film; (2) exposing the resulting cast solution very briefly to air or some other gaseous atmosphere to thereby cause the incipient formation of the requisite active layer on the exposed surface; and (3) immersing the resulting exposed still fluid film into a hot aqueous quench bath (to cause the layer of dope to set up or gel and form a porous asymmetric membrane).

The process differs mainly from conventional reverse osmosis manufacturing processes in that the water into which the exposed liquid film is immersed (immediately after the very brief atmospheric exposure) be at an elevated temperature of from about 100° to about 180°F.

Imbibing Solution

B.M. Brown and E.L. Ray; U.S. Patent 3,772,072; November 13, 1973; assigned to Eastman Kodak Company describe a single phase temperature independent imbibing solution for cellulose ester or ether reverse osmosis (RO) membranes consisting essentially of (a) 30 to 60% glycerin; (b) 25 to 55% water; (c) 3 to 10% nonionic first surface active agent such as polyethylene glycol ether of C_{11} to C_{15} linear alcohol; and (d) 0.1 to 5% of a second surface active agent which is either anionic (e.g., the dioctyl ester of sodium sulfosuccinic acid, or the sodium sulfate derivative of 7-ethyl-2-methyl-4-undecanol), or cationic (e.g., C_8-C_{18} alkyl dimethyl benzyl ammonium chloride). In the method, a cellulose ester or ether reverse osmosis membrane in a water-wet condition is immersed in the imbibing solution, and then is removed and dried under ambient conditions, or by exposing it to heated air.

In the following examples, a 2-inch diameter dry disc of RO membrane is placed in a test cell, and backed up with a 0.45 Millipore filter disc and a porous stainless steel disc having a mean pore diameter of 5, both being on the side of the RO membrane disc remote from its active layer or surface. An aqueous salt solution containing 0.5% NaCl is fed to the active surface of the RO membrane at a pressure of 600 psig and a flow rate of 800 cc/min through a flow channel 0.030 inch wide. The product water passing through the RO membrane is collected for 30 minutes (after a 20 ml wash has been collected and discarded) and salt content and flux are determined. The desalination properties of the membrane are expressed as percent salt rejection, and as product water flux in gallons per square foot of membrane surface per 24 hour day. Salt rejection is calculated as follows:

$$\%\ \text{Salt rejection} = \frac{\text{Salt in feed} - \text{Salt in product water}}{\text{Salt in feed}} \times 100$$

Example 1: Wet membranes are prepared by casting onto a glass plate an acetone-formamide solvent-liquid mixture (60:40) containing dissolved solid cellulose acetate, at a liquid-solids ratio of 3:1. The membrane is quenched in water at 34°F (2°C) after a brief exposure to moving air for partial evaporation of solvent. The wet membrane is dipped in a solution of 55% glycerin 4% Tergiton 15S7 (nonionic), 0.5% Roccal (cationic), and 50.5% water for 10 seconds, the excess is wiped off, and the membrane is dried in air under ambient conditions for 3 hours. The dry desalination parameters are 210 gal/ft²/day at 20% rejection. This is comparable to the wet membrane desalination parameters of 210 gal/ft²/day at 15% rejection.

Samples of the dry membrane are then tempered by dipping in hot water at 168°F and 180°F (79°C and 82°C). The tempered desalination parameters are 65 gal/ft²/day at 80%

rejection when tempered 1 minute at 168°F, and 48 gal/ft²/day at 86.4% rejection when tempered 1 minute at 180°F respectively, which coincides with the desalination parameters generally obtained with a wet heat tempered HT-00 membrane.

Example 2: A similar standard freshly cast wet membrane is quenched with cold water at 34°F (2°C), dipped in a solution of 55% glycerin, 4% Tergitol 15S7 (nonionic), 0.5% Aerosol OT (anionic) and 50.5% water for 10 seconds, the excess solution is wiped off and the membrane is dried under ambient conditions for 3 hours. Samples of the dry membrane are then tempered by dipping in hot water at 182°F (85°C) for 1 minute. The membrane has desalination parameters comparable to the membrane of Example 1 dried with Roccal (21.3 gal/ft²/day at 97% rejection).

Stretching Technique During Initial Gelation Period

According to a process described by *S. Manjikian and M.I. Foley; U.S. Patent 3,657,115; April 18, 1972; assigned to The U.S. Secretary of the Interior* semipermeable membranes are prepared by casting a cellulose ester from a solution of the ester, and subsequently stretching the membrane, during its initial gelation period, in a direction at a right angle to the direction of casting. The resulting membranes exhibit superior selectivity in desalination of water by reverse osmosis.

Example: Two casting solutions, having the following compositions in percent by weight, were employed.

	------Amount, wt %------	
	Solution A	Solution B
Cellulose diacetate	25	25
Acetone	45	35
Formamide	30	30
Pyridine	30	10

The solutions were cast into films by means of a doctor blade passing at a controlled distance over a latex (rubber) sheet secured to a rack type apparatus designed for stretching the sheet. The latex sheet was attached at one end to the stationary bar of a frame having side rails. The other end of the sheet was attached to a movable bar riding on these side rails. The stationary and movable bars could be moved apart or together by means of a screw thread actuating rod. The cast membrane solution had a thickness of about 0.012". The cast membrane was then allowed to gel over a 30 second dry time by exposure to the atmosphere at ambient conditions of temperature and pressure. Simultaneously, during this initial gelation period the latex sheet was uniformly stretched in directions both at a right angle and parallel to the direction of casting to give an elongation of the membrane as set forth in the data of the following table.

Direction of stretching	Casting solution	Elongation, percent	Flux, gal./sq. ft. day	Salt rejection, percent
Right angle to direction cast.	A	0	26.1	93.00
		1	30.9	92.86
		2	25.1	95.69
		3	22.5	96.61
		5	22.8	96.08
		7	11.7	96.78
		10	14.0	97.06
Parallel to direction cast.	A	0	18.85	96.46
		1	36.40	87.75
		2	36.40	88.45
		3	23.92	95.83
		5	21.32	95.94
		10	25.22	94.82
Right angle to direction cast.	B	0	24.44	92.94
		1	24.96	92.86
		2	21.84	94.78
		3	20.74	95.12
		5	19.76	96.18
		10	18.98	96.12
Parallel to direction cast.	B	0	21.36	93.84
		1	28.60	89.08
		2	29.90	82.60
		3	33.80	80.82
		5	33.40	78.30
		10	30.16	83.68

Thickness of the membrane following initial gelation and stretching was about 0.008 inch. The membrane was then placed in ice water for a period of one hour, and then heat treated by immersion in water at a temperature of 78°C, for 10 minutes. The membranes were then tested in a conventional apparatus of the type described in U.S. Patent 3,133,137. Results are given in the previous table. It will be seen from the data of the table that stretching of the membrane produced a gain in selectivity of about 2 to 4% with only a moderate reduction in flux. It will also be noted that stretching in a direction parallel to the direction of casting resulted in decreased selectivity.

Prolonging Membrane Life by Control of Water pH

C.J. Malm, M.E. Rowley and N.G. Baumer; U.S. Patent 3,342,728; September 19, 1967; assigned to Eastman Kodak Company have found that the effective life of cellulose acetate desalinization membranes is limited because the acetyl content of the membrane is reduced as the desalting procedure is carried out. This appears to be caused by the slight alkalinity of salt-containing water as occurs from natural sources, e.g., brackish water or sea water. It has also been found that the acetyl reducing effect of the water on the cellulose acetate membrane is appreciably reduced or eliminated by lowering the pH of that water to within the pH range of 4.2 to 7 either by the addition of mineral acid or by using carbon dioxide in the portion of the desalinization procedure where the water comes in contact with the acetate membrane. When the water being desalted is thus treated the effective life of the cellulose acetate membrane is measurably increased and the rate of reduction of acetyl content is substantially reduced or even eliminated so that the same membrane can be used for long periods of time without losing effectiveness.

To show the effectiveness of the method, accelerated tests were carried out showing that with the use of water without reduced pH the cellulose acetate membrane is so reduced in acetyl that acetone insolubility results whereas reduced pH reduces or eliminates this acetyl removing effect. When the cellulose acetate in the membrane is so reduced in acetyl content as to be insoluble in acetone effectiveness as a selective membrane for the desalinization of water is lost. The tests shown below were made at atmospheric pressure of various samples of saline, Sample 1 duplicating the composition of sea water and Sample 2 being a saline solution without any sodium carbonate therein. The cellulose acetate used had an acetyl content initially of approximately 40.2%.

Accelerated tests were also made in closed containers in which the membrane was in contact with simulated sea water but there was no way provided for CO_2 to escape so that the treating water was at nonalkaline pH over substantially the whole time of treatment. It may be seen that by having CO_2 present under pressure during the entire time of treatment hydrolysis of the cellulose acetate was prevented and the cellulose acetate remained acetone soluble.

Stability Tests of E-398-3 Film as Used in Water Desalinization
Tests Made at Atmospheric Pressures

	Preparation of Solutions and Quantity of Film Tested	pH Adjustment (prior to film addition)	Heating Time and Conditions	pH After Heating	% Acetyl and Film Solubility
1	500 g of 3.5% NaCl solution, 0.156 g $NaHCO_3$, 4 g film	None (initial pH of 8.0)	14 hr, refluxed	5.4	38.0, acetone insoluble
2	250 g of 3.5% NaCl solution, no $NaHCO_3$, 2 g film	4.5 by adding 1 drop of 0.1 N HCl	14 hr, refluxed	4.2	39.6, acetone soluble
3	250 g of 3.5% NaCl solution, 0.078 g $NaHCO_3$, 2 g film	6.5 by adding 0.1 N HCl	14 hr, steam bath	6.5	39.5, acetone soluble
4	250 g of 3.5% NaCl solution, 0.078 g $NaHCO_3$, 2 g film	7.0 by adding CO_2 gas	14 hr, steam bath	6.5	39.9, acetone soluble
5	250 g of 3.5% NaCl solution, 0.078 g $NaHCO_3$, 2 g film	6.4 by adding CO_2 gas	14 hr, steam bath	6.4	40.1, acetone soluble
6	250 g of 3.5% NaCl solution, 0.078 g $NaHCO_3$, 2 g film	5.4 by adding CO_2 gas (saturated)	14 hr, steam bath	5.8	39.8, acetone soluble

Also a test was run for an extended length of time using an enclosed container which would correspond to a prolonged use of a cellulose acetate membrane. The results of that test which are as follows show that there was no indication that when water is desalinated in accordance with the process the cellulose acetate membrane would necessitate replacement after use for long periods of time.

Stability Tests of E-398-3 Film as Used in Water Desalinization Tests Made for Extended Lengths of Time (Closed Containers)

Preparation of Solutions and Quantity of Film Tested	pH Adjustment (prior to film addition)	Heating Time and Conditions	pH After Heating	% Acetyl and Film Solubility
250 g of 3.5% NaCl solution, 0.078 g $NaHCO_3$, 2 g film	5.4 by adding CO_2 gas	70 hr, steam bath*	5.8	39.6, acetone soluble

*Solution was changed after each 24 hr period; pH after heating was that after each 24 hr.

Addition of Surfactant

A process described by *S. Loeb and S. Manjikian; U.S. Patent 3,310,488; March 21, 1967; assigned to The Regents of the University of California* involves a method of preparing a semipermeable membrane adapted to permit substantially selective diffusion of water from an aqueous solution and substantially improving the performance of such membrane by addition of a cationic surfactant in small concentration to the aqueous solution. The method comprises dissolving a film forming cellulosic derivative in a volatile carrier solution adapted to permit the structural organization of a permeable membrane from the cellulosic derivative by evaporation of the carrier solution.

The resulting product solution is then cast to form a thin layer of product solution and a portion of the carrier solution is then evaporated for a predetermined period of time to form a membrane. The cast membrane is then immersed in water to remove the carrier solution and then heated. Next, an aqueous solution having a high salt concentration is applied to one side of the resulting membrane under pressure and a small concentration of cationic surfactant is added to the aqueous solution. Finally, water of low salt concentration is recovered from the other side of the membrane. A variety of cationic surfactants were tested and found to improve desalinization by 35 to 60%, while other surfactants, such as nonionic and anionic surfactants, resulted in either no change or an increase in the product salt content.

GENERAL MEMBRANE PREPARATION PROCESSES

Spraying Technique

According to a process described by *J. Woo Woo, P.F. Waters, R.H. Horowitz and A.F. Hadermann; U.S. Patent 3,662,046; May 9, 1972; assigned to the U.S. Secretary of the Interior* reverse osmosis (RO) membranes are produced by spraying a film-forming solution onto a substrate in a zone into which is injected, throughout spraying, an inert gas saturated with polymer solvent vapor. Solvent is then evaporated from the film by reducing the degree of solvent saturation of the inert gas in the evaporation zone by introducing gas that is less than saturated with solvent, by lowering the atmospheric pressure in the zone or by heating the atmosphere. The film is then cured in the prior art manner by quenching and annealing. Electrostatic spraying is most effective.

Referring to Figure 1.3, reference numeral **1** designates a cylindrically shaped, closed spraying chamber. Film-forming solution **2** is supplied to the spray chamber from an adjacent reservoir **3** housed in a closed chamber **4**. A conduit **5** conveys solution to spray chamber **1** from reservoir **3**, and the solution enters the spray chamber through spray nozzle **6** at the bottom of the conduit. Most prior art spraying devices are suitable for aerosolization. For example, nozzle **6** can be of the rapidly rotating disc-type wherein solution is flung by centrifugal force to the periphery of the disc and outward as a fine spray. Temperature control is effected over spray chamber **1** with an appropriate jacket or coil. Likewise, the temperature of chamber **4** is controlled in the same manner, if desired. A spraying temperature of about minus 20° to about 60°C is suitable for most solutions heretofore employed to produce RO membranes. During the operation, to maintain control of the atmospheric composition and pressure in spray chamber **1**, and also to maintain additional control of the temperature therein, a body of solvent **7** is provided in solvent reservoir **8** housed in a temperature-controlled compartment **9**. Solvent **7** is the same as the polymer solvent in film-forming composition **2**. An inert (nonreactive) carrier gas such as nitrogen, injected into the system at conduit **10**, is diffused through solvent **7** by glass frit **11**, and then passes through conduit **12**, as solvent vapor-containing carrier gas, to spray chamber **1**.

FIGURE 1.3: MEMBRANE PRODUCTION PROCESS EMPLOYING SPRAYING TECHNIQUE

a.

b.

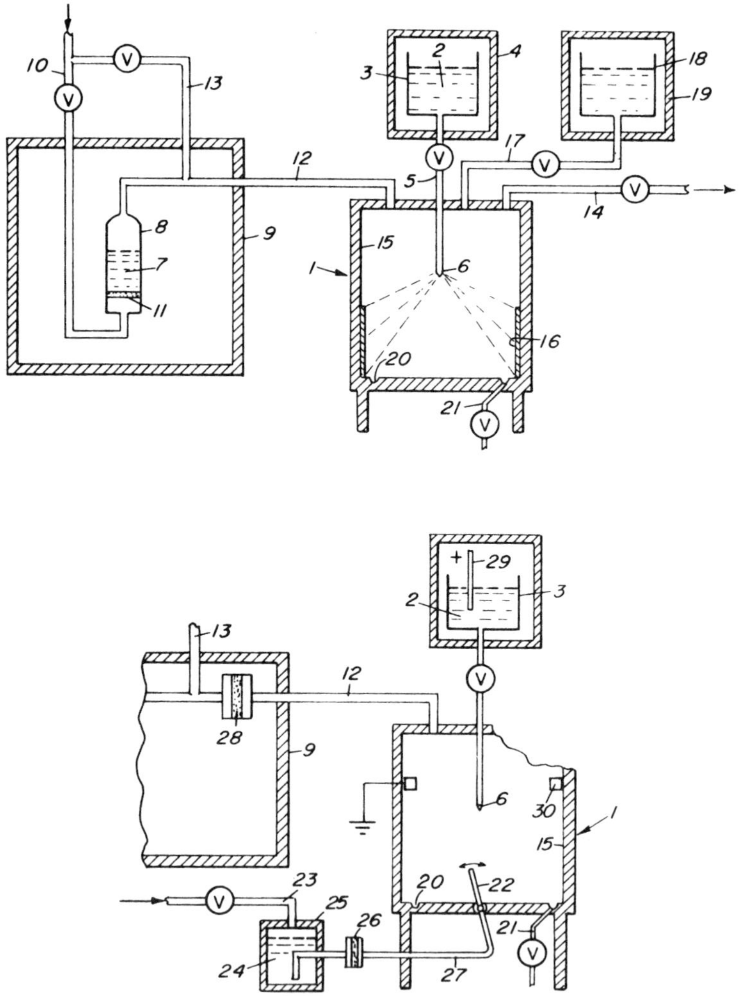

Source: J. Woo Woo, P.F. Waters, R.H. Horowitz and A.F. Hadermann; U.S. Patent 3,662,046; May 9, 1972

A by-pass conduit **13** is connected to conduit **10** to provide in spray chamber **1**, when desired, solvent-free carrier gas or a mixture of solvent-saturated and solvent-free gas. Carrier gas exits from spray chamber **1** through conduit **14**. During the spraying step, the inert gas continuously carries solvent vapor into chamber **1**, and gas is slowly and continuously withdrawn through exit conduit **14**. As to the amount of vapor in the injected gas, it is present in a saturated or near-saturated amount (e.g., within 0.5% of saturation) relative to the temperature in the spraying zone.

Inner cylindrical wall **15** of spray chamber **1** on which membrane or film **16** forms during spraying is fabricated from materials heretofore employed to produce RO membranes by casting. For example, glass, metals such as stainless steel, plastics, or ceramics are suitable. Substrates not previously employed such as fabrics, felts, porous ceramics or porous polymers can also be employed. After film **16** is deposited to a desired thickness, spraying is discontinued and conditions are adjusted to evaporate some solvent from the exposed face of the membrane. This can be accomplished by raising the temperature about 1° to 5°C within spray chamber **1**. Alternatively, evaporation is attained by reducing the pressure in the chamber, or the atmospheric composition therein is adjusted to reduce the partial pressure of solvent vapor by the addition of solvent-free carrier gas via by-pass conduit **13**.

Although the inert gas fed through conduit **12** to chamber **1** during the evaporation step may be solvent-free, the presence of some solvent vapor therein provides additional control over the rate of evaporation. Preferably, the vapor is present in an amount of about 5% to about 95% of saturation relative to the temperature in the evaporation zone. With regard to the usual polymer solvents (e.g., acetone, dioxane, or acetone/formamide), evaporation time ranges from 30 seconds to 10 minutes under the following exemplary conditions: atmospheric pressure, evaporation temperatures of about 20° to 25°C, and the presence of solvent vapor in the carrier gas in an amount of about 0 to 99% of saturation.

Next, spray chamber **1** is flooded with quenching liquid such as water, dioxane/water or acetone/water. This is supplied through conduit **17** from adjacent liquid reservoir **18** housed in temperature-controlled chamber **19**. A quenching temperature of about 0° to 15°C is preferable. Quenching liquid is drained from spray chamber **1** through annular gutter **20** and exit conduit **21**, after which the spraying chamber is disassembled, and the membrane removed and annealed, if desired, in the prior art manner.

Referring to Figure 1.3b, therein is shown a modification of the system of Figure 1.3a which enables ultra-thin RO membranes (e.g., as thin as 600 A) to be produced. First, it is necessary to carry out the spraying in a substantially dust-free environment. To accomplish this, the interior of chamber **1** is washed, prior to spraying, with a solvent such as acetone, sprayed through swivel nozzle **22** connected to the bottom of the chamber. Fluid pressure supplied through conduit **23** forces washing solvent **24** out of reservoir **25** through glass frit filter **26** and conduit **27** to the swivel nozzle. Dust-containing washing solvent flows down wall **15** and leaves the chamber by gutter **20** and conduit **21**. Dry nitrogen supplied through by-pass conduit **13** and conduit **12** then removes any traces of the solvent.

A glass frit filter **28** disposed in conduit **12** insures against particulate matter passing to the spray chamber during any part of the operation. Further, film-forming solution is passed through a filter such as a medium porosity Pyrex glass filter prior to being filled into solution reservoir **3**. Spraying is carried out electrostatically in this modification, and the necessary potential is established by a positively charged electrode **29** immersed in solution reservoir **3** and grounded ring-shaped electrode **30** adjacent and touching conductive wall **15**. A voltage range is preferably of the order of 10 to 50 kv depending primarily on the electrical properties of the solution (i.e., dielectric constant and specific resistivity), and on its surface tension. For example, solutions of low specific resistivity may lose charge rapidly necessitating higher voltages. The following example illustrates the forming of an RO membrane by electrostatic aerosolization.

A solution consisting of 26.8% (by weight) cellulose acetate, 58% acetone and 14.7% formamide was sprayed at 25 kv and 24°C onto a glass substrate in an atmosphere saturated with acetone over a period of 69 minutes to a final thickness of 0.0022 inch. The temperature was maintained at 24°C, and dry nitrogen at 0% saturation relative to acetone was added for 20 seconds to allow the film to evaporate. It was then quenched in water at 0°C for a period of one hour. Thereafter the membrane was removed from the substrate and annealed at 85°C for 5 minutes.

Using a 0.5% (by weight) NaCl feed solution at 1,000 psi flowing through a desalination cell at 250 ml/min, this membrane gave a salt rejection of 95.6%. The corresponding rate of production of desalinated water was 15 gfd. Using a 3.5% by weight NaCl feed solution of 1,500 psi flowing through the desalination cell at 250 ml/min, the membrane gave a salt rejection of 93.5%. The corresponding rate of production was 8.3 gfd. A system for carrying out the process in a continuous manner is also described.

Spraying Technique Using Discs

W.M. King; U.S. Patent 3,764,526; October 9, 1973; assigned to The U.S. Secretary of the Interior has found that dynamic membranes can be formed from suspensions of ultrathin membrane discs which are in themselves tiny reverse osmosis membranes. Thus, it has been found that these ultrathin membrane discs may be prepared from a solution of a conventional membrane material such as cellulose acetate by spraying the solution at controlled conditions onto an appropriate casting surface. Numerous droplets are formed on the casting surface and the evaporation of the solvent from these droplets forms the membranes. After appropriate post casting treatments a dilute suspension of these membrane discs can be prepared and used in the formation of dynamic reverse osmosis membranes.

One of the advantages of this process is that a wider choice of membrane-forming materials can be used than has previously been the case with continuous asymmetric membrane technology. For example, it is feasible to use polymers which possess good intrinsic osmotic characteristics, but which are either very expensive or not amenable to fabrication into pressure stable asymmetric membranes. Numerous materials may be used to form the membrane discs. In particular, the process may be used to form ultrathin membrane discs from solutions or melts of neutral polymers such as cellulose acetate, cellulose nitrate, cellulose butyrate, polyamides, polyvinyl carbonate and similar compounds.

A dilute solution of this membrane material is then sprayed on a casting surface. In this step there are a number of variables which determine whether discs will be formed and the quality of the discs formed. Among these variables are the relative speed on the spraying nozzle to the casting surface, the distance of the nozzle from the casting surface, and the nozzle pressure. It is difficult to specify conditions which will give the best results for membrane production; in any situation these operating conditions must be discovered through trial and error. From experience to date, however, the following rough guidelines can be used: relative speed of nozzle to casting surface, 8 ft/sec; distance of nozzle from casting surface, 11 to 14 inches; and nozzle pressure, 60 to 100 psi.

When selecting the proper parameters the distance of the nozzle from the casting surface and the relative velocity of these devices should be chosen so that regularly shaped individual droplets are formed. If these conditions are not properly controlled, the droplets may either coalesce or take on irregular shapes which would detract from their ability to form dynamic membranes. The nozzle pressure is adjusted to control the size distribution of the discs formed. Depending on the type of spray nozzle used, the pressure employed may make a considerable difference in both the average diameter and the size distribution of the discs formed.

Example 1: Ultrathin membrane discs were prepared from 1% solutions of cellulose acetate by spraying the solution onto a glass side. Spraying was accomplished using a nozzle with a conical spray pattern (Spraying Systems nozzle No. 60), at a nozzle pressure of 100 psi. The height of the nozzle above the casting surface casting distance was varied and the translational speed of the nozzle relative to the surface was chosen from among the alternatives of 2, 4 or 8 ft/sec. The degree and quality of disc formation was then determined by microscopic inspection. Some of the conditions at which discs were successfully formed are shown in the following table. Several of the examples are shown in which slight variations in these conditions failed to produce ultrathin membrane discs.

Test	Solvent system	Casting speed (ft/sec)	Casting distance (inches)	Description of the spray-cast particles
A	Dioxane	8	18	Discs, structured center.
B	Do.	11	24	Some discs, plus flat irregular shapes.
C	1:1 Dioxane: 1,1,2 Trichloroethane	8	18	Disc, structured center.
D	Do.	2	18	Polymer coalesced.
E	4:1 Dioxane: Methylethyl ketone	8	18	Fair discs, structured centers.
F	Do.	8	11, 8	Tendency to coalesce.
G	Do.	2	14	Polymer coalesced.
H	Do.	4	14	Some discs, also some irregular shapes.
I	Do.	8	14	Discs, structured centers.

Example 2: In Example 1 membrane discs were prepared using glass as the spraying surface. Another suitable surface is wetted chromed plated which has been prepared by flaming the chromed surface and subsequently immersing it in water. Ultrathin membrane discs were prepared by spraying 1% solutions of cellulose acetate onto this wetted chromed surface and subsequently removing the discs with warm (50°C) deionized water. A nozzle with a flat spray (Spraying Systems nozzle No. 650017) was used at a pressure of 100 psi, a height of 12.5 inches above the surface and a translational speed of about 4 ft/sec. Solutions and conditions yielding well shaped discs are indicated in the following table. The quality of the discs formed was determined by microscopic examination, and the percentage by weight of the ultrathin membrane discs which could be removed from the surface by washing was calculated.

Test	Solvent system	Results	Percentage removal by washing
A	4:1 Dioxane : methylethyl ketone	Good discs easily removed from plate.	
B	Cyclohexanone	Flat, slightly irregular shape	90
C	3:1 Ethyl acetate : acetic anhydride	Flat, slightly off-round shape	80
D	2:1 Ethyl acetate : acetic anhydride	Do.	80
E	1:1 Ethyl acetate : acetic anhydride	Do.	80
F	1:2 Ethyl acetate : acetic anhydride	Do.	80
G	1:3 Ethyl acetate : acetic anhydride	Do.	80

Room Temperature Casting

A. Sharples and W. Banks; U.S. Patent 3,439,074; April 15, 1969; assigned to Arthur D. Little Research Institute, Scotland and the U.S. Secretary of the Interior describe a method for manufacturing desalination membranes which comprises casting a solution of a film-forming cellulosic ester and a swelling agent in an organic solvent to form a film, partially evaporating the solvent from the film in order to produce a desalination layer on the side of the film in contact with air during casting, immersing the cast film in an aqueous bath consisting essentially of water, with heating if necessary, to remove the swelling agent and solvent and to consolidate the desalination layer and render it more effective, which method is characterized by selecting the concentration of the cellulosic ester and swelling agent and the temperature to which the membrane is heated from ranges which enable a satisfactory membrane to be obtained when casting, solvent evaporation and immersing are effected at room temperature.

The cellulosic ester is preferably cellulose acetate. The swelling agent may be an aqueous solution of an inorganic perchlorate salt, e.g., magnesium or sodium perchlorate, or other swelling agents such as zinc chloride may be used. It has been found that the process of desalination occurs in a pore-free layer formed during evaporation of the solvent on the side of the membrane in contact with air during casting. The remainder of the membrane may be porous but these pores must be absent in the actual desalination layer. It has also been found that a relationship exists between the useful properties of the membrane (water permeation and salt rejection) and the following process variables: (a) concentration of the swelling agent and cellulose acetate in the casting solution; (b) temperature and time of casting and evaporating; (c) temperature of immersion; and (d) temperature to which the membranes are heated. The process provides two possible methods for preparing membranes at reduced cost, namely: by working at room temperature during the casting and evaporating stage, but retaining low temperature (0° to 2°C) for the immersion step; and, by working at room temperature for both casting and immersing. Examples of the preparation of membranes according to the process are described below.

Example 1: A casting solution was prepared at 15°C from 22.2 grams cellulose acetate, 66.8 grams acetone, 10 grams water and 2 grams magnesium perchlorate (anhydrous). It was spread at room temperature with a doctor blade or knife on glass plates fitted with side runners of a height equal to the desired film thickness. The acetone was allowed to evaporate for 1 minute, and then the film was stripped. The uniform film thickness was equal in this case to 0.01 inch. By varying the side runner different film thicknesses may be readily obtained.

Partial evaporation of the solvent at about 15°C and initial organization of the film occurred during a 1 minute period. The film was then immersed in water at 15°C for 1 hour during which complete removal of the solvent was effected by diffusion into the water. Partial removal of the magnesium perchlorate also took place. The film was then heated at 73°C in water for 15 minutes to complete the organization of the film for high flow desalination of seawater.

Example 2: A casting solution was prepared from 22.2 grams cellulose acetate, 66.8 grams acetone, 10 grams water and 1.1 gram magnesium perchlorate. The solution was cast for 1 minute at room temperature, immersed in ice water and heated at 73°C in aqueous bath to yield a membrane having high throughput and low salt permeation. In the above two examples it was determined that salt rejection (decrease in salinity from the feed solution to the effluent, was greater than 94.5% for 3.5% sodium chloride solution. The water flux through the membrane for the first and second examples respectively were 1.3 and 1.4 milliliters per hour per square centimeter of membrane surface for a total applied pressure of 100 atmospheres.

Experiments to modify the casting procedure were carried out using standard cellulose acetate casting dope. Membranes were cast at about 15°C, immersed in ice water (0° to 2°C) and then heat treated at various temperatures. The properties of these membranes are recorded in the following table. The feed solution to the membranes was 3.5% NaCl at 1,500 psi.

Characteristics of Membranes Prepared by Casting at Room Temperature for 1 Minute and Immersing in Ice Water (0° to 2°C)

Heat Treatment	Water Flux, ml/hr	Salt Rejection (% Salinity Decrease)
None	780	12
Heated at 62°C	180	90
Heated at 70°C	120	95.5
Heated at 73°C	112	98.5
Heated at 75°C	104	98.6
Heated at 80°C	64	98.4

The above data were obtained using a standard membrane casting solution as in Example 1 and a standard reverse osmosis cell. The data for salt rejection and flow rate for the membrane heat treated at 73°C are virtually identical to the corresponding values for a standard Loeb-type membrane. Thus a slight decrease in the temperature of heat treatment (from 80° to 73°C) has resulted in the tremendous difference in membrane characteristics.

Ultrathin Membrane

J.E. Cadotte and P.S. Francis; U.S. Patent 3,580,841; May 25, 1971; assigned to the U.S. Secretary of the Interior describe an ultrathin semipermeable reverse osmosis membrane which may be used in combination with a porous support. The combination of ultrathin membrane and porous support gives a laminated composite membrane product which is in some ways analogous to the prior art modified membrane but which has the advantage of containing discrete laminas which perform different functions. These separate laminas, because they are made independent of each other can each be tailored for their purpose with much more precision than can the active or porous regions of the modified membrane.

The ultrathin membrane differs from prior art modified membranes in at least two aspects. First, they are not asymmetric and second, they are many orders of magnitude thinner than the prior art modified membranes. Loeb et al in U.S. Patents 3,133,132, 3,133,137 and 3,170,867 show that modified membranes are in the order of from 0.0025 to 0.0040 inch (63.5 to 101.6 microns) in thickness, whereas the ultrathin semipermeable membrane of this process is less than 1 micron in thickness. Broadly, a range of thickness of from 0.05 to 0.50 micron is desirable with the preferred range being from 0.10 to 0.25 micron.

The ultrathin membrane must be semipermeable possessing a favorable selectivity to the desired product solvent and a high degree of rejection toward the solute. Compositions which have semipermeabilities toward various solutions are well known in the art. This is especially true in the field of saline solutions where the Office of Saline Water and others have investigated a large number of materials and have determined their permeabilities to salt and water.

While the thickness of the membrane and its semipermeability are the most important characteristics, other recognized measurements may be used to establish preferred properties. Two constant terms "A" and "B" have been used to describe membrane properties. Constant A measures the permeability of the membrane to solvent and is expressed in terms of g/cm^2 sec atm. Thus, this measurement is an indication of solvent flux at a given pressure and is a particularly important consideration when the desired product is the pure solvent. This constant A has been found to be dependent on the thickness of the ultrathin membrane.

The second constant B measures the permeability of the membrane to solute and is expressed in terms of cm/sec. This constant while not pressure dependent, varies for individual solutes. Constant B by itself is not an indication of a desirable membrane property as is the ratio of A/B, which is the ratio of solvent to solute permeability and is an important indication of membrane selectivity. It should be noted that the importance of these constants depends upon the type of process in which the membrane is to be used.

This is true because the requirements of selectivity and throughput will vary in different reverse osmosis applications and also because the measurement of the constants are dependent upon the type of solution being separated. In the broad sense then, these parameters reflect not only properties of the membrane but also a desired condition. Consequently, a broad definition of the ultrathin semipermeable reverse osmosis membrane cannot be expressed in terms of these constants. However, membranes designed for a specific utility may be so defined. For example, membrane for separating solvent from aqueous saline solutions such as seawater or brackish waters should have a minimum A value of 1×10^{-6} g/cm^2 sec atm and an A/B value of at least 0.1 g/cm^3 atm in a 3.5% brine at 1,500 lb/in^2. These values may easily be determined by recognized means which include a coulometric analysis of product and a salt content determination such as obtained by a conductivity measurement.

Ultrathin semipermeable reverse osmosis membranes of the compositions and properties noted above may be prepared by one of two methods. It has been found that good ultrathin membranes may be made by casting a concentrated solution of the polymer on a liquid surface. Liquids upon which membranes have been cast include water, aqueous inorganic salt solutions, aqueous urea solutions, glycerin and mercury. One limitation on the use of any liquid is that the membrane must not be soluble in it. A number of solvents may be used to form concentrated casting solutions. The solvent should be either insoluble or sparingly soluble in the liquid over which the casting solution is to be poured. When using water as the base liquid, useful solvents include esters, ethers, ketones, alcohols, hydrocarbons, halogenated hydrocarbons, mixtures of these with each other or with small amounts of dimethylformamide or dimethylsulfoxide. To prepare the casting solution, a polysaccharide should be dissolved in the solvent to give a solution having the consistency about equal to varnish or light syrup. This viscosity will generally be obtained by solutions of 0.5 to 25% polymer concentration and usually will be found to be in the range of 3 to 15% polymer. The casting solutions spread spontaneously when poured on the liquid surface. The temperature of the casting step has not been found to be critical. However, good results have been obtained when operating at about 130°F. After the solution has been cast the solvent is allowed to evaporate leaving a polymer residue in the form of a membrane. The thickness of this membrane may easily be controlled by regulating such factors as casting solution concentration and viscosity, and in some instances by manually extending or expanding the solution.

The second method of forming an ultrathin membrane consists basically of slowly drawing a clean glass plate from a dilute solution of a polymer in a solvent. Solvents which may be used in this process include ketones, pyridine, dioxane, methyl acetate and mixtures of acetone with alcohols. As with the first method described, the concentration may be altered to vary the thickness of the resulting membrane. The operative concentration for a given polymer and solvent is a matter of choice and may easily be established by routine tests. It should be noted that the thickness of the membrane may also be controlled by the glass withdrawal rate. The ultrathin semipermeable membranes are particularly effective for reverse osmosis processes when placed in combination with a porous support. The combination is formed by separately fabricating the ultrathin membrane and a porous support and subsequently placing the two in surface to surface contact.

The function of the porous support is to protect the ultrathin membrane and to provide a solvent passage means whereby solvent may be transferred away from it downstream (low pressure) side. The porous support must accomplish these tasks under operating conditions without creating a large backup pressure and without harm to the delicate ultrathin membrane. To accomplish this end, the porous support must possess certain properties. It should have a pore size at the surface in contact with the ultrathin membrane small enough to support the ultrathin membrane, large enough to permit viscous flow.

Pore sizes should generally be smaller than about 5000 A and pores in the range of from about 500 to 2000 A are preferred. Further, the support must be able to stand the pressures of reverse osmosis operation without bursting and without compaction which restricts flow and builds up back pressure. Finally, the porous support should be relatively thin to cut down on equipment size.

Thus, porous supports will generally be in the range of from 1 to 20 mils with a preferred range of from 2 to 7 mils. There are a number of materials presently available which may serve as porous supports. These materials include filter materials such as type VF and VM Millipore filters, which are fabricated from a mixture of cellulose nitrate and cellulose acetate, commercial cellulose triacetate, metallic membranes of silver and other materials such as those made in accordance with U.S. Patent 3,287,112, and porous glasses and ceramics.

The following examples are illustrative of the process. In each example the ultrathin membrane formed was between 0.15 and 0.25 micron (1500 to 2500 A) in thickness as measured by infrared absorption techniques or by color interference patterns viewed with oblique light. The test cell was a noncirculating or batch system of the type shown in U.S. Patent 3,133,132. The membrane composite was supported by a smooth-calendered wet-strength kraft paper which in turn was supplied mechanical support by a ¼ inch porous sintered stainless steel plate.

The system was pressurized with nitrogen and runs were conducted with a 3.5 weight percent NaCl brine at 1,500 psi. The effluent from the cell was let out of the cell by a hypodermic needle and stored in pipettes. Membrane constants for the ultrathin membrane of the examples are tabulated below.

Composition	A, x10^{-6}	A/B
Cellulose acetate	2.42	0.127
9/1 mix of cellulose acetate and acetylated hydroxyethyl cellulose	1.68	0.322
Guar triacetate	1.02	0.575
Locust bean gum triacetate	1.47	0.520
Beta-glucan triacetate	2.25	0.350

Example 1: A solution of 1 part cellulose acetate E-394-60, 20 parts cyclohexanone and 1 part water was prepared and was cast on deionized water at 130°F and annealed at 176°F for 30 minutes. Meanwhile, a porous support membrane was formed by casting a mixture containing 11.1 parts E-394-60 cellulose acetate, 6.6 parts of acetone, 10.0 parts of water and 5.5 parts magnesium perchlorate on a solid substrate into a layer 10 mil thick at -10°C. This film was immediately immersed in ice water, with no evaporation step. After soaking for 1 hour in ice water, the membrane was stripped off its substrate and stored in water at room temperature. The finished thickness of this support was between 3.5 and 4 mils.

The ultrathin membrane was then laminated to the porous support membrane and placed in the test cell. Lamination was accomplished by first placing the porous support on a piece of rigid composition board, immersing the porous support and board beneath the ultrathin membrane which was floating on water bringing one edge of the porous support into contact with an edge of the ultrathin membrane, and then drawing the board out of the water at an acute angle.

Example 2: An ultrathin membrane was formed by casting a mixture of 9 parts E-394-60 cellulose acetate, 1 part acetylated hydroxyethyl cellulose and 200 parts of cyclohexanone on water at 130°F and then annealing at 170°F for 30 minutes. This membrane was laminated with a cellulose acetate support as described in Example 1 and placed in the test cell.

Example 3: Guar gum (10 grams of Guartec UF) was dispersed in a liter of water by mixing 5 to 10 minutes in a Waring Blendor. The dispersion was heated 3 hours in a pressure cooker at 120 psi, cooled and the guaran precipitated by addition of 1 liter of 95% ethanol. The precipitate was washed with pyridine on a sintered glass filter to replace the ethanol. Then 225 ml of pyridine plus 75 ml of acetic anhydride were added, and the mixture was stirred 24 hours at 100°C to give a dark solution. The solution was poured slowly into a liter of cold water to precipitate guar triacetate, which was washed thoroughly and dried at 60°C. A 1% solution of guar triacetate in 500 ml of 95/5 acetone/water was prepared, and 230 ml of 75/25 hexane/acetone was added slowly. Cooling to -10°C precipitated 63% of the polymer. Addition of hexane to the supernatant liquid gave a second fraction of 31%.

One part of a high molecular weight fraction of guar triacetate was mixed with 20 parts of cyclohexanone and cast on water at 75°F and subsequently allowed to evaporate to form a membrane. This ultrathin membrane was laminated with the type of support prepared in Example 1 and placed in the test cell.

Colloidal Thoria Sol Membrane

H.A. O'Hern; U.S. Patent 3,923,654; December 2, 1975; assigned to Owens-Illinois, Inc. describes a rigid, dimensionally stable high-flux, anisotropic, composite, ultrafiltration membrane for use in the ultrafiltration of solute from an aqueous feed stock, where thoria is filtration deposited from a dilute, aqueous colloidal thoria sol on a porous, rigid, liquid-permeable support to form the composite membrane where the initial solute rejection rate of the membrane is increased by incorporating sulfate ion into the thoria sol in the molar ratio of sulfate ion (SO_4) to thorium (Th) of about 0.9 x 10^{-2} to about 4.5 x 10^{-1}, and preferably in the ratio of about 1.35 x 10^{-1} to about 3.25 x 10^{-1}.

Figure 1.4a illustrates a schematic process flow diagram for filtration deposition of membranes from a thoria sol and use of such membrane for separation of a feedstock. In Figure 1.4a the thoria sol flows into chamber **10** under sufficient pressure to cause an aqueous permeate fraction of the sol to permeate the porous support **11** which is securely positioned in chamber **10**. The aqueous permeate is depleted in thoria as compared to the thoria sol. The fraction of the thoria sol which does not permeate the porous support leaves chamber **10** as the retentate stream. During the initial stages of the deposition, the flux rate in gfd through the membrane is very high and it gradually decreases with time as the thoria is deposited on the porous support. As the thoria deposits the pressure required to cause permeation gradually increases. Deposition pressures in the neighborhood of 500 psig to about 1,500 psig are typical to achieve these results.

FIGURE 1.4: DEPOSITION OF THORIA SOL CONTAINING SULFATE ION ON RIGID SUPPORT

a.

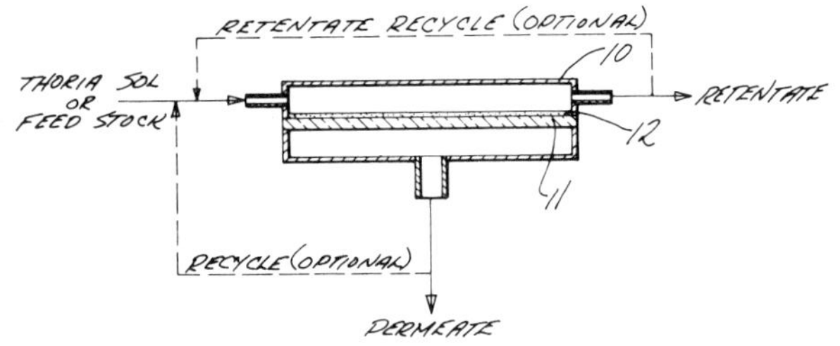

b.

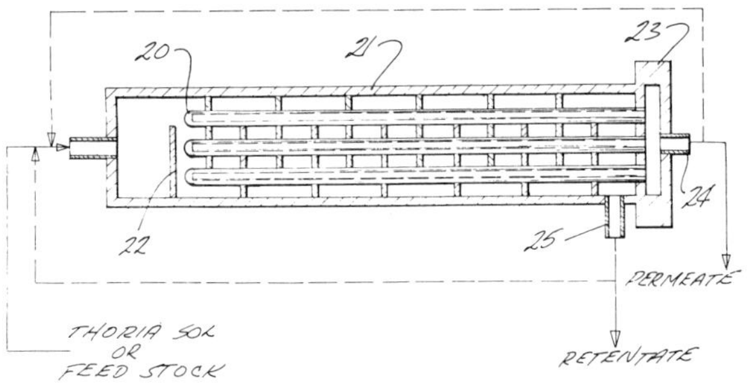

(a)(b) Flow Diagrams

Source: H.A. O'Hern; U.S. Patent 3,923,654; December 2, 1975

Pressure is also observed to be a parameter having an effect on membrane preparation. As the pressure in the deposition system is decreased, the period of time necessary to filtration deposit the thoria is increased. The flow of the thoria sol through chamber **10** is maintained in the turbulent condition to prevent concentration polarization at the membrane surface. After a few hours of deposition, thin layer **12** of a few microns in thickness of thoria is deposited in the pores and on the surface of support **11** to form the anisotropic membrane. The membrane is characterized as "anisotropic" because the thoria is deposited only on one side of the porous support and the membrane is not uniform throughout.

After the membrane has been deposited, the flow of the thoria sol is discontinued and the feedstock to be processed is introduced into chamber **10**. The feedstock separates into a permeate fraction depleted in solute and a retentate fraction enriched in solute. By using the thoria sol containing the specified proportions of sulfate ion as described above, the initial rate of solute rejection from the feedstock is improved while the membrane flux rate is also quite high. In operation with the feedstock, the permeate and/or retentate can be totally or partially recycled as desired for further purification, depending on which stream contains the product to be recovered. Several membranes can be used in various series or parallel arrangements in more complex processes.

Figure 1.4b illustrates another type of system similar to Figure 1.4a where the porous supports are in the form of tubes **20** mounted within shell **21**. This design achieves increased membrane surface area. In this embodiment, the thoria sol or feedstock enters shell **21** and flows in a tortuous path defined by baffle plates **22** while in contact with the external surface of tubes **20**. The permeate passing through the tubes is collected in header **23** and flows through outlet **24**. The retentate which does not permeate the tubes flows from shell **21** through outlet **25**. Either of these streams can be partially or totally recycled as required. The process is illustrated in the following examples where all parts are parts by weight and all percentages are weight percentages unless stated otherwise.

Examples 1 through 5: A porous alumina tube as described above is mounted in a filtration deposition apparatus as shown in Figure 1.4b with total recycle of retentate and permeate. Several aqueous thoria sols prepared from $ThCl_4$ as described above containing thorium in the concentration of 8 x 10^{-4} M and various proportions of Na_2SO_4 as indicated in the following table, flow under pressure of about 1,000 psig through the apparatus across the outside surface of the tube. Other sulfate sources such as lithium sulfate or potassium sulfate give substantially similar results. The temperature of the sols is about 85°F. The circulation of the sol is maintained for 5 to 6 hours to deposit the membrane, although similar results are obtained when the sol is only circulated for about 1 hour to deposit the membrane.

The liquid which permeates through the tube was initially at a rate of in excess of 9,000 gfd. At the conclusion of the filtration deposition, the permeation rate of the liquid through the tube is at a rate of about 90 gfd. An aqueous test feedstock having a pH of about 9 and a temperature of 85°F containing 2,000 ppm Na_2SO_4 as a solute is then circulated for several hours through the apparatus against the outside surface of the tubes as described above at a pressure of 1,000 psig. The permeate and retentate are totally recycled for test purposes. The results are set forth below as a function of time and concentration of sulfate in the thoria sol used to form the membrane.

Example No.	1			2			3			4			5		
molar ratio (SO_4/Th)	0.45×10^{-1}			1.8×10^{-1}			2.7×10^{-1}			3.2×10^{-1}			4.5×10^{-1}		
time into test (hrs.)	few minutes	¾	20	few minutes	¾	20	few minutes	¾	20	few minutes	¾	20	few minutes	¾	20
solute rejection rate (%)		24	78		42	91		68	93		77	93		10	65
flux rate (gfd)	190	225		215	217		199	176		210	173		180	257	

Examples 6 through 13: Disc type membranes are prepared from the alumina discs described above using the thoria sols of Examples 1 through 5 with various concentrations of Na_2SO_4 as indicated below. The feedstock and procedures are the same as for the above examples except that the apparatus of Figure 1.4a is employed. The results are set forth below.

Example No.	6		7		8		9		10	
molar ratio (SO_4/Th)	0.45×10^{-1}		1.7×10^{-1}		2.3×10^{-1}		2.7×10^{-1}		4.5×10^{-1}	
time into test (hrs.)	¾	20	¾	20	¾	20	¾	20	¾	20
solute rejection rate (%)	24	78	42	91	79	93	68	93	10	65
flux rate (gfd)	167	98	157	90	111	72	145	88	267	125

It appears from these examples that the higher rejection rates for the test solute at ¾ hour of testing are achieved for membranes which have been formed using 2.7×10^{-1} to 3.2×10^{-1} molar ratio of sulfate to thorium in the deposition sol.

Compression of Monomolecular Layers

E.F. Steigelmann, H.E. Ries, Jr. and R.D. Hughes; U.S. Patent 3,892,665; July 1, 1975; assigned to Standard Oil Company describe a method of forming membranes on the order of 10-200 A (0.001 to 0.01 μ) thick, particularly for use in reverse osmosis processes where solvents are caused to flow from a more concentrated solution to a more dilute solution through the membrane. The membrane is formed from a plurality of monomolecular layers of a polymer material alone, and in mixtures, which can be crosslinked, if desired.

The membranes are produced by establishing a monomolecular film of the desired film-forming material on a liquid surface such as water and compressing the film so that the molecules are tightly packed together and cover the liquid surface as a continuous layer. Thereupon, a support is passed through the continuous monomolecular film so that a monomolecular layer of the material is deposited as a continuous surface upon the support. As the support passes through the film, an essentially constant surface pressure is maintained on the film to maintain its continuous surface and uniformly apply the material to the support.

Since the molecules are transferred to the support under pressure in the film, they are under a pressure on the support which freezes in the monomolecular layer the orientation of the molecules in the film. The material used to form the monomolecular film can be monomeric, or polymeric material. Monomers can be polymerized on the surface of the liquid, or after being deposited on the support. Polymers can be crosslinked to strengthen the membrane either before, or after, transfer of each monomolecular layer to the support or after a film of several layers has been applied to the support. Normally, the support will be dipped into the liquid upon which the film is formed so that it moves through the film with an in-and-out movement. Some polymers, and monomers, prefer to transfer to the support as it is immersed whereas others transfer as the support moves in both directions through the film. With the latter materials a layer two molecules thick will be provided as the support is dipped into and removed from the liquid.

The apparatus in Figures 1.5a and 1.5b has a shallow trough **10** filled with water, or other liquid, on which the monomolecular film **15** is spread. A compression barrier **12** laid across trough **10** on the upper edges **10'** behind the monomolecular film, generally designated by numeral **15**, is driven on guide shafts **16** and **18** by a precision screw **14**, through yokes **20** and **22**. Compression barrier **12** is made to push the leading edge of the monomolecular film against a delicately suspended barrier **24** which floats on the water surface **44** above edges **10'**.

The floating barrier **24** as shown is a strip of mica attached to the sides of trough **10** by thin, flexible platinum foils **26** and **28** and to an aluminum stirrup, or yoke **30** by an unspun silk fiber **32**. Stirrup **30** is fixed to a calibrated torsion wire **34** that indicates the surface pressure on scale **36**, including dial **38** and indicator **40**. Scale **46** on the side of trough **10** is calibrated to read the area of film **15**, defined by the sides of trough **10**, barrier **24** and barrier **12**. Sweeping barriers **42**, e.g., small brass bars, are used for sweeping the water surface free of contamination prior to establishing the monomolecular film.

Monomolecular film **15** is established on the water surface by depositing a few drops of a dilute solution, e.g., containing about 0.005 mg to 0.05 mg, preferably about 0.01 mg, of the film-forming material in a volatile solvent such as benzene or hexane. Many materials such as some epoxides, cetyl alcohol, stearic acid, polyamides, etc, can be applied without benefit of the solvent; however, use of a solvent is preferred because of ease of manipulation. The amount of compound deposited is precisely measured so that the number of molecules present can be calculated from the known molecular weight of the material. Upon evaporation of the solvent, the material because of its hydrophobic and hydrophilic character forms a film one molecule thick on the surface between the barrier **12** and float **24** in which each molecule is oriented with the hydrophilic group toward the water. The orientation of the molecules in the monomolecular layer is a function of the attraction of the hydrophilic group, or groups, to the water surface and the repelling nature of the hydrophobic group. As the surface pressure is applied to the molecules on the water surface, they align themselves by reason of this attraction, and repulsion, and the surface pressure packs them together in this aligned fashion. The initial area between barrier **12** and floating barrier **24** is large enough to allow the molecules ample room to move freely on the surface.

Before the film-forming material is deposited, the entire apparatus is thoroughly cleaned and coated with a material that water does not wet, such as a high melting paraffin wax or Teflon. Since water does not wet such materials, the water surface has a downward curvature and can be raised higher, shown by layer **44**, than the edges **10'** of the trough **10** to interpose the monomolecular film **15** between barrier **12** and barrier **24**.

FIGURE 1.5: PREPARATION OF ULTRATHIN MEMBRANES

a.

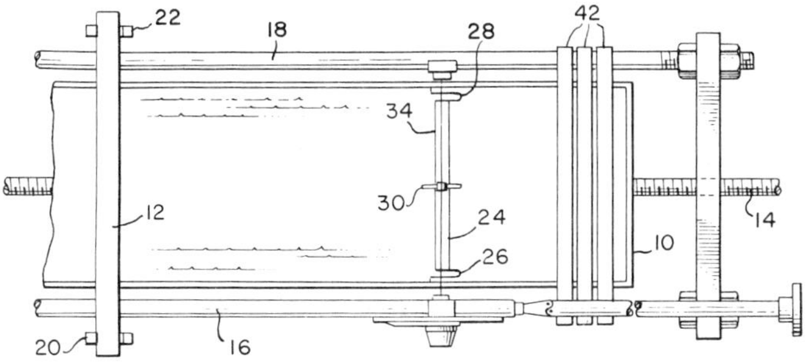

Top Plan of Apparatus

b.

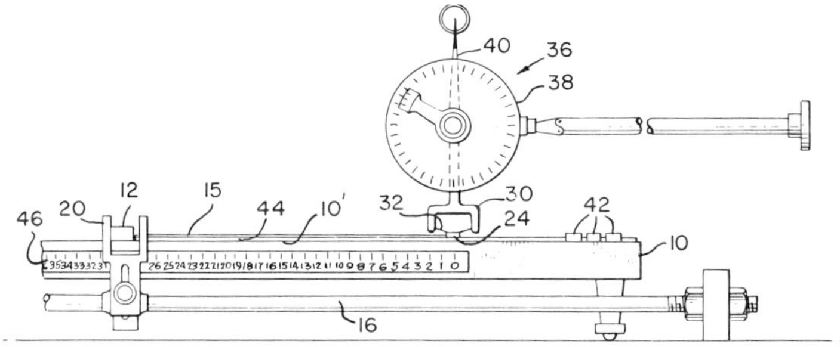

Side Elevation of Apparatus

(continued)

FIGURE 1.5: (continued)

c.

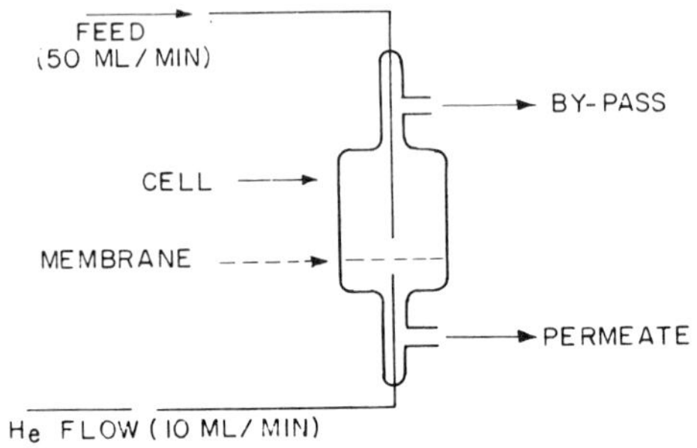

Apparatus for Testing Membranes

Source: E.F. Steigelmann, H.E. Ries, Jr. and R.D. Hughes; U.S. Patent 3,892,665; July 1, 1975

This height is necessary if sweeping barriers **42** are to be effective in removing contaminants and monomolecular film **15** is to be contained and controlled in the defined area. Since contamination and vibration can affect formation and stability of a monomolecular film, the device is mounted to minimize vibration and is contained in a temperature controlled environment.

Initially, because of the relatively large surface area of the water as compared to the amount of film-forming material deposited thereon, the molecules of the film-forming material are randomly oriented on the surface of the water and there are large open areas between molecules or groups of molecules of the material of the surface. In forming the monomolecular film **15**, however, compression barrier **12** is moved toward barrier **24** and the surface area between the two barriers is reduced, preferably in small decrements, so that the molecules begin to become oriented under the increased compression with their hydrophilic groups directed toward the water surface and the hydrophobic groups directed away from the surface. Simultaneously, the open voids in the film are reduced in size.

Area and pressure readings are taken on scales **36** and **46** simultaneously with formation of the monomolecular film. The pressure of monomolecular film **15** on floating barrier **24** is balanced by twisting of torsion wire **34** to which the floating barrier **24** is attached. The degree of twisting shown for scale **36** is a measure of the surface pressure which is expressed in dynes per centimeter. Compression is continued until the desired pressure for transfer of the monomolecular film is attained. The following examples illustrate the process.

Example 1: Two membranes of 25 and 15 monolayers, respectively, of General Electric XD-1 Resin (also known as MEM-213) on an Amicon XM-150 filter surface (pore size $\approx$35A) with an area of 4.9 cm^2 were prepared. XD-1 is a block copolymer of dimethyl siloxane and the condensation product of bisphenol A and phosgene and has a molecular weight of approximately 15,000. Transfers were carried out on a water surface as described above at 30 dynes/cm pressure on the film. To the surface was applied 0.25 ml of a 0.03 wt % solution of XD-1 in benzene for the 15 monolayer film and 0.35 ml for the 25 monolayer film. Transfer occurred in both directions. Immersion was done at about 4 in/min and emersion at 1 in/min. Between dips, the film and support were heated by a 25 watt light-bulb at 70 volts for two minutes. The bulb was 1 to 3 inches from the film. Each membrane was tested for its permeability to air.

The coated filters were less than $\frac{1}{20}$ as permeable to air as the uncoated filter. This decrease in permeability indicates that the pores in the support were bridged by the film. In a separate experiment an approximately 1% solution of trifalgic acid ($C_{131}H_{113}N_9O_{33}S_{12}$, molecular weight = 2,910) in water was passed through the coated filters under 7 to 10 psi of air. Analysis of the visible spectra of the filtrates revealed that 87% of the trifalgic acid was rejected by each of the filters. The retentate showed a corresponding enhancement in the concentration of the trifalgic acid further demonstrating the existence of a semi-permeable membrane.

Example 2: In preparing the membranes 40 monomolecular layers of XD-1 resin (0.6 ml of the solution of Example 1) were transferred to an Amicon PM-10 filter, (about 18 A pore size) with an area of 4.9 cm^2 at a film pressure of 10 dynes/cm. Transfer occurred in both directions and the film was about 200 A thick. For contrast and comparison, two thick (about 0.2 and 0.6 mil) XD-1 films were prepared using the conventional film casting technique. The 0.2 mil (50,800 A) XD-1 film was made by casting onto a clean 4" x 8" glass plate at 40% XD-1 in chloroform solution with a 0.5 mil doctor knife and allowing the film to dry. The 0.6 mil (152,400 A) film was made in a similar manner from a 10% XD-1 in chloroform solution with a 6 mil doctor knife.

The thick and the thin films were tested in the apparatus shown in Figure 1.5c. Each of the membranes was put in the test cell, and a feed consisting of 43.8% ethylene, 30% ethane, and 26.2% methane passed over the membranes. Depending on the membrane, various cell pressures were employed. To maintain a constant composition of the gas mixture in the cell, the gas mixture was flowed through the cell at the rate of 125 ml/min. Gas flowing through the membrane was picked up in a He stream. (10 ml/min). The composition of the gases passing through the membrane as well as their flow rate was determined with gas chromatography. Flow through the membrane can be calculated by knowing the amount of hydrocarbon in the He carrier gas, the He flow rate, the size of the sample and the response of the GC detector to one ml of the gas mixture. For convenience, the sensitivity of all the gases is assumed to be 1.00.

The results, shown in the following table confirm that the thin and the thick membranes both fractionate the gas mixture about equally, but the thin membrane shows a large increase in the rate of flow through the membrane.

Membrane	Permeate Composition (%) Methane	Ethylene	Ethane	Pressure on Membrane (psi)	Flow Through Membrane (ml/min)	Flow (Calculated) at 10 psi (ml/min)
0.6 mil XD-1	14.0	50.2	36.3	50	1.2	0.24
0.2 mil XD-1	18.1	47.6	34.1	75	32.2	0.43
Monomolecular layered XD-1	20.1	48.0	31.8	0.54	0.76	14.1

Example 3: In this example, 4.5 x 10^{-4} grams of Epon 872 (Shell epoxy resin) was spread on a water surface. Epon 872 is an aliphatic modification of the diglycidyl ether of bisphenol A. The resin, a very viscous semisolid, is relatively free of impurities and has a molecular weight (MW) of about 1,300. The film was compressed to 12 dynes/cm for transfer to a support. The support was a Millipore Teflon filter (2.5 cm in diameter with 5 μ holes) and transfer was made by dipping the filter in and out of the water, having the film on its surface.

With each dip a resin monolayer can be transferred to the filter. Two different experiments were used to demonstrate the effectiveness of a mild heat treatment in assisting buildup of multiple layers on the support. In one procedure the filter was dipped in and out of the film 12 times without stopping. In the other procedure, the filter was dipped in and out of the film once and then heated for two minutes next to a 25 watt light bulb. This procedure was repeated 11 times. The same filter was used in both demonstrations. Prior to these demonstrations, about 100 monolayers of the resin had been built-up on the filter surface. For this resin, transfer occurs only as the filter is removed from the layer. The following table gives the average area transferred for each condition.

Without Heating	10.0 cm^2*
With Heating	17.1 cm^2*
Calculated Area of Filter	14.1 cm^2

*A portion of the resin transferred is transferred to the support for the filter.

The results show that without heat treatment, less than a monolayer is transferred per dip whereas with heating, at least a full monolayer is transferred.

Gas Phase Desolvation

W.J. Ward III and C.H. Knapp; U.S. Patent 3,592,953; July 13, 1971; assigned to General Electric Company describe a method for the room temperature preparation of porous membranes of cellulosic derivation, as, for example, are employed in reverse osmosis systems. The process provides for a two-step preparation conducted at room temperature and consisting of casting, desolvation of the film and immersion in a room temperature bath after which the membrane is ready for use.

Gas phase desolvation with a controlled atmosphere enables (a) the room temperature preparation of such porous films even when highly volatile solvents for the film material are employed and (b) heat treatment of the cast film either during or after immersion is obviated. In spite of the elimination of the heat curing step, the porous membrane can be reproducibly prepared by this improved method to provide salt rejection of divalent ions (calcium and magnesium salts) of greater than about 90% at a water flux of at least 0.7 gal/day/ft^2/atm ΔP at pressures as low as 40 psi. Tests have established that over a period of at least 12 months, a slight decline was observed in the salt rejection, but the water flux remained substantially undiminished.

Example: A casting solution containing a mixture of cellulose acetate (39.8% acetyl), acetone, water and magnesium perchlorate (22.2, 66.7, 10.81 and 1.1 weight percent, respectively) was prepared in a covered container which was placed in the casting compartment. All the equipment (casting plate, doctor blade, immersion bath tray, weights, etc.) for conducting the casting and immersion steps were cleaned and placed within the casting compartment. A sealed container of 3% acetone solution (1,450 ml of distilled water and 50 ml of acetone) was prepared and placed within the casting compartment.

The compartment was then closed and air previously passed through a column containing an acetone-distilled water solution (12.5 mol percent acetone) was circulated through the interior of the compartment until the air within the compartment was in equilibrium with the 12.5 mol percent acetone solution. The compartment was then sealed and subsequent operations were conducted from outside the compartment so as not to change the conditions deliberately imposed within the compartment. The film is then cast by passing the doctor blade, or knife, across the glass casting plate in an inclined position with the ends of the knife resting upon side runners controlling the thickness (10 mils) of the as-cast film.

Shortly before the immersion step the container of immersion fluid is emptied into the open casting tray. At the expiration of the preselected desolvation time (4.5 minutes) corresponding to the predetermined composition of the gas phase ambient to the newly cast film the plate with its cast film was carefully submerged in the immersion bath. Care was taken to be certain that the film was completely immersed. The membrane was then left undisturbed in the room temperature immersion bath for several hours. At this point, without further treatment or conditioning the membrane was ready for use as a reverse osmosis membrane. Care should be taken to mark the side of the film that is to be oriented in contact with the saline solution.

During the desolvation and immersion stages of the process structural changes gradually occur and the membrane, which as-cast is 10 mils thick, gradually emerges about 3 mils in thickness. If the membrane is not to be placed into use at completion of the process, it may be stored indefinitely as long as it is not allowed to dry out. The properties of reverse osmosis membranes fabricated by the method are set forth in the following table.

Mol % Acetone in Aqueous Solution with Which Air in Casting Box Is in Equilibrium	Desolvation Time, minutes	Salt Rejection, percent	Water Flux, gal/day/ft^2/atm ΔP
12.5	4.5	90 - 94*	0.7
15	6.0	88 - 92**	0.8

*Range of performance of six films
**Range of performance of five films

This same process was conducted using a range of ambients within the casting compartment from a gas phase having a very low acetone content up to one which is in equilibrium with a solution of about 50 mol percent acetone. Also, at each selected gas phase concentration a series of membranes were cast using different desolvation times. It was found that an initial indication of optimum membrane fabrication conditions was provided by observation of the transparency of the resulting films. The operative range was then easily determined from the initial data, which was substantiated by actual test of the membranes. This procedure for producing a desired physical modification of membranes of cellulosic derivation may be duplicated by conduct of the same series of routine tests with other film-casting solvents with the same or other film matrices.

In the case of the specific casting solution employed those films prepared using any desolvation time and a gas phase ambient in equilibrium with a solution of less than about 12.5 mol percent acetone display too low a water flux. Those membranes prepared using any desolvation time and a gas phase ambient in equilibrium with a solution of greater than about 15 mol percent acetone possess too low a salt rejection to be useful. Having determined the desirable range of gas phase compositions the optimum desolvation time was found to be about 4.5 minutes for the ambient with the lower acetone content and about 6.0 minutes for the atmosphere having the higher acetone content. Subsequent life tests of prepared membranes established the optimum nature of the above-recited ranges of conditions.

The results shown above were obtained for city water (Schenectady, N.Y.) with a 40 psi pressure applied across the membrane. The salt rejection refers to the rejection of divalent ions (calcium and magnesium salts) which were present in the feed stream at a total concentration of about 182 ppm. Life tests conducted on a series of these membranes over a period of more than 12 months have shown that there is some slight decline in the divalent ion rejection but that the water flux remains substantially constant. The high salt rejection, high water flux and long life of such membranes make them particularly adaptable to use as home water softening membranes. These reproducible results are primarily dependent upon the relative simplicity of the casting procedure and the rigid control which is possible over the desolvation portion of the process. In another series of tests to determine NaCl removal as a function of pressure, membranes prepared by the process displayed substantial (70 to 90%) rejection of NaCl at pressures in excess of 600 psi.

The reason that rigid control over desolvation is necessary is the fact that the right amount of solvent must be removed at the right rate from the cast film to produce a membrane with useful ion rejection and water throughput properties. Also at the completion of the desolvation phase, the film must be at the right stage of development, when it is immersed because that particular structural change, which is occurring in the film during the desolvation phase, is stopped as soon as the film is immersed.

Thereafter, from the moment of immersion a different structural change occurs conditioned by the fact that water is entering the film and some of the solvent is leaving the film, the water acting as a coagulant and the solvent remaining in the film acting to prevent too rapid coagulation. Neither the rate of desolvation nor the length of the period of desolvation can be adequately controlled, if the difference between the partial pressure of the solvent in the film and the partial pressure of that same solvent in the ambient is too great because structural changes occur much too fast under such conditions. Therefore, it is a critical aspect of this process to provide in the ambient to which the newly cast membrane is exposed a partial pressure of the same solvent as is in the film, the partial pressure of the solvent in this preselected ambient being lower than the partial pressure of the solvent in the film but yet being just high enough to enable close control over the rate at which the solvent leaves the film and enters the ambient.

Critical Acetyl Content

C.R. Cannon; U.S. Patent 3,497,072; February 24, 1970; assigned to Aerojet-General Corporation describes a cellulose acetate reverse osmosis membrane characterized by high flux ability and superior salt retention properties and having an acetyl content in the range of 41.5 to 42.7% by weight of the cellulose acetate.

It has been found that there is a critical acetyl range for optimum flux and salt retention. The cellulose acetate reverse osmosis membrane of this process is characterized by having an acetyl content in the range of 41.5 to 42.7% by weight of the cellulose acetate and preferably an acetyl range of 41.9 to 42.5% with the optimum being about 42.2%. Membranes prepared of cellulose acetate compositions within the abovementioned range exhibit a remarkably lower salt permeation than compositions outside of the range and still possess adequate flux. For example, a membrane of the process having an acetyl content of 42.2% when tested against a saline water of 35,000 ppm (typical of seawater) at a pressure of 1,500 psig exhibit a flux of 11.7 gfd and a salt permeation of 164 ppm.

A cellulose acetate membrane manufactured under like conditions and tested against a water of the same salinity and at the same pressure which membrane had an acetyl content of 41.5% possessed flux of 11.9 gfd and a salt permeation of 344 ppm. A still third membrane of the process prepared under like conditions and tested under the same conditions as the foregoing membranes which third membrane had an acetyl content of 41.9% exhibited a flux of 11.1 and a salt permeation of 174 ppm. An acceptable potable water is generally considered to be a water having a salt concentration less than the range of 400 to 500 ppm.

It was found that a fourth semipermeable membrane made under like conditions and tested under the same conditions as the preceding three membranes of the process which fourth membrane had an acetyl content of 40.8% by actual analysis exhibited a flux of 9.2 and a salt permeation of 1,220 ppm which makes the membrane generally unacceptable for the treatment of seawater to recover a potable water in a one pass operation. The latter membrane having the acetyl content of 40.8 by actual lab analysis is a composition sold as a commercial diacetate material, which has an acetyl content of 39.8. It has been found by laboratory analysis that some slight variations do occur from the represented acetyl contents in commercially available cellulose acetates.

A second commercially available acetate marketed as cellulose triacetate and which has an acetyl content of 43.2 was found in one batch to have an actual acetyl content of 43.6% by laboratory analysis. When the latter commercial cellulose acetate was incorporated into a semipermeable membrane prepared in the same manner as the foregoing several membranes and tested under like conditions against a seawater salinity of 35,000 ppm a performance was found of 4.5 flux and a salt permeation of 440 ppm. It is generally accepted that a commercial flux should be in excess of about 10 gfd. Therefore, it is seen membranes having acetyl contents of 43.6% exhibit low flux rates, generally less than that commercially believed desirable.

The critical acetyl range of the cellulose acetate used in the preparation of the membranes may be obtained by blending commercially available cellulose acetate compositions to obtain the desired acetyl content or alternatively the cellulose acetate composition may be initially prepared to provide the desired acetyl content without resort to the blending technique. It has been demonstrated that blending to produce membranes of the same acetyl content employing different starting materials results in membranes having essentially the same osmotic properties.

It is frequently most convenient to prepare the membrane by blending commercially available basic types of cellulose acetate, for example, E-398, E-394, A-393 and E-383 which contain respectively, acetyl contents by weight of the cellulose of 39.8, 39.4, 39.3 and 38.3%. A-432 is a cellulose triacetate which has an acetyl composition by weight of 43.2%. The so-called triacetate composition is not fully acetyl substituted as a theoretical fully substituted cellulose triacetate would have an acetyl content of 44.4% by weight. In the data reported, laboratory analyses have been run to determine the correct acetyl content of the various materials employed.

Example 1: In this example the several membranes fabricated had average acetyl contents of 42.2% by weight and were prepared by blending equal quantities (7.7 grams) of cellulose triacetate (A-432-130B) and cellulose acetate (E-398-3). The casting solutions contained 37.8 grams of acetone, 46.8 grams of 1,4-dioxane, along with 7 grams of methanol and 3 grams of maleic acid. Water may be used in place of the methanol and other swelling agents than maleic acid may be employed. Fabrication of the membranes included casting at -10°C, immersion in a 0°C water bath and annealing for 3 to 5 minutes in a water bath maintained at a temperature of 89°C. The membranes were cast both by hand in the general fashion described above and by a machine casting apparatus. The membranes were then tested in a standard reverse osmosis test apparatus operated at 1,500 psig using a 3.5% sodium chloride water solution (35,000 ppm). The results of these tests are set forth in the following table.

Preparation Type	Number of Tests	Flux, gfd	Salt Permeation NaCl, ppm
Hand cast	18	11.7±0.7	164±35
Machine cast	12	11.1±0.9	202±58

Example 2: The work of this example was undertaken to demonstrate the osmotic properties of different blends of cellulose triacetate (A-432-130B) and cellulose diacetate (E-398-3). The several membranes were prepared in accordance with the procedure outlined in Example 1 and osmotic properties of the blends were obtained through operation of a standard test reverse osmosis cell on 3.5% sodium chloride feed solutions (35,000 ppm) at 1,500 psig. Results of the test appear in the following table. The first column gives the relative proportions of the cellulose diacetate and cellulose triacetate used in each of the blends and the resulting acetyl content.

Blends				
Parts CDA/Parts CTA	% Acetyl Content	Number of Tests	Flux, gfd	Salt Permeation
75/25	41.5	10	11.9	344
60/40	41.9	8	11.1	174
50/50	42.2	18	11.7	164
40/60	42.5	8	9.1	143
25/75	42.9	6	6.8	450

In related work, *C.R. Cannon; U.S. Patent 3,460,683; August 12, 1969; assigned to Aerojet-General Corporation* describes a reverse osmosis membrane made from a casting solution comprising cellulose acetate with a critical acetyl content, a polar solvent, water and no swelling salt.

Casting Bath Apparatus

G.F. Drake; U.S. Patent 3,721,596; March 20, 1973 describes an improved membrane and a method of forming such membrane which are particularly useful in reverse osmosis desalinization. The method includes the immersion of an apertured film support member or assembly in a liquid casting solution. The film support is withdrawn from the casting solution so that a part of the casting solution forms a homogeneous bridging film across the provided apertures. The support and the casting solution are arranged so that the film thickness is determined primarily by the fluid characteristics of the casting solution independent of the film support except at the peripheral location of film support. Once free of the solution, the bridging film should be suspended so as to allow it to form a stable film configuration.

The withdrawn portion of the film support and the homogeneous bridging film may then be provided with any subsequent curing treatments. These treatments may include exposure to a controlled humidity atmosphere and/or immersion in various aqueous wash baths. When a humidity exposure treatment is employed, the bridging film will essentially assume its final shape during or before this process. It is believed that the humidity of the atmosphere will tend to extract the solvent from the bridging film so as to form a thin skin on each surface of the film. This is believed true, since the solvents generally employed in the casting solution are miscible in water. Consequently, the subsequent aqueous baths perform a similar function until the membrane is finally completely formed. The aqueous wash baths may also be at elevated temperatures to provide a heat treatment for the membranes. These final processes determine the purification and transfer characteristics of the membrane.

A higher temperature tends to provide better desalinization, but tends to decrease the liquid transfer rate. For dialysis membranes usually such post casting treatment is either eliminated or modified appreciably. In comparison to the casting on glass method, the method of the process has many significant advantages. Initially, it may be noted that membrane thickness variations are substantially eliminated. Since the film thickness and subsequently the membrane thickness are primarily governed by the fluid characteristics of the casting solution, the surface tension will cause flow until the thickness is substantially uniform over substantially the entire area of the membrane. Thus, control of fluid characteristics, and in particular, the solution viscosity, is substantially all of the control required to assure a uniform thickness.

There will be an increase in film thickness near the periphery of the film adjacent to the film support member. However, the support member is arranged so that this area of increased thickness comprises a relatively small percentage of the entire film surface area. The natural tendency of the casting solution, which forms the bridging film to flow toward the supporting peripheries immediately after withdrawal from the liquid casting solution source will also tend to carry any impurities in the atmosphere which contact the suspended film toward the periphery. Therefore, the impurities will be localized at the peripheries instead of becoming a part of the membrane product.

A membrane produced by the process will have a symmetrical cross section structure. Since both surfaces of the film are exposed to the atmosphere, each will be an osmotic skin and the membrane is functional in both directions. The method also tends to eliminate the entrapment of air bubbles in the membrane, as compared to the coating on the glass plate or coating platform technique. In addition, the relatively low viscosity of the casting solution used in the method of this process will not tend to retain air bubbles as compared to the casting solutions used formerly.

Multilayer Membrane

K.G. Roehsler; U.S. Patent 3,855,133; December 17, 1974; assigned to Sartorius-Membranfilter GmbH describes a multilayer filter membrane for carrying out hyperfiltrations each layer being allocated to a specific function which the membrane must fulfill. One of the layers is an osmotic skin provided by an unbroken film forming polymer and the membrane also includes a porous supporting base comprising a cellulosic material. The base is in the form of a gel film which is insoluble in water, and which is backed by an impregnated sponge layer.

Example 1: For the production of the supporting base a castable solution of the following composition was used.

	% by Wt
Cellulose acetate having a 31.5% acetyl content (GU 1000, Bayer)	16
TiO_2	2
Talc	2
Acetonitrile	35
Ethanol	35
Water	10
Viscosity*	24.8 sec

*Viscosity was measured as the falling sphere viscosity using a Hoppler viscometer.

Falling distance: 100 mm
Tube diameter: 15.944 mm
Weight of steel sphere: 5.533 grams
Diameter of steel sphere: 11.10 mm
Test temperature: 30°C

The casting solution was poured continuously from a stationary pourer over a weir of adjustable height on a moving endless receiving surface (in the form of a cylindrical drum or an endless metal band) and then immersed in a water bath after a brief residence time of 5 to 15 seconds. After having been watered for 10 minutes the foil membrane was tempered in a water bath at 80°C for 5 minutes, the edges being held to prevent shrinkage. The resultant supporting base was then impregnated for 3 minutes at 50°C on that side which exhibited the sponge structure, with an impregnant of the following composition.

	% by Wt
Polyvinyl methyl ether	5
Glycerin	29
Agepon (wetting agent)	1
Methanol	35
Water	30

The supporting base was then dried in air at 50° to 60°C for 3 minutes, the edges on each side being gripped and thus fixed. This base was then coated with one or more continuous closed films of a selectively permeable inelastic, elastic or elastified high polymer.

Example 2: The procedure in this example was the same as that described in Example 1, excepting that a casting solution of the following composition was used:

	% by Wt
Cellulose acetate (GU 1000 Bayer)	16
Bentonite	4
Ethyl glycol	35
Acetone	35
Water	10
Viscosity	64.5 sec

Example 3: The procedure in this example was the same as that described in Example 1, excepting that a casting solution of the following composition was used:

	% by Wt
Cellulose acetate, GU 1000	8.5
SiO_2	1
TiO_2	1
Glass fibers	0.5
Ethyl lactate	56
Ethanol	22
Water	11
Viscosity	6.3 sec

Example 4: This example illustrates the production of an inelastic desalination film bound to the supporting base by secondary valence bonds. For the purpose of producing this layer ethyl acetate (or methyl isobutyl ketone) was poured over the gel side of the impregnated dried base (obtained as described in the preceding examples) for the purpose of flushing off particles of dust and at the same time of making this surface slightly swell. This was followed by drying which takes 1 minute in the case of ethyl acetate and 2 to 3 minutes in the case of methyl isobutyl ketone. A film of cellulose triacetate having a water absorption capacity of 10% was then poured on the still slightly swollen cellulose acetate base which had a water absorption capacity of 21%. The composition of the cellulose triacetate was as follows:

	% by Wt
Cellulose triacetate (T 1000, Bayer)	0.5
Methyl acetate	44.5
Methylene chloride	45.0
Methanol	5.0
Cyclohexanone	5.0

After the cellulose triacetate film had dried a polyamide which was soluble in a mixture of water and alcohol and which had an absorption capacity for water of 2.4% was applied in the form of a very thin film, the composition of this solution of polyamide being the following:

	% by Wt
Amilan X 8000 (a modified polyamide soluble in ethanol/water. Toyo Rayon)	0.5
Ethanol	66.5
Water	33.0

Not before the polyamide film is dry should the desalination film proper by applied. For this purpose an 0.5% by weight solution of a phenoxy resin (PKHH Brenntag) in ethyl glycol acetate was used. The high polymer of the desalination film proper (in this instance the phenoxy resin) must be dissolved in a nonaqueous solvent. Pouring should preferably proceed at 0% relative humidity.

Example 5: The production of a supporting base of hydrated cellulose is accomplished by hydrolyzing a cellulose acetate layer (obtained for instance as described in Example 1) with a hydrolyzing solution consisting of: 4% by weight of KOH, 48% by weight of ethanol, and 48% by weight of water, for 5 minutes at room temperature. After the KOH had been washed out impregnation and drying was done immediately, without prior tempering.

In order to bind this supporting base to the desalination layer by primary valence bonds the hydrophilic tendency of the hydrated cellulose must be largely suppressed on the side that is to be coated. For this purpose the gel-covered side of the impregnated and dried supporting base of hydrated cellulose was coated with a 1% by weight solution of triphenylmethane-4,4',4''-triisocyanate (Desmodur R) in methylene chloride and submitted to a warm air treatment at 50°C for 2 minutes. When provided with a film of Desmodur R the base must have been previously treated with ethyl acetate or methyl isobutyl ketone. The desalination film was applied as described in Example 4.

Polyacrylic Acid Layer

S.B. Sachs and R.L. Riley; U.S. Patent 3,676,203; July 11, 1972; assigned to the U.S. Secretary of the Interior has found that a composite membrane having superior separation characteristics may be prepared by forming a coating of polyacrylic acid (PAA) on a porous support. The support is preferably as porous as possible consistent with its primary function of supporting the PAA film in an environment where sizeable pressure differentials will exist. Generally, the support should not contain surface pores greater in diameter than the thickness of the PAA membrane. Pore size will generally range from about 100 to

5000 A. Void volume, i.e., the percentage of unoccupied space within a unit volume of the support, will generally range from 40 to 80%. Suitable support materials include porous glass, sintered metals, ceramics, etc. Preferred support materials are, however, organic polymeric materials such as homopolymers or mixed polymers of cellulose acetate, polysulfone, cellulose butyrate, cellulose nitrate, styrene, vinyl butyral, etc. A particularly preferred support material is a mixed polymer of cellulose nitrate (35 second viscosity) and cellulose acetate (40 second viscosity), hereinafter referred to as CN-CA, containing about 38.3% of the acetate.

Example: A finely porous support membrane was prepared by casting a solution of cellulose nitrate (CN) and cellulose acetate (CA). The casting solution consisted of 86% CN (35 sec viscosity) and 14% CA (40 second viscosity) in a solvent system consisting of 59.1% acetone, 21.8% ethanol, 14.7% n-butanol, 1.7% water, 2.2% glycerin and 0.5% Triton X-100 (octyl phenol polyester alcohol nonionic surfactant). Before use, the solution was pressure filtered through a 10-micron polypropylene filter at about 100 psi.

The solution was continuously cast with a knife onto glass plates (1 x 4 ft) which were then carried through a gelation chamber on a paper belt. Moist air was circulated over the surface of the membrane through a series of small chambers that served as plenums. The air flow was directed down onto the surface through an impregnated Dacron Reemay cloth and a perforated metal sheet in series, which served to distribute the flow. The air stream was first saturated with water by passage through a stainless steel sparging column consisting of a short packed section with water reflux followed by a demister section. The humidifier was maintained in a temperature-controlled water bath. The manifold and distributors leading the air to the membrane were so constructed that entrained droplets of water were not discharged onto the curing film.

The table top supporting the belt consisted of two sections of aluminum plate (6 ft and 4 ft long, respectively) that were heated independently by water circulating through copper coils attached to the bottom of the plates. The membrane was cast under standard conditions of 38°C gelling-plate temperature and 25°C humidifier temperature. Residual solvents were removed from the cured membrane by passing it over the second aluminum plate held at 45°C. The casting speed was 3 in/min; at this speed, the gelation period was long enough that the membrane was cured to a white opaque appearance when it left the gelation plate. The resulting support membrane had a thickness of 110 microns, a bulk porosity of 80% and a membrane constant of 5,000 x 10^{-5} g/cm^2-sec-atm.

A PAA film was then cast on the support membrane with a film-casting knife from a casting solution consisting of 25 weight percent solution of PAA in water. The PAA had an average molecular weight of 300,000. The cast film, which had a wet thickness of 250 microns, was allowed to dry under ambient conditions for 1 day, after which it had a dry thickness of 40 microns and was ready for use. The PAA adhered tightly to the porous support and could be removed only with difficulty.

The membrane was then placed in a high pressure reverse osmosis test cell. Flow through the composite membrane at 25°C was initially >40 gal/ft^2-day at 100 atm applied pressure, with 83% rejection of 1% NaCl. After 15 hours the water flow essentially stabilized at ~20 gal/ft^2-day and the rejection of 1% NaCl stabilized at 86%. The rejection was found to depend on pH and was a maximum at pH 7 to 9. The rejection of Na_2SO_4 was even better than that of NaCl, reaching >95% with a 1% solution of Na_2SO_4 at 100 atm applied pressure and pH 7 and with a water flow of ~20 gal/ft^2-day. The rejection of a 1% solution of $CaCl_2$ was 74% under the same conditions at pH 7.

R.L. Riley; U.S. Patent 3,648,845; March 14, 1972; assigned to The U.S. Secretary of the Interior describes a process for making thin film semipermeable membranes. A porous substrate having pores not greater than about 500 A at a surface thereof is provided, as by casting from a solution of an organic polymer. A nonporous buffer layer is coated upon that surface to fill the pores, as by applying a water-soluble material, such as polyacrylic acid. The thin film semipermeable membrane is formed in situ upon the buffer layer, as by dipping the coated substrate in a dilute solution of cellulose acetate. A less than 1 weight percent solution of cellulose acetate, having an acetyl content not less than about 39.8 weight percent in chloroform may be used. After drying, the buffer layer is at least partially leached from the composite membrane. The cellulose acetate semipermeable membrane may be less than 1000 A in thickness.

OTHER CELLULOSICS

Hydroxyalkyl Derivatives of Cellulose

E.D. Klug, H.M. Spurlin and W.L. Young, III; U.S. Patent 3,620,970; November 16, 1971; assigned to Hercules Incorporated describe a process which provides reverse osmosis membranes characterized by a salt-rejecting layer of a hydroxyalkyl derivative of cellulose. The membranes are useful for desalination and other processes involving reverse osmosis. The hydroxyalkyl derivative of cellulose from which the reverse osmosis membranes are prepared is selected from one of the following classes:

(a) hydroxyalkyl ethers of cellulose in which the hydroxyalkyl group contains from 2 to 6 carbon atoms;

(b) mixed hydroxyalkyl ethers of cellulose in which the hydroxyalkyl groups contain an average of from 2 to 6 carbon atoms;

(c) mixed hydroxyalkyl-alkyl and mixed hydroxyalkyl-aralkyl ethers of cellulose in which the hydroxyalkyl group contains from 2 to 6 carbon atoms and, respectively, the alkyl group contains from 1 to 6 carbon atoms and the aralkyl group contains from 7 to 12 carbon atoms; and

(d) mixed hydroxyalkyl-aminoalkyl ethers of cellulose in which the hydroxyalkyl group contains from 2 to 6 carbon atoms and the aminoalkyl group contains from 2 to 10 carbon atoms.

The derivative having a number average degree of polymerization (DP) within the range of from 25 to 5,000 and an MS (average number of mols of alkylene oxide combined with the cellulose per anhydroglucose unit) of at least 1 and a DS (degree of substitution) including alkyl, aralkyl and aminoalkyl groups of at least 2 (a more detailed explanation of MS and DS can be found in U.S. Patent 3,278,521 to Eugene D. Klug). Such cellulose derivatives are prepared by reacting cellulose with an alkylene oxide (class A), with mixed alkylene oxides (class B), with a mixture of alkylene oxide and alkyl chloride or aralkyl chloride (class C), or with a mixture of alkylene oxide and alkyleneamine (class D) by procedures known to the art as, for example, U.S. Patents 3,278,521 and 3,431,254. The following examples illustrate the process.

Examples 1 through 8: These examples illustrate the preparation of flat membranes comprising films of crosslinked hydroxyalkyl cellulose derivative on a microporous substrate. Two different procedures were employed. In the first procedure a thin layer (0.1 to 1.0 mil) of hydroxyalkyl cellulose derivative was deposited on a 3 x 5 inch porous substrate which was a Millipore VF/VS filter sheet composed of mixed cellulose esters having a pore size of 0.01 micron. The substrate was first moistened with a pore-filling solvent by floating the sheet dull side down on the surface of the solvent, generally deionized water. This not only filled the pores in the substrate to reduce penetration of the polymer solution during casting but also served to preswell the substrate.

Next, the moistened sheet was taped at its edges dull side down on a glass plate for rigid support during casting. This was the procedure employed in examples 1 and 3 through 8. In the second procedure the solution of cellulose derivative was cast separately on a smooth nonporous surface such as glass, and after drying was removed and transferred onto the porous substrate. This procedure was employed in Example 2.

The casting solutions of hydroxyalkyl cellulose derivative used in the examples were prepared in water except in Example 2 where dimethyl sulfoxide was employed as the solvent. After stirring for 2 to 24 hours to achieve thorough solution, the crosslinking reagent and catalyst were added and the resulting solution pressure filtered through a 14-micron nylon filter. Next, entrained air bubbles were allowed to dissipate by standing for a few hours and immediately thereafter a film was cast using a known technique such as casting with a knife edge of Meyer rod or dip coating or drain coating. The castings were subsequently dried 6 to 20 hours at room temperature in a dust-free filtered airstream and then cured to effect crosslinking at 90° to 100°C for 4 to 24 hours. Care was exercised during all preparative steps to reduce inclusion of extraneous contaminants such as dust and the like. Desalination tests were performed on 47 mm discs cut from the prepared membrane using feedwater containing 7,000 ppm dissolved sodium chloride and 3,000 ppm dissolved sodium sulfate.

Test Apparatus and Method – These were the same as those described in U.S. Patent 3,620,970 (page). Details of each example and the test results therefrom are given in the following table. In all examples cited, the cast membrane thickness was approximately 1 mil (measured dry).

		Crosslinking Agent/Catalyst, Percent of Each Based on Cellulose Derivative	----- Saline Water Test Results -----				
					--- Salt Rejection, % ---		
Ex.	Cellulose Derivative Used*		Pressure	Flux, gfd	Na^+	Cl^-	- -$SO_4^=$**- -
1	Hydroxypropyl cellulose, Type G, 2% aqueous solution	DMU/PTSA (20/1)	500	0.25	94.6	91.2	------- ---------
			1,000	0.40	97.1	95.4	------- ---------
			1,500	0.60	98.2	96.2	99.9 (99.9)
2	Hydroxypropyl cellulose, Type G, 2% DMSO solution	HMDL/DBTD (20/1)	500	2.8	41	20	------- ---------
			1,000	3.3	62	48	------- ---------
			1,500	3.2	83	76	------- (99.9)
3	Hydroxypropyl cellulose, Type G, 6% aqueous solution	DMU/PTSA (6.7/0.33)	500	0.9	77	65	------- ---------
			1,000	1.3	80	68	------- ---------
			1,500	1.4	88	77	------- (99.9)
4	Hydroxypropyl cellulose, Type G, 6% aqueous solution	DMU/PTSA (13.3/0.67)	500	-------	-------	-------	------- ---------
			1,000	0.26	92.3	83	------- ---------
			1,500	0.34	96.9	91.0	99.7 (99.9)
5	Hydroxypropyl cellulose, Type MP, 2% aqueous solution	DMU/PTSA (20/1)	500	0.58	96.2	90.6	------- ---------
			1,000	1.2	97.2	95.2	------- ---------
			1,500	1.6	99.1	98.1	------- (99.9)
6	Hydroxypropyl cellulose, Type HA, 2% aqueous solution	DMU/PTSA (20/1)	500	0.40	89	82	------- ---------
			1,000	0.65	92	87	------- ---------
			1,500	0.91	95.9	92.6	------- (99.9)
7	Hydroxypropyl cellulose, Type G (narrow mol wt distribution), 2% aqueous solution	DMU/PTSA (20/1)	500	1.2	87	82	------- ---------
			1,000	2.3	92.6	89	------- ---------
			1,500	2.7	98.3	96.1	99.7 (99.9)
8	3.0% butene-1 oxide-modified hydroxypropyl cellulose, 2% aqueous solution	DMU/PTSA (20/1)	500	0.20	92.6	86.3	------- ---------
			1,000	0.46	94.5	89.4	------- ---------
			1,500	0.68	97.4	94.9	92.6 (99.9)

*Substituent percentages based on weight of reactant added to weight of propylene oxide (68.7 g) used to react with 20.0 g of cellulose.
**$SO_4^=$ values in brackets are calculated.

Partially Hydrolyzed Cellulosic Membranes

R.E. Kesting; U.S. Patent 3,290,286; December 6, 1966; assigned half to Aerojet-General Corporation and half to the U.S. Secretary of the Interior has prepared desalination membranes for use in reverse osmosis processes for separating water from aqueous solutions. In particular, the process sets forth a treatment of cellulosic ester membranes where the membranes are particularly hydrolyzed to convert ester groups to hydroxyl groups.

It has been found that the use of cellulose esters having a higher hydroxyl content, at least in the active layer, rather than those polymers used by prior art workers results in a greater water permeation or flux rate and also results in an acceptable salt content in the product water. The usual polymers employed in the preparation of cellulosic ester reverse osmosis membranes have about 2.3 to 2.4 acyl groups per glucose unit. By hydrolyzing at least the active surface of the membrane, the benefits of this process may be utilized. The following example illustrates the process.

Example: Referring to Figure 1.6, a circular cellulose acetate membrane **1** prepared by the known process is treated with caustic solution to hydrolyze the surface of the film. The untreated membranes had an acetyl content of 39.3 weight percent prior to hydrolysis. In the preferred hydrolysis method, the membrane is placed between two hollow glass cylinders **2** and **3** in a position transverse to the mating ends of the cylinders, which are axially aligned. In this position the membranes are clamped by suitable clamping means **4** and **5** to form a tight seal, exposing the membrane surfaces to the interior of the cylinders. The active surface of the membrane is faced upward and a small quantity of hydrolyzing solution **6** is poured into the upper cylinder onto the membrane surface to a depth of 1 to 2 cm.

FIGURE 1.6: PARTIALLY HYDROLYZED CELLULOSIC REVERSE OSMOSIS MEMBRANES—TREATING UNIT

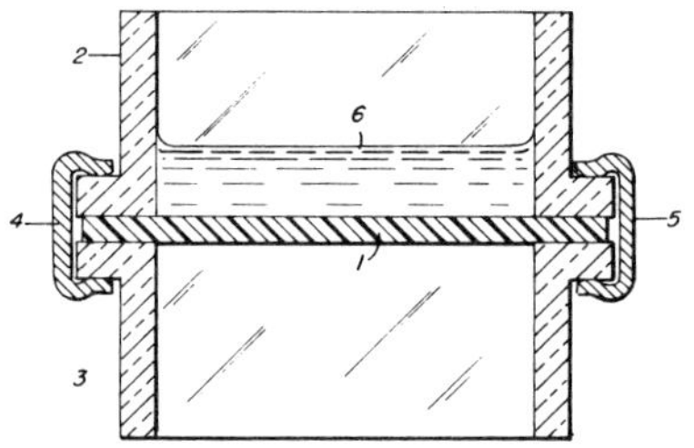

Source: R.E. Kesting; U.S. Patent 3,290,286; December 6, 1966

The solution is permitted to react with the membrane for a predetermined period of time and then is removed, for instance, by rinsing to terminate the reaction. In this example, the reagent was aqueous 0.4M NaOH and the hydrolysis temperature was 23°C. In the table below, the effect of hydrolysis on water permeation flow and salt rejection is shown for a standard reverse osmosis cell using a constant pressure differential and uniform conditions. A feed solution consisting of 3.5 weight percent NaCl was used for all tests.

Hydrolysis Time (min)	Acetyl Content (weight percent in total membrane)	Water Flux Rate (gfd)		Product Conc. (ppm NaCl)	
		Initial Rate	After 2 Hr	Initial Conc.	After 2 Hr
0	39.3	12.0	9.6	345	395
1	39.2	14.5	13.9	330	345
2.5	38.8	19.3	15.4	600	850
5	37.1	21.5	-------	4,500	------

For the particular hydrolyzing reagent used, a range of 1 to 2.5 minutes gives a desirable increase in water permeation flux rate at a low product salt content. By changing the reaction conditions, the critical time range may be changed considerably. The desired result of the hydrolysis is the conversion of a portion of the ester groups to hydroxyl groups in the active layer. The exact thickness of this layer is not known; however, results of the experiments indicate that the outer 10% of the total membrane contributes substantially all of the demineralization characteristics to the film. Only the average acetyl content of the membrane is measured. The surface acetyl content would be presumed to be smaller, since the hydrolysis would proceed from the outer molecules inward.

The desired average conversion of ester groups to hydroxyl groups is about 0.25 to 1.25% of the total ester groups in the membrane. It is not known what fraction of the surface groups are hydrolyzed, since the exact thickness of the active layer is not known, nor is the penetration of the hydrolysis easily detected. It is presumed that substantially all of the active layer is hydrolyzed under the conditions of the above table after five minutes.

This is inferred from the increase in salt content in the product water.

Cellulose Nitrate

A.S. Douglas; U.S. Patent 3,520,960; July 21, 1970; assigned half to Massachusetts Institute of Technology and half to the U.S. Secretary of the Interior has found that microporous films, or membranes, can be produced by casting solutions of amorphous glassy polymers which contain from about 7.5 to 26.5 volume percent tellurium tetrabromide, drying the film and then washing out the tetrabromide. When the polymer is a partially nitrated form of cellulose nitrate, the film is useful in separating potable water from seawater by reverse osmosis. The following example illustrates the process.

Example: In a typical preparation, 6.44 grams of tellurium tetrabromide are dissolved in 57 grams of tetrahydrofuran. After filtering the solution, 7.54 grams of cellulose nitrate are added and permitted to dissolve. A membrane cast from the resulting solution contains 24.8% tellurium tetrabromide by volume.

Cellulose Acetate Butyrate and Flux Promoters

In a related process described by *S. Manjikian; U.S. Patent 3,607,329; September 21, 1971; assigned to the U.S. Secretary of the Interior* semipermeable membranes are produced by casting a solution consisting essentially of cellulose acetate butyrate, triethyl phosphate, a flux promoter and an organic solvent into a film, and subsequently gelling the film by immersion in water. The membrane is characterized by a flux of at least 7 gal/ft^2/day and selectivity of at least 98.5% and often as high as 99.8% when tested using an aqueous feed solution containing 5,000 ppm NaCl at a pressure of 600 psig. The following example illustrates the process.

Example: The following ingredients were thoroughly mixed in a rolling jar mill, all parts are by weight.

	Parts
Cellulose acetate butyrate (EAB-171-15)	22
Triethyl phosphate	25
Glycerol	2
n-Propanol	6
Acetone	45

The mixture or casting solution was cast into 6" x 6" membranes using a flat glass plate equipped with 0.01" thick side runners and drawing the solution down the plate with a doctor blade. The film on the glass plate was exposed to the atmosphere for 60 seconds and was then immersed in water, initially at 33°F for about 1 hour, the quench water temperature rising gradually during the immersion period. The membrane was ready for use without heat treating. A 2" circle was cut from the flat membrane so produced and placed in a laboratory high pressure cell for testing.

A salt solution containing 5,000 ppm NaCl was used as feed and was pumped into the cell at a pressure of 600 psig. Purified product water passing through the membrane during a predetermined time period was collected and measured. Calculations converted the results to gfd.

The purified product was tested for conductance, and micromhos/cm converted to parts per million NaCl. Selectivity was calculated as the percentage of NaCl in the original feed solution rejected, or held back by the membrane. Results were flux, 8 gfd; product water salt content, 20 ppm; and membrane selectivity, 99.6%.

Silicone Coated Cellulosic Sheet

G.J.J. Marie; U.S. Patent 3,540,598; November 17, 1970; assigned to Ste d'Etudes pour Parois Filtrantes, France describes a filtering diaphragm intended for osmotic exchanges and molecular separations constituted by a sheet of microporous paper made of regenerated cellulose fibers which have been subjected to a marked fibrillation. The sheet is coated with a tenuous sheet of plastic material such as silicone elastomer having osmotic properties. If the coat is reduced to submicronic thickness, the diaphragm can serve for molecular separations, e.g., the removal of salt from seawater. The following example illustrates the process.

Example: A paper of the BX type produced by starting with regenerated cellulosic fibers which have been refined so as to show a very marked fibrillation, the specific weight of which paper is 30 g/m^2, while its pores have an average diameter ranging between 0.02 and 0.01 micron, is coated by a coating machine of the well known roller type with a silicone elastomer allowing selective changes between oxygen and carbonic gas. It has been found that for a like contacting area, the yield of the exchanges was clearly improved when using such a treated paper, chiefly with relation to carbonic acid gas, the actual osmotic coating having a thickness of 1 micron.

It is in fact possible to lay with a coating machine a continuous coat over the sheet of paper with a view to producing an osmotic diaphragm and according to a modification of the process, the thickness of coating may be reduced to form a porous submicronic film through which a molecular separation may be obtained through simple filtration. In particular, such a submicronic filter may serve for removing the salt from seawater.

Gelled Cellulose Triacetate

L.D. Nichols; U.S. Patent 3,846,404; November 5, 1974; assigned to Moleculon Research Corporation describes gelled cellulose triacetate (GCTA) products. In particular, the process relates to flexible, transparent GCTA products consisting essentially of cellulose triacetate (CTA) and a liquid nonsolvent for CTA, in controlled proportions, the composite material exhibiting the high tensile strength of its CTA component in conjunction with a liquid component which is freely replaceable by other liquid nonsolvents, or removable to provide shrunken and/or strengthened products composed essentially of CTA alone. This behavior, as well as permeation data, shows the GCTA products to consist essentially of two interpenetrating phases: a liquid-gellant phase and an ultramicroporous CTA phase.

The GCTA products are prepared from a solution of CTA in a nonhydrolyzing acidic solvent by forming the solution into a desired product and gelling the product with a cool liquid nonsolvent for CTA which is miscible with the acidic solvent. Suitable combinations of solvent and gelling liquid include, for example, formic acid/water; formic acid/methanol; acetic acid/water; trifluoroacetic acid/water, and meta-cresol/methanol. The CTA content of the product gels can be controlled by adjusting the concentration of CTA in the parent acidic solution. CTA contents as low as 5% are possible using dilute solutions with viscosities of a few hundred centipoises, while CTA contents of 50% or more can be obtained using solutions with viscosities of several hundred thousand centipoises, the limit being set by the maximum viscosity compatible with the chosen method of processing.

Removal of all or part of the liquid from the GCTA products, for example, by evaporation, produces stable GCTA products of reduced liquid content which can be stored indefinitely in a nonevaporative environment. Removal of the liquid component induces irreversible shrinkage, the result of complete removal being a pure CTA product of normal (about 1.3 g/cc) density which has the excellent solvent resistance and thermal stability of CTA prepared in other ways, for example, by casting.

Example 1: Various commercial grades of CTA from different manufacturers with acetyl content in excess of 42% were each dissolved in 90% aqueous formic acid in the amount of 7.5% CTA by weight. A film of the viscous solution prepared was blade-coated onto a smooth glass surface at room temperature, e.g., 20°C, with a 10 mil blade gap. The plate and film were promptly immersed in a flowing 13°C water bath.

Spontaneous separation of the film from the plate began in 15 seconds, and the gelled film was lifted from the glass after 30 seconds. The gelled film was worked in flowing water for 10 minutes to remove all indication of formic acid. The gelled film was found to be 3.7 to 4.0 mils in thickness, and did not change in appearance or dimensions upon prolonged storage in water. The gelled film so prepared was highly transparent, hydrophobic as shown by beading of superficial water on its surface, and strong enough for easy handling.

Example 2: Film specimens prepared according to Example 1 were tested in standard cells to determine their performance as semipermeable membranes. At hydrostatic pressures ranging from 200 to 600 psi, the membranes gave water fluxes between 120 and 320 gfd. Tests on Cytochrome-C (MW 12,400) solutions at pressures of 30 and 50 psi gave 85 to 88% solute retention at flexes between 30 and 40 gfd. Blue Dextran (MW 2,000,000) solutions at 50 psi gave over 99% solute retention at a flux of 56 gfd. The membranes showed little drop in flux with time during the experiments.

Microwave Curing Using Crosslinking Agent

According to a process described by *J.L. Rendall; U.S. Patent 3,864,289; February 4, 1975; assigned to Koratec, Inc.* a cellulosic semipermeable membrane is formed from a formulation containing a cellulosic membrane material, a crosslinking agent, a catalyst and a chemical blocking agent. After formation, the membrane is subjected to curing conditions preferably utilizing microwave radiation to achieve the desired crosslinking of the cellulosic structure and at the same time avoiding substantial shrinking of the membrane during curing.

Suitable crosslinking agents include those ordinarily used for crosslinking cellulose, such as noncyclic and cyclic urea type formaldehyde reactants such as dimethylol dihydroxy imidazolidone having the structure,

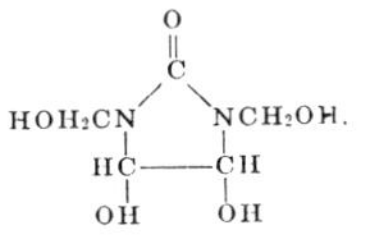

methylol ureas such as dimethylol ethylene urea (DMEU) dimethylol propylene urea, methylolated melamines, methylolated amino triazines, formaldehyde and glyoxal. One preferred crosslinking agent is a solution of dimethylol dihydroxy imidazolidone, which is available commercially as Permafresh 183.

Examples of monofunctional blocking agents are monomethylolated amides, monomethylolated ureas, cyclic ureas, and cyclic ethylene ureas which have been reacted with one molecule of formaldehyde. Monomethylol ethylene urea is a particularly useful monofunctional blocking agent. Suitable difunctional blocking agents include methylolated melamines, such as the methylolated melamine solution available commercially as Aerotex M-3. The difunctional blocking agents are believed to function not only as chemical wedges in the manner described above but also as crosslinking agents. Their ability to perform a blocking function results from their molecular size, and the ability to crosslink of course, results from their difunctionality.

One example of a suitable formulation, using ten parts of cellulose acetate having an acetyl content in the range of about 37 to about 41%, includes 40 parts of a solvent combination of acetic acid and water (approximate 4:1 ratio), 0.3 parts of the described imidazolidone crosslinking agent, 0.3 parts of the Aerotex M-3 methylolated melamine, and 0.05 parts of zinc nitrate catalyst.

After the formulation has been prepared, the membrane may be formed by casting the solution upon a smooth flat surface so that a thin layer of uniform thickness is obtained The casting operation is regulated so that the thickness of the cast membrane typically is on the order of about 100 microns. The cast membrane may remain in contact with the atmosphere for approximately one minute to permit solvent to evaporate and thereafter immersed in water at a temperature of about 1°C, where it remains for a sufficient time to set up and remove unevaporated solvent by diffusion into the water. After the membrane has been formed, it will be cured to produce the desired crosslinking. Although curing may be effected in other ways, e.g., oven curing, this is achieved preferably by subjecting the membrane to microwave radiation in a suitable environment such as a hot air oven.

Using resin catalyzed with zinc nitrate, a 5 minute microwave curing time normally is ample, and in same instances, using a hydrochloric acid catalyst system, a shorter curing time is satisfactory. In general, the microwave frequency should be at least about 900 megacycles for a period of from 10 to 300 seconds. Microwave curing is also believed to provide increased resiliency to the membrane, which enables the membrane to resist more effectively the effects of pressure in use. By the same token, it is believed that microwave curing creates more uniform blocking and crosslinking throughout the thickness of the membrane.

Persulfate-Bisulfate Crosslinking of Cellulose Acetate Methacrylate

O.S. Schaeffler; U.S. Patent 3,669,954; June 13, 1972; assigned to The U.S. Secretary of the Interior has found that a preformed cellulose acetate methacrylate (CAM) membrane may be crosslinked to a high degree of acetone insolubility, e.g., about 50 to 95% insolubility, using a persulfate-bisulfate redox system in water and that the resulting membranes exhibit superior reverse osmosis properties. The product membranes are particularly outstanding in long-term reverse osmosis because of their superior flux stability.

This superiority is evidenced by decreased flux decline slopes as compared to those of uncrosslinked membranes. Other means of crosslinking have previously been used in preparation of membranes, e.g., a redox system composed of ferrous sulfate and hydrogen peroxide and ultraviolet irradiation. However, these methods have the disadvantages of deposition of ferric oxide on the membrane in the former and dependence on exposure time and absorption in the latter. The method of this process, on the other hand, is highly advantageous since it can be performed in aqueous media, is rapid and is independent of the physical properties of the membrane.

Example: Membranes were prepared from a CAM polymer having an acetyl DS of 2.09 and a methacryl DS of 0.29 using a casting solution containing 35 parts of acetone, 10 parts of polymer, 10 parts of water and 14 parts of maleic acid. Castings of 10 mils wet thickness were made at -10°C, gelled immediately in ice water and heat treated in water for 3 minutes at 80°, 83° and 86°C. Several test circles from each of the membranes were crosslinked in a solution of 3.7×10^{-2} M potassium persulfate and 3.8×10^{-2} M sodium bisulfite at 90°C. After this treatment approximately 95% by weight of the membranes was insoluble in acetone indicating highly effective crosslinking. Three circles from each of the uncrosslinked (80°, 83° and 86°) and crosslinked (80°X, 83°X and 86°X) membranes were tested for 200 hours with 1% sodium chloride feed at 800 psi. The data are summarized below.

Membrane	Flux decline slope, m	Initial flux, gfd	Salt rejection, %	
			Initial	Final
80°	−0.042 ± 0.008	50	69.2	86.5
80°X	−0.022 ± 0.009	67	68.6	76.7
83°	−0.037 ± 0.005	35	85.1	90.3
83°X	−0.019 ± 0.003	52	86.0	82.8
86°	−0.051 ± 0.008	39	82.9	91.4
86°X	−0.013 ± 0.002	41	89.4	89.8

As is apparent from the data, the flux of the crosslinked membranes is greatly stabilized over their uncrosslinked counterparts. The 86°X membrane with a flux slope of 0.013 and a 1-hour flux of 41 gfd would be expected to have a flux of 37 gfd after 1 year operation.

Graft Polymerized or Polymer Deposited Permselective Materials

A process described by *D.J. Bridgeford; U.S. Patent 3,723,306; March 27, 1973; assigned to Tee-Pak, Inc.* is based on finding that substantial improvements in separation of mixtures by ion exchange or by permselective processes may be obtained by using ion exchange materials or permselective materials comprising a host material having a guest polymer of an olefinically unsaturated monomer deposited or graft polymerized therein. The polymer deposited or graft polymerized host material may be prepared in accordance with the properties desired in the separatory process. Thus, in producing an improved ion exchange process, a guest polymer is deposited or graft polymerized in a host material to produce a product having additional ion exchange capacity.

This is accomplished by depositing or graft polymerizing a polymer having ion exchange functional groups or having functional groups which can be converted, e.g., by hydrolysis, to ion exchange groups. The deposition or graft polymerization of a polymer into a host material may also be utilized to vary the permselective properties of the host and produce a permselective membrane or web or other separatory body having a predetermined selectivity for the particular materials to be separated or having preselected properties making it especially suitable for use in reactive solvents or in solvents which would attach the permselective host material.

In carrying out this process, an ion exchange material or permselective material in the form of a membrane, web, beads, particles, fibers, filaments, etc. is treated with a suitable monomer under conditions which effect a deposition and grafting of a polymer within and upon the host material. In many preferred applications, the host material is in a form of a film, such as regenerated cellulose, or a web such as paper, or in the form of particulate material such as sawdust or wood chips or particles or beads of an inorganic or organic polymer.

The guest polymer is deposited and grafted within and upon the host material to provide additional ion exchange sites, in the case of ion exchange materials, and to modify the permselective properties in the case of permselective materials. The polymer deposition and graft polymerization may be carried out in accordance with any of a variety of prior art techniques although the process of U.S. Patent 3,083,118 is generally preferred.

In the preparation of ion exchange materials or permselective materials for use in accordance with this process according to the process of U.S. Patent 3,083,118, a host material is selected of suitable form which has some appreciable amount of ion exchange capacity, either inherently in the material or artificially produced. The host material is treated with a catalyst component which is ionic and which is deposited in the host by ion exchange reaction. After the catalyst is deposited in the host material, the catalyst-modified host is treated with the desired monomer and any additional catalyst component which may be required, thus effecting a graft polymerization within and upon the host which is localized largely in the areas where the catalyst component was positioned by ion exchange reaction. The conditions required for carrying out this process of polymer deposition and grafting are described in considered detail in U.S. Patent 3,083,118.

Generally, in preparing materials for use in the separatory processes, any guest monomer may be used which will modify the host material to produce the desired permselective properties or which will introduce the desired ion exchange groups in the case of an ion exchange material. The process of U.S. Patent 3,083,118 is effective to deposit and to graft polymers within host materials of any desired form or shape and to locate the guest polymer is preselected regions of the host material. Thus, if desired, the guest polymer may be located uniformly throughout the host material or in selected regions of the interior or the surface of the host material. The following examples are illustrative of the preparation of ion exchange and permselective materials and their utilization in accordance with this process.

Example 1: This example illustrates the preparation of a permselective film in accordance with the process of U.S. Patent 3,083,118. Fifteen pieces of regenerated cellulose film made by the viscose process, and containing 0.025 meq of cation exchange capacity per gram of dry cellulose, about 5 mils thick in the gel state, and weighing about 1 gram were subjected individually to the following process. The film was immersed at 25°C for one minute in a solution of 0.7% ferrous ammonium sulfate at pH 5.5. The film was then washed in deionized water until no test for iron in the washings could be made, and then placed in 400 ml of 0.003% hydrogen peroxide in water at 100°C, the water containing 10 ml of pure methyl methacrylate. Polymerization was continued for 5 minutes under reflux and was terminated by pouring the contents into cold water. No clouding in the aqueous solution was observed. After extraction of the film in acetone overnight and drying at 105°C for several hours, the films were found to have increased in weight by 48%. The films were of a slight-yellowish brown in color and were quite flexible. Mullen burst tests on the films indicated no decrease in strength as compared to control films, and were slightly over 6 lb/mil/in^2 of film area. The tenacity was about 3 g/mil^2 as measured by the Scott test. Elongation at break had been decreased by about ⅓ as compared to a control film.

The films produced in accordance with this example comprise a graft polymer of methyl methacrylate on cellulose. The films when used in permselective processes are substantially less permeable to larger molecules than is untreated regenerated cellulose film. Also, when these films are treated with aqueous alkali to hydrolyze the ester groups the resulting product is a graft polymerized film having a high cation exchange capacity.

Example 2: Small sheets of manila paper were used in this experiment. They were of very closely equal weight, and contained 0.1 meq of cation exchange capacity per gram. The papers were immersed in 0.7% ferrous ammonium sulfate solution at pH 5.4 for one minute, washed, and added to a solution of 0.003% hydrogen peroxide containing 20 ml of pure methyl methacrylate per liter of solution. The polymerization was continued at reflux for five minutes. Polymerization was stopped as in Example 1.

The paper was dried and found to have increased in weight by 90%. It was a cream color, somewhat rigid, its dry Mullen strength was only about half that of the original paper, and its wet Mullen strength was about ½ times its dry Mullen strength. The graft polymerized paper web formed in this example is useful as a dialysis membrane and when saponified with aqueous alkali has substantial ion exchange capacity.

UNIT FABRICATION TECHNIQUES FOR CELLULOSIC MEMBRANES

TUBULAR

Casting Technique

A process described by *S. Loeb, S. Wasilewski, E. Selover and A. Balla; U.S. Patent 3,446,359; May 27, 1969; assigned to The Regents of the University of California* concerns the formation of reverse osmosis apparatus by dissolving a cellulosic derivative and a pore producing agent, as described in U.S. Patent 3,133,132 and 3,133,137, and casting the resultant solution as a thin tubular sheet which is cured to obtain the desired salt reduction ratio and flux by heating. Preferably the membrane is cured by heating it under pressure in situ in the perforated supporting tube or pipe to be used in desalination, from which it is separated by one or more layers of porous material providing a low resistance path for desalinized water which has passed through the membrane, to the perforations in the supporting tube or pipe.

In operation of a completed composite tubular assembly of the process, as a solution passes along the tube at the pressure necessary to obtain reverse osmosis, desalinized water continuously is removed from the stream, so that the salt content steadily increases downstream. A channel bounded by membranes of appropriately increasing salt reduction ratio may readily be obtained by making more than one composite tubular assembly in the same manner and with the same materials except that the later units are heated to a higher temperature in the curing step. Connection of the units in series, as formed, provides the desired channel.

Although any of the semipermeable cellulosic membrane systems may be used in the process upon appropriate compliance with the conditions for forming effective osmotic film membranes with the particular system, it is preferred to use the system described in U.S. Patent 3,344,214 because of the excellent salt reduction ratios and flux obtained thereby as well as the ease of handling such a system at ambient temperatures. By way of example and ease of description, the process will be described in terms of the system of U.S. Patent 3,344,214, in which a typical casting solution comprises cellulose acetate dissolved in acetone and formamide, with these components in the ratio, respectively, of 25:45:30 by weight.

Such a solution is rather viscous and may readily be handled and cast at room temperature. The casting may be performed with the apparatus shown in Figure 2.1a and 2.1b in which the casting tube **2** may be an ordinary pipe, preferably with an inner diameter such that the outer diameter of a gelled tubular membrane sheet removed from the casting tube plus the thickness of any porous wraps added to the tubular membrane sheet equals or approximates the inner diameter of the run tubes to be used in the performance of reverse osmosis. Casting could be performed in the run tube, as by blocking preformed perforations in the run tube during casting or by adding perforations after the casting process is complete, but the use of separate tubes for casting and run purposes is preferred.

FIGURE 2.1: MEMBRANE ASSEMBLY

a.

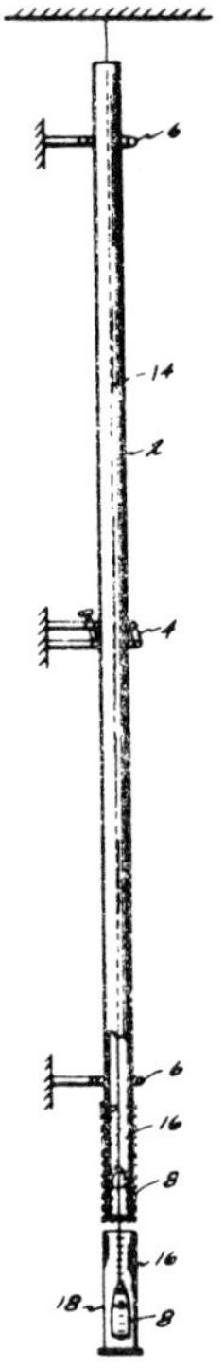

Membrane Casting Apparatus

b.

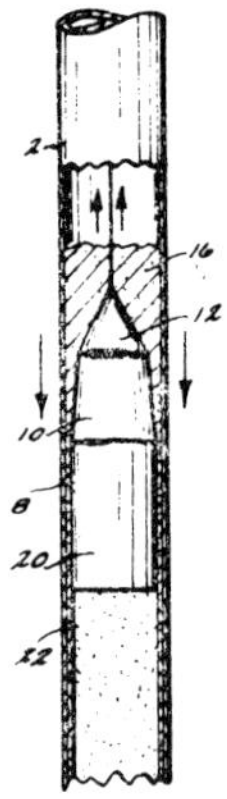

Sectional View of Casting Tube and Spreader

c.

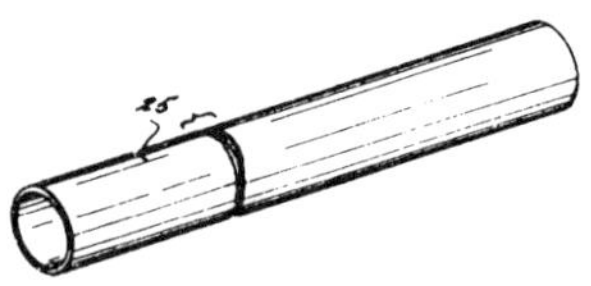

d.

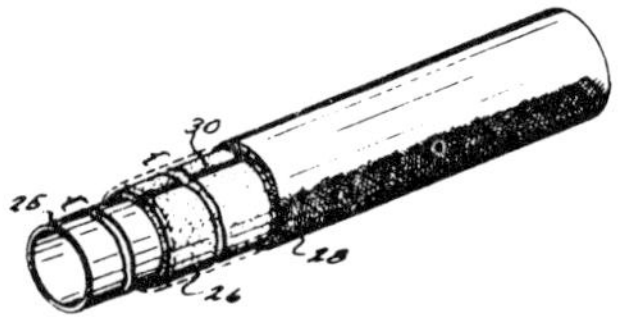

(c) (d) Tubular Membrane Sheet

(continued)

FIGURE 2.1: (continued)

e.

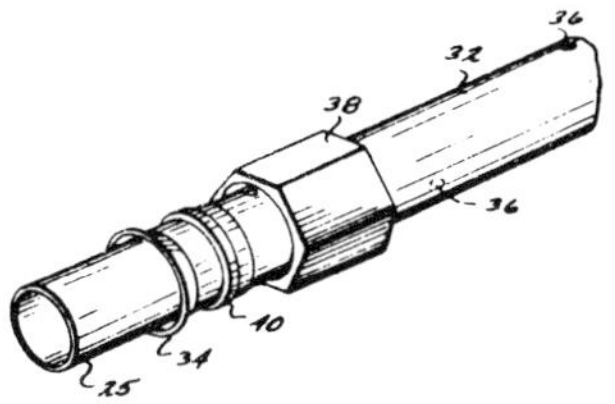

Tubular Sheet Prior to Trimming

f.

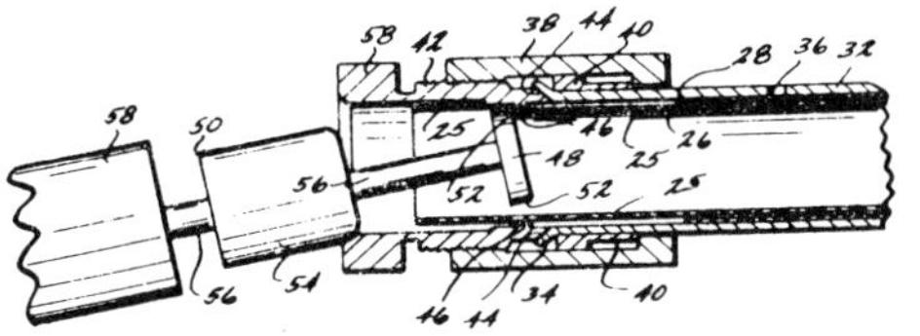

Cutting Apparatus for Trimming

g.

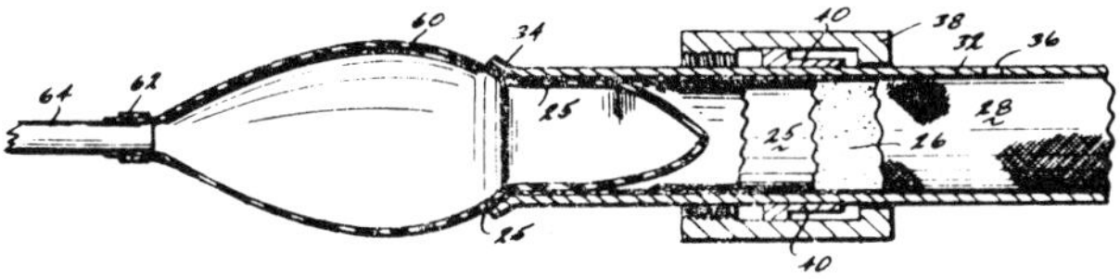

Flaring of Trimmed Sheet

h.

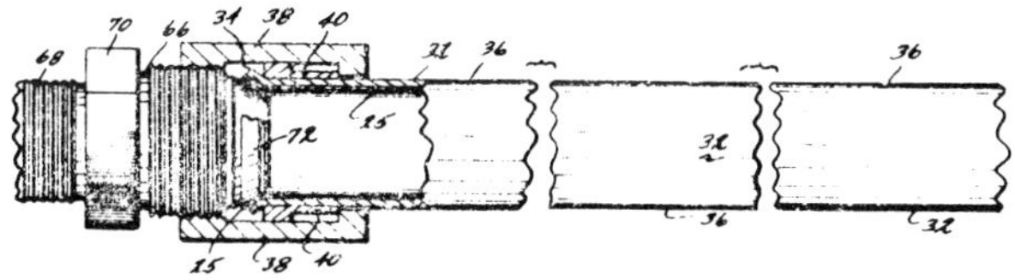

Sealing Joint, Along Line 8-8 of Figure 2.1i

(continued)

FIGURE 2.1: (continued)

i.

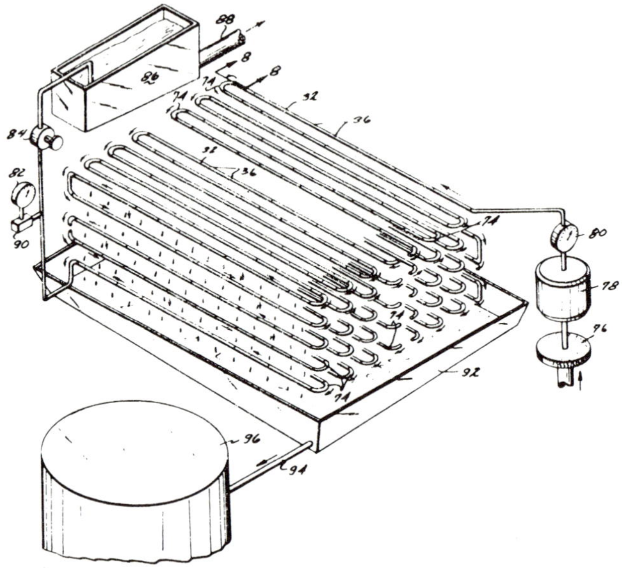

Desalination Plant

Source: S. Loeb, S. Wasilewski, E. Selover and A. Balla; U.S. Patent 3,446,359: May 27, 1969

The casting tube **2** is supported by releasable clamp **4** and held in position by guide loops **6**. A spreader **8**, preferably having two cone tapered sections **10, 12** on its leading end, is attached at its leading end to cord **14** for relative movement axially of tube **2**. The combination of an adequately long cylindrical body section **20**, (e.g., 3 inches long for use in a casting tube approximately 1 inch inner diameter) and one or more cone tapered sections **10, 12** in the spreader provides a self centering effect for the spreader in passage through the tube **2** which reduces or eliminates the need for any other centering means.

The casting solution **16** may be placed in the lower end of the casting tube ahead of the spreader by raising a graduate cylinder **18** of greater inner diameter than the outer diameter of the casting tube which contains the casting solution, over the spreader suspended below the bottom of the casting tube, and then raising it over the bottom end of the casting tube and raising the spreader into the casting tube by drawing up cord **14**, as indicated in broken lines in Figure 2.1a.

The cylindrical body section **20** of the spreader has an outer diameter smaller than the inner diameter of the casting tube by twice the desired as-cast thickness of the membrane. As-cast membrane thicknesses of 0.015 inch are preferred since membranes less than 0.010 inch thick frequently are so adhesive that they are difficult to remove from the casting tube after immersion without suffering damage. Membranes thicker than 0.050 inch are not preferred, since such added thickness appears unnecessary and merely adds to the path through which the desalinized water in reverse osmosis must pass.

The spreader may then be moved axially through the casting tube by drawing up cord **14** (indicated by the upward pointing arrows in Figure 2.1b) as by pulley means (not shown) connected to the cord above the upper end of the casting tube. Drawing the spreader through the casting tube in this manner, with the casting solution above the spreader, leaves behind the spreader a film **22** of the casting solution with the desired as-cast thickness. The viscosity and surface adhesion of the casting solution is such that the film on the inner walls of the casting tube does not sag or run appreciably. However, drawing the spreader through the casting tube is not preferred because the film is exposed to the atmosphere for varying periods of time along the length of the casting tube as the spreader moves up through the tube. Also, since the evaporating acetone solvent is heavier than air, a reverse chimney effect is obtained which accentuates greater evaporation from the film in the lower part of the casting tube than in the upper part.

The velocity of the movement of the spreader relative to the casting tube preferably is uniform, but in casting a tube of approximately 1 inch diameter it may vary between 15 and 250 inches per minute and still provide a satisfactory film. Particularly with slower relative movement, however, the casting solution must be sufficiently viscous that no appreciable sagging of the film takes place. A satisfactory casting solution for even relatively slow movement of the spreader and the casting tube has a viscosity such that a 3/32 inch stainless steel ball will drop through the solution at a rate of 1¼ inches per min.

It is preferred to hold the spreader stationary by means of cord **14** and to obtain relative movement by releasing clamp **4** so that the casting tube, held against sideward movement by loosened clamp **4** and guide loops **6**, falls into an immersion tank. The viscosity of the casting solution is such that this free fall, in the example given, with a steel casting tube approximately 11 feet long, a 1 inch outer diameter, and walls 0.049 inch thick, is completed in 45 seconds, providing a relative rate of movement within acceptable limits.

During the falling of the casting tube into the immersion tank a constant length of the film between the bottom of the spreader and the surface of the water in the immersion tank (distance preferably between 1 and 3 feet) is exposed to the atmosphere. Since the rate of fall of the casting tube is substantially uniform, each portion of the film is exposed to evaporation for a substantially uniform period of time, and since the exposed portion of the interior of the tube is sealed at the upper end by the spreader and at the lower end by the water in the immersion tank, no reverse chimney or other draft effect of significance is present. Thus, the membranes of greater uniformity are obtained since removal of solvent from the cast film has an important controlling effect upon the performance of the completed membrane in reverse osmosis; therefore, free fall of the casting tube, indicated by the downward pointing arrows in Figure 2.1b is preferred.

In any event immediately after casting the film on the inner wall of the casting tube, the tube and film are immersed in water, as in the tank. The water may be as warm as 30°C or more, but the finished tubular membrane sheet appears better if the water temperature is between 0° and 5°C. The casting tube with film, when immersed in water at 0°C, is kept in the tank for at least 1 hour. During immersion more of the solvent in the film is removed by diffusion into the water in the tank. It has been found that all of the solvent removal may be accomplished by diffusion into water, i.e., no evaporation into the atmosphere is necessary to produce satisfactory membrane. However, of course, excessive evaporation must be avoided.

The solvent removal continues until the casting solution **16** in the film gels in the sense that it achieves discrete physical strength, so that it may be removed from the casting tube as tubular membrane sheet **25** shown in Figure 2.1c. The period of water immersion required may readily be learned by feeling the sheet to determine whether it is continuing to increase in physical strength. Excessive water immersion does no harm; in the example under consideration, the period of immersion in water between 0° and 5°C should be 1 hour or more. The tubular sheet **25** is removed from the casting tube **2** simply by grasping one end and gently drawing it out of the casting tube. At this point it may easily be examined and tested, both visually and otherwise, to maintain quality control in production.

Although more expensive means for providing a low resistance path for flow of desalinized water to the exits from run tubes, and more closely spaced perforations in the run tubes (as described hereafter), may be used, it is preferred to wrap the tubular sheet **25** in one or more layers of porous material, as illustrated in Figure 2.1d. Figure 2.1d illustrates a single layer of filter paper **26** (such as that commercially available as Whatman 41H high wet strength filter paper) which has been wrapped around the sheet **25** and then covered with a double thickness of coarse porous wrap **28**, such as an ordinary woven nylon fabric or parchment (commercially available as French Fabrics 627T).

The smooth porous layer **26** and the coarse porous wrap **28** both preferably are cut as a single long rectangular sheet such that when wrapped longitudinally around membrane **25** the side edges of the sheet **26** or **28** meet exactly. However, it is more important in avoiding exercise of the membrane sheet **25** to prevent a small overlap of the wraps, particularly the inner smooth porous wrap, than to prevent a small gap, such as gap **30**, between the side edges of the wrap. When multiple layers of wrap are used they are preferably coterminous, near the ends of membrane **25**, as indicated in Figure 2.1d, in which outer wrap **28** has been cut back to better illustrate inner wrap **26**. The membrane tube **25** preferably extends beyond both ends of the porous wrap.

Although a single layer of wrap, or multiple layers of a coarse porous wrap without a smooth porous layer may be used, the combination illustrated in Figure 2.1d is preferred because the smooth porous wrap **26** minimizes membrane embossing by plastic extension of the finished membrane under high pressure, which causes exercise of the membrane structure which is deleterious to desalination performance, and the coarse porous wrap **28** provides an excellent low resistance path for the desalinized water to exit from the supporting structure such as the run tubes described below.

Formation of the desalination assembly may be radially reversed to provide assemblies to be immersed in solution under pressure, (e.g., beneath the ocean), so that the porous layer, such as wraps **26** and **28**, is positioned within a tubular membrane sheet **25** which has been cast on the outside of a casting tube **2** (to provide the essential desalinizing layer on the outside of the tubular membrane sheet **25**) which together are then placed on the outside of the run tube rather than the inside of it. Other ways of placing the low resistance porous layer between the membrane sheet and the supporting structure, such as wrapping a run tube prior to positioning the tubular membrane sheet, may be utilized.

In the preferred method, after the tubular membrane sheet has been wrapped, this sheet and its wraps **26, 28** are inserted in a run tube **32**. The run tube may be ordinary commercial pipe, of the type appropriate for the pressures at which reverse osmosis is performed, having flared lips **34** at its ends. In this example, in which the casting tube has a 1 inch outer diameter and walls 0.049 inch thick, the run tube may be ordinary steel pipe with a 1 inch outer diameter and walls 0.035 inch thick, with a length of 10 feet.

The length of the membrane **25**, when inserted, extends beyond both ends of the run tube, but the porous layer such as wraps **26, 28** falls slightly short of the ends of the run tube, such as 1 inch short in this example. The run tube has perforations **36**, spaced at intervals along the length covered by the porous layer such as wraps **26, 28**. The perforations are $\frac{1}{16}$ inch diameter holes drilled through the pipe at 3 inch intervals alternately on opposite sides of the pipe **32**.

After insertion in the run tube the membrane tube **25** is trimmed at both ends, as for example, with the cutting apparatus shown in Figure 2.1f. Nut **38** with sleeve **40** between it and the run tube (both nut and sleeve being standard fittings with commercial pipe such as the run tube), holds threaded cutter insert **42** firmly in position with end **44** (shaped to fit flared tube end **34**) centering cutter insert **42** against the tube end. Shoulder **46** around the inner circumference of the cutter insert is spaced away from the end such a distance that when membrane sheet **25** is cut at the shoulder and flared as described below it will be coterminous with flared lip **34** of the run tube. Head **48** of the cutter tool **50** has a diameter somewhat less than the inner diameter of the run tube, so that its circumferential sidewalls **52**, upon rotary movement of cutter tool **50** at an angle to run tube **32**, press and cut off the membrane sheet **25** against the edge of shoulder **46**. This action is aided by roller **54** on shaft **56** of cutter **50**, which bears against the outer end **58** of cutter insert **42**, spaced at the appropriate distance from the shoulder, upon insertion and rotation of the cutter tool, as by manipulation of handle **58**.

The clean uniform trimming of membrane **25** at the shoulder at each end of the run tube leaves a length of membrane which may be flared after removal of the cutter insert, to conform closely to the flared lips **34** at the ends of the run tube. Since flaring the end of the tubular membrane sheet to the conical shape of lip **34** requires plastic extension which imposes strain on that portion of the membrane sheet, it is preferred to soften the short length of membrane adjacent the lip either by heating it in water 70°C or higher, or by moistening it with a plasticizer softening solution such as a mixture of 9 parts by volume of n-propyl alcohol and 1 part by volume of triacetin. Also, in the example under consideration, a cellulose acetate polymer of high molecular weight material, such as Eastman Kodak E-398-10, is used in the casting solution **16**.

When the end of the membrane has been softened it is easily flared by inserting a flexible tube or balloon **60** into it and distending the balloon pneumatically or hydraulically, as by forcing air or water into the balloon through nozzle **62** and tube **64**, which may be connected to a source of fluid under pressure (such as ordinary tap water) or blown up like a balloon used as a plaything. The balloon may readily be removed by deflation and withdrawal of the portion of it inserted within the run tube. It must be kept in mind that during the flaring operation, as at all times during its preparation and operation in reverse osmosis, the membrane must not be permitted to dry out. The flared membrane in place with its porous wraps **26, 28** in the run tube in which it may be used in desalination, is now ready for the heat treatment which tailors it precisely for the service requirements it must meet.

Although this heat treatment may be accomplished by immersing the complete run tube assembly in water to obtain the conditions hereafter described, preferably it is accomplished by passing water at the desired temperature through the run tube under pressure. The run tube assembly may be joined and sealed to a separate source of heating water, and also joined and sealed in the completed desalination assembly illustrated in Figure 2.1i, with the standard commercial pipe-fitting components shown in Figure 2.1h. The seals obtained in this manner readily withstand the pressures of 1,500 psi or more which are used in desalination of sea water.

A standard threaded nipple **66**, which in turn may be connected to other piping indicated at **68** by nut portion **70**, either for treatment purposes or for reverse osmosis operation, is held by nut **38** with conical nipple section **72** bearing against flared lip **34** of the run tube. Satisfactory seals may be obtained by placing the conical nipple section directly against the flared portion of the membrane adjacent the lip. However, this generally requires the application of greater torque to the fitting in order to obtain a high pressure seal.

This greater torque increases the possibility of either plastic extrusion of the membrane at the seal, or of twisting of the membrane at the seal, or of twisting of the membrane with respect to the run tube, both of which should be avoided. Twisting of the membrane is particularly undesirable since it may cause exercise of the portions of the membrane which will perform desalination in operation. Subjection of the surface of the membrane to any stresses or strains other than its compression in use is definitely deleterious to its desalinizing performance.

It is therefore preferred, although not necessary, to place flat rubber gasket **74** over the conical nipple portion so that it bears against the membrane at the lip of the run tube. Satisfactory seals can then be obtained with a torque of 10 foot-pounds or less. Bringing the membrane to the proper temperature to complete the physical and chemical organization of the membrane layer which performs removal of solutes from the solution is essential to curing the membrane for optimum performance under service conditions.

In general, the higher the temperature at which the membrane is cured the greater is the salt reduction obtained. Correlatively, however, greater salt reduction generally results in a lower flux rate through the membrane. Therefore, the optimum membrane is one which provides solute reduction which produces the desired product, (e.g., potable water) from the feed solution at the maximum obtainable flux rate. Salt reduction performance of a membrane is most easily expressed in terms of a ratio of the salt concentration in the feed solution to the salt concentration in the desalinized product. The flux rate of production of desalinized product through the membrane is expressed in terms of quantity of product per unit area of membrane per time period, usually gallons per square foot per day.

The table below illustrates the control of salt reduction performance and flux which is obtained by variation of the temperature to which the membrane is brought in curing. The figures represent early, rough readings of a pilot plant, made with the use of the process with formamide as the pore producing agent, which the salt concentration of the product was measured by conductivity, and in which both the salt reduction and the flux measurements are not precise. The table is fully reliable, however, in illustrating the ready control of performance obtained by different curing temperatures. The pilot plant from which these readings were taken began supplying potable water to the regular drinking water system of the city of Coalinga, California on June 17, 1965.

Membrane Heating Temperature, °C	Flux, gal/sq ft/day	Salt Reduction Ratio
72	67	6
76	58	8
80	45	20
84	33	33
88	19	55

Since any plastic extension of the finished membrane is deleterious to desalination performance and since the operating pressures in reverse osmosis are quite high, optimum membrane fabrication and performance is attained (even when using relatively loose fits between the membrane, the porous wraps and the run tube to facilitate easy assembly and inspection) by performing final curing of the membrane in place in the run tube in such a manner that no further movement of the membrane and nonoperating stresses and strains are necessary.

Heating water is passed through the completed run tube assembly at a slight positive pressure. This pressure preferably exceeds the yield strength of the membrane at that temperature, so that the membrane is pressed firmly against its supporting wraps and the run tube while final membrane structuration occurs.

The high temperature yield pressure of the membrane is on the order of 4 to 6 psig for temperatures between 75° and 95°C. The heating is carried out for 4 to 10 minutes, which is sufficient time for the membrane to have reached a steady condition at that temperature. The flow rate of heated water through the run tube assembly is merely enough so that the temperature drop from one end of the tube to the other is less than ½°C. Under normal conditions, with a run tube of 1 inch inner diameter and 10 feet long, with heating to 85°C, a flow rate of 3½ gpm suffices.

Since the water to be desalinized in operation usually will be considerably cooler than the temperatures to which the membrane is brought to effect final curing, and since in any event the completed run tube assembly will be stored for varying lengths of time at ambient temperature prior to use in service, it is preferred to follow the heating step immediately by cooling the run tube assembly to the temperatures at which it will be used in service, or to ambient temperature.

This cooling, effected simply by passing water at the appropriate temperature through the run tube assembly, preferably is carried out at a pressure exceeding the yield strength of the membrane at the cooling temperature. The intent of immediate cooling at a pressure in excess of the yield strength of the membrane is to prevent the subsequent subjection of the membrane to contractile stresses and strains, relatively long after curing is complete, which harm membrane performance. The cooling is continued merely long enough to ensure that the membrane and run tube are brought to the desired temperature; 2 minutes of cooling, at a flow rate of 3½ gpm, has proven sufficient in the example given to prepare the run tube assembly for service at temperatures on the order of 15°C or less.

In practice, to ensure that the membrane is finally cured in position so that it will not later be subject to exercise in desalination performance, the yield pressure of the membrane usually is exceeded somewhat, so that a typical curing cycle comprises heating at 7 to 8 psig for 10 minutes, followed by cooling at 12 psig for 2 minutes or more. The cured run tube assembly may be stored, pending use in desalination, by filling it with water and capping its ends. Preferably the storage water is treated to prevent growth of algae and the like during storage; but so long as the membrane is kept wet and is not subjected to a negative pressure which would flex it or exercise it by moving it away from its supporting wraps, it retains effective desalinating ability even after weeks of storage.

There appears to be some tendency for the membrane to increase in flux and decrease in salt reduction ratio performance after prolonged storage; however, recuring it by passing heated water under pressure through the run tube assembly, at the temperature which provides precisely the desired combination of salt reduction performance and flux rate, will restore the membrane to precisely the desired service performance.

The heating temperatures required in curing to obtain a fixed salt reduction for a membrane which is cured under pressure usually is 3°C higher than the temperature required to obtain equivalent performance from the flat sheets discussed in U.S. Patent 3,133,132. It is believed that this is because the pressure cured membrane tube is not allowed to undergo thermal shrinkage, as is the flat sheet.

The cured run tube assemblies may be employed in desalination in a plant of the type illustrated in Figure 2.1i. A plurality of the run tubes, each containing a fully cured membrane sheet and supporting porus layer such as wraps **26, 28**, are connected in series by standard commercial U-joints **74** with the use of the seals illustrated in Figure 2.1h.

The feed (whether seawater, brackish water, or other solution) passes through filter means **76** to remove particulate and gross organic matter which might harm the pump and other parts of the assembly to pump **78** which pumps it through the assembly at the pressure necessary to obtain reverse osmosis. This pump is automatically controlled so that the pressure in the system, as read at gauges **80, 82**, remains at the level required for reverse osmosis for the particular solution being desalinized. Pressure regulator valve **84** at the brine discharge end of the assembly maintains pressure within the assembly at the desired level and discharges the brine at low pressure to brine accumulator tank **86**, which in turn, as indicated schematically at **88**, is connected to a drain for discharge of the brine.

The brine accumulator tank is preferably placed at a level above the assembly of the run tubes, and a second back pressure regulator valve **90** balanced against the atmosphere is provided at the end of the line, as indicated at **90**, so that in the event of a rupture at some point in the assembly of the run tubes which might tend to create a less-than-atmospheric pressure at some point in the run tubes which might flex the membrane away from its operating position, an input of atmospheric pressure and/or back flow of brine would be obtained which would minimize or eliminate the temporary reverse pressure causing such flexing of the membrane.

In the operation of a plant such as that illustrated in Figure 2.1i, the desalinized product flows through perforations **36** in the run tubes and may readily be collected by letting it fall into collecting means such as trough **92** beneath the assembly from which it may flow through piping as indicated at **94** to a storage tank **96** or directly into a water supply system. The table below provides an analysis of the product desalinized water produced in early operation of the experimental plant at Coalinga, California, from a brackish water feed. The figures are not precise and contain certain anomalies, (e.g., greater phosphates and iron and alumina as R_2O_3 in both the product and brine than in the feed) but the overall pattern is a fully correct representation of the performance of the desalination assembly. It will be noted that not only is the sodium chloride level reduced to well within acceptable limits for potable water, but also that the membranes have an excellent effect in softening water.

	Feed	Product	Brine
pH	7.5	7.4	7.5
Iron (Fe), mg/l	0.18	0.01	0.01
Hydroxide (OH), mg/l	–	–	–
Boron (B), mg/l	2.5	2.3	3.4
Carbonates (CO_3), mg/l	–	–	–
Bicarbonates (HCO_3), mg/l	161.7	18.3	271.5
Chlorides (Cl), mg/l	258.9	70.9	407.8
Sulfates (SO_4), mg/l	1,223.5	21.0	2,182.3
Phosphates (PO_4), mg/l	0.1	0.2	0.3
Silica (SiO_2), mg/l	44.4	11.6	70.4
Iron and alumina (R_2O_3), mg/l	3.2	4.0	5.6
Calcium (Ca), mg/l	128.5	5.1	224.4
Magnesium (Mg), mg/l	87.9	4.9	150.2
Sodium (Na), mg/l	558.0	59.0	968.0
Total hardness (as $CaCO_3$), mg/l	687.3	32.2	1,186.6
Total hardness (as $CaCO_3$), gpg	40.1	1.9	69.2
Total incrustating solids, mg/l	734.9	47.8	1,262.6
Total incrustating solids, gpg	42.9	2.8	73.6
Free carbon dioxide (CO_2), mg/l	8	1	14
Total dissolved solids (residue), mg/l	2,372.4	165.6	4,148.8

Mg/l = milligrams per liter; gpg = grains per gallon.

In many areas consumers are put to considerable expense to overcome the undesirable effects of hard water, which are largely associated with a high level of divalent ions from dissolved minerals such as limestone in the water. The membranes utilized in the process tend to have an even higher salt reduction performance for divalent ions than for monovalent ions, and therefore are particularly effective in softening the water they treat.

A very significant advantage of the process with respect to obtaining maximum efficiency in operation with simple and inexpensive manufacturing techniques lies in the fact that as the feed solution moves from the pump **78** through the assembly of the run tubes and fittings such as the U-joints it not only is constantly refreshed and mixed so that salt concentration buildup immediately adjacent the membrane is not a problem, but also the concentration of salts in the feed, at a steady flow rate of feed, increases steadily to the levels reached in the brine discharged at the end of the assembly. The level of salt concentration in each run tube may easily be calculated and determined for any given flow rate. Satisfactory operation may be achieved within very wide ranges of flow rates; in initial operation of the Coalinga plant a flow rate of between 7 and 7½ gallons per minute was used.

A membrane precisely adapted to meet the salt reduction requirement in each particular run tube, with maximum flux of desalinized product, can be provided for each run tube in the assembly by sequential manufacture of otherwise identical run tube assemblies with variation (which may be finely controlled as desired) of the temperature of the heating water used in curing the membranes. So long as the pressures used in the heating and cooling steps exceed the yield strength of the membranes at all temperatures in the range of temperatures used, nothing other than change of the heating water temperature is needed to provide the full range of varying salt reduction performance and flux of membranes for the complete plant.

Although the data given in the tables are taken from a plant designed to treat brackish water, the process is equally applicable to desalination of seawater. The use of slightly higher curing temperatures provides the membranes having salt reduction ratios well in excess of the 75 or greater level required to produce potable water from seawater. Desalination of highly saline solution such as seawater may also be achieved with the use of the process in multiple pass systems, such as a two pass system, using membranes with much lower salt reduction ratios and passing the product from an initial run through a plant as the feed solution for a second or subsequent run through the plant.

Hydrodynamic Forming

M.P. Tulin; U.S. Patent 3,647,521; March 7, 1972; assigned to Hydronautics, Incorporated describes a hydrodynamic method for forming tubular members useful in reverse osmosis apparatus. In the process, a conventional film forming liquid is forced through a tubular support by a pressurized column of gas to form a film of liquid on the inner wall of the tube. The liquid is then cured to solidify the film of liquid in tubular form.

The hydrodynamic method and apparatus make possible an advantageous reverse osmosis process and apparatus for recovering water of a low salt concentration from saline water. Tubular membranes are hydrodynamically formed within a porous tubular support to form a porous structure, and saline water is supplied under pressure to the interior of the resulting structure to recover water having a low salt concentration outside of the tubular support. When it becomes necessary to replace the tubular membrane, the supply of saline water is discontinued and a solvent is forced axially along the tubular membrane to dissolve it. Subsequently, the tubular membrane is hydrodynamically regenerated and the reverse osmosis operation can be resumed.

Continuous Process Using Mandrel

M. Marks, H.A. King and P.A. Longwell; U.S. Patent 3,601,159; August 24, 1971; assigned to the U.S. Secretary of the Interior describe a method for the continuous formation of tubular reverse osmosis elements. The process involves continuously forming the support structure together with the membrane over a mandrel. The membrane must be applied to a relatively flat, even, and not too porous surface. Because of this, it is generally preferred to first form paper about the mandrel.

The paper is directed to the mandrel having a diameter equal to the desired internal diameter of the tube. The edges of the paper are suitably scarfed and glue applied, which will set rapidly, gluing the seam of the paper together forming a flat-seamed tube about the mandrel. After this, additional pressure is applied against the formed paper, causing it to adhere and form about the mandrel. Downstream from the paper is located means for applying reinforcement fibers about the paper. This can be accomplished in various winding patterns, the purpose being to provide the necessary strength to the tubular element.

The fibers which, for example, can be resin-impregnated fiber glass, are appropriately wound about the paper to provide a porous structure having a permeability greater than that of the paper. Downstream from the means for applying the fibers are located rollers which advance the formed paper and fiber tube along the mandrel. After the rollers move the element along the mandrel, a heating means cures or melts the resin in the fibers. A cooling means is generally required and is located adjacent the heating means, to cool the tube back down to a temperature suitable for applying the membrane solution.

Inside the entire length of the mandrel is a hollow tube terminating just below the bottom end of the mandrel with an extrusion head which is provided with an opening about its entire periphery in the form of a nozzle. Surrounding the formed pipe, about its outer circumference adjacent the location of the inner nozzle, is an enclosed chamber in which slightly vacuum conditions are maintained. This serves to pull the membrane solution directed from the nozzle into tight adherence with the paper inner wall. The membrane solution, which is highly viscous for this particular application, is pumped through the center of the forming mandrel to the nozzle. The thickness of the resultant film is controlled generally by the rate of provision of membrane solution, the opening in the nozzle, and the rate at which the formed tube moves past the nozzle area, thus pulling the viscous material away from the nozzle area into a thin film. The higher the rate of tube movement, obviously, the thinner the film that is formed for a given nozzle opening and solution feed rate.

Just prior to the membrane application, the mandrel terminates. Following membrane application, the tube is directed into a water bath which gels the membrane solution. Just before or after the bath, the tube may be cut to desired lengths or can be wound or formed into spiral configurations where such is desired. The resulting process provides a membrane of extremely even and continuous thickness throughout the tubes formed.

Regardless of imperfections in the wall of the structure to which the membrane solution is applied, its thickness hardly varies. The membrane will rather follow the imperfection. This is due, as indicated, to the fact that the governing factors relating to the thickness of the membrane are the rate at which the tube moves past the nozzle, the nozzle opening, and the rate of supply of the membrane solution, and will be explained in more detail.

Referring to Figure 2.2a, there is shown the entire device utilized to form the tubular membrane structures in accordance with the process. The device comprises a support structure **11** having vertical bars **13** extending upwardly from feet suitably attached to a flooring. On a top platform **17** of the structure is mounted a motor **19**, gearbox **21** and associated pump **23**.

The pump is supplied by a reservoir **25** which contains a reverse osmosis membrane-forming solution. A thin, hollow tube **27** is directed from the pump down through the center of a mandrel **29** about which the tubular element is to be formed. The tube **27** serves to carry the membrane solution through the mandrel to a nozzle utilized to inject the membrane solution onto the formed tube, as will be further explained.

A supply of paper **31** is directed through a scarfing apparatus **33** and forming jaw **35** onto the mandrel thereby surrounding it. Glue from a supply **37** is applied to an edge of the scarfed paper prior to it being formed about the mandrel by guide rollers **39**. The scarfer device provides the necessary beveled edges on the paper so as to assure a relatively smooth-seamed construction of the formed paper while providing a large surface area to be glued at the seam. The device will be described in more detail with regard to the description of Figure 2.2c.

Guide rollers **39** are shown as idler rollers which maintain a slight pressure upon the paper placed about the mandrel so as to force it to conform into the desired tubular shape. The rollers may, however, be driven if so desired. The glue is, for example, one of several hot melts which will readily set up in a short period of time, essentially on contact with the second paper edge.

After the paper has been formed about the mandrel into the tubular shape by guide rollers **39**, fiber in the form of resin preimpregnated glass fibers **41** are wound in the desired pattern by a braiding apparatus **43** about the paper tube. The braider is a conventional device such as, for example, a Wardwellian or Butts type braider. The only modifications to the conventional braider are the use of a ring **45** over which the individual strands are directed prior to surrounding the tube passing therethrough, and the addition of stationary spools to provide longitudinal fibers.

The ring serves to direct the fibers onto the tube in a downward direction as the tube is moving through the device. After a first braider has wound the fibers in a desired pattern, idler guide rollers **46** serve to form the wound fibers against the paper tube prior to an additional layer of fibers being applied by a second braider apparatus **47**. It should be pointed out that it is not necessary in every application to utilize two layers of fibers. However, in the particular example shown, and in many instances, it is often preferable to wind the fiber reinforcing elements in two-layer configuration to reduce the roughness of the braided surface.

FIGURE 2.2: TUBULAR MEMBRANE UNIT

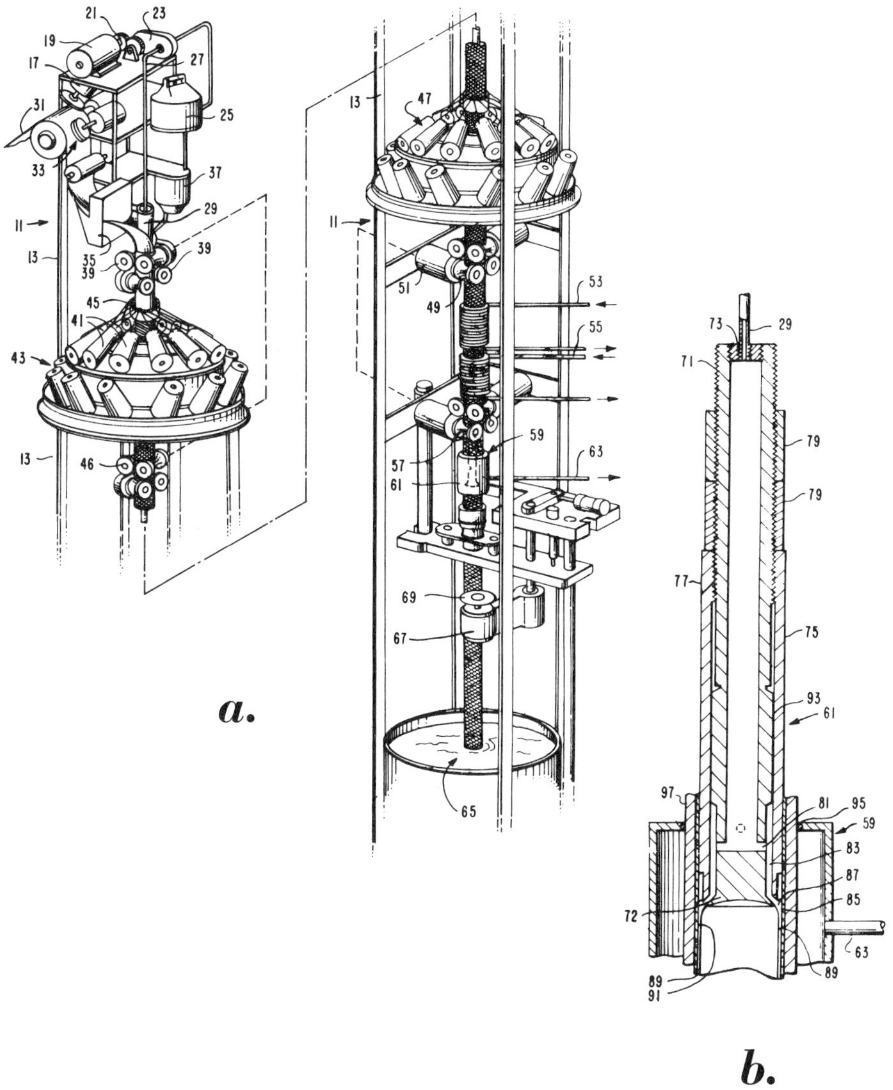

(a) Overview of Apparatus
(b) Detail of Nozzle

(continued)

FIGURE 2.2: (continued)

c.

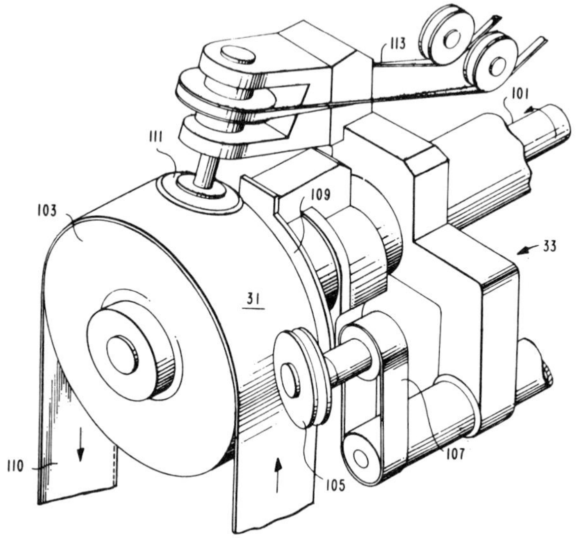

(c) Device Used to Scarf the Edges of the Paper

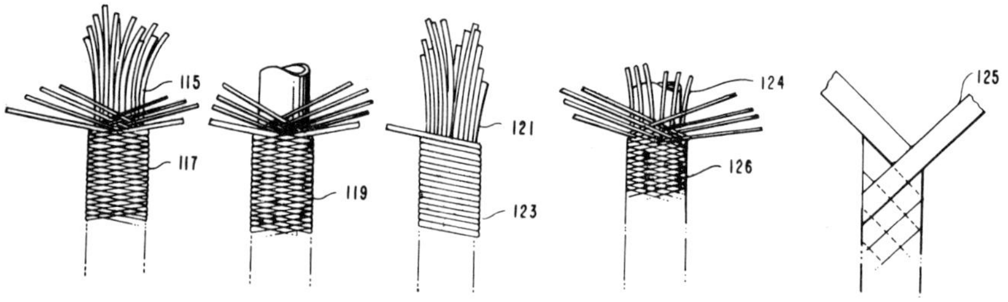

d. e. f. g. h.

(d)-(h) Winding Patterns for Forming Tube

(continued)

FIGURE 2.2: (continued)

i.

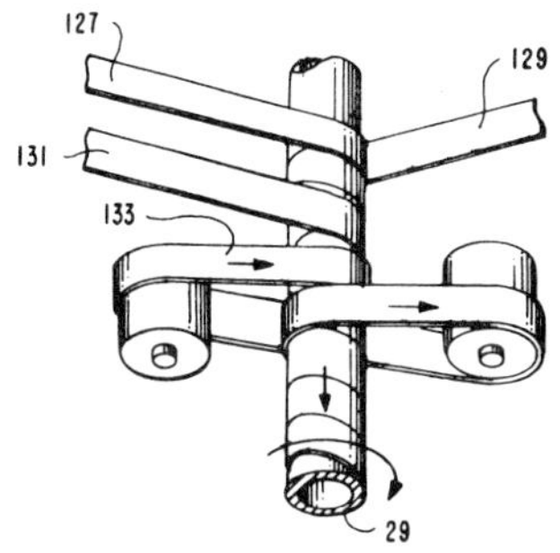

j.

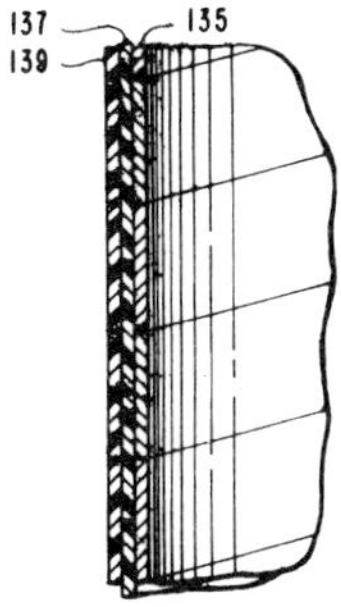

(i) Alternate Form Tube Preparation
(j) Cross Section of Tube Prepared According to Figure 2.2i

Source: M. Marks, H.A. King and P.A. Longwell; U.S. Patent 3,601,159; August 24, 1971

After the second braider, (shown on page 77) the paper with fibers wrapped about is directed to driven rollers **49** which are controlled by control motors **51** which are rigidly controlled by the same power source as the braider. The driven rollers are in tight contact with the formed element and serve to determine the speed by which the tube moves through the entire device over the mandrel and past the nozzle which applies the membrane-forming solution. Thus, the rollers **49** serve as the controlling factor for the speed of operation of the device.

Following the driven rollers, the structure is surrounded by heating coil **53** which can be a resistively heated coil that conductively heats the formed tube, melting or curing the resin material. A typical heating temperature for phenoxy thermoplastic resin is 350° to 400°F. Any other suitable means can be utilized for raising the temperature and affecting melting of the resin in both the paper and the impregnated fibers. An example of another method would be the use of ultrasonic energy or radiation at this region of the device.

Immediately following the heater coils are cooling coils **55**, which reduce the temperature of the bonded tube back to at least ambient conditions suitable for application of the membrane-forming solution. Cryogenics or ice water can pass through these coils to affect the desired cooling or cool air can be applied directly.

Following the cooling coils is a second set of drive wheels **57**, which further controls the rate of movement of the tube and which are also hooked to a master drive. Downstream and adjacent the second set of drive wheels is disposed the nozzle area **61** of the device. This area is surrounded by a vacuum chamber **59** through which a slight vacuum is maintained through line **63**. A vacuum may also be applied internally above the nozzle by plugging the entire forming mandrel and applying a vacuum to it.

The details of the nozzle are shown in Figure 2.2b. After the membrane has been applied in the region of the nozzle **61**, the tube is then directed into a water bath **65** where the membrane is gelled. Prior to entering the bath, the tube is conducted past a device **67** provided with a rotary blade **69** for controllably cutting the formed tubes to the desired length.

Referring to Figure 2.2b, there is seen the details of the nozzle portion of the device for applying the membrane to the formed tube. An inner tube **71** of machined stainless steel or the like is threadedly secured at **73** to the end of the line **29** which carries the membrane solution through the center of the mandrel. The mandrel in the device terminates just after the rollers **57**, prior to the nozzle portion. A second tube **75** is disposed concentrically about the inner tube **71**. Outer tube **75** is secured at its uppermost end **77** by being threadedly affixed to tube **71**. Two locknuts **79** secure tube **75** in place relative to inner tube **71**. The bottom end **72** of the inner tube is a solid plug.

Just above plug **72** are apertures **81** which conduct the membrane solution from the inner tube into an area **83** defined as a concentric chamber between the inner tube and the outer tube **75** and is formed by machining away a portion of the outer circumference of the inner tube in that region. The plug is flared outwardly at **85** to form a nozzle together with the canted portion **87** at the bottom of the outer tube. This provides a completely concentric nozzle area directing the membrane-forming solution to flow in a slight bubble **87** about the entire inner periphery of the formed paper tube **91**.

It is quite important that the plug portion be accurately centered relative to the outer tube so as to provide for an even flow of the membrane-forming solution about the circumference. This is accomplished by providing a tight fitting portion **93** which is comprised of a closely machined outer wall of the inner tube fitted against carefully machined inner wall of the outer tube along this area.

As indicated, membrane-forming solution **89** will tend to flow out in a slight bubble onto the paper tube. The vacuum box **59** has rubber seals **95** to ensure close fit with the outer fiber glass structure **97** so that a slight vacuum can be drawn through line **63**. A vacuum, for example, on the order of about 5 inches of water pressure, is sufficient to draw the membrane-forming solution tightly against the walls of the paper tube in an even layer. As can be appreciated, the vacuum drawn cannot be very severe or it will pull the membrane-forming solution through the paper and fiber glass windings which are porous, as previously indicated.

Thus, the vacuum is merely sufficient to cause adherence of the solution to the wall of the paper tube providing a film or membrane thereon. Further, it should be apparent that the thickness of the membrane formed will be related to the amount of solution applied by the nozzle. Thus, the thickness is controllable by the selective size of the nozzle opening and the rate of supply of solution, as well as the speed of the tube.

In order to achieve the formation of the film of membrane as shown in Figure 2.2b, the solution passing through the nozzle area must be quite viscous. The most useful solutions have a viscosity range of 25,000 to 150,000 centipoises. The final membrane must possess suitable osmotic properties, such as for example, a flux of 10 gallons per square foot per day (gfd) at less than 1,000 ppm from 3.5% sodium chloride at 1,500 psi.

Additionally, since the air space in the tube is limited, evaporation cannot be utilized as a means to rid the membrane solution of any solvent utilized. Thus, it is important that the membrane solution have a composition in which loss of solvent prior to gelation is not necessary. Additionally, since this is a continuous process, the solution should be capable of being gelled in virtually a fraction of a second after casting.

Typical compositions would comprise 10 grams of cellulose acetate, such as type E-398-3, a cellulose acetate powder of Eastman Chemical Products, Inc., containing an acetyl content of approximately 39.8% and a viscosity of 6.8 to 14.8 poises, 18 to 25 grams of acetone as a solvent, 4 to 7 grams of propionamide as a swelling agent, and 1 to 4 grams of water. A preferred solution comprises 10 grams of cellulose acetate type E-400-23 having an acetyl content of approximately 39.9% and a viscosity of 64 to 132 poises, 20 grams of acetone, 6 grams of propionamide and 3 grams of water. This produces a casting solution having a viscosity of about 80,000 centipoises. Many other forms of cellulose acetate are suitable, including E-398-6, 398-10, etc., as well as mixtures of these formulations.

Figure 2.2c shows one of the devices utilized to scarf the edges of the paper that forms the inner tube of the device on which the membrane is coated. In this view as well as the view in Figure 2.2a, only one such device is shown. However, it should be appreciated and, as will be explained, two devices are needed to scarf the opposite edges of the paper roll **31** so they will form a matting seam. The scarfing of the paper is necessary to provide a wide bond edge for the application of glue to adhere the paper together in a non-lapped seam. The motor drive is not seen in this view.

However, the motor first drives a shaft **101** which in turn rotates wheel **103** over which the paper **31** passes. A skewed drive wheel **105** driven through a belt drive **107** forces the paper against a flat guide **109** so that the paper is perfectly aligned when it comes to a skewed scarfing knife **111**. The scarfing knife is a rotary wheel driven through belt drive **113** cutting the edge of the paper at an angle determined by its skewed position.

The additional scarfing knife, not seen, will cut the opposite edge of the paper from its bottom surface **110** to form a matching joint. This is readily accomplished by doubling the paper back over a second rotating wheel so that the bottom surface then becomes the exposed portion of the roll.

Figures 2.2d through 2.2g describe the various patterns for forming the support portions of the tubular structures to which the membrane solution is applied. It should be kept in mind that it is important that the membrane be applied over a smooth surface so that it will be properly supported to avoid damage upon pressurization. For this reason the specific example illustrated has revolved around utilization of a paper inner tube. The paper inner tube can be unfilled cellulosic paper or partially resin-filled paper.

It can additionally, of course, be made from a synthetic fiber with or without a nonwater-soluble binder, such as an acrylic resin. Additionally, the paper could be further strengthened with stiffening resins, though this is not necessary. The paper used will have a thickness that will vary from 0.002 to 0.015 inch. The permeability should be generally within the range of 1×10^{15} to 1×10^{14} darcy at 750 psi operating pressure.

It is possible, however, to eliminate the paper by utilizing a plurality of longitudinal strands of fibers which have virtually no resin or very little resin content. The longitudinal fibers will serve both as a support for the membrane coating and as a structural member. Thus, as seen in Figure 2.2d, a combination of the longitudinal strands **115** and hoop braid **117** surrounding the longitudinal strands provides an adequate structure eliminating the use of an inner paper tube.

Figure 2.2e shows a hoop braiding **119** alone. In this instance the braiding would be applied over an inner paper tube. Figure 2.2f once again discloses the use of longitudinal fibers **121** wound with a simple circumferential-wound configuration **123**. This differs from that shown in Figure 2.2d in that the outer hoop fibers are not braided. Figure 2.2g shows longitudinal fibers **124** and hoop braidings **126** applied over a paper tube. The formed tubes are stationary relative to the mandrel and are merely pulled along it by the roller mechanisms previously described. Figure 2.2h shows the utilization of tapes **125** of fiber material preimpregnated with resin. Once again, these tapes would be wound over an inner paper core or longitudinal fiber.

Referring to Figure 2.2i, there is seen a method where the formed tube will be rotated on the stationary mandrel **29**. As shown, initially a roll of paper **127** is directed onto the mandrel, followed by opposed tapes of resin preimpregnated fibers **129** and **131**. A belt drive **133** is wrapped about the paper and tapes downstream therefrom, rotating and tightly adhering successive layers to each other. The tapes **129** and **131** as well as the paper **127** are automatically pulled onto the stationary mandrel and forced along in a forward direction according to the downward directed arrow due to the angle of the belt **133** relative to the mandrel.

Figure 2.2j shows a cross section of the formed underlying structure made according to

Figure 2.2i to which the membrane is applied, having an inner paper liner **135** to which the membrane solution is applied and two fiber glass layers **137** and **139**. It is noted that the fiber glass layers are in parallel directions in accord with the technique of applying tapes as shown in Figure 2.2i, even though the two tapes **129** and **131** are applied from different sides of the mandrel.

Seamless Tubular Form

E. Lowe and E.L. Durkee; U.S. Patent 3,341,024; September 12, 1967; assigned to the U.S. Secretary of Agriculture describe a molecular filtration apparatus, useful in such applications, for example, as the dewatering of fruit juices to produce juice concentrates. The apparatus for the process includes a membrane in seamless tubular form. The high flux membranes are permeable to water but essentially impermeable to nutrient, flavor, and color components of foods, and are in seamless tubular form. The process is illustrated by the following examples.

Example 1: Wool felt strip, 1 inch wide, was spirally wound on a ½ inch diameter mandrel. Woven glass fiber tape, 1.5 inches wide and 15 mils thick, was spirally wound over the felt in the opposite direction. A conventional epoxy resin (settable at room temperature) was then coated onto the glass fiber winding. Application of the glass fibers and resin was repeated, with opposite direction of winding for each layer, for a total of 10 layers of resin-coated glass fibers.

The entire assembly was placed in a chamber where it was held under a pressure of about 29 in. Hg for 10 to 15 minutes. The assembly was then removed from the vacuum chamber and allowed to stand until the resin had set. The assembly was then removed from the mandrel and the ends squared off. The ends of the assembly were made imperforate by stopping up one end, taping the side of the structure except for the last inch, dipping the untaped, plugged end into a bath of the same epoxy liquid as used before, and applying vacuum to the hollow core of the structure, whereby to suck the resin into the end of the structure. This procedure was repeated to render imperforate the opposite end.

Tapered drift pins were then inserted into each end of the assembly to compress the felt layer to a tapered configuration and the device was allowed to stand until the resin had hardened. The drift pins were removed and a seamless tubular membrane was inserted into the hollow core and sealed in place by inserting a perforated rubber stopper in each end. The tubes were threaded, and couplings were screwed on. The apparatus was then tested by plugging up one end and pumping water into the other. It was found that the particular membrane used ruptured at 2,500 pounds per square inch but the supporting tube itself was able to withstand pressures of 5,000 pounds per square inch without any damage.

Example 2: A casting solution was prepared containing 22.2% cellulose acetate, 66.7% acetone, 10.0% water and 1.1% magnesium perchlorate. A quantity of this solution was poured into a one-foot section of 0.5" i.d. precision bored glass tubing and drawn with a rod along the length of the tube to spread it uniformly in the longitudinal dimension of the tube. The tube was then capped at each end and spun about its axis at about 1,200 revolutions per minute for 20 minutes to distribute the solution uniformly in the radial sense.

The tube was then uncapped and a very gentle stream of air, at room temperature, was passed through the tube for 1 minute in one direction, then for the same time in the opposite direction. The tube was immersed in ice water and moved about in the water to promote leaching of perchlorate and residual acetone from the film.

The film was then withdrawn from the glass tube. It had a wall thickness of 4 mils; it was clear, with a tinge of milkiness and was self-supporting and quite rigid. It was heat cured for 10 minutes in water at 86°C, then utilized as the membrane in a structure fabricated as described in Example 1. The completed structure was found useful for preparing concentrated apple juice by pumping single strength (14° Brix) apple juice through the hollow core of the apparatus at a pressure of 1,000 psig.

Open Pore Tube Support System

C.H. Vandracek and A.S. Calderwood; U.S. Patent 3,598,241; August 10, 1971; assigned to Westinghouse Electric Corp. describe an open pore tube made of bonded resin-coated filler particles which serves as a support tube for cellulose acetate or other reverse osmosis membranes. The following example illustrates the process.

Example: A membrane support tube was made containing 4 axial holes. To 1,000 grams of washed, dry sand having an average particle size of 60 microns was added 9 grams of hexamethylenetetramine catalyst. This was mixed for 1 minute in a muller. Then 56 grams of the reaction product of a phenol and an aldehyde, in solution, having a viscosity of 4200 cp at 25°C and a solids content at 135°C of 67% (Durez Phenolic Resin 18115) was added and the combination mixed in a muller for 10 minutes until it was dry and free flowing.

The resin catalyst coated sand composition was poured into an 18" cylindrical mold which contained 4 axial tubes symmetrically placed. Filling was performed on a vibrating table to facilitate packing and flow. The filled mold was then cured at atmospheric pressure for 16 hours at 135°C to solidify the composition.

The mold was then stripped and a strong rigid cast cylinder of resin-coated sand particles containing 4 axial holes was obtained. It contained 3.58 weight percent actual resin, i.e., (56)(0.67)/[(56)(0.67)+1,000+9]. Tubular reverse osmosis membranes were placed in the 4 holes, which had a ½ inch inside diameter, and the device tested. Wall thickness of the tube was about ¾". During testing the cylinder was subjected to 30 psi water pressure. It withstood the pressure easily and passed the purified effluent water readily. At this wall thickness and inside tube diameter pressures of about 400 psi would be used. Using the same inside tube diameter and enlarging wall thickness of 1½", pressures of about 800 psi could be sustained.

The sand used can be round or angular. Angular sand is preferred because there is less chance of mold cracking. Any weight percent ratio is satisfactory that will provide sufficient porosity while maintaining adequate strength in the tube. The amount of catalyst will vary with the resin used and must be sufficient to obtain thermosetting properties. These tubes are corrosion resistant, should easily withstand the pressures necessary for reverse osmosis operation and can be easily fabricated in all shapes and configurations in lengths up to about 15 feet. These tubes can also be externally supported by any suitable means in order to better resist pressure.

Integrally Supported Membrane

In a process described by *W.W. Cooper IV and R.W. Pierce; U.S. Patent 3,563,889; February 16, 1971; assigned to Abcor, Inc.* an integrally supported semipermeable membrane is composed of a tubular braided fabric impregnated with a polymeric film-forming composition, and carries on at least one surface a layer of the polymer with the surface of the polymer carrying an active layer which provides the separation characteristics of the membrane.

The tubular supporting material is a fabric composed of a multiplicity of continuous monofilamentary material of relatively small diameter which is twisted together or otherwise consolidated to form strands. The strands are then woven or knitted to form a closely braided tubular structure. It is essential that the strands used to make up the material be characterized by a relatively smooth surface which is free from fuzz, that is, short fiber ends which project from the surface of the strand. Such projections would penetrate the surface of the microporous film deleteriously affecting the semipermeable efficiency of the membrane.

The use of a multiplicity of continuous monofilaments in the braided structure is critical since it permits the adjacent strands to lie close together, thereby presenting an essentially flat surface. The relative smoothness of the surface allows the application of a relatively thinner coating of polymer than would otherwise be necessary to cover protuberances, and prevents "blow-outs" at open points, i.e., failure of the polymer layer at points where the polymer is not directly supported by underlying material.

By contrast, if a weave of, e.g., wire mesh were employed, the inability of the mesh to compress would cause surface irregularities at the crossover points which would require a polymer layer excessively thick to cover these irregularities. In addition, irregularities have been found to trap air in the casting process with consequent formation of weak spots. In this process, since the strands are composed of a multiplicity of fibers twisted together, these problems are eliminated with the strands themselves becoming impregnated with polymer.

No reinforcement, as by resins, etc., of the tubular material is employed since the braided configuration provides the necessary strength and reinforcement. Further, the membranes neither require nor possess the property of rigidity, thus minimizing the possibility of damage due to shock or accident.

Direct Production of Membrane on Permeable Support

A process described by *S. Manjikian; U.S. Patent 3,544,358; December 1, 1970; assigned to Universal Water Corp.* involves the production of a semipermeable membrane directly on a supporting surface of a permeable supporting body. This is accomplished by spreading a membrane-forming composition containing at least one water-soluble constituent over such a supporting body surface to provide a film of substantially uniform thickness. Water is applied to only the surface of the film remote from the supporting surface to remove water-soluble constituents.

The resulting membrane is directly supported by the permeable body surface being uniformly conforming to the surface. The so produced semipermeable membrane is effectively employed in apparatus which comprises the semipermeable membrane having one face supported by the surface of the permeable body, and enclosure means for supplying solution to be treated at hyperatmospheric pressure to the other face of the membrane.

Means are provided to release a portion of the solution containing an increased concentration of solute from the enclosure, and to collect solution which has passed through the membrane and the permeable body and which contains a reduced concentration of the solute. The permeable body on which the semipermeable membrane is directly supported, according to this process, may be of any convenient configuration suitable for the type of apparatus in which it is employed. Preferably the body is of cylindrical form containing a tubular bore or a plurality of tubular bores containing membrane tubes as described in U.S. Patent 3,392,840.

The film-producing constituent of the membrane-forming composition is a cellulosic derivative, the solvent provides castable compositions, and the modifying agent promotes formation of a membrane having desirable flux or liquid flow-through characteristics. Cellulose acetate is preferred as the cellulosic derivative, acetone as the solvent, and formamide as a modifying agent. A preferred membrane-forming composition consists essentially of cellulose acetate, 15 to 30 parts by weight; acetone, 40 to 70 parts by weight; and formamide, 15 to 30 parts by weight. At least one constituent of the membrane-forming composition is water-soluble. In the preferred composition, above, both acetone and formamide are water-soluble constituents.

Preparation of Tubular Membranes Using Casting Bobs

Unitary tubular membranes have been made by loading a long casting tube with casting solution, gravity dropping the tube over a stationary casting bob to form a film and then immersing the tube in a bath of water to leach the solvents and swelling agents from the film and form a membrane. This method, described in *Desalination,* volume 1, No. 1, April 1966, pages 35 to 49, requires considerable time to load the casting solution, cast the membrane, clean the tube and bob and prepare the facility to cast the next membrane. It is also likely to produce variable membranes unless extremely straight precision bore casting tubes are used.

R.M. Chamberlin, J.L. Emswiller, A.S. Calderwood and R.R. Stana; U.S. Patent 3,658,955; April 25, 1972; assigned to Westinghouse Electric Corporation describe a method which involves pressure casting unitary, tubular membranes in an efficient manner. The method consists of: (1) loading membrane casting solution into one or a plurality of supported hollow smooth bore casting tubes; (2) inserting a hydraulically stable, self-centering casting bob into each casting tube; (3) driving the casting bob through the casting solution in each tube with a gas under pressure, which preferably is saturated with respect to solvent, to form a continuous film of casting solution on the inside of each tube; (4) exposing the cast film to air; and (5) contacting the cast film with a suitable leaching liquid in such a manner and for a length of time necessary to produce a satisfactory reverse osmosis membrane having an active osmotic skin.

Tubular, four-foot long, one-piece membranes produced by the method described above are consistent in structure and thickness throughout their lengths, can be fabricated in from 1 to 3 minutes, and have provided 24.5 gallons of purified water per day per square foot of membrane surface at 600 psig and 90% salt rejection. Also, since most long casting tubes are not perfectly straight, a superior membrane will be produced by this method since under pressure the casting bob becomes most self centering and will follow the hollow tube path without scraping the tube wall if it bends slightly. This method eliminates using extremely expensive quality glass tubes as casting tubes.

R.R. Stana and A.S. Calderwood; U.S. Patent 3,657,402; April 18, 1972; assigned to Westinghouse Electric Corporation describe a simplified membrane-module system by casting a unitary, tubular, reverse osmosis membrane in place, inside a stationary support module.

The method consists of: (1) placing an open pore module made of bonded, resin coated filler particles, having at least one circular bore through it, in a vertical position; (2) loading membrane casting solution, having a viscosity between 1,000 and 100,000 cp, into each of the module bores; (3) gravity dropping a hydraulically stable, self-centering casting bob, having a circular cross section at its widest point at a speed up to 1 inch per second down through the bore containing the casting solution to form a continuous film of casting solution on the inside of each module bore, the bores having an inside diameter between 0.02 and 0.12 inch greater than the diameter of the casting bob at its widest point; (4) exposing the cast in place film to air; (5) contacting the cast film with a suitable leaching liquid in such a manner and for a length of time necessary to produce a satisfactory reverse osmosis membrane, having an active osmotic skin, in each of the module bores; and (6) optionally curing the cast in place membranes in water at temperatures up to 200°F. The method is particularly suitable for multiple bore modules and eliminates several process steps by casting the membranes in place.

Hollow Teardrop-Shaped Casting Bob

W.W. Cooper IV and E.M. Shea; U.S. Patent 3,676,193; July 11, 1972; assigned to Abcor, Inc. describe an apparatus for and method of forming an integrally supported tubular membrane having an inner microporous layer by casting on an interior supporting surface a liquid polymeric composition. Polymeric compositions are applied to the interior of a tubular structure by means of a casting bob which is generally teardropped in shape and which is hollow to receive the polymer composition and further which is characterized by apertures in the sides from which the polymeric composition is discharged from the bob onto the adjacent surface.

One form of the process is illustrated in Figures 2.3a and 2.3b where tubular membrane **10** is composed of a braided fabric of fibers **12** where the fibers are made up of continuous monofilaments **14**. The braided supporting fabric is impregnated with polymer **16** which has an active layer **18**. As examples of suitable monofilamentary material which can be employed satisfactorily in the process, mention may be made of rayon, nylon, polyesters (Dacron), polyolefins, e.g., polyethylene and polypropylene, acrylics, Teflon, and the like. For very high pressure applications, high strength materials such as steel fibers and glass fibers are employed.

One integrally supported membrane of the process is formed by applying to a tubular, closely woven structure composed of strands of continuous monofilaments, a casting solution in sufficient quantity to impregnate the supporting material and to form a smooth, thin film layer on at least one surface. Evaporative drying is then applied to the structure to form the active layer on the surface of the polymer.

In one method of preparing the membranes of the process the braided supporting material is first formed into a tubular configuration, for example, around a glass tube. The supporting material is then wet with a fluid, e.g., with water and, if desired, stiffened by the application of starch, carboxymethylcellulose, polyvinyl pyrrolidone or other conventional fabric stiffening material known to the art. It should be understood that the fluid is not limited to water but can be any appropriate fluid; in the case of rayon, water is preferred; in the case of polyethylene, toluene is preferred. The supporting material is then dried, preferably at elevated temperatures.

The tube around which the supporting material was formed is then removed and the tubular supporting material is inserted into a hollow tube of slightly greater diameter to provide the necessary support and shape during the application of the polymer layer. A casting bob is then inserted inside the tubular supporting material. The dimensions of the casting bob and the supporting material are selected to provide the desired deposition of the polymeric material.

Polymeric casting solution is then provided to the upper surfaces of the casting bob either by an external feed tube or by the casting bob of the process. The casting solution is then applied to the tubular supporting fabric either by drawing the casting bob up the interior of the tubular supporting material or by maintaining the casting bob stationary and moving the tubular supporting material downwardly past the casting bob. In a third alternative both the casting bob and the tubular supporting fabric can be moved in opposite directions at the same time. The casting solution flows around the edges of the bob and onto the film. The clearance between the casting bob and the fabric walls determines the thickness of the surface layer on the fabric.

The rate of the relative motion of the casting bob and/or the tubular supporting material is not critical. The rate of motion should be such, however, that the desired amount of polymer is applied to the tubular material. A too rapid motion may result in insufficient deposition of the polymer. The polymeric material is then air dried to form the active layer. Subsequent to the drying step the thus formed self-supported membrane and the supporting tube in which it was formed is immersed in water. In the water the self-supporting membrane is released from the wall of the supporting tube and readily removed.

The rate of drying is not critical; however, the flux rate and the selectivity of the membrane can be controlled by the drying conditions selected. For example, a relatively short drying rate generally provides a less selective membrane material whereas a prolonged drying operation will produce a highly selective membrane material.

The semipermeable membranes of the process can now be handled according to conventional procedures. For example, the membranes are immersed in water to remove the solvent and nonsolvent of the casting solution by replacement with the water. The finished membrane may be surface treated thermally or by solvents and the like in order to provide a specific degree of selectivity.

FIGURE 2.3: CASTING INTEGRALLY SUPPORTED TUBULAR MEMBRANES

a.

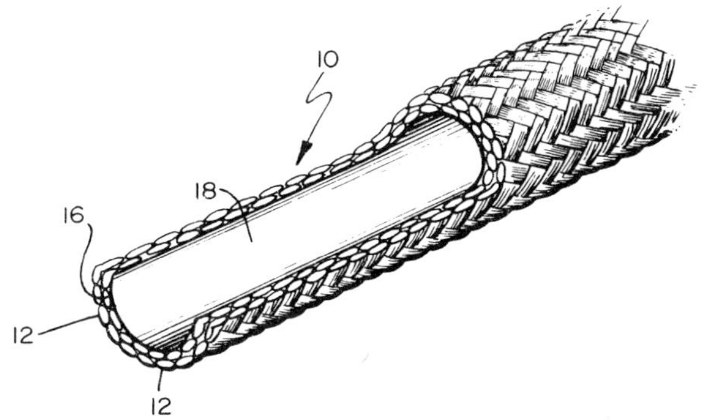

View Partly in Section of the Self-Supported Membrane

b.

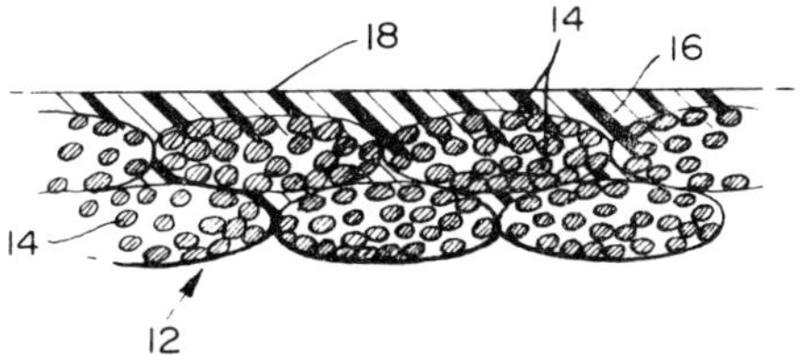

Cross-Sectional View of the Membrane

c.

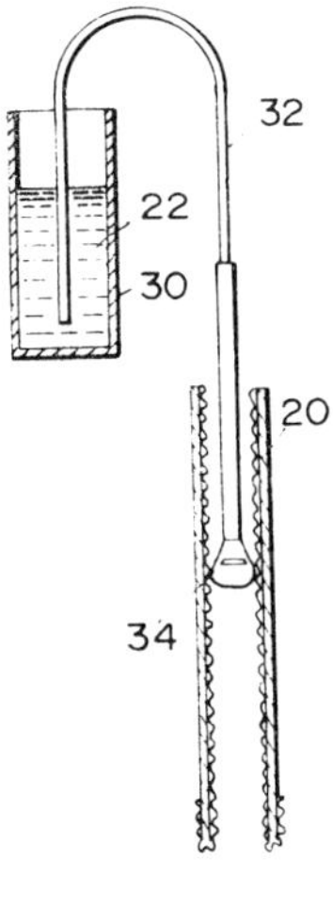

Schematic View of the Process

d.

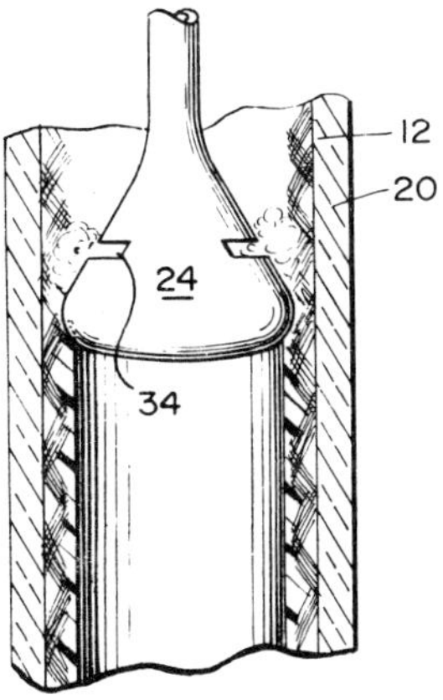

Cross-Section Showing the Casting Bob

(continued)

FIGURE 2.3: (continued)

e.

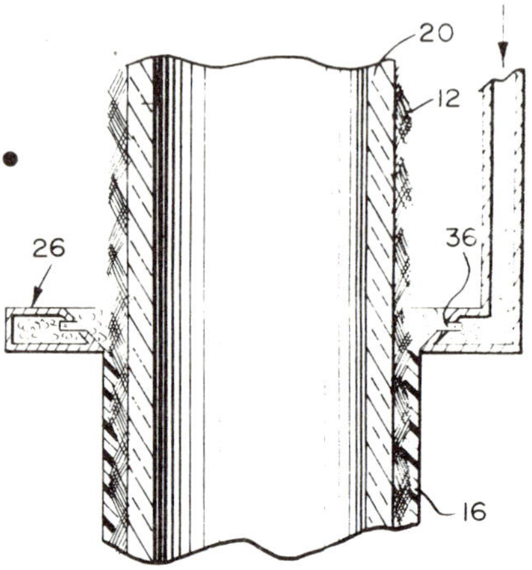

f.

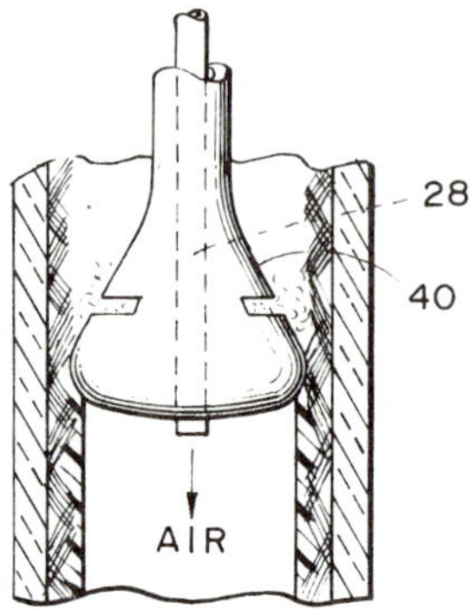

Alternative Method of Forming Membrane

Preparation of Self-Supported Membrane

Source: W.W. Cooper IV and E.M. Shea; U.S. Patent 3,676,193; July 11, 1972

The selectivity can also be controlled by the incorporation into the liquid casting solution of materials, e.g., tellurium tetrabromide or magnesium perchlorate, which are subsequently removed by the water after formation of the membrane. The characteristics of the membrane can be predetermined by the composition of the casting solution. For example, variation of the proportions of solvent, the presence of swelling agents, dopants, or diluents and the like, can effect the relative selectivity of the finished membrane.

As mentioned above, the membrane material may be prepared from any suitable film-forming polymeric material known to the art such as cellulose esters, cellulose ethers, denitrofied nitrocellulose, phenol aldehyde, urea aldehyde, acrylics, elastomers and the like. A particularly preferred class of materials are the cellulose esters, more preferably, cellulose acetate and polyion complex resins, e.g., sulfonated polystyrene. The term casting solution, as used herein, is intended to refer to solvent solutions, or aqueous polymeric dispersions, and melts. A possible casting solution includes, for example, 25% of cellulose acetate, 30% of formamide, and 45% of acetone. A membrane of this composition on a rayon braided support generally may possess a flow rate with a protein solution of about 20 gallons per day per square foot at about 10 psig as compared with 1,500 psig for prior art membranes.

A casting bob has been developed which is particularly suitable for use in forming semipermeable membranes, but which is also useful for the deposition of any polymer material on the inside of the tubular structure. The casting bob comprises a substantially circular hollow body with a lower surface and side walls sloping inwardly and upwardly from the lower surface. The side walls contain slots or apertures spaced apart from each other from which coating solution contained in the hollow cavity of the bob is expelled to flow around the bob and onto the adjacent surface to be coated.

The bob may be conventionally supported, as by a wire, with an adjacent conduit for introducing polymeric casting solution into the cavity defined by the walls and the lower surface. In a preferred form, however, the tube which carries polymeric solution to the casting bob also provides the support for the bob.

If the method of employing the casting bob involved the motion of a tube passing a stationary bob, the supply tube can be made of rigid material such as glass or stainless steel tubing. However, if a moving bob is desired, a flexible supply tube is used. Preferably, the polymeric solution is under pressure.

One method of forming the membrane of the process is illustrated in Figures 2.3c and 2.3d which show a tubular braided fabric **12** supported by the inner surfaces of a glass tube **20** to maintain the tubular configuration. Polymeric casting solution **22** is passed from reservoir **30** through conduit **32** to casting bob **24** from where it is fed to the upper surfaces of the casting bob into and onto the adjacent braided supporting fabric to impregnate the fabric and layer **16**. As illustrated, the supporting tube **20** and tubular support material **12** are maintained in a stationary position and the casting bob is drawn through the tube in the direction of the arrow.

Figure 2.3e illustrates an alternative method of forming the membrane where the active layer is on the outer surface of the fabric. Glass supporting tube **20** is inserted inside the braided fabric and annular casting knife **26** rings the outer surface of the braided material depositing casting solution through ports **36** to form polymeric layer **16**.

In an alternative case, the evaporative drying which forms the active layer is applied to the tubular supporting material. By using a moving current of drying air, greater efficiencies are achieved in the drying operation as well as providing for consistency of separation characteristics through the membrane by the uniform removal of the solvent used in the casting solution. The above mentioned high efficiency drying can be achieved by using the casting bob shown in Figure 2.3f. The casting bob, which is hollow and contains a passageway for the casting solution to pass through it and out ports **34**, also contains a tube **28** which expels air from the lower surface of the casting bob thus providing the evaporative drying conditions to polymeric film **16** which has been applied to the tubular supporting material by the passage of the casting bob.

Impregnation of Porous Substrate with Water and Controlled Heating

A process described by *J. Del Pico; U.S. Patent 3,762,566; October 2, 1973; assigned to Abcor, Inc.* relates to supported semipermeable membranes suitable for use in ultrafiltration and reverse osmosis separations. The method for preparing such membranes comprises impregnating the surface of a porous support prior to casting onto the surface a thin film of a casting solution, the surface of the porous support impregnated with a nonsolvent for the film-forming material in the casting solution, thereby providing an integrally supported semipermeable membrane of predetermined flux and rejection rates.

It has been found that at least one process variable, and preferably two process variables, should be controlled when producing the supported semipermeable membranes. One process variable is to cast the casting solution upon a porous support, the surface of which has been impregnated with a nonsolvent, such as water. The other condition is to impart thermal energy to the porous support prior to casting so as to increase the evaporation rate of the volatile solvents in the film-casting solution, thereby, in combination, providing a supported semipermeable membrane. It has been found that impregnating the porous support, such as by saturating with water prior to casting at an ambient temperature of, for example, 15° to 30°C, with a nonsolvent prior to casting, provides for increased flux rate of the membrane so prepared, whereas impregnating the porous support with a nonsolvent at an increased temperature, for example, water at 35° to 95°C, increases and controls both the flux rate and the rejection rate of the integrally supported semipermeable membrane.

A series of experiments were conducted employing varying process conditions in connection with the preparation of a supported tubular semipermeable membrane. First, a membrane-casting composition comprising a solution of about 2 grams of a cellulose acetate (a film-forming membrane material), 7 milliliters of dioxane, a solvent for the film-forming polymer, and 2 milliliters of formamide, a nonsolvent for the film-forming polymer, was prepared. A thin layer of the casting solution was then cast on the inner surface of a sintered polyethylene porous tubular support, the tube comprising a length of about 10 feet, 1 inch inside diameter, 1½ inch outside diameter, and having an average pore size of approximately 10 to 300 microns, for example, 20 to 100 microns. A thin cast membrane film was formed in the interior wall of the sintered polyethylene tubes employing a casting bob, the film having a dimension of between 20/1,000 to 30/1,000 of an inch. Immediately after casting, the films were leached in a water bath for approximately 10 minutes at 20°C. The casting operation occurred in a room having a temperature of approximately 19° to 22°C and with a relative humidity of between 30 and 55%.

After perforation, the integrally supported semipermeable membranes so prepared were tested for flux rate and rejection rate properties by using a 1% by weight Carbowax 20M water solution, which solution was passed through the interior of the supported tubular membrane at about 50 psi, 25°C, and a Reynolds number of approximately 49,000. A series of experiments were conducted by pretreating the porous support tubes illustrated in the following table with the resulting flux rates and rejection rates shown.

Example	Support Tube Pretreatment	Flux* (cc/min)	Rejection Percentage
1	None	166	82.4
2	Heated in an oven at 58°C	202	91.8
3	Immersed in a water bath of 45°C for 6 minutes	554	88.9
4	Water pumped through the porous support tube at 74°C for approximately 2 minutes	351	95.1

*Membrane surface area 2.2 feet

As illustrated, the integrally supported membrane prepared as in Examples 3 and 4 exhibited a substantially higher flux rate than those prepared in Examples 1 and 2 and higher rejection rates. The pretreatment of the porous support tube by impregnating the tube through contact with heated water is believed to prevent the rapid diffusion of the solvent from the cast film into the porous support base and to permit the diffusion of the nonsolvent into the porous support, thereby providing an open porous understructure and a high rate of flux as shown.

The rapid diffusion of the nonsolvent into the underside of the cast film, maintained relative to the outward solvent diffusion by evaporation, results in the cast film being gelled in a highly porous state. The employment of heated water imparts thermal energy and enhances the rapid formation of the active layer in the internal surface of the tube, resulting in an asymmetric membrane having a high rejection rate.

In Example 1, two active layers are formed, one at the interface of the cast film and the porous support tube, and the other at the exposed surface of the cast membrane, these layers being the same as the porous inner layer. In Example 2, the membrane was cast on a heated dry porous support tube, and again, two active inner and outer layers were formed. The layers are believed to be thicker than those in Example 1 due to the more rapid transport, that is, the high evaporation rate of the solvent from the cast membrane film.

The higher temperature conditions and rapid evaporation resulted in the higher rejection rate and flux as shown in Example 1. Examples 3 and 4 showed significantly higher flux rates and good rejection rates. Example 4 showed an even higher rejection rate where the support tube was heated to an even higher temperature.

Multilayer Braided Tubular Reinforcement

J.L. Richardson, A.O. Brodie and C.H. Lewis; U.S. Patent 3,891,556; June 24, 1975; assigned to Oxy Metal Industries (Intra) Inc. describe a method of continuous single or multilayer braiding to form a high pressure support for a thin cast membrane tube, which tube can be of greater length than those previously used and can be afterwards confined in a small volume, for instance, wrapped into the form of a spool of tubing, without detriment to the membrane.

The actual braiding of a membrane tube is performed with machines readily available (New England Butt Company, Wardwell Braiding Machine Company, etc.). Braiding is best accomplished with the membrane tube in a horizontal position. In addition, precautions are taken to ensure adequate lubrication of the machine parts without leakage of the lubrication onto the product. It has been found to be mandatory, in the case of cellulose acetate membrane tubes, to keep the membrane tube immersed in water before braiding and to maintain both it and the braid in a wet state continuously.

Highly successful reinforced semipermeable tubing capable of resisting internal pressures of as high as 2,000 psig and exhibiting a high rejection of sodium chloride from dilute solutions has been prepared employing as the semipermeable membrane cellulose acetate tubing having a diameter of approximately ¼ to 1 inch and a wall thickness of about 0.008 inch. This reinforced tubing is capable of being bent around a very small radius when it is pressurized with moderately hot water.

It is essential that this cellulose acetate semipermeable membrane tube be kept wet at all times subsequent to its formation by casting. The reinforcing braided layer immediately adjacent the cellulose acetate semipermeable membrane is braided from a 1,100 denier Dacron polyester yarn, zero twist. This braiding is accomplished on a conventional braiding machine with each yarn being made up of two individual 1,100 denier fibers. This inner low denier reinforcing layer is applied with the fibers at an angle of approximately 69° with the axis of the semipermeable tube. There is a distinct advantage to having the inner layer of the fine denier yarn braided at an angle greater than the lock angle. Such braiding has been found to minimize any tendency for the membrane tube to buckle or distort if afterwards coiled or bent to a radial configuration.

Braided Glass Fiber Support Prepared Using Mandrel Technique

A process described by *A. Shaines, R.L. Kalish and E.A.G. Hamer; U.S. Patent 3,695,964; October 3, 1972; assigned to Abcor Water Management Company, Inc.* provides a cylindrical rigid porous support for an osmotic or ultrafiltration membrane. The support is made of braided fiber sleeves, which are intimately conformed to a mandrel by axial lengthening and consequent radial shrinkage of the braided fibers around the mandrel and which are then permanently cemented together in such conformed position with a limited amount of cement to leave open interstices between the fiber crossover points.

Example 1: A mandrel is prepared by cutting to the desired length a polytetrafluoroethylene (Teflon) rod 0.51 inch in diameter. A single thickness of fine filter paper is moistened with water and wrapped around the mandrel, dried and 4 layers of braided glass sleeving, nominally ½ inch in diameter, are conformed over the mandrel and the paper line in the manner previously described. The prepared mandrel is then saturated with an epoxy-polyamide-solvent formulation consisting of 60 parts by weight of epoxy resin (ERL 2795), 40 parts by weight of polyamide (Versamid 125) and 47 parts by weight of trichloroethylene.

The saturated lay-up is then subjected to peristaltic action by wiping it down with a sponge compressed around the circumference of the glass sleeving, to remove as much as possible of the impregnated adhesive. The solvent is then evaporated and the formulation is cured in an oven at 212°F for 2 hours, after which the completed porous support tube is easily removed from the mandrel.

Example 2: A mandrel is prepared by placing a 34 inch length of ½ inch outside diameter by ⅜ inch inside diameter Teflon tubing on a ⅜ inch steel drill rod 3 feet in length. On one end of the steel rod is threaded a round stop nut ½ inch outside diameter. The other end of the steel rod is left bare for gripping. A 2.1 inch by 22½ inch strip of Whatman No. 1 filter paper is cut from a sheet 22½ by 18¼ inches. The strip of filter paper to serve as liner is wet with water and carefully placed on the Teflon mandrel and shaped to conform to it with about ½ inch overlap. It is then left to dry. With one end of the steel rod gripped in a vise, a 2½ foot length of braided glass sleeving ½ inch in diameter (style EG 530, Electra Insulation Corp.) is slipped over the filter paper, taped at the free end, (i.e., the end that is not in the vise) with an adhesive tape, and then pulled taut. This is repeated for a total of 5 layers of braided sleeving, pulling each taut to tightly conform the layers to the mandrel. The lay-up (consisting of the liner and layers of reinforcement) is then removed from the vise and placed on a sheet of polyethylene film.

An impregnating formulation is next prepared consisting of 6 parts ERL 2795 (Union Carbide), 4 parts Versamid 125 (General Mills) and 40 parts trichloroethylene. This is mixed thoroughly. The impregnating formulation is poured over the lay-up and allowed to soak for 5 minutes while covered with a plastic sheet to prevent evaporation of the solvent. The lay-up is then placed vertically and the excess resin is squeezed out by wrapping a paper towel around the top end (the end at which the glass sleeving had been taped), grasping it tightly while drawing downward. This "milking" procedure is repeated twice. The finished lay-up is placed in a hot air circulated oven at 210°F for 2 hours in order to cure. The lay-up is removed from the oven and allowed to cool. The mandrel is next pulled out while the stop nuts prevent the Teflon from coming off the steel rod. The resulting porous tube is then cut to the desired length.

Textile Sheath for Tubular Units

M. Pages; U.S. Patent 3,853,757; December 10, 1974; assigned to Rhone-Poulenc SA, France describes a support element for a tubular membrane having particular utility in reverse osmosis and ultrafiltration comprising a support tube and a shrunk textile sheath in gripping contact with the outer surface of the support tube. The tube is manufactured by placing a shrinkable textile sheath around a support tube whose external diameter is smaller than the internal diameter of the sheath and shrinking the textile sheath so that it grips the outer surface of the support tube.

Example: A woven sheath with a reinforcement of the 5-on, 2-off satin type and 30 millimeters wide when flat, in which the warp consists of a polypropylene yarn of 200 deniers, with 340 turns per meter, and the weft consists of bulked yarns of 70 deniers spun from a mixture of 83% by weight of polyvinyl chloride and 17% of postchlorinated polyvinyl chloride, is produced using 30 cop strokes to the centimeter per face, on both faces.

A portion of this sheath 145 centimeters long is placed around a steel tube (1.35 centimeters inside diameter by 1.8 centimeters outside diameter by 150 centimeters long) provided with perforations 0.08 centimeter in diameter every 8.5 centimeters. The sheath and tube are then immersed in water at 85°C for 5 minutes. The sheath shrinks along its diameter, thus bringing the sheath into contact with the tubular support over its entire surface; the sheath is thus held fixed of its own accord and stretched uniformly over the tubular support. No unevenness or discontinuity of the surface of the sheath is observed.

A collodion based on the cellulose acetate is deposited on the element thus formed, according to the techniques described in French Patents 1,426,548 and 1,506,402. The element thus treated is kept at a temperature of 0°C for 30 minutes and is then subjected to a heat treatment which consists of bringing it to a temperature of 85°C for 20 minutes. The membrane thus formed on the element is of even thickness and adheres well to the sheath.

Polyvinyl Chloride Support Using Water-Soluble Surfactant and Polyvinyl Methyl Ether

According to a process described by *L.A. White, B.O. Baum, W.H. Holley and H.E. Podall; U.S. Patent 3,763,055; October 2, 1973; assigned to the U.S. Secretary of the Interior* a microporous support structure for reverse osmosis membranes is prepared by compounding: (1) a water-insoluble thermoplastic resin from the group consisting of polyvinyl chloride, vinyl chloride-propylene copolymer and a blend of polyvinyl chloride and vinyl chloride-propylene copolymer; (2) polyvinyl methyl ether; and (3) a surfactant, forming the resulting composition into a suitable shape and extracting with water to remove a substantial portion of the polyvinyl methyl ether and surfactant, thereby forming a microporous structure.

Example: The following materials were used in preparation of a microporous support structure.

	Quantity (solids) Parts by Weight
Geon 103EP (B.F. Goodrich Chemical Co.), a suspension polymerized, rigid PVC homopolymer having a weight average molecular weight of about 150,000	100
Gantrez M-154 (G.A.F. Corporation), linear PVME of 43,000 cp viscosity as a 50% solution in water, specific viscosity, 0.47 (1 gram per 100 milliliters solvent, 30% solution)	15
Triton X-100 (Rohm & Haas Co.), isooctyl phenoxy polyethoxy ethanol containing 10 mols of ethylene oxide	10
Acryloid K120N (Rohm & Haas Co.), acrylic/methacrylic polymer of molecular weight about 1,000,000 (processing aid)	3
Thermolite 831 (M&T Chemicals, Inc.), di(n-octyl)tin S,S'-bis(isooctylmercapto acetate), heat stabilizer	2

This formulation was milled at 160°C. The PVC was fused and the PVME was then added gradually to the banded resin to avoid breaking up the sheet. The compounded material, as sheeted off the mill rolls, was white, tough, and somewhat flexible. It was then compression molded into films of about 6.5 mils thickness at 175°C and a pressure of 1,000 pounds per square inch, and the films were then leached in distilled water at room temperature overnight. The weight loss on leaching was found to be 10%.

Following leaching, the films were dried for 2 hours in a 50°C vacuum oven at 30 inches vacuum and were then tested for flow rates of water through the films and for pore size of the films. Flow rates of water were 173 gfd at 40 psig pressure, and the average pore size was about 0.35 micron with 80% of the pores falling between 0.05 and 0.8 micron.

Lap Seam Adhesive

E.A.G. Hamer; U.S. Patent 3,715,036; February 6, 1973; assigned to Abcor Water Management Company, Inc. describes a reverse osmosis purification system in which the osmotic membrane is formed from a flat membrane which is curled and lap seamed. An adhesive is employed to seal the lap seam. The adhesive does not permanently bond together the lapped surfaces of the membrane until after the membrane is positioned in contact with the support tube and fluid pressure is applied to force the membrane into firm and conforming contact with the support member. Three techniques can be used to effect the sealing of the overlapping membrane area without destroying membrane properties.

(1) An adhesive can be applied to the overlapped area which will not permanently bond during either its application or the seating of the membrane against the porous tube. The adhesive may or may not form a permanent bond after full brine pressure is applied. All that is required of this type of adhesive is that it prevent the brine solution under pressure from seeping between the layers of the overlap. An example of such an adhesive is a viscous solution of aqueous polyvinyl alcohol.

(2) An adhesive which forms a permanent bond to the membrane before it is put under pressure may be used, if the mechanical properties of the adhesive are such that it will stretch under the applied pressure, in preference to any strain undergone by the membrane itself. Low modulus elastomers may be used as this type of adhesive.

(3) Adhesives which bond permanently before the brine pressure is applied may also be used if, in seating the membrane against the porous tube, a pressure is applied which is equal to or greater than the brine pressure to be subsequently used. If this is done, the overlap will open up sufficiently during the seating process so that there will be no further strain either on it or on the membrane when the brine pressure is applied. An example of such an adhesive is E.I. du Pont de Nemours Company's Super Cement.

Example: For an internal membrane, the sheet material was prepared by casting a 10-mil film with a doctor knife from a solution containing 23.15% cellulose acetate (Eastman E398-3), 69.4% acetone, 1.65% magnesium perchlorate, 0.33% hydrochloric acid and 5.47% water. This operation was carried out with apparatus and solution precooled to -9°C. Part of the solvent was allowed to evaporate by exposing the freshly cast film to air for 2 minutes in the same temperature range. After immersion in water at 1°C for 1 hour, the film was removed from the casting plate and wrapped around a mandrel comparable in size to the bore of the support tube before being submerged in a hot water bath maintained at 73°C for 15 minutes.

During this step, the air exposed surface of the film faced the mandrel. In the course of this annealing step, the membrane acquired a permanent curl, so that it tended to retain a cylindrical, rather than a flat configuration, even when unsupported. A sufficient width of film was unwound from the mandrel and slit along its length resulting in a section which overlapped when curled to the desired diameter. The membrane was wrapped around an elastomeric tube, the outside diameter of which was slightly smaller in diameter than the inside diameter of the permanent support tube into which the membrane was to be placed. One of the overlapping surfaces of the membrane was coated with a thin layer of aqueous polyvinyl alcohol.

The elastomeric tube and membrane assembly were inserted into the porous support tube and one end of the elastomeric tube clamped shut. Air pressure was applied to the other end of the elastomer tube, and the elastomeric tube was expanded sufficiently to bring the membrane into intimate contact with the porous support tube. The elastomeric tube was depressurized and removed, and the membrane remained in the porous support tube in intimate contact with it since the adhesive in the overlapped area of the membrane fixed the membrane into the larger diameter. The ends of the tube were then fitted with attachments which permitted it to be used in a desalination system.

The support tube with its membrane lining was then coupled into a high pressure brine line circulating brackish water at 0.5% NaCl concentration under a pressure of 600 psi. Purified water containing only 300 ppm (0.03%) NaCl was collected at a rate of 7.3 gallons per day per square foot of membrane area.

The composition of the casting solution may be varied to some degree. Thus, the casting solution is preferably about 23.15% cellulose acetate, but may be from about 15 to 25%. The preferred composition also contains about 69.5% acetone, but this may vary between about 60 and 80%. Similarly, the content of magnesium perchlorate, preferably about 1.65%, may range from about 1 to 2%.

Hydrochloric acid is not an essential component of the casting solution, but may be used in amounts up to about 1%, preferably about 0.33%. Water is also not essential, but is preferably present in an amount of about 5.47%. It may be present, as desired, in amounts up to about 20%. Variations outside the ranges just stated tend to produce membranes which are either deficient in physical properties such as tensile strength, or lacking in their functional properties such as selectivity and/or permeability to water.

Cellulose Ester Adhesive

R.S. Stana, E.Mitchell and J.B. Coffay; U.S. Patent 3,791,526; February 12, 1974; assigned to Westinghouse Electric Corporation describe a method of sealing a reverse osmosis membrane to a contacting porous support which comprises applying a sealant composition, comprising a cellulose ester, an organic solvent and an alcohol to the membrane and its contacting support.

Example: A reverse osmosis apparatus was made, containing 18 parallel ½-inch diameter membranes containing axial bores in a porous support and a stainless steel perforated outer shell, having means to supply feed, exhaust feed and collect the purified water. The support was made from foundry sand, having an average particle size of about 180 microns, and the resinous reaction product of a phenol and an aldehyde, in solution, having a viscosity at 25°C of 400 cp and a solids content at 135°C of 67%. The phenolic resin along with hexamethylenetetramine catalyst was mixed with sand to uniformly coat it.

The mixture was then poured into the shell, which acted as a mold and contained removable rods, and cured at about 175°C in a Norman kiln oven. The resulting module contained resin in the range of about 5 weight percent of the coated filler particle weight. The module was 48 inches long and 4 inches in diameter.

Osmotic membranes were cast in place, within the axial bores through the module, from a casting solution containing about 25 weight percent cellulose acetate, 45 weight percent acetone and 30 weight percent formamide (methanamide). The membranes were leached in place by immersing the module containing them in cold water, to form an osmotic skin. The module was then allowed to stand vertically for 5 minutes to allow most of the water used in leaching the membrane to drain from the module pores and the module end face was cut off and filed flush with the outer shell.

The membrane-module sealant composition was mixed, containing 200 milliliters acetone (160 grams, 68 weight percent), 50 milliliters n-propanol (40 grams, 17 weight percent), and 35 grams cellulose acetate powder (15 weight percent). The cellulose acetate was added to thicken the mixture to the consistency of a thick varnish. The mixture had a postgelation shrinkage of about 2%.

Membrane sealant was applied in several coats on the membrane and ends of each bore with each coat being allowed to dry for about 5 minutes. The sealant overlapped the membrane end and welded it to the bore end and module end face. The module was then inverted and the same procedure followed as above to seal the other end of the module. End flanges were bolted on and the module tested. The seals successfully operated at 300 psi for over 500 hours continuously with no leakage or failure.

Diepoxide Resin and an Amidopolyamine Sealant

A.S. Calderwood and R.G. Black; U.S. Patent 3,773,181; November 20, 1973; assigned to Westinghouse Electric Corporation describe a method of sealing reverse osmosis membranes to a porous support which comprises applying a sealant composition, comprising diepoxide resin and amidopolyamine, to the membrane and support in an atmosphere having a relative humidity over 85%, the sealant contacting both the membrane and the support.

Example: A reverse osmosis apparatus was made, containing 18 parallel ½-inch diameter membranes containing axial bores in a porous support and a stainless steel perforated outer shell, having means to supply feed, exhaust feed and collect the purified water. The support was made from foundry sand, having an average particle size of about 180 microns, and the resinous reaction product of a phenol and an aldehyde, in solution, having a viscosity at 25°C of 400 cp and a solids content at 135°C of 67% (Durez Phenolic Resin 18115, Hooker Chemical Corporation). The resin was mixed with sand and hexamethylenetetramine catalyst to uniformly coat it.

The mixture was then poured into the shell, which acted as a mold and contained removable rods, and cured at about 175°C in a Norman kiln oven. The resulting module contained resin in the range of about 4 weight percent of the coated filler particle weight.

The module was 4 feet long and 3½ inches in diameter. Osmotic membranes were cast in place, within the axial bores through the module, from a casting solution containing about 25 weight percent cellulose acetate, 45 weight percent acetone and 30 weight percent formamide (methanamide). The membranes were leached in place by immersing the module containing them in cold water, to form an osmotic skin.

The porous support module, having the semipermeable reverse osmosis membrane therein, was placed in a humidity chamber. This was a plastic box 14 inches square and 5½ feet high. Air was circulated in a closed loop by a fan through the box and over a packing of flexirings in a column 5 feet high and 4 inches in diameter. The packing was kept wet by a spray of water at room temperature and evaporation of the water maintained the relative humidity at more than 95%. The sand module end face was cut and filed flush with the outer shell. The ends of the bores were chamfered at an angle of 45° to the end face of a 1/16 inch depth. The module was then allowed to stand vertically in the closed humidity chamber for 1 hour to allow most of the water used in leaching the membrane to drain from the module pores.

The humidity chamber top was opened and the ends of the bores dried with cotton swabs. The end of the membranes, i.e., the inner sides near the module terminal end face, were thinly coated to a depth of ½ inch down the bore with a polyurethane-solvent primer (TPR-421 primer, Thiokol Chemical Corp.). The chamber was then closed and the primer allowed to set for 5 minutes in the 95% humidity atmosphere.

The diepoxide resin membrane-module sealant composition was mixed, containing 10 grams of a liquid diglycidyl ether of bisphenol A resin (Araldite 6005 Epoxy Resin, Ciba Products Co.), 1.5 grams of a liquid diglycidyl ether of propylene glycol (DER 732 Epoxy Resin, Dow Chemical Co.) and 4.3 grams of an amidopolyamine curing agent (Epicure 870, Celanese Corp.). This gave a composition having a 63 weight percent bisphenol A type resin, 9 weight percent epoxy-glycol type resin, 28 weight percent curing agent, and a postgelation shrinkage of about 2%.

This resin was heated under a heat lamp for a few minutes at a 6-inch distance until all the bubbles were dispensed and the resin sealant had a watery consistency. The top of the humidity chamber was opened and the resin sealant brushed liberally over the sand end face. The end face was then recoated with the warm sealant which penetrated about 1/8 inch into the module end. A thin coat of room temperature resin sealant was then applied with a narrow brush in a thin coat on the membrane ends to a depth of about ½ inch from the end face and chamfered ends of each bore so that the sealant contacted both the membrane and the support and glued them together.

A membrane-module sealant composition containing filler particles was then applied about 1/32 inch thick with a spatula to the end face. This second coat had a higher viscosity and was thixotropic due to the filler. It contained, for each 5 grams of the diepoxide-curing agent composition described above, 0.3 gram (5.6 weight percent) fumed (pyrogenic) silicon dioxide having an average particle size of 0.007 micron (Cab-O-Sil EH-5, Godfrey L. Cabot, Inc.). The use of the filled sealant was optional and may be substituted with another coat of the unfilled room temperature sealant. The humidity chamber, at 95% relative humidity, was then closed and the sealant allowed to cure for 18 hours.

The chamber was then opened, the module inverted, allowed to drain for ½ hour, and the same procedure followed as above to seal the other end of the module. Finally, the epoxy and faces were smoothed off with light filing, end flanges bolted on and the module tested. The seals successfully withstood pressures of 450 psi and operated at 350 psi for 700 hours continuously with no leakage or failure.

Staged Desalination Unit

According to a process described by *E.F. Miller; U.S. Patents 3,898,158; August 5, 1975; and 3,839,201; October 1, 1974* a reverse osmosis separator unit is comprised of a plurality of reverse osmosis separator stages, each stage including a semipermeable membrane. At least one stage of the unit is comprised of a plurality of semipermeable membranes and at least one stage preferably includes generally annularly arranged membranes. A feed solution under pressure is fed to at least one of the stages, and at least one outlet is connected to each of the stages for drawing off the solution from each stage after the feed solution has passed through at least a longitudinal portion of the respective stage.

The pressure of the input (feed) solution is sufficient to maintain the pressure in the first stage substantially above the osmotic pressure in the stages to permit a portion of the feed solution to permeate and pass through the respective membranes, without requiring additional pressurization. The unit is preferably enclosed by an outer casing, and the solution which permeates and passes through the final stage, which preferably is the space between the outer membrane(s) and the outer casing, is drawn off as separated output. Valves are incorporated in each of the outlet lines, including both interstage concentrated solution drawoffs in the direction axial to the permeate flow, and in the permeate outlet line from the last stage.

This process is directed to substantially reducing or eliminating the need for interstage repressurization and to permit the utilization of a series of low to intermediate salt rejection semipermeable membrane stages, instead of a single or more highly salt rejection membrane, thus materially reducing the overall energy requirements of a reverse osmosis plant.

Referring to Figure 2.4a, the reverse osmosis separation unit comprises an input feed line **10** for feeding a solution to be treated, such as seawater containing 36,000 ppm of dissolved salts to the unit. The feed line feeds the input solution to a stage **11**, which may be tubular, annular or spiral annular depending upon construction of the system. The outer boundary of the stage is defined by a tubular or spiral annular member **12** which is comprised of a first semipermeable membrane and a structural matrix or backing material. The structural matrix or backing material, which physically supports the unit, is not shown.

Around tubular member **12** is mounted a second tubular member **14**, also made of a semipermeable membrane and a structural matrix or backing material. An annular stage **13** is defined between the first and second membrane. Around the second membrane and spaced therefrom is an outer shell member **16**. A second annular stage **15** is defined between the second membrane and outer shell **16**. The membranes and outer shell are maintained in relative position by means of structural elements.

An output line **17**, with a valve **18** therein, removes the unpermeated solution from the first stage **11**; an output line **19** with a valve **20** removes solution from the middle annular stage **13**, and an output line **21** with a valve **22** therein removes solution from the outer annular stage **15**. In accordance with the process, the solution removed from the outer stage is the purified output solution. The valves **18, 20** and **22** are adjustable and may be used to vary the internal pressures in the various stages **11, 13** and **15**.

The inner stage **11** may be annular, spiral annular or generally tubular. If an internal, centrally axial, support structure for the apparatus is used, then this will result in an annular inner stage **11**. Alternatively, the interior of the stage **11** may be kept free, thereby resulting in a generally tubular including spiral annular interior stage **11**.

The membranes **12** and **14** and the outer shell member **16** need not be circular in cross section. If desired, the cross-sectional configuration of the apparatus may be oval, rectangular, polygonal, spiral, helical, etc. Various shapes may be mixed in a given separator unit. The particular configuration will depend upon the particular system requirements. For ease of description the term tubular is deemed to encompass all of the above configurations of the membranes and the outer shell.

A preferred technique for removing salts from seawater is illustrated by reference to Figures 2.4a and 2.4b. Seawater, containing 36,000 parts per million of dissolved salts is charged through line **10** at a pressure of about 1,000 more or less depending upon the desired amount of driving force in excess of the osmotic pressure of seawater (approximately 870 psi). The charged seawater is to a first inner stage **11** comprised of first semipermeable membrane **12** and a structural matrix or backing material. The capacity of membrane **12** for rejection of dissolved sodium chloride, the predominant component of the seawater, can be, for example, about 90%. Unpermeated solution is removed from stage **11** through line **17** containing valve **18.** The output from this line, when concentrated by a factor of 2, would contain approximately 72,000 ppm of dissolved salts.

FIGURE 2.4: REVERSE OSMOSIS SEPARATOR UNIT

a.

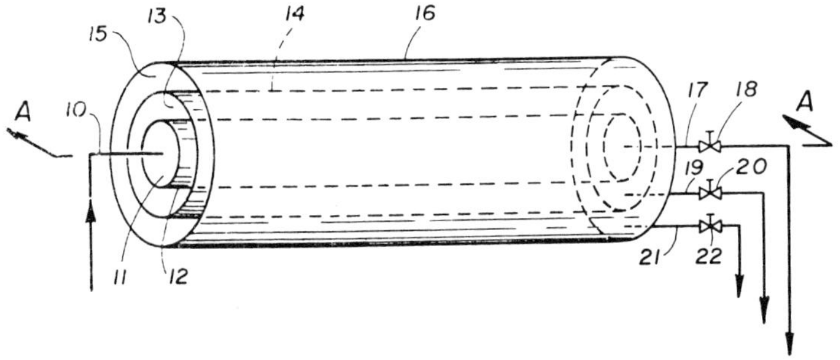

Reverse Osmosis Separator Unit Having 2 Semipermeable Membranes, 1 Inner Zone and 2 Annular Zones

b.

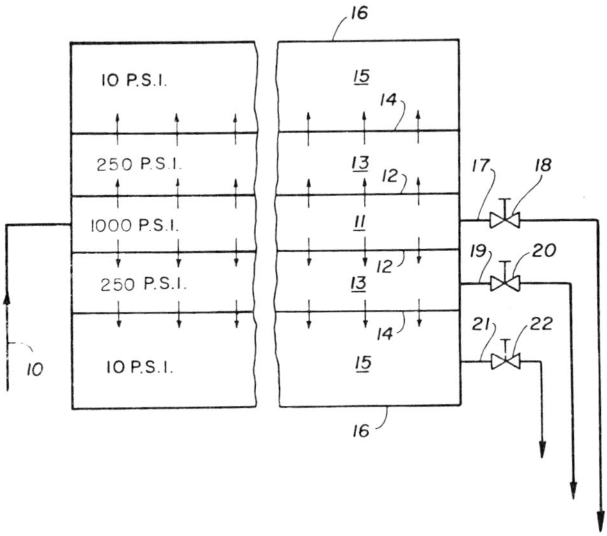

Cross Sectional View of Figure 2.4a Taken Along Line A-A

Source: E.F. Miller; U.S. Patent 3,898,158; August 5, 1975

The pressure of the solution charged into the first stage may range from as low as 20 psi for brackish water feeds to as high as 4,500 psi for highly saline water feeds. The permeate from stage **11** is obtained in a second annular stage **13** and will contain about 5,400 ppm of dissolved salts. This stage is bounded by second semipermeable membrane **14** and a structural matrix of backing material. This membrane can be the same as membrane **12** or can be more or less permeable than membrane **12**. Preferably, for seawater desalination, it is in the order of 90%. In this illustration, membranes **12** and **14** are the same.

Thus, the permeate from the first stage **11** (or the first or extreme inner tube) becomes the feed for the annular stage **13** (or the annulus space between the extreme inner and next outer tube). The reverse osmotic pressure requirement of the permeate in stage **13** is provided by maintaining a sufficient back pressure on line **17** by controlling valve **18**. In this illustration, the pressure maintained in stage **13** is approximately 250 psi.

Unpermeated solution in stage **13** is removed therefrom through outlet **19** containing valve **20**. As indicated above, this will contain about 11,000 ppm of dissolved salts when concentrated by a factor of 2. The permeate from stage **13** is obtained in a further annular stage **15** and contains about 800 ppm of dissolved salts. This permeate is removed from the system in stage **15** through line **21** containing valve **22**. Pressure in this annular product water stage **15** may be negligible or positive to force the product water to a storage tank, for example, a collection zone.

As illustrated above, initial pressurization of the seawater, or other aqueous solution to be purified, and subsequent control of the permeated and unpermeated (spent) solution fractions on each side of a reverse osmosis semipermeable membrane for pressure control, eliminates interstage repressurization. It is to be noted that staged desalination accomplished by back pressurization in the manner described minimizes the pressure drop across the membrane and provides some support to maintain the structural integrity. Thus, it is possible to use thinner membranes to accomplish the described separations than without such pressure operations.

HOLLOW FIBERS

Density Gradient in Wall Structure

A process described by *R.L. Leonard and J.D. Bashaw; U.S. Patent 3,724,672; April 3, 1973* is directed to hollow filament membranes derived from esters of cellulose and characterized by a density gradient in the filament wall structure. When employed in reverse osmosis operations the filaments exhibit extraordinary flux and rejection properties. A continuous fabricating method is also provided which comprises continuously extruding filaments into a controlled evaporation zone thence directing the filaments into a cold water coagulation bath, thereafter passing the same to a wash bath and finally to an annealing zone.

A unique feature of the fibers which contributes very substantially to the water transport properties is the asymmetric structure of the fiber walls. That is, the walls have a density gradient running progressively less dense from the outer surface to the surface facing the hollow area. This morphology can best be described as a skin core structure. It is in contrast to the essentially homogeneous structure of common hollow fiber membrane systems, and accounts largely for the improvement in the flux rate property.

The process flow is described using cellulose acetate for purposes of illustration by reference to Figure 2.5. After forming a viscous solution of cellulose acetate (the spinning dope), the cellulose acetate dope is extruded through a hollow filament shaping orifice in spinneret **1** into an evaporative gap or zone **3**. The evaporative zone is most critical to the formation of high flux fibers. The filament must be extruded into an atmosphere rich in a cellulose acetate solvent.

FIGURE 2.5: HOLLOW FIBER MEMBRANES

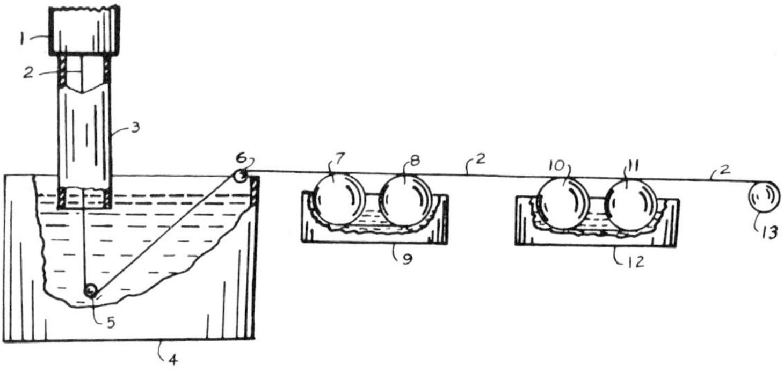

Source: R.L. Leonard and J.D. Bashaw; U.S. Patent 3,724,672; April 3, 1973

This is accomplished by enclosing the extrusion orifice in what may be likened to a chimney. The chimney rapidly becomes saturated with solvent evaporating from the hot dope. Once saturated at chimney temperature, further evaporation is greatly slowed. The rapid cooling during the evaporative phase causes gelation and the rapid viscosity buildup aids in maintaining the hollow fiber structure which has been shaped by the extrusion orifice.

From the evaporative zone, the fiber is passed to coagulation bath **4**, which is a water bath maintained at ice water temperatures. By immersion in this cold water bath, the filament structure becomes set. The filament then passes over roller **5**, which is driven at a greater speed than the extrusion velocity to effect a stretching and consequent reduction of the filament diameter to bring it within an outside diameter of from 100 to 700 microns and preferably from about 200 to 400 microns in outside diameter.

From roller **5** the filament is forwarded over guide **6** to revolving wash drums **7** and **8** which are partially immersed in water tank **9**. In this washing zone remaining traces of solvent are removed. From passage around the wash drums, the filament is advanced to revolving drums **10** and **11** which are positioned in hot water bath **12**. The filament passes around drums **10** and **11** and through the hot water in bath **12**. The water in this zone is maintained at a temperature of from about 65° to 95°C. The purpose of this heat treatment or annealing operation is to effect a contraction or a tightening of the filament structure which still permits the transfer of water, but restrains the passage of salt. Finally, the filament is forwarded to a take-up bobbin **13**. Alternatively, the filament bypasses the hot water bath and is taken up on the bobbin for subsequent annealing. In forming the spinning solutions, it has been found that a combination acetone-formamide in ratios in the range of 3:2 or 2:3 is the solvent of choice.

Example: In the 6 runs made in this example, the dope used was 30% by weight cellulose acetate in 40:60 acetone-formamide mixture; the extrusion was through a single hole tube in orifice hollow filament spinneret at 1.22 cc/min; the evaporative chimney was ¾" by 3½" outside diameter; the spinning rate was 70 ft/min; and the fiber had an outside diameter of 296 microns and an inside diameter of 100 microns.

The annealing times and temperatures and flux rates were varied as shown in the table on the following page. In each of the runs the feed water presented to the membrane contained 3,000 ppm of sodium chloride and was at ambient temperature. The pressure employed was 250 psig, and each of the runs was of 2 days duration.

- - - - - - - Annealing - - - - - - -		Flux	
Time	Temperature	gal/sq ft/day	Rejection, %
5	75°C	6.30	93.8
15	75°C	5.91	95.0
5	80°C	2.90	97.1
15	80°C	2.34	97.8
5	85°C	1.32	98.4
15	85°C	1.03	98.5

Foraminous Separators for Hollow Filaments

P.R. McGinnis and G.J. O'Brien; U.S. Patent 3,690,465; September 12, 1972; assigned to E.I. du Pont de Nemours and Company describe an element for the selective permeation separation of components of fluid mixtures comprising thin layers of substantially parallel hollow filaments separated by thin foraminous materials which restrict movement and prevent nesting of the hollow filaments during use of the element, and a resinous tube sheet structure encapsulating a portion of the hollow filaments, with open ends of the filaments extending through the tube sheet structure.

A permeation separation device incorporates the improved element in a surrounding shell with means for introducing into and removing a fluid mixture from the shell and means for removing separated components of the mixture from the open ends of the hollow filament membranes. Preferably the device also contains a foraminous tube in the center with the foraminous material and hollow filaments convolutely wrapped around the tube.

Commonly, permeation separation devices have generally required that hollow filaments be packed tightly into a jacket with means for introducing a fluid feed such as brackish or saline waters at a point near one end of the filament bundle under pressure. In such instances water passes through the walls of the hollow filaments more rapidly than dissolved impurities and purified water is obtained from the open ends of the filaments while the remaining solution, enriched in impurities, is withdrawn from the devices through exit pores.

In such devices the ideal flow of the fluid feed is that which permits for the greatest contact between fluid feed and filament surface. In practice, however, considerably less than ideal conditions prevail. The flow of the fluid feed has a persistent tendency to cause the filaments to migrate and nest slightly together in compact bundles. This greatly reduces the filament surface available to the feed fluid as the bundles are practically impenetrable and the result is marked decrease in the efficiency of the permeator device. These difficulties are largely overcome by this process by inserting between layers of the hollow filaments one or more layers of foraminous web-like material which both separate and stabilize the position of the hollow filaments but do not materially restrict the flow of fluid feed.

These foraminous separatory materials restrict movement and nesting of the adjacent hollow filaments by preventing their contact and by reducing their freedom of movement perpendicular to their length. The foraminous materials impart friction between the filaments and the materials and are resistant to such distortions as wrinkling and tearing. Preferably the foraminous material has uneven surfaces so as to touch only a small fraction of the adjacent filament surfaces and thereby to reduce only insignificantly the effective membrane area. In addition, the movement and nesting of filaments within a given filament layer, which further removes the filaments from a separating material, can be restricted by compression of the filament layer. The compression is caused by the foraminous separating material and the resulting increased friction between filaments which cross within the layer.

The foraminous separating materials must be thin so that they do not excessively reduce the fraction of the volume of the separatory elements which is available for the hollow filament membranes and correspondingly the membrane area which can be achieved in a given volume. They can range in thickness between the diameter of a single filament and about 20 mils. Preferably they are between about 2 and 20 mils in effective thickness. Such thin foraminous materials are also desirably flexible and can conform to the shapes of the hollow filament layers and to the textures of their surfaces.

These foraminous materials can be of any form within the general principles of the process. Thus, they can be porous flexible solids such as sintered metals or porous plastic sheets; pervious fibrous materials such as paper, cloth, or nonwoven cloth-like materials; perforated films; or any other similar material. The structures should preferably have appreciable tensile strength in one direction to resist forces applied during fabrication of a permeator assembly but need not have appreciable tensile strength in a direction at right angles to the first direction.

The preferred materials are in general web-like, in that they are flexible structures of relatively large area compared to their thickness, composed of fibrous elements, and with the general properties of cloth and paper. Particularly preferred structures include the spunbonded structures described by Shealy et al in the *Textile Research Journal,* volume 35, pages 322 to 329 (1965) and volume 38, pages 7 to 15 (1968). Representative of suitable materials for use as foraminous separating materials are those listed in the table below. Complete details of the assembly of the separation units and related performance parameters are provided in the patent.

Composition and Construction	Aperture	Basis Weight oz/yd^2	Thickness mils
Polyethylene spunbonded sheet	10 microns*	2.2	8
Polypropylene spunbonded sheet	300 microns*	2.5	10
Nylon knitted tulle cloth	20 x 20 per inch hexagonal	–	5
Nylon knitted tulle cloth	9 x 9 per inch hexagonal	–	9
Nylon knitted tulle cloth	27 x 27 per inch square	–	5
Nylon woven cheese cloth	32 x 32 per inch	1.5	–
Nylon woven tobacco cloth	20 x 20 per inch	1.0	–
Nylon nonwoven crosslaid skrim	10 x 10 per inch	0.5	–

*Approximate maximum pore size

Nonwoven Fabric Comprising Hollow Fibers

In a process described by *A.L. Kohl; U.S. Patent 3,557,962; January 26, 1971; assigned to North American Rockwell Corporation* an improved reverse osmosis fabric is provided having warp fibers and hollow woof fibers woven into a stable fabric configuration; each of the hollow woof fibers has a permeable wall defining a longitudinally extending fluid channel. The permeable wall is in fluid contact with a pressurized feed fluid and is adapted to pass a product fluid forced from the feed fluid in a predetermined direction through the wall.

Other forms of woven and nonwoven fabrics are also provided having selected hollow fibers. Referring to Figure 2.6a, one form of reverse osmosis fabric **10** having a relatively stable fabric configuration is woven in a predetermined arrangement or weave pattern with warp or support fibers **12** and woof fibers **14**. The reverse osmosis fabric can have a weave pattern such as a right angle weave as shown by Figure 2.6b, which is similar to a plain cloth weave where the warp fibers provide the lengthwise support for the crossing woof fibers.

The reverse osmosis fabric as contemplated can have other weave patterns. For example, the bias weave fabric **16** as shown by Figure 2.6c, or the mesh weave fabric **18** as shown by Figure 2.6d, or other conventional cloth weave patterns. In the right angle weave pattern **10** of Figure 2.6b, the woof fibers **14** of the reverse osmosis fabric are formed from a hollow working fiber **20** as shown by Figure 2.6e and described below. The woof fibers can be continuous hollow fibers that are bent in a generally serpentine arrangement through about 180° bends at the fabric edge into similar bights **22** during a weaving process.

FIGURE 2.6: NONWOVEN FABRIC COMPRISING HOLLOW FIBERS

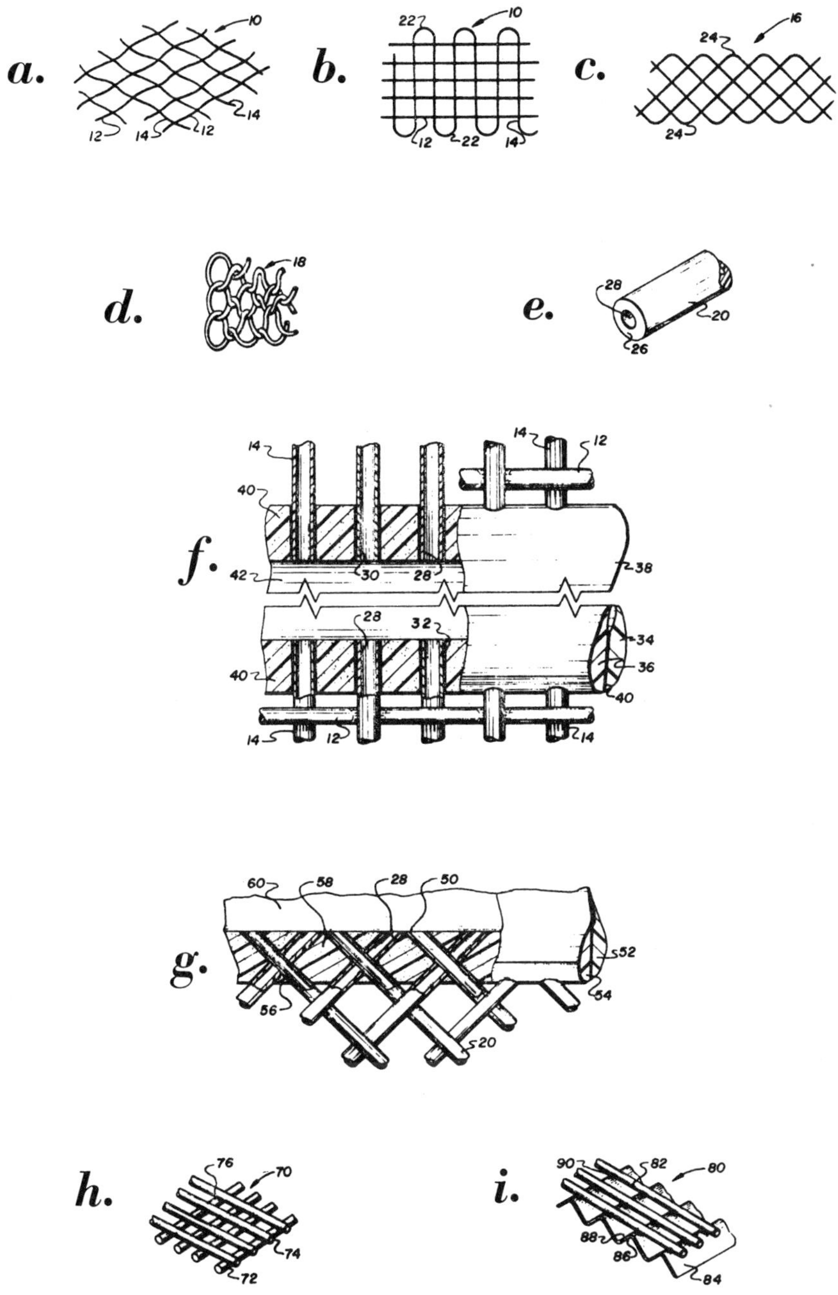

Source: A.L. Kohl; U.S. Patent 3,557,962; January 26, 1971

The warp and woof fibers **12** and **14** are interwoven with interconnected adjoining fiber portions in the weave pattern of Figure 2.6b. It is contemplated that the woof fibers **14** can be individual hollow fibers positioned in a parallel array rather than the serpentine arrangement with the raw or open ends suitably sealed as required. It is also contemplated that the warp fibers **12**, in addition to the woof fibers **14**, can be formed from hollow fibers, such as hollow fiber **20**.

In the bias weave pattern **16** of Figure 2.6c and in the mesh weave pattern **18** of Figure 2.6d the interwoven fibers are preferably all hollow fibers like fiber **20** of Figure 2.6e. However, it is contemplated that selected ones of the fibers in the bias weave pattern or in the mesh weave pattern could be solid so that functionally the solid fibers would serve as support fibers. The fibers in the bias weave pattern are bent through 90° bends into similar bights **24** at the fabric edge during a weave process. Any raw edges that remain after a weaving process in either the right-angle weave **10** of Figure 2.6b, the bias weave **16** of Figure 2.6c, or the mesh weave **18** of Figure 2.6d can be suitably sealed, thus forming a finished edge.

The hollow reverse osmosis fiber **20** that is used in the fabrics is shown in Figure 2.6e. This fiber can be a conventional hollow reverse osmosis fiber formed from a variety of polymeric materials. In one commercially available form, the hollow fiber has an outside diameter range from 25 to 250 microns with a permeable wall **26** that has a thickness of 5 to 50 microns. The inner surface of the hollow fiber wall **24** defines a longitudinally extending fiber channel **28**. Operatively, the reverse osmosis fabric can find a variety of uses as a stable fabric in reverse osmosis systems. It will be beneficial for a clear understanding of the process to briefly consider, for example, how the right angle weave fabric **10** of Figure 2.6b and the bias weave fabric **16** of Figure 2.6c can be used in such a system.

In Figure 2.6f the reverse osmosis fabric has the right angle weave pattern as shown by Figure 2.6b with the warp fibers **12** formed from solid fibers and the woof fibers **14** formed from hollow fibers such as the hollow fiber shown in Figure 2.6e. The woof fibers of the fabric are cut and the exposed fiber ends **30** and **32** are positioned between the complementary parts **34** and **36** of a longitudinally split tube **38**. The exposed fiber ends and the complementary tube parts are sealed together by a suitable bonding material such as an epoxy resin **40** so that the ends and the tube parts form a fluid-tight structure. Thus sealed, the similar fiber channels **28** of the hollow woof fibers are in fluid communication with a tube channel **42** as defined by the assembled parts **34** and **36** that form the tube **38**.

A product fluid is recovered from the operational arrangement of the reverse osmosis fabric as shown by Figure 2.6f when, in one operating mode, a pressurized feed fluid contacts the outer surfaces of the warp and woof fibers. A product fluid forced from the feed fluid passes through the similar walls **26** to the fiber channels **28** where the product fluid collects and is directed through the cut fiber ends **30** and **32** to the tube channel **42**. It is contemplated that a reverse flow pattern could be used where the feed fluid passes from the tube channel through the cut fiber ends into the similar fiber channels **28**. The product fluid then passes through the similar walls **26** to the outer surface of the woof fibers where the product fluid collects. The product fluid is in fluid contact with the outer surfaces of the warp and woof fibers.

In Figure 2.6g, the reverse osmosis fabric has the bias weave pattern **16** as shown by Figure 2.6c with all or selected ones of the fibers formed from hollow fibers such as the hollow fiber **20** shown in Figure 2.6e. The fabric is cut and the exposed fiber ends **50** are potted between the complementary parts **52** and **54** of a tube **56** by a suitable epoxy resin **58** that seals the cut fiber ends and the tube parts into a fluid-tight structure. The similar fiber channels **28** are exposed by the cut fiber ends and are in fluid communication with a tube channel **60** that is defined by the assembled parts **52** and **54** that form the tube **56**. A product fluid is recovered from the operational arrangement of the reverse osmosis fabric as shown in Figure 2.6g in a manner similar to that described for the operational arrangement as shown in Figure 2.6f.

Concentration polarization is minimized by the reverse osmosis fabric in the operational modes as described since the pressurized feed fluid flow is maintained in a turbulent flow condition rather than in a laminar flow condition by the particular fabric weave pattern. The right angle bias, and mesh weave patterns as shown in Figures 2.6a through 2.6d, respectively, and as described, do not permit the feed fluid to flow in parallel with a fiber outer surface for any extended distance without encountering a cross fiber that interrupts such a parallel or laminary flow pattern. This interruption ensures that the undesirable concentration polarization effect is substantially minimized if not totally eliminated.

By design, hydraulic losses of a product fluid can also be substantially reduced dependent upon the length of the hollow working fibers, for example, the relatively short length of the hollow woof fibers **14** in the right angle weave pattern **10** of Figure 2.6b when placed in the operating mode as shown by Figure 2.6f minimizes hydraulic losses because of the relatively short product fluid flow path through the hollow working fibers to the tube channel. Thus, a reverse osmosis fabric has the advantages, inter alia, of minimizing the concentration polarization effect during operation of a reverse osmosis system using the fabric as described and minimizing hydraulic pressure losses, particularly of the product fluid. Both of these desirable advantages enable the reverse osmosis system to operate with lower system pressures.

The tubes **38** and **56** as shown in Figures 2.6f and 2.6g, respectively, can be formed where desirable by filament winding or otherwise building up the tube configuration so that a resulting tube channel is in fluid communication with the hollow fibers of the particular reverse osmosis fabric, or can be formed by coring a solid rod that secures the reverse osmosis fabric so that the cored rod defines a tube channel that is in fluid communication with the hollow fibers of the fabric.

Nonwoven Reverse Osmosis Fabrics – Referring to Figure 2.6h, one form of nonwoven reverse osmosis fabric **70** has warp or support fibers **72** and woof fibers **74**. It is contemplated that a nonwoven reverse osmosis fabric can also have a plurality of fibers suitably connected together in a predetermined nonwoven arrangement or array where the warp and woof fibers are hollow fibers such as hollow fiber **20** shown in Figure 2.6e. The warp fibers and woof fibers are bonded together with a stable fabric configuration by a suitable bonding material such as an epoxy resin that forms similar bond joints **76** at the adjoining portions of the warp and woof fibers, i.e., where the warp and woof fibers cross.

It is contemplated that the warp fibers can be arranged in separate and adjacent planes spaced apart by the crossing woof fibers; similarly, the woof fibers can be arranged in separate and adjacent planes. Another form of reverse osmosis fabric **80** having a nonwoven fiber array is shown in Figure 2.6i. Hollow reverse osmosis fibers **82** are bonded by a suitable bonding material such as an epoxy resin to a generally nonplanar corrugated member **84** that has similar ridges **86** and grooves **88** oriented transverse to the hollow fibers. The hollow fibers **82** are bonded to the ridges by the epoxy resin at similar bond joints **90**. It is contemplated that the corrugated member can be formed from a flexible plastic, or polymeric material.

Melt Extrusion of Plasticized Polymers for Hollow Fibers

In a process described by *E.A. McLain and H.I. Mahon; U.S. Patent 3,423,491; January 21, 1969; assigned to The Dow Chemical Company* permselective hollow fiber membranes are melt extruded from a molten intimate mixture of a thermoplastic polymer and a plasticizer for the polymer, the plasticizer characterized in having a boiling point above the temperature at which the molten mixture is melt extruded and further characterized in being soluble in a nonsolvent for the polymer.

The resulting solid, plasticized hollow fiber can then be leached with the solvent that is a solvent for the plasticizer but a nonsolvent for the polymer and, preferably, kept wet prior to being used as a separatory membrane or element in a separatory cell or apparatus. Or, the solid plasticized hollow fiber may be stored for a length of time prior to being leached after which it is kept wet or put in use immediately as a separatory membrane. Another and desirable alternative is to fabricate the solid plasticized hollow fiber in a separatory cell by securing or potting the ends or whatever other fabrication techniques may be employed, and leaching the plasticizer with a suitable solvent that is a nonsolvent for the polymeric constituent of the hollow fiber. Any of the known thermoplastic polymers that can be suitably melt extruded from a plasticized composition into a hollow fiber shape can be utilized in the process.

Example 1: A uniformly blended mixture of cellulose triacetate (43.2% acetyl content) and sulfolane (tetramethylene sulfone) is formulated according to the following procedure. The cellulose triacetate is ground to fine powder in a ball mill. A mixture of 300 milliliters of benzene-pentane containing 20 volume percent pentane is prepared. Forty grams of the ground cellulose triacetate is added to the benzene-pentane mixture while stirring with a propeller type agitator and the resulting mixture is then cooled to about 0°C.

While stirring, 40 grams of sulfolane is added to the cooled mixture. The cooled mixture is heated while stirring to 50° to 60°C until a gel is obtained. 500 milliliters of pentane is placed in a blender and, while agitating, the polymer gel is poured in which forms a precipitate which is separated by filtering through a coarse glass frit. The precipitate is twice more repulped in 500 milliliters pentane and filtered. The final precipitate is dried in a vacuum oven for 16 hours at about 60°C. The dried powder resulting from the above procedure is found to be a uniform mixture free from gels. The proportions of constituents are determined by analyzing for sulfur content and the sulfolane content is thus computed. The weight ratio of sulfolane to cellulose triacetate is approximately 0.77:1.

The finely divided cellulose triacetate-sulfolane mixture, to remove entrained air, is compacted under vacuum and a pressure of about 20,000 psi. A portion of the essentially air free mixture is then spun into hollow fibers according to the following procedure. The mixture is placed in a stainless steel sample tube having an internal diameter of ½ inch. The tube is approximately 9 inches long and the external diameter tapers from ¾ inch at the base to about ⅝ inch at the top. The bottom of the tube is plugged except for a ⅛ inch diameter opening through which the molten mixture is able to flow.

The sample tube is inserted into a 4 inch diameter melt block which is electrically heated and is controlled at a temperature of about 203°C. The sample tube is forced into the bore of the melt block (which is tapered to match the tube) by a spinneret holder which is also electrically heated and controlled at about 235°C. The spinneret holder contains a ⅛ inch diameter passage which conducts the molten mixture to the spinneret. In order to assure that the mixture attains a uniform melt temperature, it is allowed to remain in the melt block for about 15 minutes.

The mixture, in a uniform molten condition, is forced through an annular orifice in the spinneret formed between the outer walls of the orifice and a fine center core tube by the action of a hydraulically actuated piston moving down into the sample tube at a controlled rate. The internal diameter of the extruded hollow fiber is maintained by applying nitrogen under controlled pressure in the center core tube.

After the hollow fiber is passed through air at room temperature, a sufficient distance to solidify it, the hollow fiber is taken up on a 10 inch diameter plastic drum driven at a rate of peripheral speed greater than the rate of extrusion such that the external diameter of the final hollow fiber is less than that of the extruded fiber. Excellent smooth-surfaced hollow fibers in the size range of about 30 to 200 microns outside diameter are prepared in accordance with the above procedure.

A bundle of 50 hollow fibers, prepared according to the foregoing, about 30 cm long and each having an outside diameter of about 190 microns and an inside diameter of about 108 microns, is leached in water at room temperature to remove the sulfolane and then dried. The leached and dried bundle of fibers is mounted in a glass U-tube, equipped with an outlet valve at the bottom of the U, by sealing the space between and around the fibers at both ends of the U with an epoxy resin. The ends of the hollow fibers extend slightly beyond the seal and are kept open.

Water under a pressure of about 188 psig is admitted to the interior, i.e., the hollow cores, of the fibers through a fitting attached to one end of the fiber bundle for about 16 hours. During this time about 0.38 cc of water permeates the fiber walls and is recovered through the outlet valve at the bottom of the U-tube, indicating the fibers are excellently suited for use in permeability separatory apparatus and processes. The permeation coefficient (PC) for this test is calculated as follows.

$$PC = \frac{\text{permeation rate x thickness}}{\text{transfer area x driving force}}$$

The transfer area = πDLN, D being the mean diameter of the fiber, L being the length of the fiber, and N being the number of fibers. The permeation rate is the amount of water per given time and the driving force is water pressure minus osmotic pressure. Substitution in the equation gives a PC of 2.9×10^{-14} cm/sec.

Example 2: A bundle of 104 hollow fibers prepared according to the procedure of Example 1, about 50 cm long and each having an outside diameter of about 104 microns and an inside diameter of about 69 microns, was leached in water at room temperature and then air dried. The bundle was formed in the shape of a loop with all of the open ends together and the bundle was strung through a test cell made from copper tubing. The open ends of the fiber were first capped over with a cement to prevent plugging with epoxy and then that section of the bundle was potted in the copper tube with epoxy resin.

After the resin was cured, the end of the potting was cut off to expose the open ends of the fibers. A reverse osmosis test was then conducted by pumping seawater (total solids, 35,000 ppm) through the copper tubing over the outside of the hollow fibers at 600 psig. Water containing 189 ppm total solids permeated the fiber walls and flowed from the open fiber ends at a rate of 0.20 gram per hour. Using the appropriate calculations as described above, the PC in this case is 3.8×10^{-14} cm/sec.

Example 3: A bundle of 200 hollow fibers spun according to the procedure in Example 1, about 48 cm long and each having an outside diameter of 63 microns and an inside diameter of 32 microns, was assembled in a copper tube test cell as in Example 2, except that the sulfolane in the fibers was not leached from the fibers and the fibers were not dried before potting. In this case the bundle was potted before leaching. The sulfolane was leached with water at 25°C after the epoxy resin had cured and the bundle was kept wet.

A reverse osmosis test was conducted under the same conditions as Example 2. Water containing 132 ppm dissolved solids permeated the fiber walls and was collected from the open fiber ends at the rate of 0.42 gram per hour. The PC in this case was 7.8×10^{-14} centimeters per second.

Thermally Gelled Membranes

According to a process described by *B.M. Riggleman, M.E. Cohen and M.A. Grable; U.S. Patent 3,745,202; July 10, 1973; assigned to the U.S. Secretary of the Interior* membranes are developed by changing the temperature of a melt of the membrane material. These thermally gelled asymmetric membranes are prepared in two steps. First, a porous thermally gelled substrate is prepared. This is followed by the formation of a dense semipermeable skin on one side of the substrate.

Initially, the membrane material is mixed with plasticizers at a temperature of about 100° to 200°C to form a melt. Typically this melt will contain from about 15 to 75 weight percent polymer depending on the porosity and strength desired. It is significant that even at polymer concentrations as low as 25%, the hollow fibers produced can withstand external pressures of up to 1,000 psi without collapsing. Suitable plasticizers which may be used alone or in combination with one another include benzyl acetate, triethylene glycol, ethylene glycol, diethylene glycol, tetraethylene glycol, polyethylene glycol, propylene glycol and dipropylene glycol.

The melt is then formed into a hollow fiber. Although the melt could be formed into flat sheets or any other appropriate shape, the design of hollow fibers is preferred because it allows the largest possible membrane surface area per volume of equipment.

Hollow fibers are spun from the melt by pumping or in some other suitable manner pushing the melt through an orifice designed to shape the melt into a hollow fiber. This can be achieved by having a needle extending into the circular opening through which the melt is forced. By pumping a gas through this needle, the drawn fibers will be hollowed. The specific device used in the experiments is illustrated and described in Office of Saline Water Research and Development Progress Report No. 518 entitled "Second Report on Development of Improved Cellulose Acetate Membranes for Reverse Osmosis," on pages 62 to 65.

Typically, the outside diameter of the hollow fiber will be from about 15 to 41 mils with the wall thickness being approximately one-third the outside diameter. These dimensions are subject to variation, however, since in the next step, the hollow fiber is drawn by stretching to the proper size. Depending on the draw ratio, the hollow fibers may have a final outside diameter as small as from 3 to 10 mils up to the original diameter of the fibers. The wall thickness will continue to be about one-third of the outside diameter.

The drawn fiber is then gelled by exposure to either air or water to cool the melt and consequently harden the fiber. Gelation may be characterized as any such cooling technique which transforms the melt into a more firm structure. Typically, this treatment is performed for less than 1 second, although the duration may be as much as several days. The temperature of the bath is usually room temperature as is the final temperature of the fiber. A dilute aqueous solution containing about 10% or less of the melt plasticizer may also be used as a gelation bath. The use of aqueous solutions may be particularly useful in tailoring the properties of the thermal gel. For example, incorporation of plasticizer in the gelation bath produces a more permeable gel.

The product of these steps is a porous hollow fiber which is suitable for supporting a semipermeable membrane but has no selectivity of its own. There are problems, however, in trying to forming a composite asymmetric membrane by coating these fibers with a dense layer of polymer. These dense polymers tend to have low flux properties. Furthermore, there are physical difficulties in attempting to cast a membrane on the porous surface of the fibers; polymer solution may seep into the pores clogging the substrate or the membrane may be unsupported at spots, tending to rupture under the force of osmotic pressure.

It has been found that the porous thermally gelled hollow fiber may be treated to transform the surface into a membrane skin, thereby avoiding all the disadvantages of attempting to apply a separate membrane to the porous substrate.

Since the structure of a thermal gel is induced by heat transfer rather than mass transfer, its porosity is expected to be relatively homogeneous in contrast to that of Loeb type membranes in which the rapid evolution of a volatile solvent from the surface of the membrane during casting forms a latent skin. Examination of thermally gelled fibers by electronmicroscopy reveals, however, that the pore size on the air exposed surface of the thermally gelled fibers is smaller than that in the substructure. Although plasticizers are nonvolatile, apparently during spinning some removal of plasticizer occurs, affecting either a reduction in the pore size at the membrane surface or surface tension forces which promote skin formation.

Although the thermally gelled hollow fibers are not selective, since a gradation in porosity exists, it should be possible to tighten the latent skin by appropriate treatment without fully densifying the porous substructure. A technique was developed to form a membrane skin on the support surface without impairing the qualities of the porous fibers. The steps involved are as follows.

After the gelation of the hollow fiber, it is leached in water to remove any remaining plasticizer. Since the gelation may be carried out in water, a separate leaching step may not be necessary in all situations, i.e., water gelation will simultaneously effect leaching. Generally, however, a treatment in water at ambient temperature for at least about 1 hour should be effective in removing the plasticizer.

Subsequently the hollow fiber is exposed to hot water at a temperature of from 40° to 100°C for from 2 seconds up to 20 minutes. Typically, this treatment is conducted for 10 minutes at 60°C. Although the fibers do not exhibit selectively after this procedure, hot water treatment significantly increases their crystallinity and permeability.

Finally, the fibers are exposed to a hot solution of p-dioxane, formic acid, nitromethane or any other material having a high solvent action for the polymer used. The concentration of the solution is generally less than 50 weight percent of the active ingredient and preferably from 5 to 20 weight percent. The treatment time ranges from less than 1 second to 10 minutes and the temperature of the solution is from 40° to 100°C. Typically, a treatment of about 1 minute in a 10% solution at 65°C is quite effective. This last treatment develops a thin selective membrane skin. Membranes produced by this method are believed to be asymmetric because of their dye staining characteristics, their permeability and salt selectivity (permselectivity).

While the exact mechanism for the formation of this permselective skin is unknown, several factors in the manufacture of the hollow fibers have been isolated which favor the formation of a porosity gradient and the subsequent formation of the skin. These favorable factors are as follows: (1) plasticizers of greater solvent action (kinetically or thermodynamically) toward the polymer and those of increased volatility; (2) hotter spinning temperatures; (3) increased temperature difference between the melt and the gelation bath; (4) increased velocity of air past the membrane during spinning; and (5) water gelation rather than air gelation. Optimization and control of these variables can enable one to vary the pore size of the membrane for the particular application.

Example 1: In connection with this example, see the accompanying table on the following page. In each test given in the table, a melt was prepared consisting of 30 weight percent cellulose acetate and 70 weight percent plasticizer. In series A tests the plasticizer was triethylene glycol, in series B tests it was a 3/1 mixture of triethylene glycol and methyl Carbitol, and in series C tests the plasticizer was a 1/1 mixture of triethylene glycol and methyl Carbitol. The melt was spun into hollow fibers of the dimensions given in the table, and without further drawing were gelled in water at 25°C.

Tests A-1, B-1, and C-1 show the properties of the hollow fiber tubes at this time. All tests of the osmotic properties throughout this example were conducted at an applied pressure of 600 psig on the saline solutions indicated.

Tests A-2, B-2, and C-2 represent hollow fibers which having been prepared as mentioned were then further treated by leaching in water for about 1 hour at ambient temperature. Subsequently the fibers were heat treated in water at 60°C for 10 minutes. The examples show that this further treatment increased the flux of the fibers but did not impart any selectivity.

Tests A-3, B-3, C-3, and C-4 represent experiments on fibers which have been treated by all the steps indicated above. Thus, having been heat treated in water some of the membranes were further treated in a 10% aqueous solution of p-dioxane at 66°C for 1 minute. The osmotic properties indicate that the hollow fibers now have selectivity and in three of the four cases the salt rejection is greater than 91%. Although the membranes are less permeable than the nonselective fibers, the flux of the membranes might be further improved by drawing the spun fibers into thinner wall membranes. Reducing the thickness of the fiber would lessen the resistance of the membrane to the passage of water.

Test No.	Plasticizer composition	Fiber dimensions (mils)			Heat treatment (water)	Solvent treatment [1]	Feed solution (wt. percent NaCl)	Water flux (g.f.d.)	NaCl rejection percent
		OD	ID	Wall					
A-1	TEG [2]	15	5	5	No	No	3.5	3.6	0
A-2	TEG	15	5	5	Yes	No	3.5	5.2	0
A-3	TEG	15	5	5	Yes	Yes	3.5	4.5	16.3
B-1	TEG/MeC,[3] 3/1	15	5	5	No	No	3.5	8.1	0
B-2	TEG/MeC, 3/1	17.5	6.0	5.25	Yes	No	0.5	14.2	0
B-3	TEG/MeC, 3/1	15	5	5	Yes	Yes	3.5	1.0	91.9
C-1	TEG/MeC, 1/1	15	5	5	No	No	3.5	6.9	0
C-2	TEG/MeC, 1/1	15	5	5	Yes	No	3.5	10.6	0
C-3	TEG/MeC, 1/1	15	5	5	Yes	Yes	3.5	0.45	96.1
C-4	TEG/MeC, 1/1	15	5	5	Yes	Yes	0.5	5.6	96.0

[1] Exposure to a 10% aqueous solution of p-dioxane at 66° C. for 1 minute.
[2] TEG=Triethylene glycol.
[3] MeC=Methyl Carbitol (diethylene glycol monomethyl ether).

Example 2: A melt was prepared consisting of 30% cellulose acetate and 70% of a 3/1 mixture of triethylene glycol-methyl Carbitol. By the method of Example 1 this melt was spun and gelled into hollow fibers with an outside diameter of 20 mils. The fibers were treated in deionized water at 60°C for 10 minutes followed by annealing in a 10% aqueous solution of nitromethane at 70°C for 10 seconds.

The resulting membranes were stained by exposure to a 0.1 weight percent aqueous solution of Acridine Red for 1 minute. The membrane surface showed a very light stain, comparable to that of a commercial Loeb type membrane capable of at least 90% salt rejection. After slight scraping of the surface, an intense stain resulted by the same procedure, illustrating that the selective portion of the membrane was very thin and that the structure was asymmetric.

Hollow Fibers Supported at Terminal Portions

H.I. Mahon; U.S. Patent 3,228,877 and U.S. Patent 3,228,876; both dated January 11, 1966; both assigned to The Dow Chemical Company describes a permeability cell in which the multicomponent fluid composition or solution is passed around the exterior wall surfaces of a multiplicity of fine hollow fibers, supported only at their terminal portions. Where the process is applied to reverse osmosis, the fluid on the exterior of the hollow fibers is maintained at a pressure in excess of the specific osmotic pressure.

Referring to Figure 2.7a, there is shown the overall structure of one example of the basis permeability cell of this process. This is generally designated, in its entirety, by reference number **10**. The cell **10** includes an inlet **11** for untreated fluid having, where reverse osmosis is utilized, means such as a pump to apply a desired fluid pressure to fluid entering this inlet, and an outlet **12** for the discharge of the treated fluid, this outlet being positioned on the opposite side of the cell casing from the inlet, it being noted that there may be more than one outlet, spaced along the axis of the casing. Additional outlets **13, 13** are formed for discharge of the fluid component after it has passed through the permeable membranes of the cell.

At the exit end the outlets **13, 13** are fed by header chambers **14, 15**. Situated between these headers and supporting them is a casing **16**. Advantageously, this casing is tubular, although it may have other desired configuration. The casing is sealed at its end from the headers by means of header end plates **17** and **18** (**18** not shown). Outlets **13** provide an exit for fluid accumulated within the interior of the hollow fibers, while outlet **12** carries off the treated fluids. Positioned within the casing are a multiplicity of bundles **19** of hollow, fine fibers.

As is shown in Figures 2.7c and 2.7d, each of the bundles has its terminal portions **21, 22** sealed to each other and to the walls of openings or passages in plates **17** and **18** by means of a suitable adhesive or cement, resin or the like, indicated generally at **23**.

FIGURE 2.7: TUBULAR UNIT UTILIZING HOLLOW FIBERS

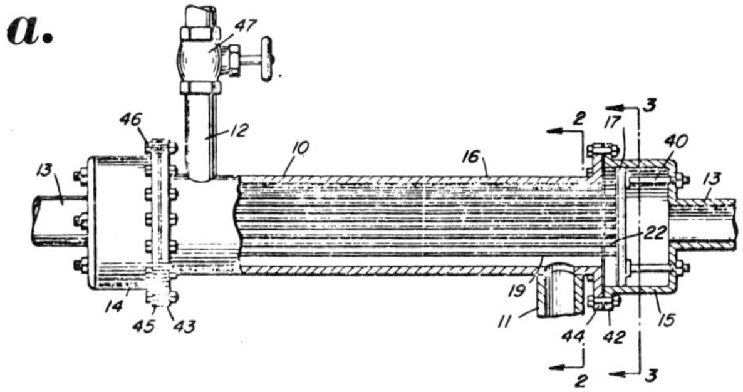

Overall View of Permeation Cell

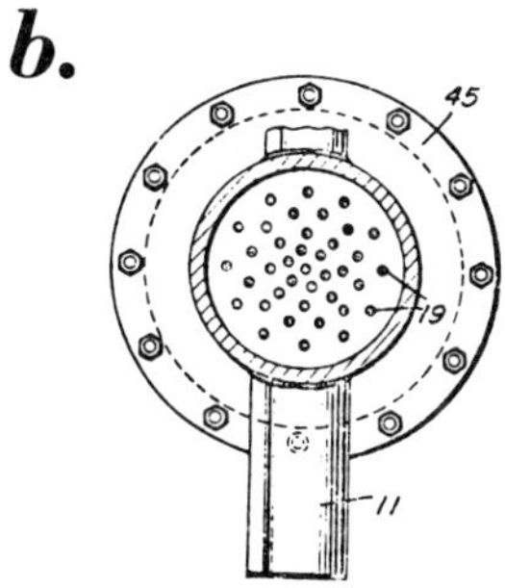

Cross-Section Along Line **2-2**

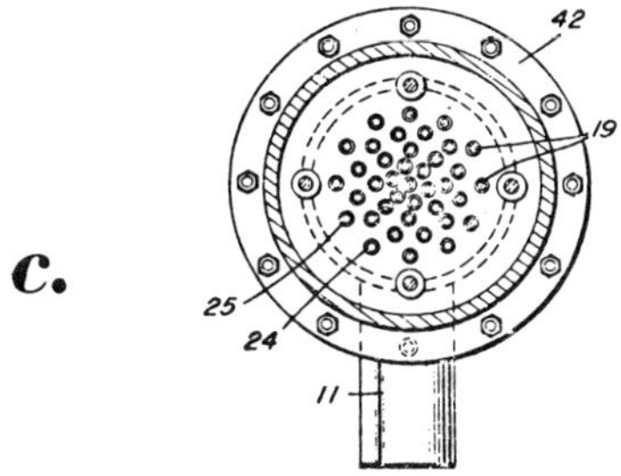

Cross Sectional View Along Line **3-3**

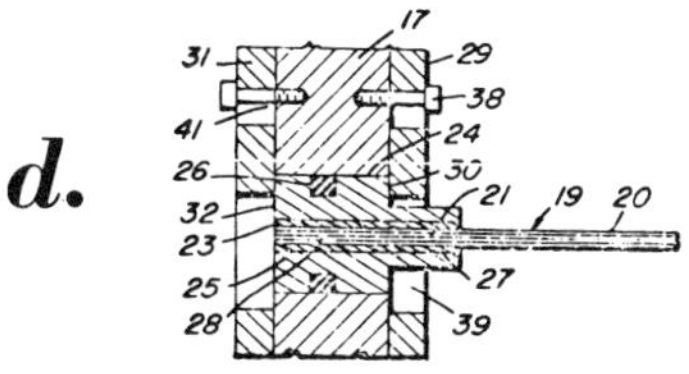

End Seal Unit for Hollow Fiber Bundle

Source: H.I. Mahon; U.S. Patent 3,228,877; January 11, 1966

These openings or apertures are designated **24**. The adhesive seals the exterior terminal portions only of the fibers to each other and to the interior walls of the apertures. The bundles are thus supported at their terminal portions only by the header plates. In this way, the opposite open ends of the many small hollow fibers are all in fluid communication with the header chambers **14** and **15**.

Each of the hollow fibers **20** in each bundle has the same or about the same length. Likewise, it is generally desirable for the other dimensioned parameters of all the hollow fibers in each bundle, (i.e., outside diameter and wall thickness) to be the same in each fiber although, if desired, bundles of fibers having different individual outside diameters and wall thicknesses may be employed. Similarly, each of the bundles advantageously has the same dimensional parameters as the other bundles; i.e., the same length and outside diameters, although the latter of these can, if desired, also be varied in individual bundles.

Thus it will be seen that each of the fibers has a uniform bore. There is thus formed a permeability cell that includes a bundles multiplicity of parallel, fine, hollow fibers, each of which is supported only at its terminal portions, these terminal portions being in substantial contiguous relation to each other, as are the remainders of the fiber lengths, and substantially unsupported throughout its length. Such a cell, as indicated, provides hundreds of pounds per square inch and their wall thickness is so small that they present minimum resistance to passage of a permeable fluid component.

Referring to Figure 2.7d, there is shown a preferred form of an end seal for a hollow fiber bundle. The header end plate **17** has a plurality of apertures of which one is shown at **24**. Fitted slidably within each of the apertures **24** is a coupling fitting **25**. An O-ring **26** seals the joint between the end plate **17** and coupling **25**. A bundle of hollow fibers **19** is positioned within the bore **27** of coupling **25**. The individual hollow fibers **20**, shown on exaggerated scale, in the bundle are sealed on their exterior terminal portions **28** only to each other and to the interior of bore **27** by means of a suitable cementitious material **23**, shown on exaggerated scale for purposes of illustration. Epoxy resin compositions are especially well suited for such purpose.

A sliding gate lock plate **29** having a plurality of fitting apertures **39** is slidably positioned parallel to the end plate **17** within header **14**. The gate lock plate is shown in locked position where it engages a projection or shoulder **30** on coupling **25**. A similar slidable gate lock **31** is preferably positioned within header chamber **14** and on the side of header end plate **17** opposite the sliding gate lock plate. Gate lock **31** coacts with the end **32** of the fitting.

Gate lock plates **29** and **31** are slidingly secured to header end plate **17** by headed studs **38** which extend through suitable slots **41** and are fixedly threaded into end plate **17** as shown in Figure 2.7d. The heads of the studs engage the lock plates at each of the slot sides to secure the plates in position. Preferably the couplings **25** are inserted in apertures **24** and locked by gate lock plates **29** and **31** during assembly of the permeability cell, although it is contemplated that operating means could extend outside of the cell through suitable seals so that the locks could be actuated externally of the cell.

The lock plates **29, 31** may be formed with a cam or other means to longitudinally adjustably position the coupling **25** in the end plates **17, 18**. Or the entire end plate may be longitudinally adjustably positioned by any suitable means, such as positioning screws **40**, shown in Figure 2.7a. In this way, if desired, the tension upon the individual fiber bundles **19** may be varied although it is generally preferable to avoid application of positive drawing force on the fibers in the bundle in order to avoid difficulties from fiber breaks.

It will be understood that there is a similar aperture **24** and a similar coupling **25** for each of the hollow fiber bundles employed, and that the end seal and support structure is duplicated at the other header **15**. This preferred form presents a simple means for supporting and sealably positioning the individual bundles of hollow fibers. Such an arrangement allows each bundle to be removed individually in the event of rupture or other damage to one of the bundles. Each bundle of fibers together with its cementitiously terminally adjoined couplings **25, 25** in fluid-sealing relation forms a unitary permeability separator element. Such an element is described in U.S. Patent 3,228,876.

Each of the header chambers **14** and **15** are formed with external circumferential flanges indicated at **42** and **43**. These flanges coact in sealing relation, when the overall unit is assembled, with mating flanges **44** and **45** positioned on the exterior of casing **16**. The flanges are held in sealing relation by means of bolts **46**. The outlet **12** is desirably provided with a suitable valve **47**. This valve may be utilized to control the rate of withdrawal of treated fluid from the apparatus in order to regulate the concentration of the solution within the casing.

In assembling the apparatus, as thus far described, the individual bundles are sealed, as previously delineated, to their individual end fittings **25**, and these end fittings are assembled within the apertures in one of the end plates **17** or **18**. This end plate is then positioned within its appropriate header chamber in sealing relation, and the bundles with their end portions **25** are led through chamber **16**. With the end plates held toward the casing within the header chambers, the opposite ends of the bundles are positioned within the apertures of the other of the end plates.

The sliding gate locks are reciprocated to engage the shoulders on the fittings to lock them to the end plates. Subsequently, the header chambers are sealed to the casing by means of the above described flanges and bolts and, if desired, final adjustment of the tensioning of the bundles is accomplished by means of the positioning screws **40** or their counterpart. The permeability cells of the process, when utilized in pluralities in any given separating installation, may be connected in parallel relation or may be serially connected to further purify the separated component passed sequentially through the units. In such a serial battery, the outlets **13, 13** of the first cell are connected to the inlet **11** of the second cell, etc.

There is shown in Figure 2.8 a preferred form of a seawater conversion plant incorporating one or more of the cells described above used in the process. In the schematic flow diagram of Figure 2.8, the numeral **33** indicates a pretreatment plant of any suitable construction. Such pretreatment plants are usually comprised of various settling tanks, other clarifiers, filters, bacterial treatment units, etc. The raw seawater is fed to the pretreatment plant **33** and the treated effluent from the plant is led to a first stage pump **34**. The first stage pump is preferably a low pressure unit connected directly mechanically to the output shaft of an impulse turbine **35** which is motivated by the energy from the moving stream of waste brine solution after its travel through the permeability apparatus.

FIGURE 2.8: SEAWATER CONVERSION PLANT

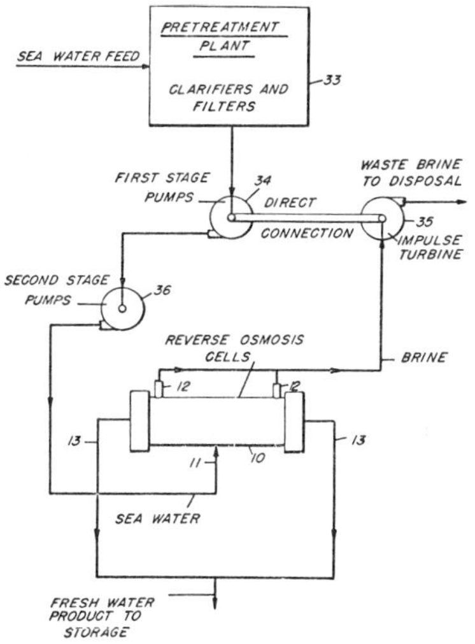

Source: H.I. Mahon; U.S. Patent 3,228,877; January 11, 1966

From the low pressure pump **34**, the saline solution is fed to a second stage high pressure pump **36** which is fluid connected to the output of pump **34**, driven by an electric motor or other suitable means. The saline solution under high pressure from the output second stage pump is fluid connected to the inlet side **11** of a permeability cell **10** or to the inlets **11** of a battery of parallel connected cells **10** similar to those described above. Fresh water is discharged at the outlets **13, 13** and, if desired, is further cycled through a battery of additional, serially connected permeability cells. The treated saline solution is led from the outlets **12, 12** of the cell or cells to drive turbine **35**. The waste brine solution from the turbine is led to a disposal area which can be the ocean or other source of saline water.

In a typical seawater conversion plant patterned in accordance with the process along the lines above illustrated, an output of 10 million gallons per day of fresh water contained in the seawater feed can readily be achieved. To attain this output, approximately 16,000 gallons per minute of seawater containing about 3.5% dissolved salt are fed to the pretreatment plant. This solution, after passing through the pumps, is fed to the inlet of the permeability cells at a pressure of about 835 pounds per square inch.

Brackish or incompletely desalted water, from which 95% of the initial salt content has been removed, is withdrawn from the parallel connected cells at a rate of approximately 7,000 gallons per minute. The remaining treated brine issuing from the cells at the rate of about 9,000 gallons per minute and at a pressure of about 700 pounds per square inch is used to drive the impulse turbine. This remainder is then led to the brine disposal outlet. If desired, the waste brine solution may be used as the starting material in an evaporative plant for saline solutions to further recover fresh water and to recover the salt content.

Taking into account the amortized cost of the permeability cells and other components of this system, as well as the energy losses due to the pumping system and to the conversion of mechanical energy into thermal energy resulting from membrane forced flow and from friction, such a plant can produce fresh water from seawater on an economically attractive basis which is substantially less than can be afforded by previously known procedures.

Where the permeability apparatus is to be utilized in the treatment of aqueous fluid compositions, particularly in procedures pertaining to osmosis, reverse osmosis, or dialysis, a variety of materials capable of formation into hollow fibers are suitable for use as membrane materials. Such materials are generally hydrophilic in character and include cellulose esters such as cellulose acetate, triacetate, formate, propionate, nitrate or the like, including mono-, di- or tri- esters and mixtures of such esters; cellulose ethers such as methyl, ethyl, hydroxy-alkyl, carboxy-alkyl, or the like, including mixed cellulose ethers; regenerated cellulose ethers, regenerated cellulose, polyvinyl alcohols, polysaccharides, casein and its derivatives, and similar hydrophilic polymeric materials such as certain forms of acrylonitrile polymers.

Certain of these materials are more readily formed into fine, hollow fibers than others. A highly advantageous material for conversion into hollow fibers is cellulose triacetate. Cellulose triacetate in hollow fiber form has excellent qualities of permeation, cost, strength, and resistance to chemical or solvent action. Furthermore, fiber formation is relatively easy with cellulose triacetate.

Fine, continuously hollow fibers of the type suitable for use in the process can be prepared, depending upon the fiber-forming material employed, by melt, dry (or evaporative) and wet spinning procedures using spinnerets of a design that is apt for the purpose. Such fibers may be made in accordance with the teachings of British Patent 514,638. For example, fine hollow fibers of cellulose triacetate are readily produced by a wet spinning process in which the cellulose triacetate is first dissolved in a suitable solvent to form a viscous spinning solution, after which the spinning solution is extruded into a coagulant bath through the annular space in a small annular orifice (or spinneret).

The extruded fiber coagulates or gels in the desired form of a continuously hollow, uniform-walled filament. Prior to use, the cellulose triacetate hollow fiber is washed free of solvent and of any soluble core material, if used, in a suitable bath for the purpose, advantageously one that is aqueous in nature or which consists of water. It is usually expedient to take up the hollow fibers on a reel or other suitable means for collection prior to their assembly in the cells.

In general, the uniform wall thickness of the fibers should be held to the limits of between about one-third to one-eighth of the outside diameter of the hollow fiber. Also, in general, the hollow fibers should be able to withstand an external pressure of at least about 100 pounds per square inch without collapsing. In accordance with the previously noted parameters of rate transfer, the maximum outside diameter of the hollow fibers should, advantageously, not exceed 300 microns. In fact, a more advantageous range of outside diameter is from about 10 to 50 microns.

The uniform wall thickness of the hollow fibers may range from a minimum of about 1 micron to a maximum of about 80 microns. Decrease or increase from the specified limits will lead to an increased chance of collapsing on the minimum side and to a decrease in mass transfer rate on the maximum side.

The most advantageous range of wall thickness of the hollow fibers for most purposes is from about 2 to 20 microns, with a generally suitable and desirable working range of from about 1 to 50 microns. Regardless of the specific dimensional parameters of the fibers involved in any hollow fiber bundle used in the process, the transfer area of the bundles (or of a plurality of bundles assembled in cells) may be reckoned in ways analogous to that illustrated in the following calculations, using for purposes of illustration hollow fibers having an outside diameter of 13 microns, an inside diameter of 9 microns, a wall thickness of 2 microns, a length of 1 foot and a cylindrical bundle size of 1 inch diameter.

Transfer Area	Installation A	Installation B
Fiber spacing in bundle	2 x outside diameter	1.5 x outside diameter
Number of fibers in bundle	8.65×10^5	1.54×10^6
Transfer area per bundle	80 square feet	142 square feet
Bundle spacing in cell	2 x outside diameter	1.5 x outside diameter
Total transfer area per unit volume of cell	3,300 square feet per cubic foot	10,500 square feet per cubic foot

As has been indicated in the foregoing, the total transfer area per unit volume available from a flat membrane assembly (assuming a 1-millimeter unsupported membrane spacing) is about 305 square feet per cubic foot.

Examples 1 through 4: The system and separatory procedure of the process is successfully applicable to a wide range of separating operations. A few such applications actually accomplished are illustrated in the following tabulation. In each instance, for purposes of uniform operation and to aid in comparison, the tests were conducted using a single 350 fiber bundle of cellulose triacetate fibers, each of which had an outside diameter of about 45 microns and a wall thickness of about 10 microns. The yield of permeate obtained in each of the listed illustrations is that obtained from the entire fiber bundle. Permeation tests were also conducted on aqueous sodium chloride solutions using both a bundle of relatively large, thick walled fibers having a relatively low permeation rate (Example 3) and a larger bundle of small, thinner walled fibers (Example 4).

Example 1

Material	Apple cider
Osmotic pressure	380 to 400 psi
Applied pressure	425 psi
Permeation rate	0.04 cc/hr
Permeate	White, negative for sugar, refractive index-distilled water

Example 2

Material	Coffee
Osmotic pressure	90 psi
Applied pressure	210 psi
Permeation rate	0.06 cc/hr
Permeate	Substantially only water; no odor; no residue on evaporation

Example 3

Number of fibers	9
Outside diameter	290 microns
Inside diameter	150 microns
Length	8 inches
Feed solution	1% NaCl
Osmotic pressure	100 psi

(continued)

Example 3 (continued)

Applied pressure	300 psi
Permeation rate	0.0015 cc/hr
Permeate	97% rejection of NaCl

Example 4

Number of fibers	1,500
Outside diameter	48 microns
Inside diameter	32 microns
Length	30 centimeters
Feed solution	1% NaCl
Osmotic pressure	100 psi
Applied pressure	520 psi
Permeation rate	1.28 cc/hr
Permeate	97% rejection of NaCl

Permeability Cell

A process described by *N.S. Strand; U.S. Patent 3,342,729; September 19, 1967; assigned to The Dow Chemical Co.* takes advantage of the efficacious features of hollow fiber membranes by means of a highly efficient arrangement of the fibers in the form of a separatory cell. This cell is comprised of a relatively flat mesh-like membrane of a plurality of woven or interlaced line, filamentary hollow fibers, much like a fabric. The network usually involves a plurality of the fibers running essentially perpendicular to a plurality of others. The edges of the mesh membrane are sealed in a frame with the ends of the fibers, which are open, terminating in openings around the side of the frame and separated from the central opening.

One fluid can be passed through the central opening and over the outside of the fibers. Usually, a second fluid is then passed through the openings around the side of the frame, which fluid traverses through the hollow cores of the fibers and the required material transfer takes place through the hollow fiber membrane wall. The material transfer, or more specifically, permeation, can be made to proceed from outside to inside, or inside to outside of the fibers, depending on the system. Moreover, it is not always necessary to have two fluids since in some cases, permeation can be made to occur from one side to another without relying on the driving permeation force developed between two fluids.

Figure 2.9a illustrates a relatively flat mesh separatory membrane, generally referred to by reference numeral **10**, of hollow fibers **11**. Usually, this web or mesh membrane is only one layer thick of hollow fibers, although if desired, each of the ribs of the web, which are represented by an individual hollow fiber, can be a small bundle of hollow fibers. Advantageously, the mesh membrane is made up by interweaving or interlacing the hollow fibers much on the order of a woven fabric. This provides a relatively dimensionally stable membrane, holding the fibers in place.

In Figure 2.9b is shown an exploded view of a fragmentary cross section of a woven mesh membrane like that illustrated in Figure 2.9a. Thus, as shown, fibers **11** are interlaced between one another. Described in textile terms, those fibers in one direction being called the warp and those fibers in a general perpendicular direction to the warp being called the weft. Other means for laying up the webs can be employed, including hand lacing or the like. Each of the fibers have a wall **12** and a continuously hollow bore **13** capable of having a fluid pass through. The actual separation of components actually is facilitated by the wall which is of a selectively permeable material such as a normally solid synthetic polymeric material that can be extruded into a fine, hollow fiber shape.

Hollow fibers suitable for the process can be made by techniques known in the art, as shown for example in British Patent 514,638. Advantageously, the hollow fibers can be made by the method described in U.S. Patent 3,423,491. In general, such fibers are spun by melt, dry or wet spinning techniques depending upon the particular fiber-forming materials being used.

FIGURE 2.9: PERMEABILITY SEPARATING CELL

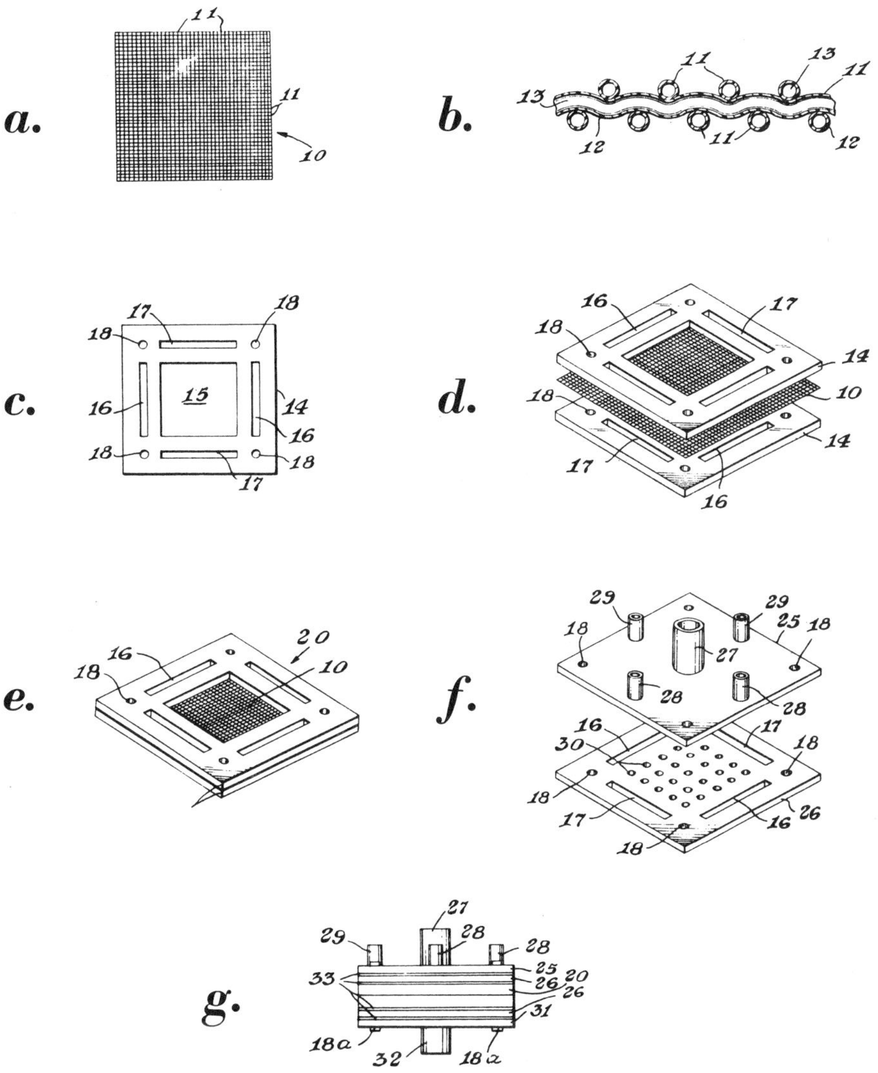

(a) (b) Mesh Membrane Detail
(c)-(g) Fabrication Details

(continued)

FIGURE 2.9: (continued)

h.

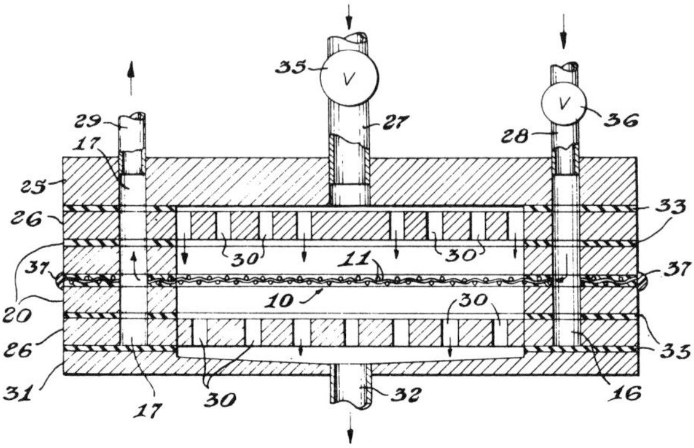

i.

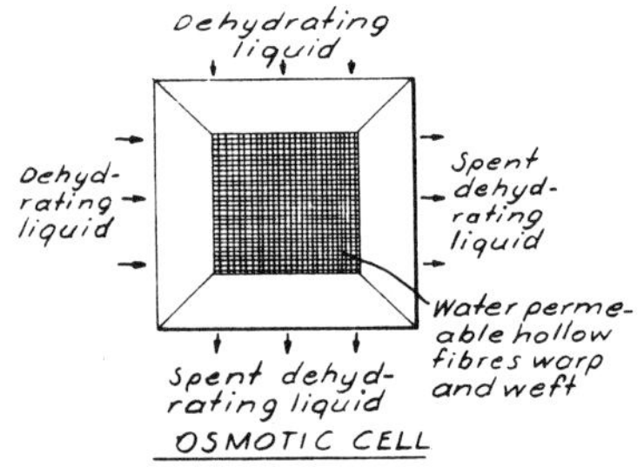

j.

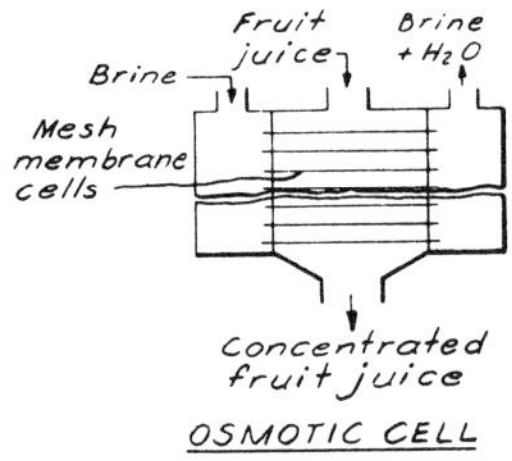

k.

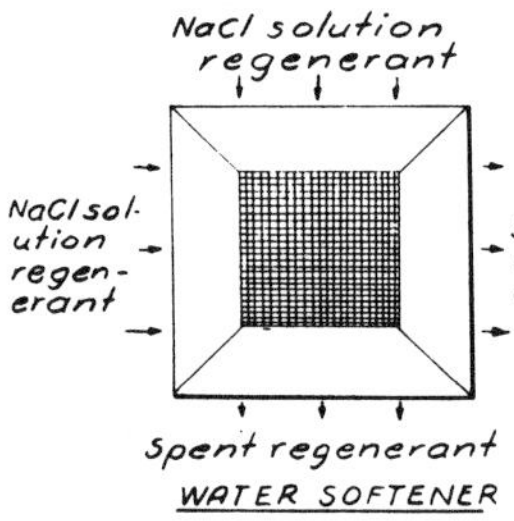

l.

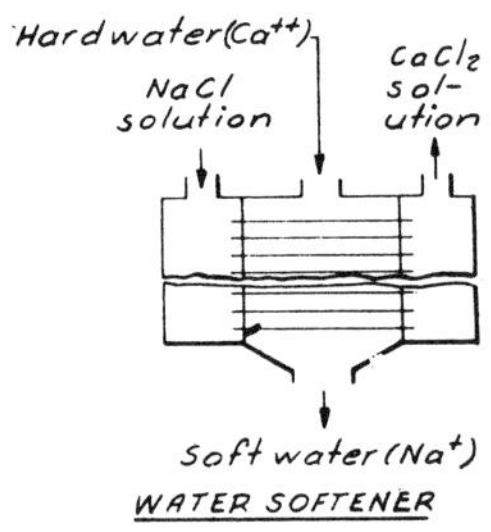

(h) Separatory Apparatus
(i)(j) Osmotic Cell
(k)(l) Water Softener

(continued)

FIGURE 2.9: (continued)

m.

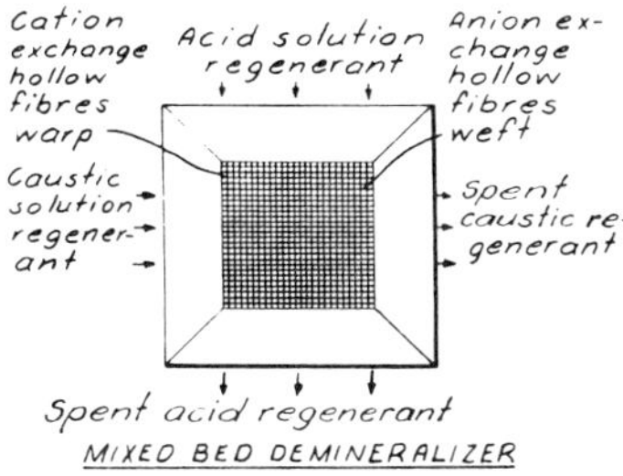

n.

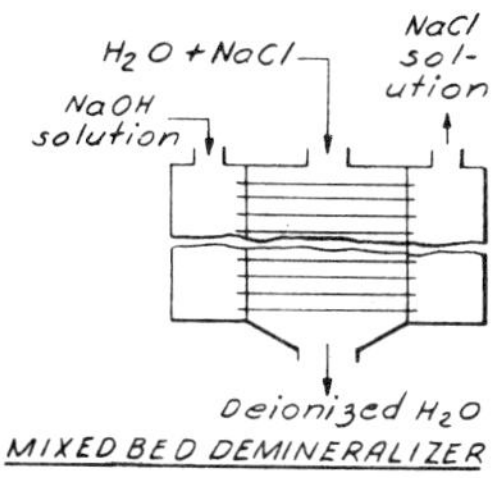

(m)(n) Demineralizer

Source: N.S. Strand; U.S. Patent 3,342,729; September 19, 1967

The spinneret is selected according to the type of spinning procedure used and the particular dimensions desired in the hollow fiber. For the production of the hollow fiber, the spinneret has a small annular opening in the orifice through which the spinning composition is extruded.

The hollow fibers employed in mesh membrane are relatively fine filamentary structures. The wall thickness of the fibers is desirably sufficient to withstand the pressure that will be exerted in the subsequent permeability separation utilization of these fibers. Generally, a capability of withstanding pressures of 100 psi or more is desired. It is found that the small diameters of these fine, hollow fibers permit the self-supporting membrane walls of the fiber to withstand considerable pressures.

It is generally preferred that the outside diameter of the hollow fibers does not exceed 350 microns, advantageously no more than 300 microns. Preferably the outside diameters are in the range of about 10 to 50 microns. Advantageously, a wall thickness to outside diameter ratio of about 1/8 to 1/3 is employed in the hollow fibers. Profitably, the wall thickness of the fibers is in the range of about 1 to 80 microns, preferably from about 2 to 15 microns. Wall thicknesses below this range may result in an inability to withstand the desired pressures, whereas thicknesses above this range increase the resistance to permeation through the fiber wall.

In Figure 2.9c there is shown a plan view of one member of a frame **14** suitably designed to be combined with mesh membrane **10** of Figure 2.9a. This frame member is provided with a central opening **15**, oppositely spaced peripheral openings **16** and oppositely spaced peripheral openings **17**, and bolt holes **18**.

In Figure 2.9d mesh membrane **10** is shown about to be sandwiched between a pair of frame members **14**, and in Figure 2.9e a cell, generally referred to by reference numeral **20**, is perspectively illustrated, which is the product of combining the illustrated elements of Figure 2.9d. When this particular means of fabricating cell **20** is employed, the mesh membrane **20** is generally made slightly larger than frame **14**. The sides of frames **14** that are to be sealed to membrane **20** are coated with a suitable adhesive material, such as an epoxy resin, that will securely and permanently bind the members together as well as seal the areas between fibers **11** of membrane **10** so that there is no leakage between the peripheral openings **16, 17** and the central opening **15**.

Membrane **10** is then sandwiched between frame members **14** in a manner such that the central, peripheral and bolt hole openings of the frame members coincide. After sufficient drying or setting, those portions of mesh membrane **10** spanning peripheral openings **16** and **17** are then trimmed out leaving a clear passage through the openings. The excess fibers of the membrane extending past the exterior edge **21** of the frame are trimmed off and the ends of the fibers at the exterior edge are sealed with a suitable sealant such as an epoxy resin.

Frame **16** can be made of any suitable material that is relatively inert to the fluid materials to be processed in the cell. In order to take full advantage of the available fiber surface in the mesh membrane, however, at least four peripheral openings, i.e., two pairs of oppositely spaced openings should be employed no matter what the frame configuration. This arrangement permits delivering a fluid to the central bore of the fibers from at least two directions and removal of the fluid from the oppositely spaced exits.

The reasons for this are apparent when consideration is taken of the fact that the fibers in the mesh membrane form a criss-cross angular pattern, and usually half of the fibers are perpendicular to the remaining half in the plane of the mesh membrane. The peripheral openings need not necessarily be in the top and bottom of the frame. Equally satisfactory results can be obtained if the openings are along the edge of the frame coupled with suitable passage openings in a separatory apparatus.

A particularly advantageous and beneficial means for fabricating the cell **20** is to first extrude or cast the desired frame member from a suitable thermoplastic polymeric material. A mesh membrane can be sandwiched between two such members and the assembly subjected to heat sealing conditions whereby a unitary, integral cell member is provided. This means has the added feature of readily and securely bonding the members into an intimate joined relationship, but additionally avoids the need for any adhesive and sealant material and the attendant setting or drying time. Means can also be provided to simultaneously heat seal the ends of any fibers protruding beyond the outer edge of the joinder of the two frames by causing the material of the frame to flow over the joinder forming a smooth surfaced seamless edge.

Care must be exercised that the hollow fibers are not materially altered in any portion where flow therethrough is desired. The frames can be made in pairs with mating male and female fittings such as lugs and indents to facilitate and assure alignment of the various matching openings. Rapid production of the cells can thus be achieved.

Cell **20** can be assembled into any suitable separatory apparatus adapted to take advantage of the two-way flow through the central bore of the fibers of the membrane and/or flow over the exterior of the fibers. One such apparatus is assembled with the aid of the members illustrated in Figure 2.9f. Thus, as perspectively illustrated, a header plate **25** and distributor plate **26** can be employed in a separatory apparatus assembly.

The header plate is provided with a central part or passage **27** which is so adapted and arranged as to be suitably coupled with means for supplying a fluid thereto and further to guide the fluid to the central opening of a cell like that illustrated in Figure 2.9e. Entrance passages **28** are so adapted and arranged as to be suitably coupled to means for supplying a fluid and further to guide the fluid to the peripheral openings of a cell like that illustrated in Figure 2.9e.

Exit passages **29** oppositely spaced from entrance passages **28** are adapted and arranged so as to receive the fluid that has traversed the entrance passages and the core of the hollow fiber of the mesh membrane and/or the permeate from the outside of the hollow fibers through the walls of the hollow fibers. Of course, it is not required or necessary to flow a fluid through the hollow core of the fibers, in which case the hollow cores would serve only to collect the permeate; then entrance passages would actually be employed as exit passages. Bolt holes **18** in header plate **25** are designed to mate with the bolt holes in frame member **14**.

Distributor plate **26** has a pair of openings **16** and a pair of peripheral openings **17** designed to mate with like numbered openings in frame member **14**, bolt holes **18** designed to mate with bolt holes **18** in frame member **14** and header plate **25**. When a fluid flows through entrance passage **27** in the header plate, it becomes distributed before contact with the mesh membrane of the cell by passing through perforations or holes **30** which are generally randomly distributed to prevent channeling of the fluid through the cell.

Figure 2.9g illustrates a simplified assembled separatory apparatus. The top member is a header plate **25**, then a distributor plate **26**, a cell **20**, another distributor plate **26**, followed by a bottom plate **31** equipped with central exit passage **32** adapted to flow away fluid that has been introduced to the cell through central entrance passage **27**. The various members are held together by nuts and bolts **18a**. The members are sealed at the junctures by gaskets **33** of a suitable material such as a modified rubber.

Figure 2.9h is a more detailed view, partly in cross section, of a separatory apparatus. The assembly comprises a header plate **25** with central entrance passage **27**, equipped with valve **35**, a peripheral entrance passage **28**, equipped with valve **36**, and a peripheral exit passage **29**. The ends of hollow fibers **22** at the exterior edge **21** of the junction in cell **20** are sealed with an epoxy resin **37**. A fluid is delivered to the central passage, down over distributor **26**, through holes **30** and over mesh membrane **10** of hollow fibers **11**, through holes **30** in distributor **26** and the fluid then exits through central exit passage **32** in bottom plate **31**.

A second fluid, if desired, is flowed through the peripheral passage, through peripheral openings **16**, through the hollow core of the fibers, whose open terminal ends terminate in opening **16**, to peripheral opening **17** and out through peripheral passage **29**. Valving means can also be employed on exit passages **29** and **32** to aid in governing the fluid pressures developed. If the apparatus were rotated 90° and a cross section taken through it, a view similar to Figure 2.9g would be seen.

Figures 2.9i through 2.9n illustrate schematically and diagrammatically the operation of a separatory apparatus for some of the intended uses. Thus, in Figures 2.9i and 2.9j, hollow fibers that are permeable to water are used for concentrating a fruit juice.

A dehydrating liquid, such as a brine, is fed in two directions through the hollow cores of the fibers in a mesh membrane and the fruit juice (as illustrated in Figure 2.9j) is fed over the mesh membrane on the outside of the fibers. A diluted brine (spent dehydrating liquid) is flowed out the opposite ends of the hollow fibers and a concentrated fruit juice is recovered at the bottom of the apparatus.

In similar fashion, as illustrated in Figures 2.9k and 2.9l, a water softener is provided using hollow fibers having cation exchange properties, and as illustrated in Figures 2.9m and 2.9n, a mixed bed demineralizer is provided for deionizing water.

In order to illustrate the process, two cells were fabricated having a configuration and following the general procedure discussed. The two cells were stacked upon each other, separated by rubber gasketing material, and assembled in a separatory apparatus.

A dilute brine solution, (i.e., an aqueous solution of about 5% NaCl) was fed through two of the peripheral passages spaced about 90° from each other, which brine passed through the cores of the fibers and exited through exit passages oppositely spaced from the entrance passages. Water, containing about 200 parts per million $CaCl_2$, was percolated down through a central passage over the mesh membrane and out through a bottom central passage.

The effluent water obtained from the bottom passage was found to be significantly reduced in calcium ions and the effluent water could be made to be completely "soft" with successive cycling through the mesh membrane or by utilizing more cells in series in the separatory apparatus or sequential separatory apparatus.

Polyepoxide Potting Formulations

P.G. Schrader; U.S. Patents 3,619,459; November 9, 1971 and 3,703,962; November 28, 1972; both assigned to The Dow Chemical Company describes an improved method of bonding a thermoset resin header tube sheet member to plasticized, acetylated cellulosic hollow fibers secured therein. The process is directed to the preparation of the tube sheet by potting the plasticized fibers in a thermosettable resin formulation comprising a polyepoxide resin, from 10 to 100 parts per 100 parts of the resin (phr) of a reactive diluent, from 30 to 200 phr of a polycarboxylic anhydride, from 0.5 to 5 phr of a bond promoter, based on total formulation from 12 to 200 phr of a long chain flexibilizing agent and from 0.5 to 5% by weight of total resin solids of a thixotropic agent.

Examples 1 through 5: A simple test used for screening resins comprised a small mold made from a ⅜-inch x 1¼-inch piece of polyethylene tubing having a cork inserted in one end. The mold was set upright on the cork end and a looped bundle of sulfolane plasticized acetylated cellulosic hollow fibers were inserted, loop end, into the mold cavity. The potting resin was then poured in and cured for up to 16 hours at about 50°C.

After curing, the ends of the tubing were cut to expose open fiber ends and the exposed tube sheet surface examined under a microscope. A poor resin can be detected by a visible separation between the fiber wall and the resin. If the fibers appeared to be well bonded the potting was soaked in distilled water. After 4 hours of soaking the exposed ends were examined again under the microscope. Most of the unsatisfactory formulations are detected by the 4-hour soak and practically all by a 24-hour soak.

The screening tests were usually made without the thixotropic agent in order to check the effectiveness of the bond first. Formulations which passed the screening test were then formulated with a thixotropic agent and used to manufacture actual separatory units.

A series of formulations were prepared with maleic anhydride as the curing agent, and tested as above. The formulations contained stoichiometric amounts of maleic anhydride (i.e., 1 anhydride equivalent weight per 1 epoxide equivalent weight) and the additional components shown in the following table. Resin A is a glycidyl polyether of resorcinol having an epoxide equivalent weight of 124 to 140 and available commercially as Kopox 159. Resin B is a glycidyl polyether of bisphenol A having an epoxide equivalent weight of 186 to 192. The formulations were cured for an additional 1 to 1½ hours at 100°C.

Example	Resin	Reactive Diluent	Bond Promoter	Flexibilizer
1	A	RD-2 50 phr	TMBDA 4 phr	Epoxide G 50 phr
2	B	RD-2 50 phr	TMBDA 4 phr	Epoxide G 50 phr
3	A	RD-2 50 phr	TEDA 4 phr	Epoxide G 50 phr
4	A	RD-2 50 phr	N-MM 4 phr	PGR 33 phr
5	A	RD-2 50 phr	N-MM 4 phr	PGR 33 phr

Note: The following abbreviations are used
phr = parts per 100 parts of resin
RD-2 = diglycidyl ether of 1,4-butanediol
TMBDA = N,N,N',N'-tetramethylbutanediamine
TEDA = Triethylenediamine (Dabco)
N-MM = N-methyl morpholine
Epoxide G = diglycidyl ether of a polyglycol having an epoxide equivalent weight of 175 to 205
PGR = propylene glycol monoricinoleate

All of the above formulations, when cured, had excellent water resistance and bonding of the fiber to the resin after soaking. The addition of colloidal silica to the above formulations provides good thixotropic application characteristics.

Hollow fiber separatory elements can be prepared in a variety of ways and one method for preparing small laboratory test elements consisted essentially of winding a tow of hollow fibers spirally around a cylindrical core (about 6 inches long) in a plurality of layers and simultaneously applying the resin in a longitudinal strip during the winding operation. After the resin is cured the longitudinal strip may be cut or holes drilled in a staggered overlapping pattern to expose open fiber ends in a tube sheet.

The above resin formulations containing 2% of colloidal silica (Cab-O-Sil) were used to prepare a hollow fiber separatory element in a similar manner to that above and by drilling a pattern of holes in the longitudinal tube sheet. After leaching the sulfolane away from the fiber the resin tube sheet had an excellent bond to the fibers and the tube sheet showed excellent water resistance.

Example 6: The resin formulation of Example 2 was modified to contain butyl glycidyl ether in place of the RD-2 as a reactive resin diluent. Similar results were found in that a hard tough resin was formed with excellent water resistance and bonding of the fiber to the resin. Similar results are also found if the maleic anhydride is replaced in equivalent amounts by phthalic anhydride, chlorendic anhydride or other well-known anhydride curing agents.

In related work *P.G. Schrader; U.S. Patent 3,728,425; April 17, 1973; assigned to The Dow Chemical Company* describes an improved method for potting plasticized, acetylated cellulose hollow fibers. An improved bond between the cured resin and the fiber results when the resin comprises a polyepoxide resin, from 6 to 100 phr of an aromatic amine and from 0.5 to 5 phr of a thixotropic agent. The thixotropic agent imparts desirable application characteristics to the uncured resin and, particularly, minimizes wicking.

SPIRAL FLOW

Apparatus Utilizing Wound Membrane Module

D.T. Bray; U.S. Patent 3,504,796; April 7, 1970; assigned to Desalination Systems, Inc. describes a reverse osmosis apparatus employing a membrane pack or module wound around a perforated tube inside a pressure container, and from which container brine is released through a conduit formed by a spiral groove in the face of a plug which is maintained in contact with another surface. The spiral groove is of such length and diameter so that the conduit offers resistance to flow of brine to regulate its release from the container and to maintain liquid pressure. A plurality of two or more grooved face plugs may be employed to form an extended conduit for desired control. The perforated tube around which the membrane pack is wound may contain a water-purifying material and the outlet of such tube may be equipped with a collapsible plastic tube valve to shut off water flow if membrane rupture occurs.

Referring to Figures 2.10a through 2.10e, the purification unit **10** comprises an outer casing **12** which is pressure-resistant and may conveniently be fabricated of high-strength plastic. The right-hand or inlet end of casing **12** is closed by end piece **14** having a circumferential groove **16** into which fits O-ring **18** to seal the end piece to the wall of the casing. Snap ring **20** fitting into its groove **22** in the interior wall of the casing holds the end piece in position as shown. An inlet connection **24** is provided which passes through the end piece and communicates with the interior of the casing. Plate **26** is also positioned inside the casing and is sealed to the inner wall thereof by O-ring **28** in groove **30**.

Plate **26** is provided with extending rim **32** which forms a socket into which is fitted a short tubular filter element **34**, which may be fabricated, for example, of porous cellulosic material. The hollow space inside the tubular filter may advantageously contain some granular, slow-dissolving water treatment or purifying chemical, for example, a polyphosphate for pH control, indicated at **35**. Passing through plate **26** in the way of the edge of the filter element are transfer holes **36** through which feed solution entering through connection **24** and after passing through the filter element is transferred to the interior of the casing.

FIGURE 2.10: REVERSE OSMOSIS APPARATUS

a.

b.

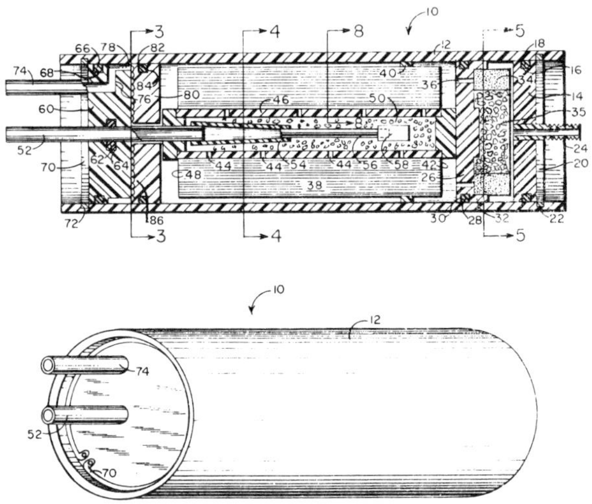

(a)(b) Purification Unit

c. d. e.

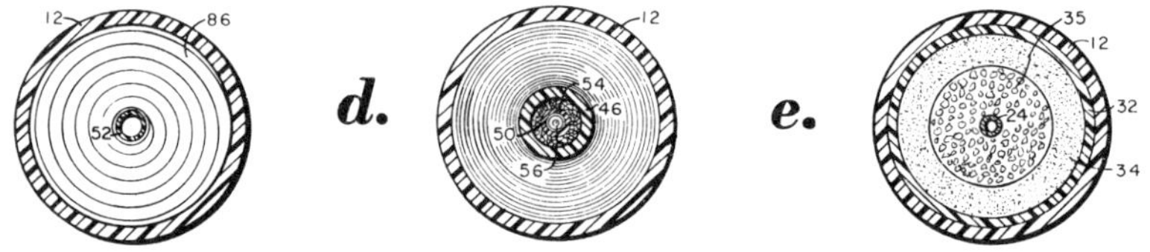

(c)(d)(e) Vertical Cross Section Along Lines **3-3, 4-4** and **5-5** respectively

f. g.

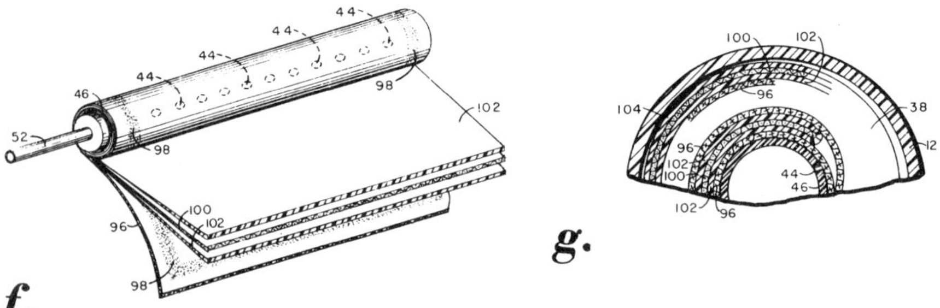

(f)(g) Membrane Module Construction

Source: D.T. Bray; U.S. Patent 3,504,796; April 7, 1970

A membrane module **38** is disposed inside the casing as shown, passage of feedwater around its sides being prevented by gasket **40** which may be of felt or other soft plastic material. The membrane is composed of a spirally wound membrane, backing sheet and screen sandwich whose construction and operation has been referred to in *1965 Saline Water Conversion Report* of the U.S. Department of Interior, Office of Saline Water. Impure feedwater enters the module through its right-hand edge **42** and as it passes longitudinally through, purified water is separated by the membrane action and is transferred through holes **44** in tube **46** around which the membrane pack of the module is wound. Brine, or water concentrated in the impurities or salts separated by the membrane in the module flows out of the module through its left-hand end **48**.

The interior of tube **46** is arranged advantageously to contain a water-purifying agent, which may for example be activated carbon in granulated form as at **50**. The ends of tube **46** are sealed by conventional end pieces maintained in place by a suitable adhesive, the left-hand end as shown being pierced by product water delivery pipe or tube **52** which projects some short distance into the interior of tube **46**. Fastened around the interior end of tube **52** is a length of soft plastic tubing **54** which will normally remain open but which will collapse under pressure to act as a shut-off valve.

Preferably at the end of tubing **54** is affixed a length of relatively stiff, small-diameter tubing **56**. The end of tubing **56** is capped with a filter closure as at **58**. For illustration of suitable materials for fabrication of these elements, tubing **54** may be of plasticized vinyl, tubing **56** may be of nylon, and filter **58** may be of the same type and construction as employed in a cigarette filter.

Thus it will be seen that water having passed through the membrane in pack **38** and into the interior of tube **46** through holes **44** is further purified by contact with granulated purifying agent. It then passes through the filter which functions principally to keep detached particles of activated carbon or purifying agent from plugging the entrance of tubing **56**. Passing through tubing **56**, then through tubing **54**, the water flows out of product water delivery tube **52** which may be connected in any convenient manner, not shown, to piping or plumbing for delivery to point of use.

The left-hand end of the casing is sealed by end piece **60** which, like other elements of the apparatus described, may be fabricated of strong plastic. Thus the casing with end pieces **14** and **60** form a pressure-resistant container in which is contained membrane pack **38**. Product water delivery tube **52** is sealed where it transfixed the end piece by O-ring **62** set in groove **64**, and the periphery of the end piece is sealed to the interior of unit **12** by O-ring **66** in groove **68**. The end piece is maintained in position in closing the casing by snap ring **70** in groove **72**.

The end piece is provided with brine outlet pipe **74** which communicates with an interior conduit **76** which in turn communicates with the annular edge space **78** between the end piece and the interior of the casing. The interior vertical surface of the end piece is plane as shown. Abutting against the end piece is plug **80** whose outer edge is sealed to the inner wall of the casing by O-ring **82** in groove **84**. Product delivery tube **52** passes loosely through a central bore in plug **80**.

The left-hand face of the plug is formed having a spiral groove **86** as shown more clearly in Figure 2.10c which, when contacting the plane back face of the end plate forms a spiral conduit which communicates at its inner end with the bore in the center of the plug where delivery tube **52** passes through, and at its outer end with outer edge space **78** between the end piece and the adjacent wall of the casing. The diameter of groove **86** has been shown somewhat enlarged for purposes of illustration and will, in practice, be relatively small to produce a suitably small-diameter, elongated conduit capable of providing the required resistance to regulate flow and maintain water pressure within the container.

In the case of the example illustrated in Figure 2.10a, the brine flowing from the edge **48** of membrane pack **38** flows through the central space between the product tube **52** and the internal bore of plug **80**, then through the elongated conduit formed by the spiral groove **86** and the contacting face of end plate **60**, out through the annular **78** space between the inside of the case and the edge of the end piece, and then through connecting channel **76** to brine outlet pipe **74**. The pressure of water against the back of the plug holds its grooved face tightly against the plane back face of the end piece.

Construction of the membrane module **38** will be more clearly understood by reference to Figures 2.10f and 2.10g. As will be seen in Figure 2.10g, central tube **46** extends through the length of the pack and is perforated at intervals as at **44** to provide porosity in the area under the membrane pack. Around the central tube is first wound the end of a length of porous fabric **96**, at least one full turn being placed around the tube. The fabric is advantageously formed of a base of porous sheet material with fine glass beads attached to its surface. The extending length of fabric has a band of suitable adhesive laid around or near the edges as at **98**, this band also running adjacent the turn of the fabric around the tube. The adhesive in the band soaks through the porous fabric to provide glueing surfaces on both sides.

Then a sheet of plastic screen **100** is placed between the inner surfaces of length of doubled-over membrane **102** as seen in Figure 2.10g. The inner surfaces of the membrane, that is, those next to the screen, are the active membrane surfaces and will face the incoming water to be purified. Then the crease of the folded membrane with the interleaved plastic screen is inserted in the nip between the extending porous fabric and the central covered tube, and the assembly wound up to form a spiral pack.

As the pack is wound and formed, the porous fabric becomes glued on each side to faces of the membrane. As the final turn is taken, the end of the outer layer of the membrane is overlaid on the end of the porous fabric, and the plastic screen is wound over to form the outer layer of the pack, the complete module organization being maintained intact by outer plastic tape winding **104**.

In Figure 2.10g is shown an enlarged fragmentary view of the membrane module as it appears in vertical section. Tube **46** is shown with one of its perforations **44**, and around this a layer of porous fabric **96**, then a layer of membrane **102**, then a layer of plastic screen **100** and then another membrane layer followed by another layer of porous fabric with the sequence repeated to form the spiral-wound pack of desired membrane area. As previously described the rolled pack is arranged with screen **100** as its outer layer and this all held in place by tape winding **104**.

The semipermeable membrane **102** is fabricated of material possessing the requisite properties, principally adequate flux or flow through, and selectivity, or salt rejection. Materials which have been found advantageous are cellulose acetate films such as described in U.S. Patents 3,133,132; 3,133,137 and 3,344,214.

In operation of a purification unit, an impure water supply line may be connected to inlet fitting **24**. This may be a hose or plumbing connection to a domestic water supply if it is desired to purify such domestic supply to provide a purer grade of water, for example, for drinking and cooking. Product outlet **52** is connected, as for example, by a plastic tube, to a suitable container for accumulating a supply of purified water, and brine outlet **74** is connected in any suitable manner to drain or sewer. The pressure of a normal domestic water supply, that is, from about 50 to 100 lb per square inch, will be sufficient to cause a flow of water through the membrane pack **38** with purified water passing through the membrane **102**, flowing through holes **44** in tube **46** and into the interior of tube **46** where it contacts purifying carbon granules **50** and then flows through filter **58**, tubes **56** and **54**, to product water outlet **52**.

The brine outflow is controlled as described above by directing its passage through one or more elongated conduits formed by a spirally grooved plug face or faces as **86** in plug **80** in Figure 2.10a.

Thus pressure is maintained inside case **12** to provide proper driving force for reverse osmosis and purification of the impure feedwater, while a regulated amount of water more concentrated in removed salts, and termed brine, is allowed to outflow through brine outlet **74**. A small unit as described is capable of producing several gallons per day of water in the range of 50 to 200 ppm total dissolved solids from a somewhat unpalatable household supply containing, for example, up to 800 to 1,000 ppm.

Spirally Wound on Mandrel

G.A. Newby and A.J. Navoy; U.S. Patent 3,397,790; August 20, 1968; assigned to Gulf General Atomic Incorporated describe semipermeable membrane separation devices particularly designed for production line manufacture. In the use of semipermeable membranes for the separation of fluid components from fluid mixtures, one of the operating parameters which affects the amount of throughput which is obtained using a given membrane is the total surface area of membrane which is in contact with the high-pressure fluid feed mixture; all other operating conditions being equal, a larger area membrane will pass more of the permeating fluid component than a membrane of lesser area.

Accordingly, for a separation apparatus employing semipermeable membranes to have a high capacity, it is important to incorporate a large amount of surface area of semipermeable membrane which will be in contact with the feed fluid mixture in a given volume. The thinner the semipermeable membranes that can be operatively employed, the greater the amount of surface area that can be incorporated in a given volume.

To obtain large amounts of semipermeable membrane surface area in a compact separation apparatus, it is known to employ a central mandrel containing an axially extending collection passageway and to spirally wrap around this mandrel a series of overlying sheets, such as two sheets of semipermeable membrane material sandwiching between a sheet of backing material (which backing material provides a flow passageway for the fluid component permeating through the semipermeable membranes) and to dispose another flow-passageway providing sheet between spiral windings of the semipermeable membranes, which sheet provides a passageway in the completed spiral wound assembly through which the feed mixture can be conveniently supplied to the surfaces of the semipermeable membranes, as by pumping the fluid feed mixture through in a direction generally parallel to the axis of the central mandrel. This assembly of central mandrel and spirally wound sheets is sometimes termed a membrane module.

In operation, the fluid component which permeates from the feed mixture through the semipermeable membranes enters the backing material and flows in its plane spirally inward to the axial passageway in the central mandrel, with which passageway it is in fluid communication.

This process provides an improved membrane module for use in separating a first fluid component from a fluid mixture of the first fluid component and a second component employing semipermeable membranes and apparatus for utilizing such a membrane module. In addition, an improved method for making such modules is also provided. Briefly, the improved membrane module comprises a mandrel having an axial passageway or a plurality of axial passageways, which mandrel is spirally wrapped with sheets of semipermeable membranes and associated sheets of porous backing material and sheets of feed passageway-providing means.

Inasmuch as the spirally wound sheets are variously connected to or disposed adjacent the mandrel before they are spirally wound, these sheets are variously referred to as leaves and as extending generally radially outward from the mandrel. It should be understood, however, that a single sheet may pass through the mandrel and thus constitute two or more leaves because, in its assembled condition, it emanates from a point at or near the periphery of the mandrel at a plurality of locations.

One of the features of the module is that the porous flow-passageway-providing leaves extend into the axial passageway means in the mandrel to provide good fluid communication between the axial passageway means and the generally radially or spirally extending flow passageways, which in operation are the passageways for the fluid component that is separated from the fluid mixture via permeation through the semipermeable membranes.

Illustrated in Figure 2.11a is a separation unit **110** embodying various of the features of the process which unit includes a high-pressure generally cylindrical chamber **111** fabricated from an outer shell **112** having a cap **113** secured to its upper end, an improved membrane module **114** disposed within the chamber **111**, and a fluid takeoff assembly **122** at the bottom end of the chamber **111**. The cap **113** is formed with an inlet **115** through which the feed solution is supplied to the chamber **111**. A side outlet pipe **124** permits outflow from the chamber **111** of the feed mixture after it has passed through the module **114**.

FIGURE 2.11: MEMBRANE MODULE CONSTRUCTION

a.

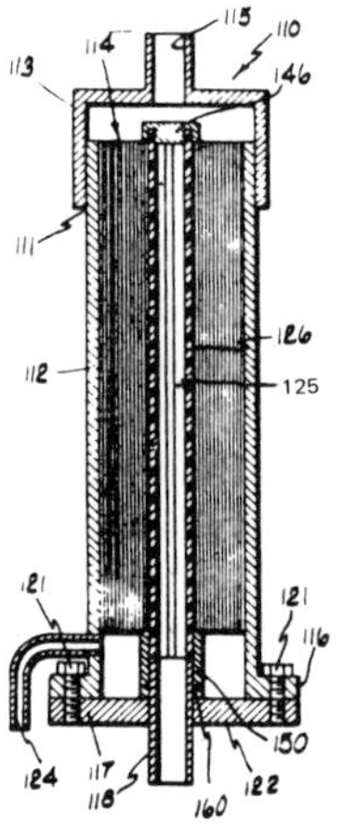

b.

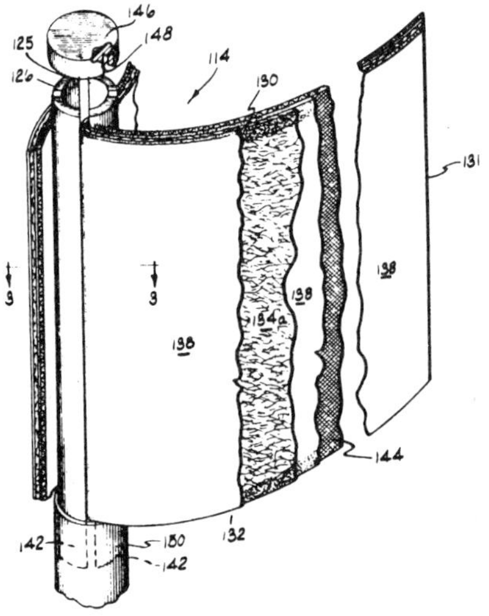

Separation Unit

Membrane Module in Unwound Position on Mandrel

c.

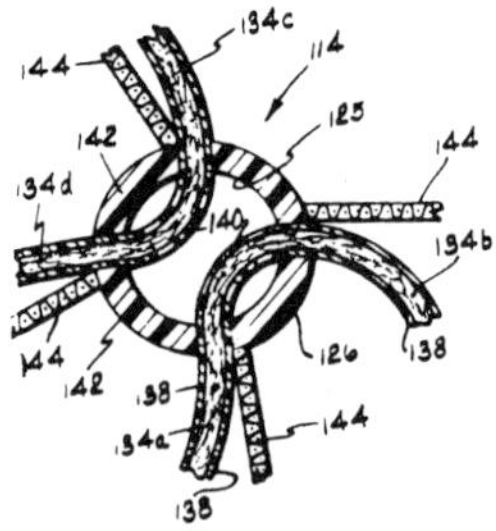

Initial Fabrication Step, View Along Line **3-3** of Figure 2.11b

Source: G.A. Newby and A.J. Navoy; U.S. Patent 3,397,790; August 20, 1968

The hollow cylindrical shell **112** has a peripheral flange **116** at its lower end by which connection is made to the fluid takeoff assembly. The fluid takeoff assembly includes a circular plate **117** having a pipe **118** extending through a central hole. A coupling ring **150**, such as a piece of plastic tubing, connects the upper end of the pipe **118** to the membrane module. A means is provided to seal the chamber. One such arrangement employs bolts **121** to secure the lower flange **116** to the circular plate **117**, and a suitable peripheral gasket may be employed to assure a fluid-tight seal.

Generally, the module includes a central mandrel **126** having an axial passageway **125**. As best seen in Figures 2.11b and 2.11c, the module includes a plurality of sheets or layers of backing material **134**, a plurality of sheets of semipermeable membranes **138** and a plurality of separator grids **144**. In the particular illustrated example, two sheets of backing material are employed, both of which extend into and out of the central mandrel, thereby traversing the axial passageway. Each sheet of backing material is disposed in the mandrel at a location intermediate the ends of the sheet, generally in the middle of the sheet, so as to provide two leaves of generally equal length. In the illustrated example, four leaves, **134a**, **134b**, **134c** and **134d** are used.

Each of the leaves of the backing material is flanked by and enclosed by two adjacent sheets of semipermeable membranes **138**. In the example, it can be seen that the sheets of semipermeable membrane extend along the entire length of sheets of backing material even along the portion of the backing material which is disposed within the mandrel, residing in the axial passageway **125**. This particular arrangement facilitates economical fabrication of the sheets of backing material and semipermeable membrane in three-ply groups or laminates.

Obviously, the semipermeable membrane could be stripped from the backing material along the portion that resides within the mandrel so as to facilitate transfer of the fluid from the flow passageways provided by the backing material into the axial collection passageway in the mandrel. However, from a production standpoint, it is contemplated that a plurality of perforations **140** would be made in the portion of the three-ply laminate within the mandrel, an operation which can be performed more simply than stripping.

As best seen, in Figure 2.11b, the three edges of each backing material leaf are sealed between adjacent edges of the pairs of flanking semipermeable membranes **138**. The seal is simply made by applying a sufficient amount of a suitable adhesive to the three edges, **130, 131, 132** to assure that a seal is achieved and that no fluid can enter the edges of the backing material leaves **134** and thus must pass through the semipermeable membranes before it can reach the passageways provided in the backing material. In mass production techniques, this adhesive is economically applied before assembly of the three-ply laminates with the mandrel **126**; however, it can also be applied after the membrane layup (Figure 2.11c) is assembled.

Between each three-ply leaf, a sheet or layer of separator grid material **144** is disposed. This separator grid material, in assembled condition of the wound membrane module **114**, provides feed passageways between adjacent spiral windings of adjacent three-ply leaves. The innermost edge of each sheet of separator grid material is preferably bonded in a suitable manner to the periphery of the mandrel **126** so as to facilitate orderly spiral winding of the membrane module. In Figure 2.11c, it can be seen that four sheets of separator grid material are provided, one adjacent each of the four three-ply leaves. The upper and lower edges of the separator grid sheets are completely open so as to facilitate free passage of fluid through.

In the spirally wound membrane module **114**, the feed mixture is fed into the top of the separation unit **110** and enters only into the upper edges of the separator grids **144** because the adhesive bonds prevent any flow into the three-ply laminates of semipermeable membrane **138** and backing material **134**. The fluid feed mixture which enters the top edges of the separator grids flows generally downward therein, parallel to the axis of the central mandrel **126**, and exits through the open bottom edges, finally leaving the separation unit via the outlet pipe **124**. Because, in the wound spiral membrane module **114**, each sheet of separator grid material **144** lies between two sheets of semipermeable membrane, the incoming fluid feed mixture is carried to the entire surface area of all eight leaves of semipermeable membrane.

In the form illustrated in Figure 2.11b and 2.11c, the central mandrel **126** could be made from a tube of suitable length, diameter and wall thickness which is slotted to provide four slots of suitable width extending downward to a depth of slightly greater dimension than the lateral dimension of the sheets of backing material **134** and semipermeable membrane **138**. In such an instance, the three-ply laminates would simply be threaded through the slotted tube, perhaps from the open end. However, it has been found that mass production assembly of the membrane module **114** is facilitated by using, instead of a single tube with slots, a plurality of mandrel segments **142** (which in Figures 2.11b and 2.11c are arcuate sections of the sidewall of the tube). The mandrel segments are disposed in the desired position in association with the sheets of backing material, thereby dividing each sheet into two leaves.

By clamping or fastening the mandrel segments together, both at the top and at the bottom, a composite, stable central mandrel is provided about which the plurality of leaves can be wound to form the membrane module.

An upper cap **146** is employed which has an upward extending annular passageway **148** into which the tops of the mandrel segments **142** are interfitted. A suitable peripheral coupling ring **150** is employed to hold the mandrel segments together near the lower ends, beneath the lower edges **132** of the multiple-ply laminate sheets. The lowermost portions of the ends of the mandrel segments extend down inward into the coupling ring, which extends from the base of the membrane module to some point below where the outlet pipe **118** abuts the mandrel segments, preferably till it abuts the circular plate **117**. Any suitable means may be used to provide a good seal between the coupling ring and the outlet pipe; one such means would employ a standard O-ring **160** at some point below the mandrel outlet pipe junction as shown in Figure 2.11a.

Accordingly, the axial passageway **125** through the central mandrel **126** is thus in fluid communication with the outlet pipe **118**. The selection of the semipermeable membrane material is dependent of course upon the intended use of the separation unit **110**. If separation of a gaseous mixture is to be employed, a suitable semipermeable membrane is selected which affords high permeation of the desired gaseous component. The situation is similar when a liquid is to be separated from a mixture of liquids. For the separation of a solvent from a solution, an osmotic membrane is used. When it is desired to separate water from an aqueous saline solution, the cellulose acetate membranes are considered to have particular economic advantage because they exhibit excellent salt rejection characteristics while permitting above average flow rates.

The backing material **134** is made of relatively thin sheets of a material having a sufficiently high porosity to permit ready flow of fluid, and being able to withstand substantial pressure in a direction generally transverse to the plane without collapse or undue creep. The backing material should be sufficiently flexible so that it may be wound in the spiral configuration without fracturing. If compaction of the backing material should occur upon subjection to relatively high pressures, it is usually accompanied by a corresponding reduction in porosity.

Such compaction and reduction in porosity increases the resistance of the backing material **134** to fluid flow therethrough, reducing the average pressure differential on opposite sides of the semipermeable membranes **138** and thus reducing the efficiency of the unit **110** at a specified set of operating conditions. Graphite cloth is one material which exhibits good qualities for backing material; however its fairly expensive cost makes it economically unsuitable for certain operations. Silicon carbide grit, properly sized, or properly sized particles of sand, held on the surface of a sheet of plastic felt with a suitable binder is considered especially useful in high-pressure operations, for example where the feed pressure may reach 1,500 psi. Various types of felt or glass fibers fabricated in the form of thin flexible sheets also display good advantages as backing materials for certain separation processes.

In instances where a pressure of not more than about 500 psi is to be employed, various synthetic plastic fibrous materials may be used as the backing material **134**. Examples of such synthetic materials include nylon, polyester, rayon, rayon viscose, and acrylic fibers. These fibers are generally unaffected by exposure to water and thus may be used in the separation of aqueous solutions as well as gases. Whereas the backing material provides the flow channels for the fluid which permeates through the semipermeable membranes, leading to the axial collecting means, the separator grid material **144** provides the passageways for the input or feed mixture.

In operation, the feed mixture is pumped through the separator grid material in a direction generally parallel to the axis of the mandrel **126**, and therefore generally at right angles to the direction of flow of the permeating fluid, which flow is generally laterally, spirally inward through the backing material to the central mandrel passageway **125**. The separator grid material is preferably fabricated of thin, very porous, flexible sheets which may be readily wound into spiral configuration. Because the separator material is on the high pressure side of the semipermeable membrane **138**, it need not provide support, nor particularly resist compaction.

In general, a relatively inexpensive woven structure, such as plastic screening material made of polyethylene, can be employed as the separator grid material **144**. The mandrel **126** can be fabricated of any relatively corrosion-resistant material which is unaffected by the particular components of fluid mixtures being treated. For example, for the separation of fresh water from saline water, the mandrel may be fabricated of synthetic plastic materials, such as cellulose butyrate or an extruded acrylic, both of which materials have good dimensional and structural stability in a high pressure and relatively moist environment.

The particular adhesives employed, of course, depend upon the particular semipermeable membranes used. In general, and in particular when cellulose acetate membranes are used, a modified epoxy resin is found to be a very suitable adhesive for bonding the edges of the semipermeable membranes **138** and backing material leaves **134** so as to prevent any inflow of the feed mixture.

The high-pressure chamber **111** can be made of any corrosion-resistant material which is unaffected by the particular feed mixture intended to be treated. In general, examples of suitable materials include copper, coated mild steel, stainless steel, fiber glass-reinforced epoxy, and polyvinyl chloride. After the membrane module has been wound into its spirally wound condition, an outer wrap of plastic film may be disposed thereabout so as to hold the membrane module **114** in its tightly wound condition. In order to assure that there is a good seal between the outer lateral surface of the membrane module and the interior lateral wall of the shell **112**, a few turns of plastic tape are often wrapped about the module at one or more locations to provide suitable sealing bands.

Alternatively, a circumferential seal may be provided at a suitable vertical location on the interior wall of the shell **112**. These sealing bands prevent the feed mixture from bypassing the membrane module **114** in the separation unit **110** and leaking from the inlet **115** to the outlet **124** thereby reducing the effectiveness of the separation unit **110**. The upper cap **146** and the lower coupling ring **150** are suitably sealed to the central mandrel **126** to prevent leakage of the feed mixture.

Although the separation unit may be employed for treating a variety of feed mixtures, depending upon the particular semipermeable membranes employed, it is considered particularly adapted for the treatment of aqueous solutions, such as sea or brackish water; and its operation is described with reference to sea water. Sea water is supplied to the upper end of the chamber via the inlet pipe and flows downward through the various spiral windings of the separator grid material. After completion of its downward passage through the membrane module the sea water is depleted of a portion of its original water content and has a higher salt concentration. Exit of the sea water with its increased salt concentration is via the outlet pipe.

Because in the separation of fresh water from sea water pressure is maintained above the osmotic pressure of the particular semipermeable membrane used, a suitable pump is employed to pump the sea water into the separation unit **110**. Likewise, a suitable adjustable pressure control valve is provided in connection with the outlet pipe **124** so that the desired pressure can be maintained throughout the entire membrane module **114**. Preferably, the pump is adjusted to provide a relatively turbulent flow of input feed mixture through the separation unit so that, when a solution such as sea water is employed, the boundary layer salt concentration in the flow stream adjacent the membrane **138** is minimal.

The semipermeable membranes permit the diffusion of water through into the backing material where it flows spirally inward until it reaches the axial passageway in the central mandrel. Of course, the water experiences a drop in pressure in its flow spirally inward through the backing material sheets. This pressure drop is generally proportional to the square of the distance the water must flow, and inversely proportional to the square of the effective hydraulic diameter of the pores in the backing material, so long as the volumetric flow per unit area of membrane is substantially constant.

It is accordingly desirable to provide a backing material having relatively large-sized pores to minimize such pressure drops. But, because the back material also serves to support the sheets of semipermeable membranes, the pores cannot be so large as to permit the membrane to be forced into the pores. The use of the plurality of backing material leaves shortens the average distance which the separated fluid component must travel to reach the axial passageway, and the entry of the backing material sheets directly into the axial passageway means in the central mandrel eliminates the possibility of a further pressure drop at this point which might reduce the efficiency of the unit, and facilitates its manufacture.

If desired, more than one membrane module may be employed within a single separation unit by connecting the top of the central mandrel of a lower module to the bottom of the central mandrel of the upper module. This arrangement may be of particular interest where the overall desired height of the separation unit exceeds the width of the sheets of semipermeable membranes which may be conveniently fabricated, so that sheets of a width about one-half the height of the unit may be employed.

In related work, *J.C. Westmoreland; U.S. Patent 3,367,504; February 6, 1968; assigned to Gulf General Atomic Inc.* describes a reverse osmosis separation apparatus employing an envelope of semipermeable membrane sheets spaced apart by a layer of porous backing material and spirally wound with a continuous length of separator grid material about a hollow mandrel. An aqueous feed solution is introduced through the separator grid material, permeated water passes through the membranes and into the backing material by reverse osmosis, and the backing material discharges the permeated water into the hollow mandrel. Porous walled tubes are provided in the membrane envelope and serve as further discharge passageways for the permeated water.

Spirally Wound Membranes—Three-Ply Layup

U. Merten; U.S. Patent 3,386,583; June 4, 1968; assigned to Gulf General Atomic Inc. describes the fabrication of a membrane module. An elongated sheet of semipermeable membrane material is flanked on opposite surfaces by a sheet of porous backing material and a sheet of spacer material to provide a three-ply layup which is convolutely folded and arranged around a central porous mandrel with the backing material layer in fluid communication with the center of the hollow mandrel. The individual folded flaps are spirally wound about the mandrel and wrapped with tape about the exterior circumferential surface. The spacer material provides axially extending feed solution passages between adjacent flaps while the permeating fluid travels spirally inward in the plane of the backing material to the central mandrel.

Referring to Figure 2.12a, a water purification unit **10** is illustrated, which includes a membrane module **14**, a pressure-tight container **18** for containing the module and a product water take-off assembly **22**. The construction of the membrane module is illustrated in greater detail in Figures 2.12b and 2.12c. The module generally includes a tubular central mandrel **26** which also serves as the product water collection means, or product water take-off tube. A membrane layup **28**, comprising a plurality of radially extending flaps **29**, is disposed about the circumference of the mandrel and the flaps are spirally wound on the mandrel forming a convoluted configuration, prior to placing the module within the pressure-tight enclosure **18**.

The membrane layup comprises adjacent elongated sheets of spacer material **30**, semipermeable membrane material **34** and backing material **38**. The manufacture of a membrane module constructed in accordance with the principles of this process entails the fabrication of a membrane layup, which is plurally folded upon itself a plurality of times to define a number of flaps, thereby providing a relatively large area of membrane surface through which product water may diffuse. At the same time, since each of the flaps is of relatively short longitudinal length, the distance through which the product water flows is minimized, thereby also minimizing pressure drop.

Preferably the semipermeable membrane material, which comprises the plurally folded membrane layup, is in the form of a relatively thin sheet of material, which has predetermined permeability properties. When desalination operations are intended, it is advantageous to fabricate the membrane material of a substance which exhibits substantial salt rejection properties. An example of one such material which has been found to be highly advantageous for use in such applications is cellulose acetate, such as is described in U.S. Patents 3,133,132 and 3,133,137.

The backing material is preferably fabricated of a relatively inexpensive material, having sufficiently high porosity properties to permit unencumbered flow of product water therethrough. The backing material is also capable of withstanding substantial applied pressures without collapse or undue creep. In addition, the backing material is preferably sufficiently flexible so that it can be initially supplied from supply rolls, folded upon itself, and then wound in a generally convoluted configuration without being fractured, since it is desirable that its continuity be maintained for efficient operation.

The backing material serves as a flow channel for the product water which has diffused through the membrane material, and accordingly extends substantially through the membrane layup. Since the membrane layup is plurally folded upon itself, the portion of the backing material within each of such separate folds or flaps in effect serves as an individual flow channel for the purified water to follow to the take-off tube **26**. Thus, a plurality of relatively short product water flow channels are provided.

On the other hand, the spacer material, which separates adjacent folds of the membrane layup, generally serves as a passageway for the feedwater which is to be purified. During operation of the apparatus, the feedwater is generally supplied through the spacer material in a direction generally parallel to the axis of the tube **26** and at right angles to the direction of flow of the product water, which generally flows spirally through the convoluted backing material toward the central tube **26**. The spacer material is preferably in the form of a relatively inexpensive grid-like structure, such as woven screen, and generally provides a low-resistance path for the feedwater to follow in passing through the module.

The central mandrel **26** is tubular and is provided with a plurality of slots or apertures **40** positioned on its outer circumferential surface and extending through its wall. The apexes **42** of each of the folds of the membrane layup are maintained in position, seating tightly against the central tube by suitable position maintainers or posts **46**, as is subsequently explained. Thus, fluid communication is provided between the slots **40** and the backing material, which carries the product water, so that the product water may flow through the slots into the hollow interior of the tube **26** from which it may be appropriately removed through the take-off assembly **22**.

FIGURE 2.12: MEMBRANE MODULE

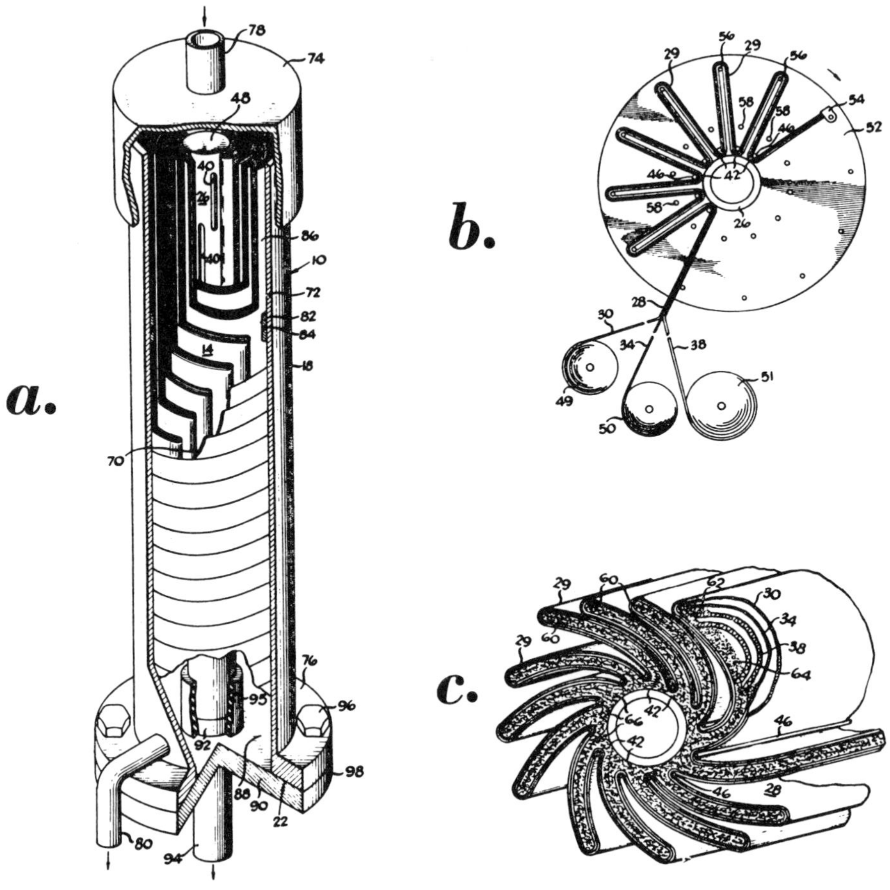

(a) Separation Unit

(b)(c) Fabrication Steps for Membrane Module

Source: U. Merten; U.S. Patent 3,386,583; June 4, 1968

Preferably, the tube **26** is plugged at its upper end by a suitable plug **48** so that the purified water, supplied to the interior of the tube may be directed toward the other end, while precluding feedwater from entering the tube. The tube is generally fabricated of a corrosion-resistant material, such as cellulose butyrate or extruded acrylic, since it is to be disposed in a relatively moist environment.

The particular structure described on the previous page is generally quite convenient to manufacture. The elongated sheets of material which comprise the membrane layup are preferably initially contained on supply rolls **49, 50** and **51**, as shown in Figure 2.12b so that they may be conveniently supplied during the manufacture of the membrane module. The respective sheets of material are simultaneously fed from the supply rolls, with the flaps being initially held in position, so as to form a fin-like configuration about the circumference of the mandrel. This is preferably accomplished by initially fabricating the layup on a suitable fixture of jig **52**, having a centrally located central opening adapted for receiving the mandrel.

In this connection, one end of the adjacent sheets of material, comprising the membrane layup, is maintained in position by a removable clamp **54** secured on the jig adjacent its outer periphery. The layup then is extended radially to the hollow mandrel at which point it is folded upon itself with the apex of the fold being maintained in position relatively tightly against the mandrel by the position maintainer **46**. Preferably the position maintainers comprise thin plastic rods which may be conveniently secured in position on the jig, so as to maintain the respective apexes of the flaps in position. The layup then extends radially outwardly from the mandrel, and is folded over a pin or post **56**, which is inserted on the jig at the same radial distance from the mandrel as is the clamp. The pin preferably comprises a readily insertable and removable metal rod. A readily insertable and removable roller **58** adapted for axial rotation is mounted on the jig spaced radially from the position maintainer.

The jig is then rotated and the layup is extended radially inwardly to the mandrel and is folded upon itself with the apex of the fold being secured to the mandrel by another position retainer **46** secured on the jig so as to maintain the apex tightly against the periphery of the mandrel. The jig is again rotated and the layup similarly is extended radially outwardly from the mandrel and is folded over another pin. Another axially rotatable roller is then rotatably mounted on the jig and similarly spaced radially from the position maintainer.

In this fashion, the layup is initially disposed on the jig about the circumference of the mandrel, forming a continuous elongated fin-like structure, including a relatively large amount of flaps generally limited in number by the thickness of the sheets of material and the size of the casing in which the module is to be disposed. Thus the membrane layup is fabricated of a plurality of flaps, each of which is relatively short in comparison to the overall length of the membrane layup.

When a desired number of flaps have been formed in the above manner, the respective sheets of material comprising the membrane layup are separated from the supply rolls, such as by cutting. As shown in Figure 2.12c, the starting end and the terminal end of the layup are preferably sealed together with a suitable epoxy sealant, which strongly adheres to the materials after curing so that the clamp may be removed. Alternatively, if desired, the starting end and portions of the layup may be maintained in sealed relation by the clamp.

After completion of the formation of the plurality of flaps, the pins are removed and the central tube together with the position maintainers are rotated on the jig relative to the rollers. Since the rollers are adapted for axial rotation, the flaps readily slide under the rollers, which define the circumference of a cylinder. In this fashion the membrane layup is wound spirally about the central mandrel to form the desired convoluted configuration. The spirally wound structure is generally cylindrical in shape and has a diameter approximately equal to the diameter of the cylinder defined by the rollers. Generally, for the sake of convenience, the position maintainers are merely allowed to remain in the completed structure, since they take up very little space and have substantially no effect on the operation of the apparatus.

In order to insure that the backing material only communicates with solutions which have diffused through the membrane material, it is essential to provide suitable seals so as to provide separate feed passageways through which the feedwater is supplied and product passageways through which the product water flows.

In this connection a suitable epoxy sealant is applied on the outer edge **60** of the backing material within each of the flaps. The epoxy sealant is applied relatively copiously along the edges so as to substantially saturate this portion of the backing material. As a result of this saturated condition, the epoxy sealant flows between the outer edges **62** and **64** of the adjacent sheet of backing material and membrane material along the outer edges of each of the flaps, thereby sealing substantially the entire outer edge of the membrane material to the backing material in each of the flaps so as to define a plurality of product water passageways, extending through the respective flaps comprising the membrane module.

When the flaps are convolutely wound on the mandrel, a slight spacing **66** remains between the apexes of adjacent flaps. This spacing **66** extends along substantially the entire axial length of the tube and if left unsealed would permit feedwater to enter the interior of the tube through the apertures, thereby contaminating the purified water being produced. This problem is substantially alleviated by the liberal application of the epoxy sealant along the edge of the flaps.

Thus, sufficiently ample quantities of the sealant are applied so as to saturate the edges, as previously mentioned, as well as to cause portions of the excess to flow into the spacings **66**, thereby sealing it from communication with the feedwater. The edge of the spacer material **30**, which extends between adjacent flaps is not sealed in position on the layup. Accordingly, the spacer material defines a feed passageway, extending between adjacent flaps, so that the feedwater may flow through the module **14** in an unemcumbered fashion from one end to the other.

By thus providing separate feed passageways, the feedwater may be supplied to the module through the spacer material without communicating with the backing material prior to having diffused through the membrane material so that only water from which the desired impurities have been removed reaches the product passageways.

Since it is contemplated that relatively high pressures are to be employed when utilizing the apparatus, it is generally desirable to employ suitable sealant means so as to obtain adherent bonds which have sufficient strength to withstand such pressures. For example, when cellulose acetate is utilized as the membrane material, an appropriate adhesive which adequately seals the edges of the membrane material to the backing material and also seals the edges of the backing material from communication with feedwater comprises a modified epoxy resin. Such an adhesive has been found to be compatible with the materials being used and adequately wets them, forming a high strength bond after completion of curing.

The spirally wound module is then suitably taped, as shown in Figure 2.12a, about substantially all of its outer circumference. This is efficiently achieved by feeding a suitable plastic tape **70**, which is tension-sensitive so as to maintain the convoluted flaps in position and which is substantially unaffected by moisture, between the rollers **58** and the outermost convolution of the module. Preferably a plastic tape such as that commonly designated by Mylar is employed. The module may then be disposed in the pressure-tight enclosure **18**.

Since each of the plurality of flaps comprising the membrane layup in effect serves as a separate path of travel for purified water, it can be seen that the distance of travel of the purified water in flowing to the central mandrel through the backing material within the respective flaps is relatively short. At the same time the total surface area of membrane material available, through which purified water diffuses, is maintained relatively large by forming a plurality of flaps. The pressure-tight enclosure is preferably fabricated from a relatively strong corrosion-resistant material such as copper, polyvinyl chloride, coated mild steel, stainless steel, fiber glass-reinforced epoxy, etc.

The enclosure generally includes a cylindrical outer shell **72** having a cap **74** secured to its upper end and having a flange **76** secured at its lower end. The cap and flange may be secured to the shell by various operations such as welding, brazing, etc. A suitable feed inlet pipe **78** is disposed in a generally central position on the cap, and a suitable waste outlet pipe **80** is further disposed in the shell adjacent the flange.

Often the interior diameter of the enclosure **18** is approximately equal to the outer diameter of the spirally wound, taped module **14**, so that a sufficiently tight fit is provided to direct the flow of feedwater through the feed passageways, defined by the spacer material extending through the module, rather than around the exterior of the module. If desired, additional sealing may be provided in some instances by wrapping several turns of a relatively narrow plastic tape about a suitable location on the exterior of the module.

In certain instances, where the spacing between the module and the interior of the enclosure is too large to permit adequate sealing by the above-described means, a sufficient low-pressure seal may be advantageously provided by disposing a grooved ring **82** about a suitable region on the exterior of the module. A suitable adhesive material, preferably a silicone rubber compound, is employed to cement the ring to the module. An O-ring **84** is disposed in the groove provided on the ring **82** so as to complete the seal.

As a result of the seal proided by the above-described means, a feed compartment **86** and a concentrate compartment **88** are defined within the pressure-tight container and exterior to the membrane module. The seal means prevents direct communication between the feed compartment and the concentrate compartment. In operation the fluid which it is desired to process may be supplied through the inlet pipe **78** and directed from the feed compartment through the feed passageways, defined by the plural folds of spacer material between adjacent flaps of the module, into the concentrate compartment.

The purification unit **10** also includes the product water take-off assembly **22**. The product water take-off assembly preferably includes a flange **90** having a nipple **92** disposed on its interior surface and a product water outlet pipe **94** on its exterior surface communicating with the nipple. A suitable connector **95** which preferably comprises a short piece of rubber tubing is disposed over the end of the nipple at one end and disposed over the lower end of the take-off tube **26** on its other end.

Thus, fluid communication is provided between the interior of the tube and the product water outlet pipe. Preferably the take-off assembly **22** is secured to the flange **76** by securing the flange **90** to the flange **76** by a plurality of bolts **96**. Generally, a gasket **98** is also disposed intermediate the flanges **90** and **76** for providing a fluid-tight seal.

The apparatus described above may be employed for treating a wide variety of solutions, but is particularly adapted for the treatment of saline solutions, such as seawater. In operation, the seawater is supplied through the feed inlet pipe **78** as shown by the arrow, to the feed compartment **86**, from which it flows through the passageways, defined by the spacer material extending between the adjacent flaps, comprising the membrane module **14**, into the concentrate compartment **88**. Upon completion of its passage through the purification unit, the seawater, which has been processed, and is more highly saline than originally, is discharged from the concentrate compartment **88** through the waste outlet pipe **80**. This discharged material is referred to as "seawater residue."

Throughout its residence time in the purification unit **10** the seawater is preferably maintained at a predetermined pressure. The pressure is in excess of that required to cause reverse osmosis to occur. At pressures in excess of the osmotic pressure purified water diffuses from the seawater through the semipermeable membrane into the backing material. The requisite pressures are advantageously obtained through utilization of an appropriate pump, which supplies the seawater to the purification unit, and by further providing an appropriate pressure control valve in the waste outlet pipe, thereby adjusting the pressure of the seawater present in the feed passageways, defined by the spacer material extending between adjacent flaps of the membrane module **14**, as desired.

In addition, the pump is preferably adjusted to provide a relatively constant flow of seawater through the feed passageways. Such a provision is desirable because it promotes the maintenance of an equilibrium salt concentration in the solutions being processed, thereby obviating the need for diluting more concentrated solutions prior to processing, or for increasing the applied pressure.

The purified product water diffuses through the plurality of flaps, defined by the plurally folded sheet of membrane material **34** as the seawater flows through the feed passageways, defined by the spacer material extending between adjacent flaps. This product water diffuses into the backing material disposed within the plural flaps of membrane material, and flows spirally inwardly toward the mandrel **26** through the backing material, comprising the product passageway, until it reaches the apertures **40**. Since the apex **42** of each of the flaps is disposed in contact with the outer circumference of the tube **26**, providing fluid communication between the backing material and the hollow interior of the tube, purified water flows into the tube.

Upon reaching the hollow interior of the tube, one end of which is plugged, the product water flows out through the opposite end of the tube through the take-off assembly **22** and out through the product water outlet pipe **94**. If desired, the purification unit may be disposed in a vertical position so that the purified water may flow out through the open end of the tube by gravity.

In order to operate at high levels of production and efficiency, it is desirable to employ membrane modules having as large a membrane surface area as possible so as to substantially increase the amount of purified water which is produced. The structure described here achieves this objective, while maintaining the pressure drop developed within acceptable values from an engineering standpoint, by minimizing the distance through which the product water must flow. At the same time the apparatus may be fabricated in a highly convenient and efficient manner so as to substantially decrease the unit costs of manufacture.

Similar membrane modules based on the use of coextensive leaves of sheet-like backing material attached to a hollow mandrel is described by *D.T. Bray; U.S. Patent 3,417,870; December 24, 1968; assigned to Gulf General Atomic Inc. U. Merten; U.S. Patent 3,429,957; February 25, 1969; assigned to Gulf General Atomic Inc.* also describes a process for continuously casting a semipermeable membrane.

Casting on Highly Pressed Woven Fabric Support

B.M. Brown and W.D. Slowig; U.S. Patent 3,894,166; July 8, 1975; assigned to Eastman Kodak Company describe the preparation of an integral cellulosic reverse osmosis membrane having a permeable fabric support. The process comprises casting a concentrated membrane-forming dope and forming a reverse osmosis membrane on a smooth surface of a permeable casting and membrane support fabric comprising substantially unsized continuous multiple warp and fill strands of fiber-forming crystalline organic thermoplastic resin.

Thus, it has been found that the prior art disadvantages found in attempting to use a woven fabric, such as a fabric woven from Dacron polyester, e.g., D-601 fabric, as a support member for a semipermeable polymeric, preferably cellulosic, film membrane in an integral membrane suitable for use in reverse osmosis processes, are substantially avoided and overcome by a process for forming an integral membrane which comprises

(1) subjecting at least one surface of a substantially unsized, fluid-permeable fabric comprising multiple substantially continuous warp and full multifilament strands of fiber-forming crystalline organic thermoplastic resin to high pressing forces at temperatures and pressures sufficient substantially completely to close permanently the interstices at the strand intersections while leaving the pressed fabric permeable to fluids, such as gas and desalinated water, and sufficient to smooth the surface(s) of the fabric to a high degree of smoothness,

(2) cooling the pressed fabric to a temperature at least sufficiently cooler than the pressing temperature to set the fabric in its permanently pressed form,

(3) casting onto a smooth surface of the fabric a concentrated polymeric, preferably cellulosic, film-forming dope, and

(4) subjecting the dope on the fabric to subsequent steps of a known reverse osmosis membrane forming process.

In particular, it has been found that the pressing of the unsized woven fabric of thermoplastic resin must be done under conditions of temperature and pressure sufficient to close as completely as practically possible the interstices in the fabric so that membrane ruptures will not occur at those parts of the fabric when the integral membrane is subjected to reverse osmosis membrane operating pressures, e.g., of 600 to 800 psig or higher, without, however, causing the filaments of the strands to fuse together to render the fabric impermeable, or to rupture from excessive pressure applied during pressing.

The filaments thereby undergo sufficient plastic flow to cause the interstices to close. Simultaneously, under the conditions existing at the time of pressing, the surfaces of the fabric are rendered plastically flowable and can be pressed with a smoothing means to make one, or preferably both, flat surface(s) highly smooth, thereby reducing the dimensions of substantially all ridges and valleys in the faces of the fabric. Absence of the usual fabric stiffening or texturizing sizing from the fabric permits enhanced adherence of the cellulosic membrane composition to the support fabric and avoids interference with the membrane-forming process by the sizing.

Because of the highly pressed and smoothed form of the membrane-facing surface of the support fabric, penetration of the surface of the fabric by the dope is minimal and permits the fabric to be used as a casting surface. No need for any filler material, such as melamine resin, discussed above, is found to exist.

The membrane layer of the resulting integral membrane can be made superthin in thickness and highly flexible while still providing a salt rejection capability and water flux rate which is the same or better than that previously obtained with a much thicker membrane (e.g., 4 to 5 times thicker) supported on an as-received D-601 fabric. The integral membrane made according to the process is particularly useful in a spiral-wound module for use in water desalination applications, because of its flexibility and the greater surface area of active membrane which can be compacted into the same module space formerly used.

Pulsed Water Flow

E.L. Gibbs; U.S. Patent 3,846,295; November 5, 1974; assigned to Ultrascience Incorporated describes a reverse osmosis process which proceeds contrary to the general accepted proposition that a predetermined minimum continuous water flow is essential to efficient operation whereby such water flow is sufficient to avoid the undesired buildup of boundary layers and to maintain particulate material in suspension.

This is accomplished by suitably pulsing the concentrate water flow, wherein such flow is divided into periods of maximum flow and intermediate periods with at least a lesser flow, which may, particularly where small output volume is involved, be reduced to no flow. With this method, the operating parameters are so determined that, contrary to accepted theory, the periods of maximum flow involve a sufficient flow to provide adequate flow turbulence over the membrane surface with sufficient flushing to prevent excessive boundary layer formation at the membrane and to maintain particulate material suitably in suspension in the concentrate water with the intermediate periods of a lesser or no concentrate flow being so dimensioned as to duration that any boundary layer formation or particulate disassociation resulting during such period is materially countered by the established flow and period duration of the maximum flow period.

The apparatus may be very simple in construction, with the bulk of the components being commercially procurable items. A very efficient unit for small operations may be produced, utilizing commercially procurable membrane units, as for example tubular modules employing membrane elements of spiral configuration, in conjunction with suitable valve means in the concentrate water discharge line adapted to be operated by suitable timing means to provide desired durations of maximum and intermediate flow periods. The flow-reducing orifice in the discharge line may be of any suitable construction, being formed either as a separate element, or as a part of the valve structure, i.e., either as a separate port in the valve body or being formed by the valve orifice itself. The apparatus may conveniently include additional equipment as for example, a suitable filter unit, a recovery water circulating pump, means for effecting a monitoring of the conductivity of the reverse osmosis recovery water, etc.

CLEANING AND PRESSURE CONTROL

Cartridge-Type Cleaning Apparatus

K. Ishii, S. Konomi and Y. Ishiguro; U.S. Patent 3,883,431; May 13, 1975; assigned to Daicel Ltd., Japan describe a cleaning apparatus for use in connection with fluid-flow tubing systems, particularly for cleaning tubular semipermeable membranes. The unit comprises a cartridge which is adapted to supply scrubbing and cleaning objects such as sponge balls into the tubing system, and which is adapted to receive and contain the objects when the direction of fluid flow in the system is reversed. A by-pass conduit and a check valve are provided to permit fluid to flow around the cartridge when the flow direction is reversed.

Referring to Figure 2.13, the process provides a casing **4** serving as a cartridge or container for receiving the rubbing and cleaning members **1**. The casing **4** is open at its rightward end and is adapted for connection to the tubing system so that the members can move from the casing through the tubing system and vice versa. At its leftward end the casing **4** has a fixed porous barrier **7** which permits passage of fluid but does not allow passage of the rubbing and cleaning members.

The casing is disposed within a housing **8** and is spaced therefrom to define an annular by-pass passage **6** surrounding the casing. The by-pass passage extends from the vicinity of the rightward end of the casing to a position beyond the barrier. A tube **20** extends axially from the barrier into the fitting **21** attached to the leftward end of the housing. The tube has openings **5** so that fluid can flow from the fitting into the by-pass passage, and vice versa. A check valve **3** is slidably movable within the tube and is adapted to open and close the opening in the tube. The check valve is automatically opened and closed by reversals of the direction of flow through the housing.

One or more spherical elastically compressible members, preferably sponge balls, are used as the rubbing and cleaning members in the apparatus. The rightward end of the housing has an opening **9** adapted to be connected, as by a threaded or soldered connection, to the tubing system, which preferably is a reverse osmotic semipermeable membrane module. By the provision of such structure, connection of the reverse osmotic semipermeable membrane module (tubing system) to the housing can be accomplished very easily. The opening is sized so that the balls **1** can move therethrough from the casing into the membrane module and vice versa. The fitting on the opposite end of the housing has an opening **2** for connection, as by soldering or a threaded connection, to a pipe connected to a valving system, such as a system containing valves.

FIGURE 2.13: CARTRIDGE-TYPE CLEANING APPARATUS

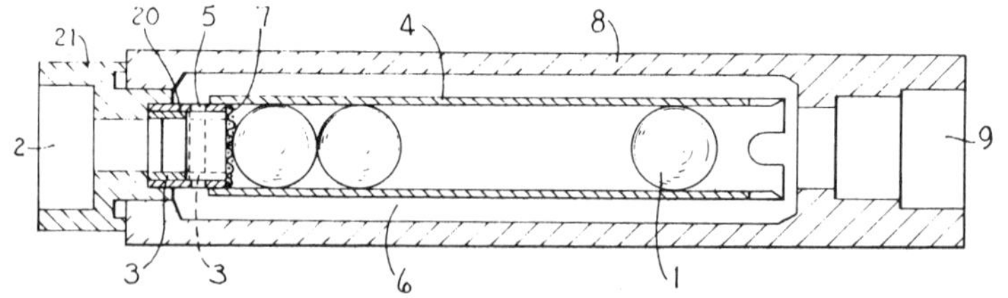

Source: K. Ishii, S. Konomi and Y. Ishiguro; U.S. Patent 3,883,431; May 13, 1975

Membrane Cleaning Solution

R.R. Stana and J. Markind; U.S. Patent 3,827,976; August 6, 1974; assigned to Westinghouse Electric Corporation describe a process that will quickly and economically clean most fouling materials from a membrane contained within a reverse osmosis liquid treatment system. The method consists of (1) contacting the fouled reverse osmosis membrane, under pressure, with a cleaning solution having a temperature between about 32° to 55°C and containing about 0.25 to 2.5 ml of an organic nonionic surfactant, about 20 to 75 grams of at least one salt selected from the group consisting of Na_2SO_4, K_2SO_4 and NaCl, per liter of water, and at least one mineral acid selected from the group consisting of hydrochloric acid and sulfuric acid in an amount effective to provide a cleaning solution having a pH between about 1.5 to 4.0, (2) releasing the pressure on the reverse osmosis membrane, and (3) flushing the contaminated solution from the membrane surface.

The release of pressure, after the cleaning solution contacts the membrane, causes liquid backflow through the membrane by natural osmosis. The backflow removes gross fouling attached to the osmotic skin side of the membrane and fouling residue trapped inside the membrane and its support. The acid dissolves most of the foulants, the hot water increases the rate of cleaning as well as the solubility of the fouling materials, the organic nonionic surfactant acts as a wetting agent, which easily penetrates the membrane, softens the foulants and keeps nonsoluble foulants suspended in the cleaning solution until they can be flushed away from the membrane surface, and the salt provides a solution with a high osmotic pressure to insure proper backflow through the membrane.

Reverse Pressure Cleaning Technique

According to a process described by *R.R. Stana; U.S. Patent 3,853,756; December 10, 1974; assigned to Westinghouse Electric Corporation* foulants are removed from the surface of a supported semipermeable membrane by (A) contacting the feed side of a semipermeable membrane, supported by a wall of a module comprising resin-bonded filler particles, with a continuous stream of liquid feed solution under a driving pressure of up to 1,000 psi, (B) applying a back pressure, on the support side of the membrane, of up to 50 psi greater than the driving pressure, to cause a backflow of liquid through the membrane, from the support side to the feed side, and (C) flushing the solution containing the concentrated foulants from the feed side membrane surface.

Example: The liquid purification system used consisted of a driving feed pump, a cylindrical reverse osmosis module containing 18 axial feed bores, the inside walls of which supported semipermeable ultrafiltration membranes, and a permeate reservoir tank located 17 feet vertically above the module.

The membranes were cast in place in the module bores from a silica-filled casting solution, containing about 20 weight percent cellulose acetate, 45 weight percent acetone and 35 weight percent formamide liquid component, and were of substantially uniform cross-section having an average pore diameter of about 40 to 60 A. Preferred ultrafiltration membranes will have an average pore size diameter of between about 10 to 150 angstrom units. The membranes used had no NaCl rejection and about 5 to 10% rejection on sucrose.

The membranes in tubular form, were supported by the interior walls of 18, ½-inch diameter axial bores through a cylindrical support module. The module was 48 inches long with a 4-inch outside diameter, and was made of phenolic resin-bonded 180-micron foundry sand particles, with about 4 weight percent resin content based on resin-coated filler particle weight. The sand module contained about 9 square feet of active membrane surface area. The sand module in turn was enclosed in a stainless steel shell having permeate outlet holes.

Using this liquid purification system, the following data were obtained using a feed pump pressure of 70 psi and a temperature of 24°C. The feed solution contained protein and lactose fraction from ultrafiltration of a whey protein lactose concentrate.

Feed	Operating Cycle	Product Flux (ml/min)	Back Pressure (psi)	Reverse Flux (ml/min)	Time (min)
tap water	continuous	2800 ml/min	—	—	—
0.5% lactal-bumen	continuous	1000 ml/min	—	—	2
0.5% lactose					
do.	continuous	420 ml/min	—	—	45
do.	continuous	450 mil/min	—	—	60
do.	10 sec. pump off 50 sec. pump on	820 ml/min	7.5	150	120
do.	5 sec. pump off 55 sec. pump on	880 ml/min	7.5	150	180
do.	10 sec. pump off 90 sec. pump on	850 ml/min	7.5	150	300
do.	continuous	506 ml/min	—	—	360

As can be seen from the above, the product flux decreased from 2,800 to 1,000 ml/min, 2 minutes after the lactalbumen and lactose were added to the tap water. After operation for 60 minutes, the product flux was down to 450 ml/min, whereupon a back pressure was applied, by shutting off the feed pump and releasing reservoir back pressure, causing the net product flux to rise to 820 ml/min after total operation of 120 minutes.

Periodic back pressure maintained this level. Subsequent continuous operation decreased product flux to 506 ml/min. The back pressuring is made possible by the integral bond of the membrane to the sand walls which permits continuous cycling at suitable pressures without damaging or causing separation of the membrane. The back pressure in this case caused a reverse flux of about 6 gal/day for each sq ft of membrane surface area.

When the pressure is increased, the flux does increase instantaneously, but so does the fouling rate which returns the flux to its original level in a few seconds. Increases in applied pressure to the feed also extends to compact the foulant gel on the membrane. Thus, low-driving pressure operation of from about 15 to 250 psi is preferred, since it saves energy costs and prevents shearing of the whey solution.

Back-pressuring can also be used in low driving force ultrafiltration of such other materials as blood, enzymes and other proteinaceous solutions which can be damaged by the high shear velocities that some ultrafiltration membranes require for efficient operation.

Periodic Relaxation of Pressure

E.R. Watson, H.W. Heidsman and B. Keilin; U.S. Patent 3,250,701; May 10, 1966; assigned one-half to Aerojet General Corporation and one-half to the United States Secretary of the Interior describe the stabilization of desalination membranes.

It has been found that the flux rate may be maintained at a high level by intermittently relaxing the pressure applied to the feed solution on the upstream side of the reverse osmosis cell and then repressurizing the system to resume continuous operation. The exact nature of the mechanism for stabilizing the membranes by such interruption of continuous operation is not completely understood.

However, it is believed that the degradation of the membranes is related to the plastic flow of the polymer under applied pressure which tends to compress the highly porous material, thus causing the fresh water flow through the membranes to decrease. The periodic relaxation of upstream pressure for a period of about one hour for each day of operation is believed to permit elastic rebound of the structure before a permanent plastic deformation takes place.

Automatic Pressure Control of Flow

D.T. Bray; U.S. Patent 3,887,463; June 3, 1975; assigned to Desalination Systems, Inc. describes a reverse osmosis system which includes a pressurized purified water storage container and an automatic valve. The valve has a sealing member spanning a first internal cavity and movable responsive to the pressure of purified water in the storage container. A central internal cavity is of effective cross-sectional area smaller than that of the first cavity spanned by a sealing member operatively connected to the member sealing the first cavity. A third cavity has an opening of cross-sectional area smaller than the effective cross-sectional area of the central cavity and leading to the central cavity which may be opened or shut off by a sealing member operatively connected to the sealing member in the central cavity.

When the pressure in the storage container rises above a predetermined value, it acts through the sealing members in the first and central cavities to shut off the pressurized feed solution introduced into the third cavity from flowing through the opening of the third cavity into the second cavity and then to the input of a semipermeable membrane-containing module.

When the pressure in the storage container drops below a lower predetermined value, it acts through the sealing members in the first and central cavities to open the opening of the third cavity leading to the central cavity, thus restoring flow of feed solution through the central cavity and then to the input of the semipermeable membrane-containing module.

Referring to Figures 2.14a through 2.14c, the control valve **10** comprises a cylindrical body **12** having, near one end, a first internal cavity in the form of circular bore **14**, a central internal cavity or circular bore **16** communicating with the lower part of bore **14** and being of smaller effective cross-sectional area or diameter than bore **14**, and a third internal cavity or circular bore **18** having an opening **19** of smaller effective cross-sectional area of diameter than bore **16** and communicating with it. The juncture of bores **16** and **18** may be formed as a tapered shoulder **20**. Perforating body **12** and communicating with the lower part of bore **16** is conduit **22** into whose outer end may be threaded feed solution output connector **24**. Also perforating body **12** and communicating with the lower part of bore **18** is conduit **26** into whose outer end may be threaded feed solution input connector **28**.

The top of first bore **14** in body **12** is closed as by plug **30**, which is sealed to the inner side wall of bore **14** by resilient O-ring **32** sealed in groove **34** in the outer edge of the plug. Snap ring **36** seated in groove **38** in the inner side wall of bore **14** holds the plug in bore **14** against internal pressure from solution or liquid introduced thereinto. Connector **40** is threaded into conduit **42** through the plug to communicate with bore **14** inside body **12**; then connector is adapted to connect the top of upper bore **14** with a storage container for purified water.

The bottom portion of bore **14** communicates with vent conduit **44**, for example as shown through the shoulder between it and bore **16**, a vent tube outlet **46** being threaded into the outer end of conduit **44**.

Floating so that it may move up and down in bores **14, 16** and **18** is a unitary double-step piston **48** having a top or large end **50** of diameter slightly smaller than the diameter of first bore **14** and to which it is slidably sealed as by resilient O-ring **52** seated in a conventional circumferential groove around its outer edge. The central section **54** of the step piston is of diameter slightly smaller than the diameter of central bore **16** and to which it is slidably sealed as by resilient O-ring **56** seated in a conventional circumferential groove around its outer edge.

The lower or small end **58** of the step piston is of diameter slightly smaller than lower bore **18** and to which it may be slidably sealed as by resilient O-ring **60** seated in a conventional circumferential groove around its outer edge. The juncture of central portion **54** and lower portion **58** of the step piston is formed as a tapering end **62** with degree of taper roughly corresponding to the taper of shoulder **20** between bores **16** and **18**.

FIGURE 2.14: REVERSE OSMOSIS SYSTEM WITH AUTOMATIC VALVE FOR MODULE OPERATION CONTROL

a.

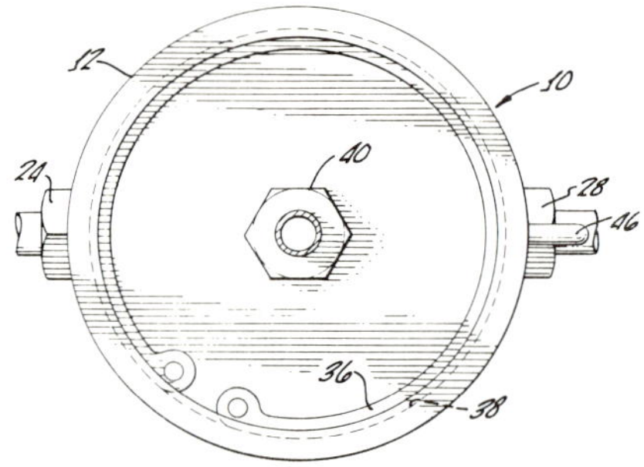

Top View of Control Valve

b.

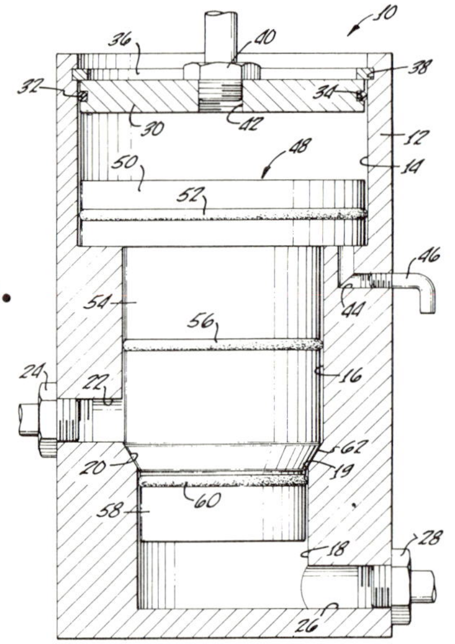

Central Vertical Section of the Valve of Figure 2.14a with the Feed Solution Supply Shut Off

c.

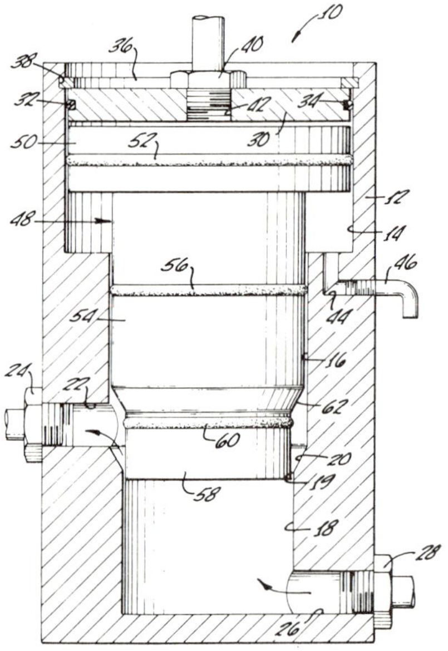

Same View as Figure 2.14b, with the Feed Solution Supply Open

(continued)

FIGURE 2.14: (continued)

d.

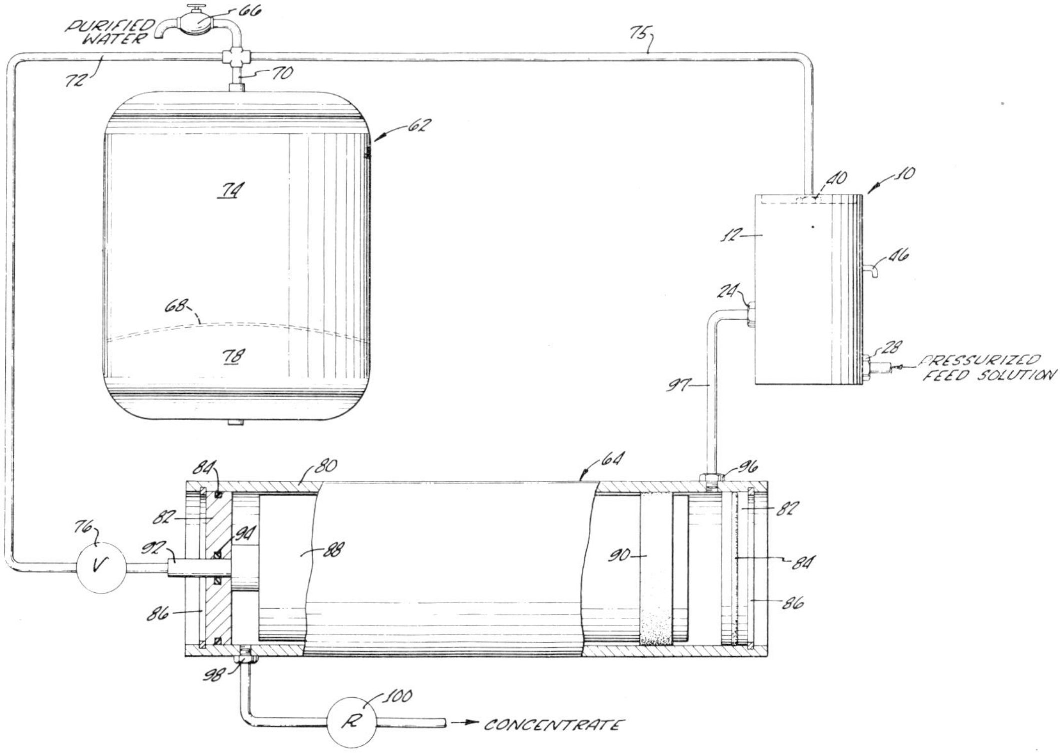

Partly Diagrammatic Representation of the Valve of Figure 2.14a

Source: D.T. Bray; U.S. Patent 3,887,463; June 3, 1975

When the step piston is in lower position as seen in Figure 2.14b, piston tapering end **62** will rest on bore shoulder **20**, with O-ring **60** sealing lower piston end **58** to the opening **19** of bore **18** and shutting off communication and flow of solution from connector **28** through conduit **26** and lower bore **18**, to central bore **16**, conduit **22** and connector **24**.

When the step piston is in raised or upper position as seen in Figure 2.14c, resilient O-ring **60** is raised up and out of bore **18**, thus providing communication and restoring flow of solution from connector **28** through conduit **26** and lower bore **18**, through the aperture between shoulders **62** and **20**, which are separated, and thus to central bore **16**, conduit **22** and connector **24**.

Referring to Figure 2.14d, a reverse osmosis system is illustrated in which a control valve **10** of the type described and illustrated in Figures 2.14a through 2.14c is employed. This system includes a storage tank **62** of the type in which purified water delivered from a reverse osmosis module **64** is maintained under pressure for delivery to dispersing means such as faucet or spigot **66**. The storage tank may be of the type in which a flexible, impervious diaphragm **68** separates the interior of the tank into two compartments. Purified water from module **64** is introduced through lines **70** and **72** from the module into the top compartment **74** of the storage tank, check valve **76** being placed in line **72** to prevent any back flow of purified water back into module **64** under any conditions of pressure relationship.

The bottom compartment **78** of storage tank **62** is filled with fluid such as air under slight initial pressure, which will maintain through diaphragm **68**, delivery pressure of desired magnitude in purified water stored in top compartment **74**, this pressure being communicated by line **75** connecting the purified water compartment **74** through line **70** to connector **40** at the top of valve **10**. Storage tanks of this general design are described in more detail in U.S. Patents 3,568,843; 3,452,199; and 3,493,496.

Semipermeable membrane module **64** may be of a commercially available type having a pressure-resistant container **80** closed at each end by plugs **82** sealed to the adjacent container walls by resilient O-rings **84** and maintained in position against internal pressure by snap rings **86** in grooves in the pressure-resistant container wall. Inside the pressure resistant container is semipermeable membrane cartridge **88** of well-known spiral-wound type of which examples are shown in U.S. Patents 3,367,504; 3,397,790; and 3,568,843, among others.

The cartridge is sealed to the inner wall of pressure-resistant container **8** by circumferential seal **90**. Purified water delivery tube **92** projecting from one end of the cartridge passes through one of end plugs **82** through a perforation therein, being sealed thereto by resilient O-ring **94**. Feed solution inlet **96** in the pressure-resistant container supplies feed solution from line **97** to one end of the cartridge and brine or concentrate outlet **98** communicates with the other end of the cartridge to provide for release of brine or concentrate after the feed solution has passed longitudinally through the cartridge and purified water which has passed through the semipermeable membrane has been delivered through tube **92**.

Concentrate outlet **98** is connected to a restrictor **100** through which a small flow of concentrate may flow while maintaining suitable high operating pressure within pressure resistant container **86**. The restrictor may be in the form of a restricted orifice, a length of small diameter tubing, a back pressure control valve of commercial type, or other equivalent device.

Preferably, the relationship of the areas of the cavities in the control valve and the cross-sectional areas of the seals spanning them should be as follows: The cross-sectional area of the central cavity and its seal is from one-third to two-thirds of the area of the cross-sectional area of the first or largest cavity and its seal, and the cross-sectional area of the opening from the third or smallest cavity into the central cavity is from one-half to slightly less than the cross-sectional area of the central cavity and its seal. An advantageous ratio is 10 to 5 to 4, that is a relationship in which the cross-sectional area of the central cavity is about one-half that of the first or largest cavity and the cross-sectional area of the opening from the third or smallest cavity is about four-fifths that of the central cavity.

In operation, the effective cross-sectional areas of the bores **14** and **16** and the opening **19** to bore **18** will have been chosen to provide the required areas which, when the appropriate pressures are applied thereto, will provide force to move piston **48** to open or close the valve, that is to shut off or restore feed solution flow from connector **24** to module **64**. For example, it may be desired to shut off feed solution supply to the module when the purified water pressure in line **75** exceeds a first predetermined pressure of 50 psi and to restore flow of feed solution to the module when the pressure in line **75** has dropped below a second predetermined pressure of 40 psi caused by withdrawal of purified water from storage tank **62** through faucet **66**.

Assume a feed solution supply pressure of 100 psi. Under these conditions, the effective cross-sectional area of bore **14**, and the corresponding surface area including O-ring **52** of the top surface of piston **48** could be one square inch, the effective area of opening **19** to be sealed by the bottom section of piston **48** would be 0.4 square inch and the effective cross-sectional area of central cavity **16** and the corresponding cross-sectional area of the central section of piston **48** with its O-ring **56** would be 0.5 square inch.

With these related dimensions, when the pressure in line **75** (essentially the pressure of the purified water in the top portion **74** of storage tank **62**) exceeds 50 psi, then piston **48** would be moved downward by the force exerted on its top, which would amount to 50 psi over an area of 1 square inch or 50 pounds to shut off feed solution flow. When the pressure in line **75** and on top of piston **48** drops to less than 40 psi, then the force exerted on the bottom of the piston through opening **19** of 40 pounds (an 0.4 in^2 area times 100 psi feed liquid pressure) would exceed the less than 40 pounds now exerted on the top of the piston and it would move upward to open opening **19** to restore flow of feed solution from connector **28** through opening **19**, conduit **22** and line **97** to inlet **96** of module **64**.

As soon as valve **10** is opened, the line pressure, 100 psi, is exerted against the greater cross-sectional area of central bore **16** and an upward force of 50 pounds on the piston results. A pressure of over 50 psi acting against the effective top surface area of the piston is required to again move piston downward to close opening **19** and valve **10** to shut off the feed solution supply to module **64**.

Therefore, the system operates with valve **10** maintaining a pressure of purified water in storage tank **62** between slightly less than 40 psi and slightly greater than 50 psi, with minimum pressure swings from 50 psi to 40 psi and back. This prevents undue opening and closing of the valve under small pressure changes, while still maintaining a pressure lag so that the turn-on pressure is less than the turn-off pressure.

When valve **10** closes at above 50 psi storage tank pressure and shuts off feed solution to module **64**, no additional purified water is produced by the reverse osmosis module to replace any used from storage tank **62** by opening faucet **66** until the pressure in the storage tank has dropped below 40 psi. Then valve **10** opens again as has been described to restore feed solution flow to the module and this on-off cycle is repeated as the pressure in the storage tank fluctuates between pressures of 50 psi to 40 psi or less. The pressure lag, in this case 10 psi, between opening and closing of valve **10** is of utmost importance. It provides a suitable turn-on pressure, in this case 40 psi, below which the pressure of purified water in the storage tank will be insufficient for adequate or desired pressurization of the purified water supplied as from faucet **66**.

As the reverse osmosis module operates when feed solution is supplied thereby by open valve **10**, purified water is produced and introduced into the storage tank through line **72** and **70** under pressure which will gradually rise in compartment **74** of the storage tank, due to the compression of the gas in compartment **78** and, of course, initially pressure exerted by the feed solution pressure less its pressure drop through the membrane in cartridge **88** while producing purified water. As the pressure in the storage tank rises, the pressure differential, which is the driving force across the membrane in cartridge **64**, decreases and osmotic efficiency decreases. It is advantageous therefore, to shut off the feed solution supply to the reverse osmosis module before the pressure differential across the membrane reaches too low a value caused by high pressure on the purified water side. Thus, the feed solution supply will be shut off when the pressure in the purified water storage tank reaches a predetermined pressure higher than the predetermined lower pressure in the storage tank **62** at which the feed solution supply to module **64** is restored.

The system of this process is advantageous in providing a pressure lag between predetermined storage tank pressures for turn-on and shut-off of the feed solution supply to the semipermeable membrane module. This results in a range of pressures between which pressurized purified water can be supplied and operation of the semipermeable membrane module shut down when its production is not at that time needed. Shutting down module operation for such periods will save an appreciable amount of water which would otherwise be wasted by continual flow of feed solution into the semipermeable membrane module, and possibly by discard of overflow purified water from the purified water storage tank.

Low-Pressure Recycling

A process described by *M.C. Gross, R.R. Stana, F.D. Klem and R.E. Shaffer; U.S. Patent 3,836,457; September 17, 1974; assigned to Westinghouse Electric Corporation* involves concentrating feed solutions, having osmotic pressures up to about 600 psi using a staged and recycle liquid purification fluid concentration system, without employing pressures significantly above the irreversible compaction pressure for cellulose ester or cellulose ether reverse osmosis membranes.

The fluid concentration system is operated by (1) pumping a feed solution of low concentration, having an osmotic pressure below about 600 psi, into a first-stage reverse osmosis membrane separation apparatus operating at a driving pressure of up to about 800 psi, to provide a first lower concentrate solution and a first high concentrate solution; (2) pumping the high concentrate solution into a second-stage reverse osmosis membrane separation apparatus, operating at a driving pressure of up to about 800 psi, to provide an intermediate low concentrate solution and a second high concentrate solution, having an osmotic pressure of between about 800 to 3,000 psi; and (3) recycling at least one of the low concentrate solutions back through at least one of the reverse osmosis membrane separation apparatus.

Figure 2.15a illustrates a typical tubular-type reverse osmosis system. Feed solution is pumped through a battery of support tubes **1** constituting the reverse osmosis apparatus. The feed pump **2** can operate as high as 1,500 psi. The tubes can be mounted in batteries of about 150 each. They serve to support the reverse osmosis membrane **5** contained within the tube wall **6**. The membrane is usually made of an ether cellulosic derivative such as cellulose acetate, cellulose acetate-butyrate or cellulose propionate, among others, or ester cellulosic derivative such as ethylcellulose, among others.

Generally, modified cellulose acetate is used. The membranes are generally cast from a solution of cellulose acetate, acetone and formamide, and reference can be made to U.S. Patents 3,170,867; 3,310,488; 3,344,214; 3,446,359; 3,593,855 and 3,657,402 for detailed information concerning these materials and their method of manufacture. These membranes have a dense osmotic skin layer which faces the feed solution and a highly porous spongy sublayer which contacts the membrane support.

Three reverse osmosis fluid purification apparatus were made, shown in Figure 2.15b as first-stage apparatus **41**, second-stage apparatus **42** and third-stage recycle apparatus **43**. Each apparatus contained series-connected reverse osmosis modules, about 4' long and 4" in diameter having 18, ½"-diameter, cellulose acetate reverse osmosis membrane-containing axial bores. The supports were tubular and made from foundry sand, having an average particle size of about 180 microns, with phenolic resin bonding the sand filler together,

FIGURE 2.15: SYSTEM FOR CONCENTRATING SOLUTIONS BY LOW-PRESSURE RECYCLING

a.

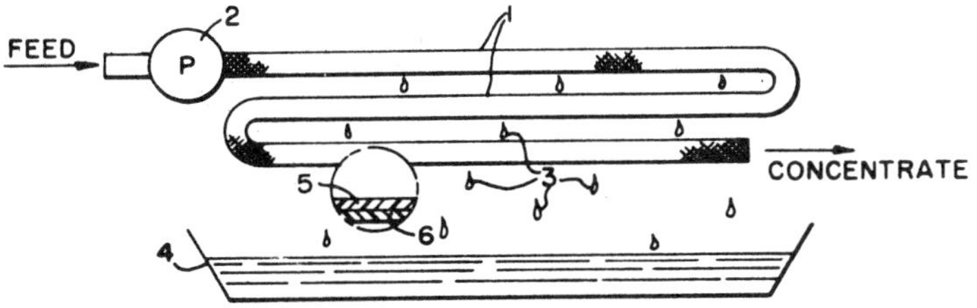

Concentrating a Feed Solution by Reverse Osmosis.

b.

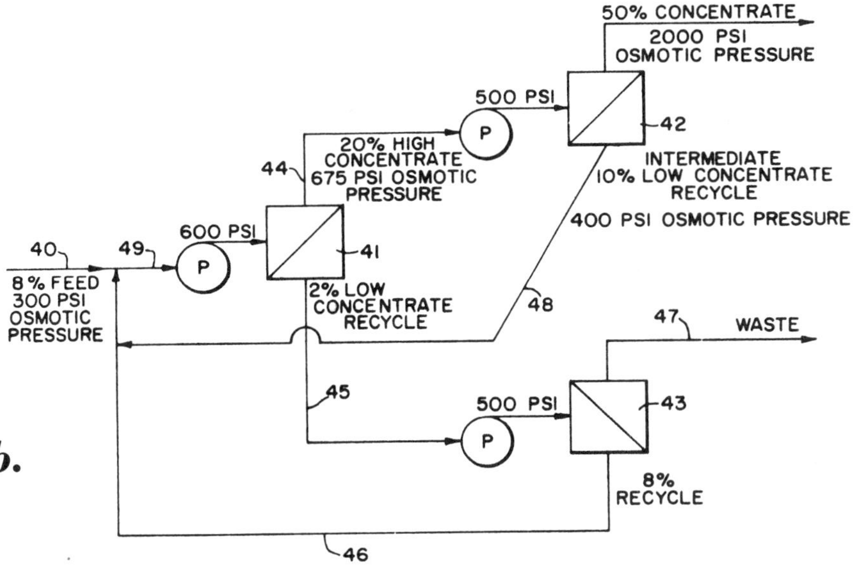

Concentration of the Feed Solution Using the Recycled Low-Pressure Method

Source: M.C. Gross, R.R. Stana, F.D. Klem and R.E. Shaffer; U.S. Patent 3,836,457; September 17, 1974

the resin constituting about 5 weight percent of the coated filler particle weight. Initial first-stage reverse osmosis membrane apparatus **41** contained about 9.25 square feet of membrane area. Recycle reverse osmosis membrane apparatus **43** contained about 9.25 square feet of membrane area and concentrate reverse osmosis membrane apparatus **42** contained about 9.25 square feet of membrane area.

The membrane in apparatus **41** had a NaCl rejection rate of 40%, the membrane in apparatus **42** was more porous than the other membranes in the system and had a NaCl rejection rate of 20%, and the membrane in recycle apparatus **43** had a NaCl rejection rate of 80%. Average fluxes were 15 gfd for apparatus **41** and **42** and 30 gfd for apparatus **43**. Apparatus **41** was driven by a 600-psi pump, apparatus **42** was driven by a 500-psi pump and recycle apparatus **43** was driven by a 500-psi pump. A low concentration feed containing 8 weight percent solids tobacco leaf extract solution **40**, having a total osmotic pressure of 300 psi, was pumped into the apparatus at 600 psi. The feed contained a mixture of inorganics such as NaCl and KCl, and organic materials. The inorganic fraction comprised about 25% of the total solids and contributed about 50% of the osmotic pressure. The organics were almost all soluble organics in water such as soluble cellulosics, esters and ethers. Reverse osmosis apparatus separated its input into a first high concentrate stream **44** and a first low concentrate solution stream **45**.

The high concentrate solution was pumped into second-stage concentrate reverse osmosis apparatus **42** at 500 psi and the solution was separated by reverse osmosis into the final high concentrate and an intermediate low concentrate solution stream **48**. The intermediate low concentrate solution was recycled back and combined with the feed solution **40** to form blended stream **49**. The low concentrate solution **45** from the initial apparatus **41** was pumped into the third-stage recycle reverse osmosis apparatus at 500 psi, and the solution was separated by reverse osmosis into a concentrated stream **46** which was pumped back and blended into the feed solution and stream, and a waste stream **47** which had only 0.3% solids and was discarded.

The 8 weight percent solids feed solution blended with recycled 8 weight percent solids, 400 psi osmotic pressure solution **46** from recycle apparatus **43** and recycled 10 weight percent solids, 400 psi osmotic pressure solution **48** from final concentrate apparatus **42** was separated into high concentrate 20 weight percent solids, 675 psi osmotic pressure and 2 weight percent solids, 100 psi osmotic pressure components, in streams **44** and **45** respectively.

The initial 8% feed produced an initial 1% permeate. As the solution passed through the apparatus and was concentrated to 20%, it produced a 4% permeate. The low concentrate recycle solution in stream **45** averaged about 2 weight percent solids. The 2 weight percent component was pumped into recycle apparatus **43** at 500 psi and was separated into 0.3 weight percent solids waste and 8 weight percent solids component. The 8 weight percent solids component was recycled into the feed stream.

The 20 weight percent component **44** from initial apparatus **41** was pumped into second-stage concentrate apparatus **42** at 500 psi and was separated into an intermediate low concentrate 10 weight percent solids component stream **48**, which was recycled into the feed stream, and the final concentrate which was concentrated into a 50 weight percent solids solution, having an osmotic pressure of 2,000 psi.

The osmotic pressure difference in this final concentrate separation apparatus was 100 psi. The feed in stream **44** produced an initial 7% permeate. As the solution passed through the apparatus and was concentrated to 50%, it produced a 30% permeate. The combined permeate averaged about 10 weight percent solids with an osmotic pressure of 400 psi.

Thus, it can be seen that an 8 weight percent feed of 300 psi osmotic pressure was processed to yield a 50 weight percent concentrate of 2,000 psi osmotic pressure and a low pollutant 0.3 weight percent waste stream of 30 psi osmotic pressure, without employing operating pressures above 600 psi, thus saving on membrane life, pump and valve maintenance and high-pressure components and providing a nonpolluting waste stream of very low solids concentration.

Back Pressure Treatment

J. Kopecek and S. Sourirajan; U.S. Patent 3,536,612; October 27, 1970; assigned to Canadian Patents and Development Limited, Canada describe a method of increasing the permeability of a reverse osmosis membrane without substantially reducing the separation factor. The method comprises contacting one side of the membrane, from which liquid passing through the membrane is withdrawn in the reverse osmosis separation, with a liquid at a pressure and for a time sufficient to effect the increase in permeability.

The back pressure treatment, it is believed, reopens and enlarges the smaller pores on a selective basis without substantially opening the larger pores. Thus, the back pressure treatment causes the liquid under pressure to permeate from the back surface to the top surface opening up the pores on the top surface and this treatment results in a new distribution of pores on the top surface with a higher average pore size. Further, it is believed that other advantageous structural changes probably also occur in the spongy porous mass underneath the top surface and it is possible to adjust the conditions of the back pressure treatment so the smaller pores are opened more than the larger ones because of the greater amplifying effect of the back pressure on the area of the smaller pores.

Thus, it is believed that the porous cellulose acetate reverse osmosis membrane has an asymmetric and amorphous structure with extensive physical cross links in the membrane structure and the initial temperature and pressure treatments create additional cross links which contribute to the rigidity of the surface layer. During the back pressure treatment according to this process the pores on the surface layer are widened and/or opened, the change in pore size being more with respect to the smaller pores than the larger ones.

Further, it is believed the back pressure treatment brings about significant viscoelastic deformations in the macromolecular structure of the surface layer; the extent of such deformation depends upon the secondary or tertiary gel structure existing in the membrane prior to the back pressure treatment. Due to such deformation, the polymer flows and the macromolecules change their spatial distribution including the rearrangement of the physical cross links and the most probable distribution of end to end distances of polymer segments finally obtained is time dependent.

Due to the long duration involved in the back pressure treatment the creep recovery is reduced, and a degree of permanent deformation is set in the membrane structure. Thus, the product rate, i.e., permeability of the solvent through the membrane by the reverse osmosis separation, can be increased by the back pressure treatment according to the process without decreasing the separation factor under a given set of operating conditions. The following examples illustrate the process.

Example 1: Six asymmetric porous cellulose acetate membranes were initially shrunk in hot water at different temperatures and subsequently normally pressure-treated for one hour at 700 psig. The membranes were then subjected to a reverse osmosis separation in a cell such as that described in U.S. Patent 3,133,132 in which they were subjected to feed stream of an aqueous sodium chloride solution containing 0.5% by weight sodium chloride at a feed rate of 250 cc per minute, operating at a pressure of 600 psig on a membrane area of 7.6 square centimeters.

The membranes were then subjected to a back pressure treatment which comprises pumping pure water at ambient temperature past the back side of the membrane at 400 psig for 85 hours. The back-pressure-treated membranes were then subjected to the same reverse osmosis separation as the untreated membrane. A plot of the product rates obtained in the reverse osmosis separation against the solute separation factor was made as shown in Figure 2.16a.

In Figure 2.16a line I is the plot for the untreated membranes and line II is the corresponding plot for the same membranes which have been subjected to the back pressure treatment. It will be seen that line II is considerably above line I and the arrows indicate the change in performance of the individual membranes tested.

It will be seen that as a result of the back pressure treatment the product rate increased by 15 to 30%, the solute separation either remained the same or was better in all cases except one in which the solute separation decreased by 0.6%. Thus it will be seen that the back pressure treatment according to the process improves membrane performance considerably. The reverse osmosis separation with the back-pressure-treated membranes was continued for a period of 72 hours.

FIGURE 2.16: BACK PRESSURE TREATMENT OF REVERSE OSMOSIS MEMBRANE

a.

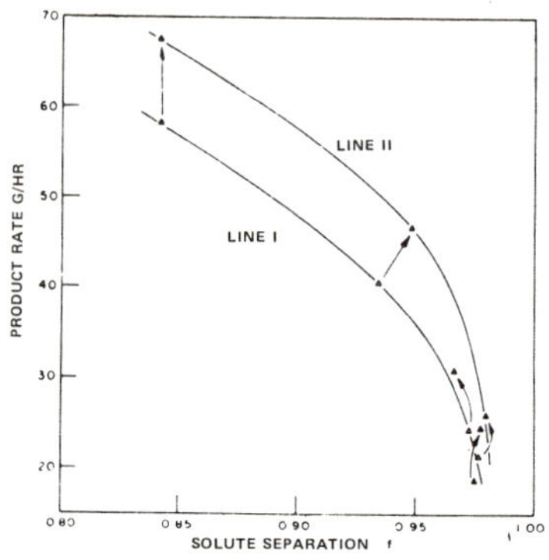

Product Rate for Untreated (I) and Treated (II) Membranes

b.

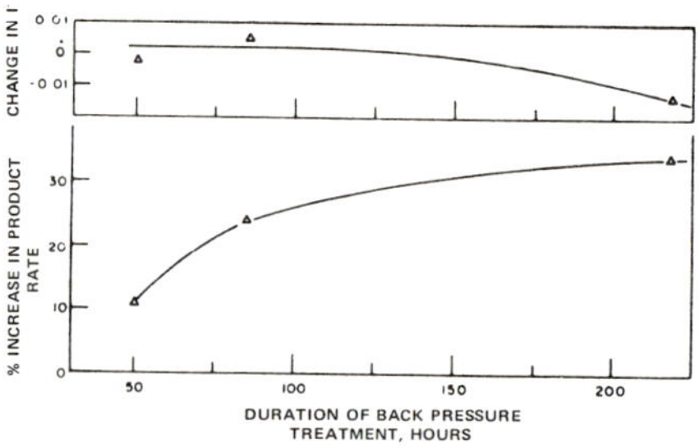

Effect of Duration of Back Pressure Treatment

Source: J. Kopecek and S. Sourirajan; U.S. Patent 3,536,612; October 27, 1970

The performance of porous cellulose acetate membranes before and after back pressure treatment was tested. The test data show that the structure of the surface layer of the back-pressure-treated membrane remains stable under the reverse osmosis conditions and it can be concluded that the reduction in product rate obtained due to membrane compaction during the extended operation is no more than that obtained with membranes not subjected to back pressure treatment. Thus the 15 to 30% increase in product rate obtained by the back pressure treatment is a net gain in the producitivity of the membrane for saline water conversion and similar applications.

Example 2: A set of six asymmetric porous cellulose acetate membranes which had been preshrunk and normally pressure-treated as in Example 1 giving solute separation in the range 85 to 99% were subjected to a back pressure treatment as in Example 1 under pressure of 400 psig for 50 hours. A similar set of six asymmetric porous cellulose acetate membranes were subjected to a back pressure treatment at 400 psig for 85 hours and a third set of six asymmetric porous cellulose acetate membranes were subjected to a similar back pressure treatment at 400 psig for 218 hours. The membranes were then subjected to a reverse osmosis precisely in the same manner as in Example 1 and the variation in product rate and solute separation were determined.

The results obtained are given in Figure 2.16b from which it will be seen that the product rate increased significantly and that the solute separation changed less than 1%. Thus Figure 2.16b gives the average change obtained in each set of six membranes and it will be seen that a duration of 145 hours at that pressure treatment at 400 psig gave an average increase in product rate of 30% with substantially no change in the separation factor in the reverse osmosis separation. Further, it will be seen that with 85 hours' back pressure at 400 psig the average increase in product rate was 23% and there was, in fact, a slight increase in solute separation and this back pressure and period of treatment was considered optimum for use in the demineralization of salt water.

OTHER PROCESSES

Grooved Rod Assembly

I.D. Aitken, R. Gayler, J.R. Grover and R.A.P. Spencer; U.S. Patent 3,819,056; June 25, 1974; assigned to United Kingdom Atomic Energy Authority, England describe a membrane assembly for reverse osmosis apparatus, which assembly comprises a rod having at least one longitudinally extending groove therein, a permeable substrate in the form of a sleeve covering the rod, a reverse osmosis membrane deposited on the permeable substrate, there being an opening providing for direct communication with the groove.

Preferably the groove extends to at least one end of the rod, thereby to provide the opening at the end of the rod. Very desirably, at the end of the rod remote from that at which the groove reaches the end of the rod, the rod, substrate and membrane are sealed together. In this way the product or permeate passing through the membrane is collected by the permeable substrate and channeled into the groove to flow along the groove and emerge at that one end of the rod.

Although metal rods could be used, it is preferred to use rods of a plastics material and preferably of polypropylene or an unplasticized grade of polyvinyl chloride. A convenient length of rod is 8 feet (2.44 meters) and a convenient diameter is ⅛ inch (0.318 cm). Such a rod can be made by continuous extrusion using conventional extrusion materials and machinery, and the grooves can be made either during the extrusion process or by machining subsequent to extrusion.

It should be explained that the grooves do not need to be straight although this is convenient and indeed preferable. A convenient size of groove for an arrangement as above described is a groove of square section and 0.020-inch (0.051 cm) deep and 0.020-inch (0.051 cm) wide. Three such grooves are conveniently provided on each rod and calculations show that they will permit the flow of all the permeate that is expected to flow through such a membrane assembly. However, in one arrangement three larger grooves 0.030-inch (0.076 cm) deep and 0.030-inch (0.076 cm) wide have been employed. In yet another arrangement six grooves 0.020-inch (0.051 cm) deep and 0.020-inch (0.051 cm) wide have been employed.

Alternatively to providing a plurality of small grooves, a single large groove may be used. In one form this may be of keyhole section so that the rod could be considered as a tube with a slit therein communicating the surface with the bore.

The most convenient form of substrate is a winding of thread and this thread does not of itself need to be permeable, the permeability of the substrate being provided by the gaps between the threads. Threads of glass fiber, polyester fiber, nylon or polypropylene are typical examples and a polyester fiber is preferred. The threads may be applied to the rod by winding or braiding and, depending upon whether it is convenient to apply the threads immediately after the rod has been formed, the thread may be applied either directly or at a later stage. It is particularly convenient to apply the threads by a braiding machine operating directly following the output of the extrusion machine.

In the case of a single large groove, particularly a keyhole groove, part of the groove may be filled with a tape or strand of fibers to prevent the substrate penetrating into the groove.

The apparatus shown in Figures 2.17a through 2.17c comprises a plurality of reverse osmosis rods **10** of which seven are shown in Figure 2.17a and a section through one of them in Figure 2.17b. Each rod **11** is in the form of an 8-ft-long, ⅛-inch-in-diameter rod of plastics material which in the preferred construction is polypropylene. The rod is provided with three longitudinally extending grooves **12** of dimensions 0.020 inch square. Over the grooved rod is a permeable substrate **13** which is in fact a braiding of polyester fibers. Over the substrate is a semipermeable membrane layer **14** derived from cellulose acetate in the known way.

FIGURE 2.17: MEMBRANE ASSEMBLY COMPRISING A GROOVED ROD COVERED WITH PERMEABLE SHEATH

a.

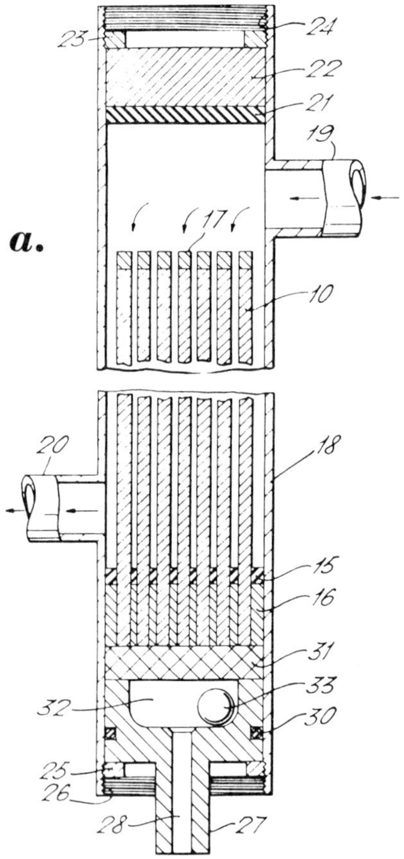

Cross Section Through Part of a Reverse Osmosis Apparatus

b.

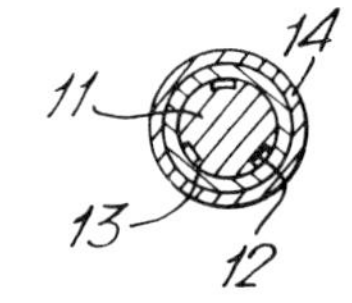

Cross Section of a Rod

c.

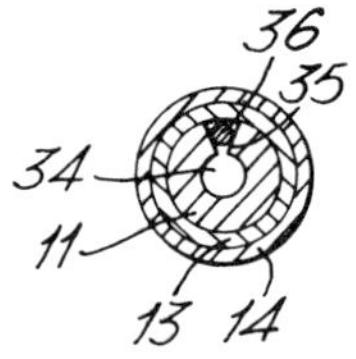

View Similar to Figure 2.17b Through an Alternative Rod

Source: I.D. Aitken, R. Gayler, J.R.Grover and R.A.P. Spencer; U.S. Patent 3,819,056; June 25, 1974

After being made as above described in continuous lengths the rods are cut into the 8-ft lengths required and then one end of each rod is pushed through a molded rubber seal **15** which has the necessary apertures to receive the rods. The ends of the rods are then pushed through a locating disc **16** which is of apertured form similar to the molded rubber seal. The effect of the locating disc is to provide an adequate support for the ends of the rod.

If the rods protrude through the locating disc, the ends are trimmed, and it will be appreciated that any damage done in inserting the rods into the locating disc or in trimming the ends will be on the side of the rubber seal remote from the high-pressure brine to be treated. The complete bundle of rods is dipped at its other end in a sealing medium, for example polyvinyl alcohol, followed by a dip in formalin to polymerize and seal the ends of the rods remote from the locating disc. This sealing is shown diagrammatically at **17**. It is, however, preferred to seal the ends of the rods at **17** with molded rubber end caps which fit tightly over the ends of the rods over the top of the membrane and braiding.

In this case, however, it is most desirable for the rods to be cut with differing lengths. This avoids possible binding together of the rubber end caps and consequent blocking. The complete bundle as above described forms a replacement unit for insertion into the reverse osmosis apparatus if necessary.

As shown in Figure 2.17a, this reverse osmosis apparatus comprises a tubular shell **18** made of stainless steel or the like and of a sufficient thickness to withstand the pressure which must be applied in reverse osmosis. Adjacent each end is a side-limb **19, 20** for the inflow and outflow of the brine, and the arrangement is desirably such that the side-limb **19** is located above the sealed end **17** of the rods, while the side-limb **20** is located close to the seål **15**. Above the side-limb **19** is a molded rubber seal **21** which engages an end piece **22** held in position by an externally threaded nut **23** engaging a screw thread **24** on the inside of the pressure shell **18**.

At the other end of the pressure shell a similar screw-threaded nut **25** engages a screw thread **26** and holds in position an end cap **27** which is provided with an internal bore **28** for the outflow of the product fresh water. This end cap is sealed to the pressure shell by means of an O-ring **30**, and its inner end engages against a porous support member **31** made of sintered metal or the like which in turn engages against the end of the locating disc. The function of the porous support is to provide a channel for the product fresh water from the substrate on each of the rods to convey the same to the channel **28**. The support **31** together with the end cap **27** also absorbs the end load on the rods due to the pressure of the brine within the pressure shell.

The inner end of the end cap is recessed as shown at **32** to accommodate a ball **33** functioning as a ball valve. In normal operation the ball is located as shown. However, should the membrane on any of the rods become defective and fail, the flow of water through the apparatus will increase very markedly, for example by a factor of 10, and since the ball is located so as to be flow-sensitive, it will be moved from the position shown to seal the exit from the end cap and thus cut off all flow from the module. This has the advantage that in the case of failure it is not necessary to shut down the plant, since the failed module automatically shuts itself down and can be withdrawn at some convenient time when the plant is shut down for maintenance or the like. The O-ring is necessary because under failure conditions the high-pressure brine may be in contact with the end cap.

The alternative form of rod shown in Figure 2.17c has a single keyhole-shaped groove and is conveniently described as a ⅛-inch-diameter rod having a $\frac{1}{16}$-inch axial bore **34** with a V-slot **35** communicating the bore with the surface. In this V-slot is located a woven or braided strand of fibers **36** designed to permit the passage of water but prevent the deformation of the substrate into the groove.

With the 37-rod configuration, the internal diameter of the shell should be about 1½ inches. The output using 8-ft-long rods is about 200 gpd. With the 91-rod configuration, the internal diameter of the shell should be about 2½ inches. The output using 26-ft-long rods can approach 1,000 gpd.

Helical Configuration

G. Segovia; U.S. Patent 3,873,665; March 25, 1975; assigned to Oxy Metal Industries Corporation describes a process for producing a reverse osmosis module in which a delicate reverse osmosis membrane is wrapped about a coil form. The rejection characteristics of the reverse osmosis membrane are developed by a heat-treat process. Upon termination of the heat-treat process and while conditions of the heat-treat process are maintained within the membrane, the membrane is coiled about a coil form in order to form a reverse osmosis module in which the membrane is in a helical configuration. The membrane is cooled in its helical configuration.

Flexible Webs Having Enhanced Releasability

A process described by *M.E. Rowley, W.D. Slowig and C.F. Smith; U.S. Patent 3,636,150; January 18, 1972; assigned to Eastman Kodak Company* relates to improved procedures for casting and subsequently (immediately after casting) handling films of cellulose ester, such as cellulose acetate, that are intended for use as "reverse osmosis" membranes.

It has been found that the problem of excessive adherence between the dimensionally stable polymeric films useful as webs upon which the cellulose ester dopes can be cast is caused by the fact that the surfaces of such polymeric films are almost always very hydrophobic in nature. As a result, the problem can be solved by providing a fairly hydrophilic surface (on the dimensionally stable polymeric web) upon which the cellulose ester dope solution is cast.

The degree of hydrophilicity to which the surfaces of the polymeric webs useful in this process must be converted is represented by the well-known measurement of "contact angle" (of a drop of distilled water with the surface of the polymeric web). Those that are useful in the process must have a contact angle (at 25°C) of at most about 55°C.

The polymeric webs that are particularly useful for the purpose of providing a smooth base upon which the layer of dope solution can be cast in accordance with generic processes for manufacturing high-quality, uniform reverse osmosis membranes must be extremely resistant to losses in dimensional stability such as wrinkling, stretching, softening and the like when in contact with the dope solutions themselves, the aqueous "leaching" baths, and temperature changes of as much as 30°C or more during the various processing steps.

Typical examples of useful polymeric films are the biaxially oriented, heat-set polyesters such as poly(ethylene terephthalate) and poly(1,4-cyclohexylenedimethylene terephthalate) and related copolyesters, and biaxially oriented polyolefins such as high-density (0.94 to 0.97 g/cc) polyethylene, polypropylene, and the like, as well as films of heat-treated cellulose triacetate (having at least 43% acetyl). Unfortunately, these "dimensionally stable" films all have naturally very hydrophobic surfaces. Thus, in order to be used in the process at least one of their surfaces must be made more hydrophilic (by comparison with its natural hydrophobicity).

Examples: In a conventional stainless steel mixer are blended for several hours the following ingredients:

100 parts cellulose acetate (39.8% acetyl, I.U. 1.5)
180 parts acetone
120 parts formamide

The result is a typical cellulose acetate dope solution. This dope solution is cast in a layer 0.10-inch thick onto the dope surface of a flat, moving web of polymer that is being tested. The moving polymeric web is run continuously, at a rate of 2 to 20 feet per minute, under a hopper from which the dope solution is cast, and then, immediately thereafter, through an area in which the layer is exposed to air for a short time (less than 30 seconds). Then the layer is run directly into a large pool of cool (34°F) water, whereupon the layer "sets" almost immediately. After moving in the water bath for 50 seconds, the set layer of cellulose acetate film is stripped from the moving polymeric web. The ease of such stripping is observed and recorded. Results of this test with various polymeric webs, some treated in accordance with the process, are listed in the table below. Contact angles are measured in accordance with the procedure of H. Schonhorn as published in *Nature,* volume 210, page 896 (1966).

No.	Polymeric Web Material	Surface Treatment	Amount of Coating (mg/ft^2)	Strippability	Contact Angle, °
1	Poly(ethylene terephthalate)	None	-	Very poor	71
2	Poly(ethylene terephthalate)	Rough finish (mechanical)	-	Very poor	------
3	Poly(ethylene terephthalate)	Gelatin	-	Excellent	40
4	Poly(ethylene terephthalate)	Alkali-treated[1]	-	Good	------
5	High-density polyethylene	None	-	Very poor	110
6	High-density polyethylene	Electron bombardment[2]	-	Excellent	55
7	Polypropylene	None	-	Very poor	------
8	Polypropylene	Electron bombardment[2]	-	Good	50
9	Cellulose triacetate	None	-	Very poor	60
10	Cellulose triacetate	Alkali hydrolysis[3]	-	Excellent	31
11	Cellulose triacetate	Gelatin	90-150	Excellent	40
12	Cellulose triacetate	Sodium ethyl cellulose sulfate	30-60	Excellent	------
13	Cellulose triacetate	Polyvinyl alcohol	90-150	Excellent	------
14	Cellulose triacetate	1-gelatin/cellulose nitrate:50/50 2-gelatin	60-60 90-150	Excellent	40
15	Poly(ethylene terephthalate)	MaVelt terpolymer[4] plus gelatin overcoat	60-90	Excellent	40

[1]Treated for 600 seconds with 10 weight percent NaOH in water at 70°C
[2]By passing film between a highly charged bar electrode and a grounded metal roll
[3]Treated for 1 minute with 5 weight percent of NaOH in methyl alcohol and water (60-40) at 180°F
[4]83% vinylidene chloride, 15% methyl methacrylate and 2% itaconic acid

Continuous Fabrication of Elongated Web

E.R. Watson, G.V. Rowley and C.R. Wunderlich; U.S. Patent 3,432,585; March 11, 1969; assigned to Aerojet-General Corporation describe a method and apparatus for continuously fabricating a desalination membrane material of cellulose acetate in an elongated web form. The membrane material is of high uniformity and is suitable for use in desalination processes operating on the principle of reverse osmosis. In practicing the method, a film-forming casting solution of cellulose acetate is continuously deposited along the surface of a moving elongated belt which is directed into a refrigerated water bath to subject the coating of casting solution on the belt to the refrigerated water bath. The coating gels to form a film which is automatically stripped from the belt and directed into a heated water bath, with the film being thereafter wound on a take-up roll and stored under water in a continuing sequence.

Figure 2.18a diagrammatically illustrates the method and apparatus for continuously fabricating a desalination membrane in accordance with the process. Reference numeral **10** broadly designates a casting box assembly which comprises a housing **11** having a hopper or receptacle **12** serving as a casting box adjustably mounted therein and a table **13** supported on the floor of the housing. The forward end of the table is equipped with a delivery roll **14** on which an elongated belt **15** is wound, the belt being guided longitudinally across a platform defined by the top of the table within the housing by suitable guide means, such as a guide roll **16** and out through a slot **17** provided in the rear wall of the housing to a take-up roll to be subsequently described. A refrigeration unit **20** is associated with the housing so that the interior of the housing may be maintained at a predetermined temperature, preferably below ambient temperature.

FIGURE 2.18: CONTINUOUS FABRICATION OF DESALINATION MEMBRANE

a.

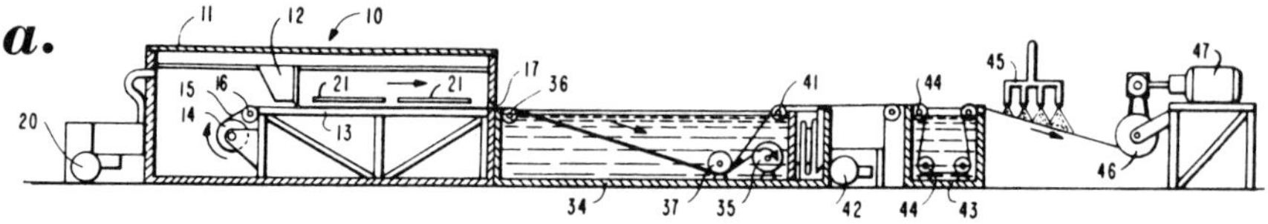

Schematic View of Apparatus

b.

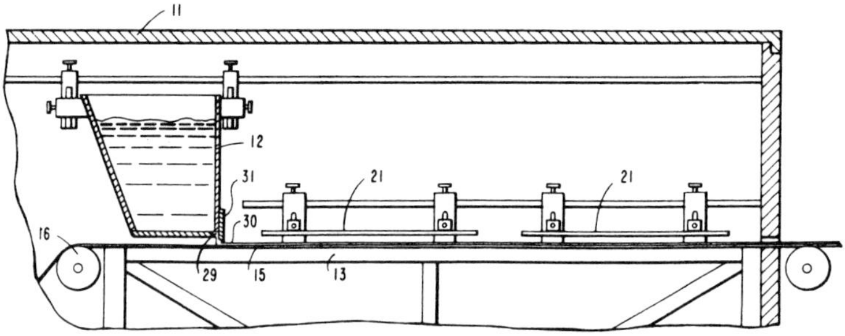

Enlarged View of Casting Box

c.

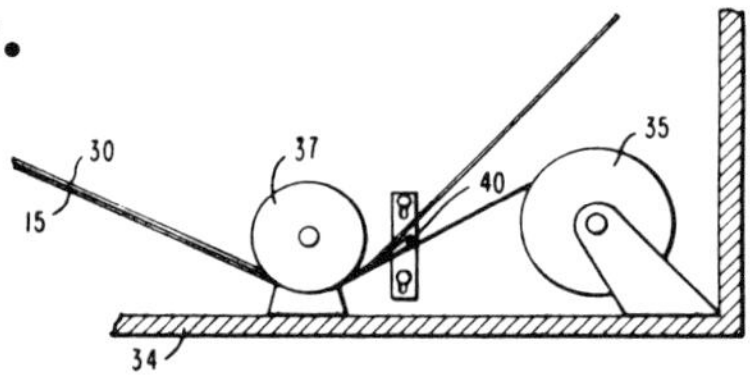

Refrigerated Water Bath Treatment

d.

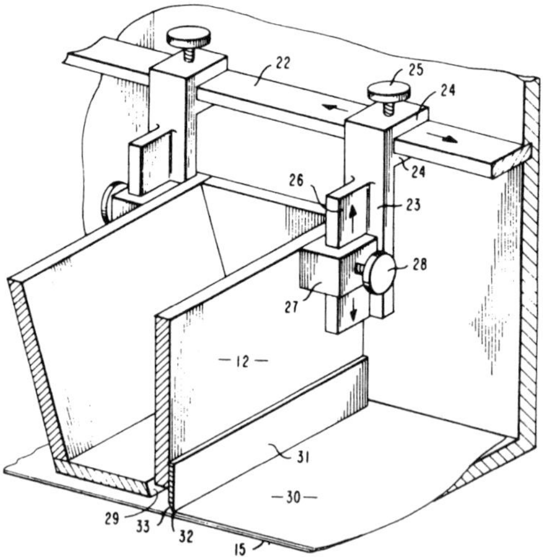

Casting Box and Mounting Assembly

Source: E.R. Watson, G.V. Rowley and C.R. Wunderlich; U.S. Patent 3,432,585; March 11, 1969

As seen more clearly in Figure 2.18b, the housing further contains a series of vapor baffles **21**, two such vapor baffles being illustrated. The vapor baffles comprise a plurality of plates which are arranged in longitudinal alignment so as to be disposed in spaced, superimposed, parallel relationship to the platform defined by the top of the table and the belt extending across. The vapor baffles may be independently adjusted vertically so as to vary the space between the respective vapor baffles and the belt extending across the top of the table, such vertical adjustment of the vapor baffles being accomplished by suitable means.

The hopper or receptacle serving as the casting box is mounted for longitudinal and vertical adjustment within the housing so as to be disposed above the table at the forward end. Referring to Figure 2.18d, it will be observed that the opposite side walls of the housing are respectively provided with an internal elongated track **22** running the length of the housing, only one such track being illustrated in Figure 2.18d. The tracks are disposed adjacent the top of the housing such that the casting box can be dependingly supported therefrom above the top of the table. Means are provided to mount the casting box in depending relation from the tracks so as to permit both longitudinal and vertical adjustments of the casting box with respect to the top of the table.

The mounting means comprises four mounting assemblies of identical construction disposed respectively at the four corners of the casting box. Each of the mounting assemblies includes a bracket member **23**. The bracket member is provided with a pair of spaced parallel legs **24** extending transversely outwardly from the upper end and defining a track-receiving channel. The legs serve as runners which are respectively disposed for slidable engagement with the upper and lower surfaces of the track, the track being received in the channel defined by the space between the legs. A thumb screw **25** is mounted in a suitably threaded bore provided in the upper leg and is adapted to extend through the engagement with the upper surface of the track so as to fix the bracket member at a selected location along the track when the thumb screw is tightened.

The bracket member further includes a vertical flange **26** which extends transversely inwardly from the inner surface, the vertical flange acting as a bracket track along which the casting box may be slidably adjusted vertically with respect to the top of the table. To this end, each of the mounting assemblies further includes a channel-shaped guide member **27** which is affixed to the casting box at a respective corner for slidably receiving the bracket track of the bracket member to permit vertical adjustment of the casting box. Each channel-shaped guide member is provided with a thumb screw **28** which is received in a suitably threaded bore formed in the outermost leg of the channel-shaped guide member and is adapted to extend through into engagement with the opposed surface of the bracket track **26**. The thumb screw when tightened fixes the channel-shaped guide member in a selected vertical position on the bracket track.

The casting box is of the gravity flow type in that the film-forming solution to be contained therein is dispensed by gravity through a transverse slot **29** provided in the bottom of the casting box. As shown, this transverse dispensing slot is defined in the casting box by spacing the bottom wall from the rear wall, wherein the space between the bottom and rear walls of the casting box has a transverse extent equal to the internal transverse width of the casting box as measured between the internal surfaces of its side walls. A transversely extending doctor blade **31** is fixedly mounted across the outer surface of the rear wall of the casting box so as to depend partially below the bottom wall of the casting box. The lower leveling end of the doctor blade comprises a flat lower edge surface **32** and a beveled surface **33**, rather than a knife edge.

The casting box is adapted to contain a film-forming casting solution which is gelable to form a desalination membrane **30** suitable for use in desalination processes involving the reverse osmosis technique. The film-forming casting solution is dispensed by gravity flow through the transverse slot **29** defined in the bottom of the casting box and onto the belt which extends across the top of the table such that the belt forms a substrate for the film-forming casting solution which is applied as a uniform coating along the length of the belt. In this respect, the belt is caused to undergo continuous movement across the top of the table as the film-forming casting solution is applied from the casting box. The doctor blade **31** performs a leveling function on the gelable casting solution as it is dispensed from the casting box onto the belt where the flat lower edge surface **32** of the doctor blade conditions the casting solution so that the coating applied to the belt is of substantial uniformity along the length of the belt.

The film-forming casting solution may comprise any suitable solution which is gelable to form a desalination membrane for use in processes involving the principle of reverse osmosis to desalinate water. Typical casting solutions of this type are described in U.S. Patent 3,444,286.

Following application of the film-forming casting solution to the moving belt, the moving belt with its membrane coating **30** travels through the slot **17** provided in the rear wall of the housing and into a cold water bath. In the latter connection, a large vat or tank **34** is located adjacent to the housing and downstream so as to be in longitudinal alignment with the slot **17**. The tank is adapted to be filled with water which is maintained at a temperature lower than ambient so as to provide a cold water bath treatment for the membrane coating **30** on the moving belt. To this end, it will be observed that a belt take-up roll **35** is mounted on the bottom wall of the tank **34** at the end thereof remote from the housing.

The take-up roll is positively driven and receives an end of the belt thereon so as to wind the belt while rotating, thereby imparting longitudinal movement to the belt between the delivery roll **14** and the take-up roll. Between the take up roll and the casting box assembly **10**, guiding means are provided to direct the belt with the membrane coating **30** into the tank **34** and below the surface of the cold water for subjecting the membrane coating to a cold water bath treatment.

As shown, such guiding means may take the form of a first or upper idler roll **36** located in the upper portion of the tank **34** at the end thereof adjacent the slot formed in the housing of the casting box assembly. A second or lower idler roll **37** is located between the upper guide roll and the take-up roll at the bottom of the tank **34**. Thus, the belt **15** with its membrane coating **30** emerges from the housing through the slot at the rear of the housing and is directed downwardly by the upper guide roll to the lower intermediate guide roll for immersing the membrane coating on the belt in the cold water contained in the tank **34**. The cold water treatment sets the membrane and washes out or leaches out excess solvent material or salts from the membrane.

As the membrane-coated belt passes by the lower intermediate guide roll, means are provided for stripping the membrane coating which has been gelled and set as a solidified sheet. To this end, a wedge-shaped stripping blade **40** is mounted just to the rear of the guide roll so as to extend transversely across the tank **34**, as best shown in Figure 2.18c. While the coating on the belt has coagulated into a continuous membrane and will readily strip from the belt substrate, the stripping blade or wedge **40** facilitates such stripping action.

Following stripping of the membrane from the substrate belt, the belt continues on its path of movement to the take-up roll **35** on which it is wound, while the film or membrane is directed upwardly onto a guide roll **41** located at the top of the tank **34** adjacent the rear end wall thereof. A suitable refrigeration unit **42** is associated with the tank for maintaining the water to be contained at a predetermined temperature below ambient.

From the tank **34** containing the refrigerated water for subjecting the membrane **30** to a cold water bath treatment, the membrane is thereafter directed to a second tank or vat **43** which is adapted to be filled with water maintained at a temperature higher than ambient. The membrane is entrained about a series of guide rolls disposed at upper and lower levels within the tank **43** so as to immerse the membrane within the heated water contained in the tank. The water within the tank is maintained at a temperature higher than ambient by suitable heating means. The length of time during which the membrane is immersed in the heated water within the tank is governed by the speed of the moving membrane and the arrangement and number of the guide rolls located within the tank. The heat-treatment both anneals the membrane causing contraction of its cellulose acetate structure to give the membrane its ability to pass water therethrough while restraining the passage of salt.

A series of spray nozzles **45** are located rearwardly of the heat-treatment bath provided by the tank and the heated water, the spray nozzles being adapted to direct water sprays onto the membrane as it emerges from its heat-treatment bath for cooling purposes. Additionally, the water from the spray nozzles moistens the membrane to prevent the membrane from drying out, wrinkling, or becoming heat-set, thereby guarding against an irreversible loss of water content in the membrane.

After cooling by the water sprays released from the spray nozzles, the membrane is directed onto a positively driven membrane take-up roll **46** located downstream of the spray nozzles. The speed at which the take-up roll is driven may be varied, as desired, by a variable speed drive **47** which rotates the take-up roll. The variable speed drive is synchronized with a similar variable speed drive for the belt take-up roll. In this way, the necessary parameters as to length of treatment of the membrane during various stages of the method may be observed.

A membrane take-up roll **46** upon receiving the desired number of windings of the membrane entrained thereabout is then placed in a storage tank where the elongated membrane is immersed in water at all times during storage to prevent loss of water content which is irreversible and adversely affects the performance of the membrane in desalinating water.

In practicing the process, a film-forming casting solution is prepared which may comprise, for example, 1,200 grams of acetone (CP grade), 500 grams of Eastman E-398-3 cellulose acetate, 250 grams of water and 125 grams of tartaric acid (NF powder grade). In the example given, the water and acetone were thoroughly mixed and the solids, cellulose acetate and tartaric acid, were added to the water-acetone solvent solution. The resulting mixture was thoroughly stirred for 20 to 30 minutes to dissolve the cellulose acetate, thereby forming a viscous casting solution which may be placed in the casting box so as to be subsequently deposited upon the moving belt.

A portion of the belt is unwound from the delivery roll and threaded onto the belt take-up roll **35** so as to extend across the platform defined by the top of the table. The belt may be made of any suitable material which will not be chemically affected by the application of the casting solution or by immersion in the cold water bath. Stainless steel or polyethylene terephthalate film, known as Mylar, are examples of suitable belt material. The belt take-up roll is then driven to wind the belt, which moves the belt across the top of the table. The belt speed was adjusted to 2 ft/min. The casting box was adjusted above the moving belt so as to set the lower flat edge **32** of the doctor blade **31** at a distance 0.012 inch above the belt.

The film-forming solution is then continuously cast onto the moving belt which passes out of the housing for the casting box assembly through the slot. During the time that the casting of the film-forming solution onto the belt was being carried out, the temperature within the housing was maintained at −11°C by the refrigeration unit **20**. The membrane coating on the belt was within the housing of the casting box assembly over a 3-minute period. The vapor baffles **21** are instrumental in maintaining a saturated atmosphere of acetone above the membrane-coated belt and in preventing convective currents from passing across the membrane-coated belt. This further aids in preventing air-drying of the membrane coating during the casting procedure.

The membrane-coated belt is then directed into the tank **34** where membrane coating **30** is immersed in the cold water which is maintained at 1.5°C by the refrigeration unit **42**. The membrane required 2.25 minutes to traverse the cold water bath treatment. It will be understood that initially the membrane is manually stripped from the belt and processed through the heated water bath treatment and past the spray nozzles **45** so that the membrane can be threaded onto the membrane take-up roll. Thereafter, the membrane take-up roll is positively driven by the variable speed drive **47** to wind the membrane thereabout and thereby impart movement to the membrane independent of the movement of the belt but in synchronization.

The membrane is then automatically stripped from the belt while undergoing the cold water bath treatment, such stripping being facilitated by the action of the wedge **40**. The membrane is then directed to the tank **43** containing the heated water which is maintained at a temperature of 85°C. An appropriate thermostatic means, not shown, is preferably associated with the heating unit for treating the water contained in the tank **43** so as to maintain a constant temperature of such water while the membrane is traveling therethrough, which in the given example was 85°C. The speed of the membrane through the heated water bath was adjusted such that the membrane was immersed in the heated water bath for a period of 3 minutes.

Emerging from the heated water bath, the hot membrane was cooled and moistened by water from the spray nozzles **45** to prevent drying out, wrinkling, and heat-setting thereof which might otherwise occur. A cellulose acetate desalination membrane prepared in accordance with the process and as related in the foregoing example was found to have a flux of 12 gfd and a salt permeation of 1.4% against 35,000 ppm sodium chloride feedwater at 1,500 psi.

A method and apparatus is thus described for continuously fabricating a desalination membrane which does not require solvent evaporation from the membrane as a separate step following the casting of the membrane from a film-forming solution. The membrane is therefore exposed to air for a minimum time only following the casting before it is immersed in a cold water bath treatment. This prevents loss of water content from the membrane which is an irreversible process and which adversely affects the proper performance of the membrane, while also permitting flexibility in the speed at which the membrane may be cast.

Desalination membranes offering superior performance have been found to have a water content ranging between 54 and 60%, and the method and apparatus consistently produce continuous desalination membranes whose water content lies within this preferred range of percentages. The method and apparatus provide for effective control over the various parameters necessary to produce a satisfactory desalination membrane so as to yield a continuous membrane exhibiting a high degree of homogeneity.

Control of Thickness of Boundary Layer at Interface

A process described by *R.E. Lacey and M.S. Mintz; U.S. Patent 3,676,335; July 11, 1972; assigned to Southern Research Institute and the U.S. Secretary of Interior* involves increasing the efficiency of reverse osmosis separation processes by reducing the thickness of the boundary layer formed at the membrane-solution interface. This is accomplished by placing a second membrane, which is substantially permeable to the solution, in a close-spaced relationship to the reverse osmosis membrane. A portion of the solution is withdrawn from the space between the two membranes.

Figure 2.19 illustrates a reverse osmosis system incorporating the process. The solution **200** to be demineralized is pressured and introduced into a cell similar to those used in conventional reverse osmosis, and the solution flows to the top surface of a porous membrane **201** that is positioned the desired distance from the selective side of a reverse osmosis membrane **203** by a net or mesh spacer **202**. The spacer material may be a net or screen-like material which provides a multiplicity of support locations for membranes **201** and **203**, and which is the desired thickness and which allows solution flow parallel to membranes **201** and **203**. An example of a suitable material is a thin net of polyethylene.

Porous membrane **201** may comprise a felted mat of fibrous materials such as paper, polyethylene, polypropylene, Dynel, Acrilan, or Nylon; it may consist of porous plastic film such as thin sheets of open-cell polyurethane foam, or cellulosic films and parchment.

FIGURE 2.19: DOUBLE MEMBRANE REVERSE OSMOSIS SYSTEM

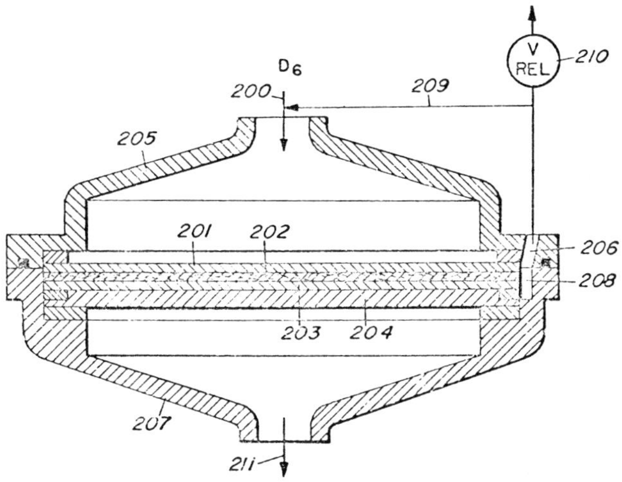

Source: R.E. Lacey and M.S. Mintz; U.S. Patent 3,676,335; July 11, 1972

The material out of which the porous membrane is made is not critical. However, the material must have the ability to pass solution fluxes of from about 10 to 650 ml per minute per square inch of membrane surface under a hydraulic pressure drop of about 2 inches of water.

The reverse osmosis membrane **203** rests on and is supported by a member support **204** of the type used in conventional reverse osmosis. Slots at the edges of the cell **208** at one edge of the net spacer **202** (or part way around the perimeter of a circular cell) are interconnected with withdrawal ducts **206** so that the feed solution flows through the porous membrane **201** toward the reverse osmosis membrane **203** and then through the solution channel formed by the net spacer **202** and the two membranes to the slots at the edges of the cell **208** and is withdrawn through the ducts **206.**

This solution may be recycled and mixed with fresh feed (via **209**) or it may be discarded to a waste stream through a pressure reducing valve **210.** The demineralized water **211** that is transferred through the reverse osmosis membrane **203** goes through the membrane support **204** and is withdrawn through conventional means **207** of enclosing and withdrawing the product.

POLYAMIDES AND OTHER NITROGEN-CONTAINING MEMBRANE MATERIALS

POLYAMIDES

Crosslinked Aromatic Polyamides

According to a process described by *R. McKinney, Jr. and W.L. Hofferbert; U.S. Patent 3,904,519; September 9, 1975; assigned to The U.S. Secretary of the Interior* reverse osmosis membranes of improved flux or flux stability are prepared by crosslinking aromatic polyamide membranes. Crosslinking may be achieved by reaction with crosslinking reagents such as aldehydes, polyamines, polycarboxylic acids, polyisocyanates, oxidizing agents, peroxides, etc. It may also be accomplished by means of irradiation, e.g., with gamma radiation.

Linear polyamides are conventionally prepared by condensation of diamines with dicarboxylic acids, or by condensation of amino carboxylic acids. The aromatic polyamides, from which the membranes of the process are prepared, are characterized by the presence of aromatic groups in the chains between the functional groups, i.e., the amide groups. The aromatic groups will generally be phenylene groups, but may also include naphthylene, 2,5-oxadiazolylene, 3,5-triazolylene, etc., and may be ortho, meta or para substituted.

The aromatic polyamides are generally prepared by low temperature, i.e., about 0° to 10°C, solution or interfacial polymerization, in which the reaction is direct amidation, i.e., reaction of amine groups and carboxyl groups with elimination of water. For this type of reaction dicarboxylic acid chlorides, rather than the corresponding acids, are usually employed and the condensation by-product is hydrochloric acid. Optimum molecular weights and viscosities of the aromatic polyamides may vary considerably, depending on the specific type of polymer, type of subsequent crosslinking employed, etc., but will usually have a molecular weight of about 20,000 to 70,000 and a viscosity of about 6,000 to 200,000 poises. Following polymerization, hydrochloric acid formed in the reaction is neutralized with a base such as $CaCO_3$, Li_2CO_3, $Ca(OAc)_2$, LiOAc, etc.

Examples of linear aromatic polyamides that may be employed are shown below in terms of the structural formula of the characteristic group of which the polymer chain is composed. As illustrated in structures 7 and 8 below, the group may include a hydrophilic or polar group, such as a carboxylic acid group, in the polymer backbone. Such a group has been found to impart increased flux level to the membrane, as well as providing for ionic crosslinking, as described below. Other suitable polar groups include amine groups, hydroxyl groups, $O(CH_2CH_2O)_xH$ groups, etc.

(1) 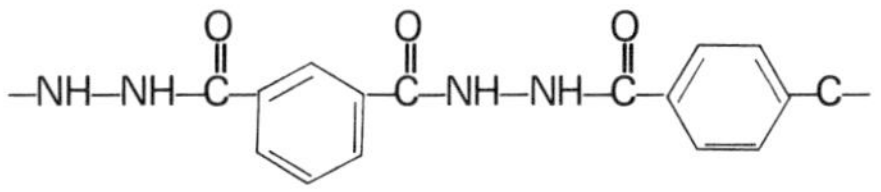

(2) 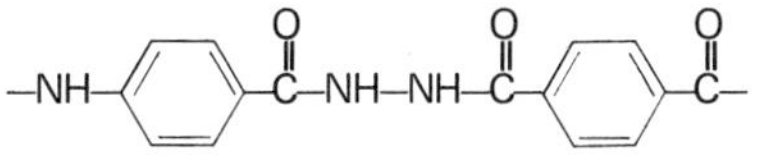

(3) 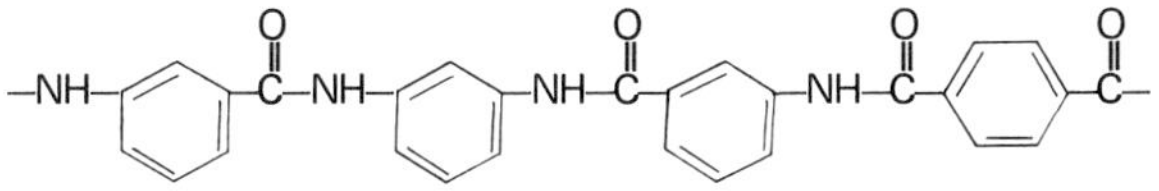

(4) 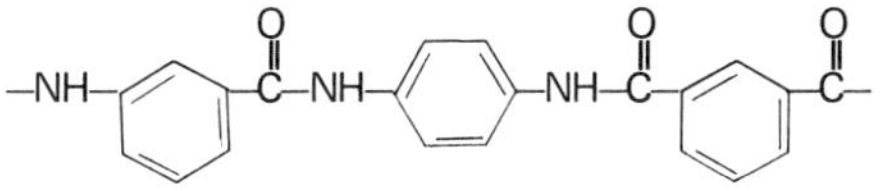

(5) 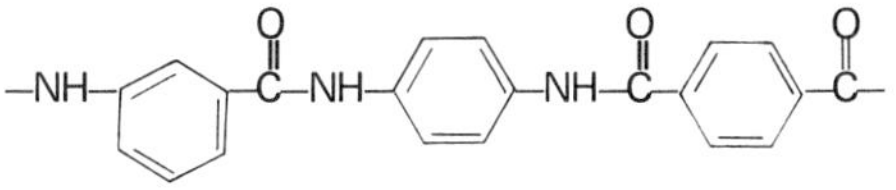

(6) 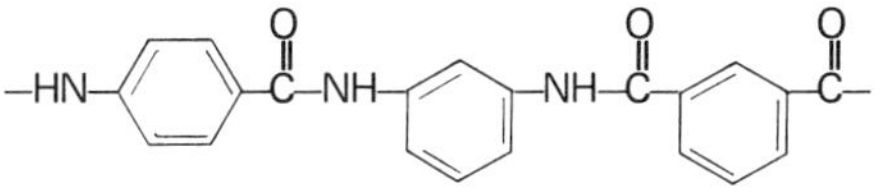

(7) 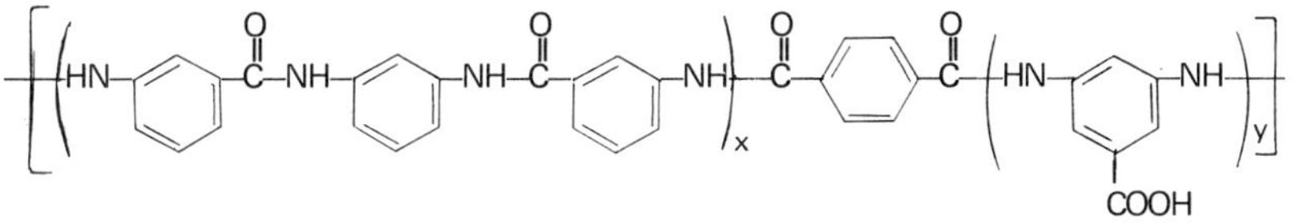

(8) 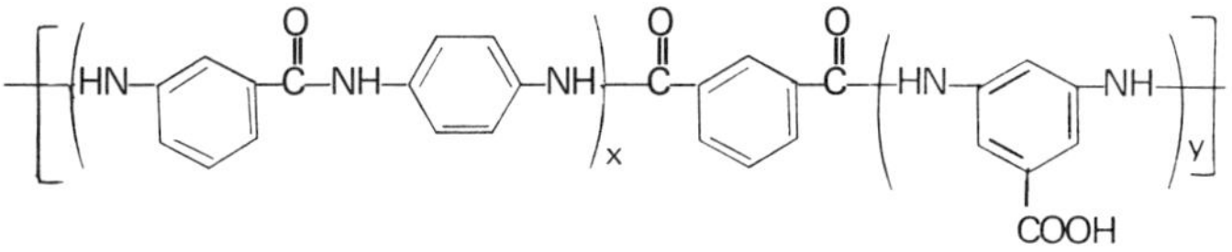

Example 1: In this example crosslinking was accomplished by means of post treatment of the linear polyamide membrane with a 1.0% aqueous solution of formaldehyde. This solution also contained 1.0% sulfuric acid. The linear polyamide was the type shown in structure No. 3 above. It was prepared by copolymerizing 1,3-bis(3-aminobenzamide)benzene and terephthaloyl chloride in equal molar amounts, at a temperature of -15°C, in solution in dimethylacetamide containing 5% LiCl, to a molecular weight of 30,000. The membrane was prepared by casting mils on a glass plate, followed by oven drying at 105°C for 180 minutes and coagulation in a 24°C water bath for a period of 16 hours.

The membrane was treated in the formaldehyde solution at 24°C for a period of 18 hours to effect crosslinking. It was then removed from the formaldehyde solution, drained, and washed with water, and was then tested in a conventional reverse osmosis test cell at 1,500 psi with 60,000 ppm NaCl feed. An untreated (linear) control membrane was also tested under the same conditions. Salt injection for both the treated (crosslinked) and untreated membrane was found to be 99.6. Initial flux (gfd) and flux stability (ϵ) for the control were 8.7 and -0.120, respectively, while the corresponding values for the crosslinked membrane were 5.8 and -0.050. Thus, crosslinking resulted in some decrease in initial flux, but the more significant flux stability was greatly improved.

Example 2: In this example crosslinking of a linear polyamide of the type shown in structure No. 4 above was accomplished by the same procedure as that employed in Example 1. This polyamide was prepared by copolymerizing 3,4'-diamino-benzanilide and isophthaloyl chloride in equal amounts. Formation of the polyamide and the membrane was accomplished by the same procedure as that employed in Example 1. It was found that the crosslinked membrane had an initial flux of 10 and a flux stability of -0.12, while the control showed corresponding values of 13 and -0.18. In each case rejection was 99.7%.

Example 3: In this example crosslinking was accomplished by means of post treatment of the linear polyamide membrane with a 1.0% aqueous solution of glyoxal di(sodium bisulfite). This solution also contained 1.0% sulfuric acid. The linear polyamide was the type shown in structure No. 3 above, and was prepared as in Example 1. The membrane was also prepared by casting, drying and coagulation as in Example 1.

The membrane was treated in the glyoxal di(sodium bisulfite) solution at 24°C for a period of 24 hours to effect crosslinking. It was then removed from the treating solution, drained and washed with water, and was then tested in a reverse osmosis cell, as in Example 1. A control membrane was also tested under the same conditions and gave a salt rejection of 99.6, an initial flux of 8.7 and a flux stability of -0.120. The treated membrane showed a salt rejection of 99.7, an initial flux of 5.3, and a flux stability of -0.060. Again, flux stability was greatly improved by crosslinking.

Example 4: In this example crosslinking was accomplished by means of irradiation. The linear polyamide membrane was the same as that employed in Example 1, except that the formaldehyde treatment was eliminated and drying was at 105°C. It was exposed to gamma radiation at a dose rate of 8.75×10^5 rads/hr for a period of 10 hours, and tested as in Examples 1 and 2. Initial flux and flux stability for the control (no irradiation) were 7.57 and -0.119, respectively, while the values for the irradiated membrane were 5.01 and -0.034, respectively, with both showing 99.7 salt rejection. Thus, flux stability was again greatly improved over that of the linear polymer membrane.

Example 5: In this example ionic crosslinking was employed. The linear polyamide was the type shown in structure No. 7 above. It was prepared by copolymerizing equal molar amounts of 1,3-bis-(3-aminobenzamide)benzene and terephthaloyl chloride with 3,5-diaminobenzoic acid (DABA), the mol ratio of either of the first two ingredients to DABA being 75/25. Polymerization was carried out at -15°C in solution in dimethylacetamide containing 5% LiCl, to a molecular weight of 10,000 to 30,000.

The membrane was prepared by casting the polyamide solution to a thickness of 20 mils on a glass plate, followed by oven drying at 105°C for 180 minutes and coagulation in 24 ± 1°C deionized water. The membrane was allowed to remain in the deionized water for a period of 12 hours to remove all residual solvent and salt. This membrane (linear polyamide) was tested in a conventional reverse osmosis test cell at 1,500 psi with natural seawater at a simulated 45% recovery. Salt injection was 98.7%, with an initial flux of 18.5 and flux stability of -0.100.

A control membrane was also prepared in the same manner except that DABA was not employed in the polymerization reaction mixture and the resulting linear polyamide, therefore, did not contain the polar carboxylic acid groups. This membrane showed a salt rejection of 99.6, an initial flux of 8.0 and a flux stability of -0.120. Thus, the incorporation of the polar groups resulted in a substantial increase in initial flux, and some improvement in flux stability, but in a decrease in rejection.

A membrane containing the polar groups, prepared as above, was placed in an HCl-water mixture adjusted to pH 2.5, for a period of 12 hours at ambient temperature of 23 ± 1°C, in order to exchange the membrane and provide free carboxyl groups. The exchanged membrane was then washed for an additional 12 hours in deionized water to remove all residual HCl.

The membrane was then placed in aqueous 0.5% p-phenylene diamine solution and allowed to equilibrate for 24 hours at ambient temperature to effect crosslinking. The crosslinked membranes possess a uniform blue coloration resulting from the p-phenylene diammonium benzoate linkage. The crosslinked membrane was washed in deionized water, changing the water frequently to remove any unreacted diamine. The membrane was then tested in the reverse osmosis test cell under the same conditions as those employed in testing the control and noncrosslinked, polar group-containing membrane and showed a rejection of 99.6, an initial flux of 11.2 and a flux stability of -0.070. Thus, the initial flux is greater than that of the control and less than that of the noncrosslinking membrane, while the flux stability is better than either.

Aromatic Polyamides

J. Ikeda, R. Bairinji and N. Dokoshi; U.S. Patent 3,878,109; April 15, 1975; assigned to Toray Industries, Inc., Japan have found that an excellent permselective membrane of high water-flux and low salt passage, which is particularly useful for desalination of seawater and brackish water, can be prepared from synthetic organic aromatic polyamide represented by the formula:

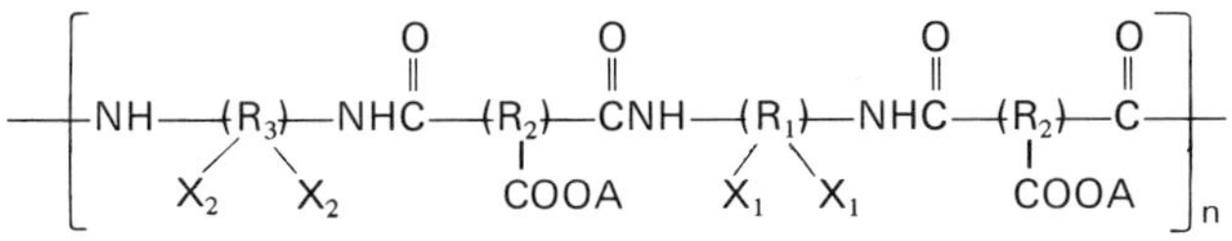

where

(a) R_1 is a tetravalent benzoid group;

(b) R_2 is a trivalent benzoid group;

(c) R_3 is a tetravalent benzoid group;

(d) X_1 and X_2 are groups selected from the class consisting of —R, —OR, —SR, —COOR, —NHR, —NRR and halogens, where R is hydrogen or a hydrocarbon radical of 1 to 6 carbons;

(e) —COOA represents —COOH or an ionic group selected from the class consisting of carboxylic salt of alkali metal, alkaline earth metal, organic amine salt and the mixtures thereof, the concentration of the ionic group in the polymer (PI) being 1/490 to 1/215 where

$$PI = \frac{\text{total number of pendant ionic groups in the polymer}}{\text{polymer molecular weight}}$$

(f) n is an integer sufficiently large to provide film-forming molecular weight; and

(g) the polymer has a solubility of at least about 10% by weight in a solvent selected from the group consisting of dimethylformamide, dimethylactamide, dimethylsulfoxide, N-methylpyrrolidone, hexamethylphosphoramide, and mixtures thereof at 25°C, and the membrane has a water permeability of at least 1,000 and a solute passage through the membrane of less than 20%.

The following examples illustrate the process. All parts and percentages are by weight except where otherwise specified. All polymerization reactions were carried out in solution, usually cooled below room temperature, employing standard preparation techniques, e.g., as described in U.S. Patent 3,598,786. In the following, the logarithm viscosity (inherent viscosity) η inh is a value represented by

$$\frac{\text{natural logarithm } \dfrac{\text{flowing down time of the solution}}{\text{flowing down time of the solvent}}}{\text{concentration}}$$

where concentration is the number of grams of the polymer in 100 ml of the solution. The ordinary viscosity in these examples is based on a measurement at 30°C, in N-methyl-2-pyrrolidone at a concentration of 0.5 g/100 ml.

Example 1: A 10.52 gram portion of purified 4-chloroformyl phthalic anhydride was dissolved into a 35 gram portion of N-methylpyrrolidone and 14.5 gram portion of propylene oxide in a nitrogen-blanketed vessel with vigorous stirring. After the solution was cooled to -20°C, a 6.6 gram portion of 2,2'-dichloro-4'-methylene bis-aniline dissolved in a 22.9 gram portion of N-methylpyrrolidone was added dropwise to this solution at -20°C over a period of 30 minutes, and the stirring was continued for 2 hours.

To this reaction mixture at -10°C, a 5.0 gram portion of fully dried 4,4'-diamino diphenyl ether was added, followed after 15 minutes by addition of an 18 gram portion of N-methylpyrrolidone with vigorous stirring. The solution gradually became viscous, and the polymerization proceeded smoothly at 0°C over the period of 30 to 60 minutes. Further, the vigorous stirring was continued for another 2 hours at 25°C to obtain an extremely viscous polyamide solution. The inherent viscosity of the polymer thus obtained was 1.10. The unreacted propylene oxide or by-product, propylene chlorohydrin, produced from the reaction of propylene oxide with hydrogen chloride included in the polymer solution was removed under vacuum.

To this 22% of polyamide solution, a 6.7 gram portion of lithium chloride and a 31.9 gram portion of dimethyl formamide was added to obtain a 15% polymer solution, which is a dope. The composition of the dope was as follows: 15% polyamide, 5.2% lithium chloride, 25% dimethyl formamide, 49.6% N-methylpyrrolidone, 2.0% propylene chlorohydrin, and 2.0% water. A portion of the dope was poured onto a glass plate, at 29°C, 68% humidity by using a 250 μ blade doctor knife at a rate of 3 cm/sec, followed by drying the dope on the glass at 80°C for 10 minutes in a dryer having a circulating hot air system.

The partially dried membrane "protomembrane" on the glass was dipped into a 10% aqueous calcium chloride solution at 29°C for 1 hour. The membrane was peeled off when rinsing the protomembrane. The membrane obtained was treated with hot water at 80°C for 1 minute with tension. This heat treatment of the membrane having a thickness of 50 μ is preferred for the membrane applied in a saline water conversion, if the rough side of the membrane, or skin layer, is facing the feed solution under the reverse osmosis condition. This membrane was fixed in a porous metal.

The permeation tests were carried out by using seawater (collected at Wakasa Gulf in Japan Sea) with 32,000 ppm, measured by a conductometer, at the pressure of 105 kg/cm^2. The water permeability was 2,860, the water flux was 0.30 m^3/m^2/day and salt passage was 0.4% which meant the salt concentration of the passed water was 128 ppm measured by the conductometer. The permeated water was essentially useful for drinking water and industrial water.

Example 2: As described in Example 1, the following copolymerization was conducted. From a 5.9 gram portion of 4-chloroformyl phthalic anhydride and a 3.7 gram portion of 2,2'-dichloro-4,4'-methylene bis-aniline, an acid dianhydride was obtained. To this solution, a 3.2 gram portion of 4,4'-diaminodiphenyl ether was added followed by addition of a 2.3 gram portion of 3,4,4'-triaminodiphenyl ether, and finally a 4.7 gram portion of pyromellitic dianhydride. Thus 20% of copolyamide in N-methylpyrrolidone was obtained.

The inherent viscosity of this polymer was 0.94. The powder of this polymer was obtained after reprecipitating the polymer with ice water thereby providing a 90% yield. The dope was made from the combination of 20% polyamide thus obtained, 75% dimethyl formamide and 5% lithium chloride. The membrane was made according to the same procedure shown in Example 1 at 27°C and 72% humidity except that the heat treatment was omitted.

The permeation test of this membrane having 71 μ thickness was carried out by using a sodium chloride solution containing 5,000 ppm of chloride ion at an applied pressure of

40 kg/cm^2 with the same testing cell shown in Example 1. The water permeability was 17,500, the water flux was 0.70 m^3/m^2/day and the chloride passage was 3.2%. The permeated water was useful for both drinking water and industrial water.

Type-8 Nylon

H.H. Hoehn and D.G. Pye; U.S. Patent 3,497,451; February 24, 1970; assigned to E.I. du Pont de Nemours & Co. describe a process for separating the components of fluid mixtures, particularly the desalination of seawater, by the use of membranes formed from type-8 nylon. The process of treating fluid mixtures and solutions by reverse osmosis or ultrafiltration utilizes a semipermeable membrane which is a substantially linear polyamide having an inherent viscosity of at least 0.4 and containing as an integral part of the main polymer chain recurring groups of the formula:

$$-\underset{\underset{O}{\parallel}}{C}-\underset{\underset{CH_2OR}{|}}{N}-$$

which are separated by an average number of carbons of at least two and where R represents the organic radical obtained by removal of OH from (a) compounds of the formula: $R'-(O-CH_2-CH_2)_n-OH$ where R' is hydrogen or a lower alkyl group and n is 1 to 100 (or more preferably, 1 to 4), (b) lower alkyl glycolates, or (c) glycerin, the groups containing at least 10% of the amide groups in the main polymer chain. The modified polyamides of which the permeator membranes of this process are made are described in U.S. Patent 2,430,860. The polyamides from which these modified materials are prepared are of the general type described in U.S. Patents 2,071,250; 2,071,253 and 2,130,948.

Such polyamides having an inherent viscosity of at least 0.4 as defined in U.S. Patent 2,130,948 and having hydrogen-bearing amide groups as an integral part of the polymer chain are treated at a temperature from 0° to 150°C with formaldehyde and a formaldehyde-reactive compound of formula ROH as defined above in the presence of an oxygen-containing acid catalyst having an ionization constant of at least 9.6×10^{-6} and an equivalent conductance, measured at 25°C in 0.001 N concentration, no greater than 370 $ohms^{-1}$ cm^2.

Formaldehyde may be used in monomeric form or prepared by depolymerization of one of its condensed forms such as paraformaldehyde, α-polyoxymethylene and trioxane. Operable alcohols are those compounds of formula ROH, noted above, in which H is a reactive hydrogen as determined by the Zerewitinoff test. Acid catalysts as defined above include formic, trimethylacetic, trichloracetic, oxalic, chloroacetic, benzoic, maleic, p-toluenesulfonic, and phosphoric acids and substituted phosphoric acids such as $(CH_3)H_2PO_4$ and $(C_4H_9)H_2PO_4$. All polyamides (nylons) employed in the examples which follow had inherent viscosities in excess of 0.4 and when modified had $-CH_2OR$ groups as shown above on 10% or more of the amide groups in the main polymer chain. Explanations of the terms used in the examples are given below.

66 Nylon Film — 66 nylon film in thicknesses of 0.5, 1, and 2 mils was prepared from 66 nylon flake (Zytel 43, Du Pont Company) by extruding at temperatures in the range of 288° to 296°C through a slit die onto a polished quench roll, controlled at 66° to 77°C, at a rate of 100 to 150 linear feet/minute with an air gap between the die and the quench roll of 4" to 7".

Permeation Flow Measurements — The permeation flow rate W through the film membrane was calculated from the test data by the formula:

$$W = \frac{Q \times t \times 10^6}{A \times \Delta P}$$

Q is the permeate flow in gallons per day;
t is the thickness of the permeation membrane in square feet,
A is the area of permeation membrane in mils;

ΔP is the pressure of the feed solution in psig minus the difference in the osmotic pressures of the feed solution and the permeate in psig.

Feed Solutions – The examples below refer to tests with two types of saline solutions. One of the saline solutions used was a 3.5% NaCl solution (35,000 ppm NaCl), a concentration of salt about equal to the mineral content of seawater. With this solution, the permeation data were not complicated by unknown factors regarding the effect of the various ions present in seawater or synthetic seawater. The salt content of the permeate, being a pure NaCl species, was easily monitored by conductivity measurements. Tests with 3.5% NaCl solution were run at 30°C and 1,500 psig unless otherwise indicated.

A brackish sulfate mixture containing 700 ppm of calcium sulfate (0.07%), 400 ppm of magnesium sulfate (0.04%), and 400 ppm of sodium sulfate (0.04%), making a total solids content of 1,500 ppm of mixed sulfate (0.15%), was also used for desalination tests. This mixture was formulated to simulate many midcontinent groundwaters. Tests with the mixed sulfate solution were run at 30°C and 500 psig unless otherwise indicated.

Salt Passage – Salt passage (SP) is defined as the percentage of the salt in the feed solution passing through the membrane with the permeate. It is determined from the conductivity of the permeate and the salt content in the feed solution. In the typical permeator used for these experiments, a 1 mil unmodified 66 nylon film that had been prepared by extrusion showed a flow rate (W) of 3.6 and a salt passage of 7% (2,500 ppm) for 3.5% NaCl feed at 1,500 psig. With the mixed sulfate feed at 500 psig, unmodified 66 nylon showed a W of 6 and a SP of 13% (196 ppm).

Example 1: Part A – Flake polymer of 66 nylon of the type used for textile fiber spinning as described in U.S. Patent 2,163,636 (105 grams) was dissolved in 420 grams of a mixture prepared from 900 ml of 90% formic acid and 675 ml of acetic anhydride. To the solution were added 129 grams of paraformaldehyde and 798 grams of methyl glycolate. The reaction mixture was stirred and heated at 65°C for 30 minutes. The product was then poured into water with vigorous agitation. The precipitated polymer was washed with water and dried to obtain 56 grams of N-methoxycarbonylmethoxymethyl nylon in the form of a white fluff.

Part B – A sample of N-methoxycarbonylmethoxymethyl-modified nylon prepared by the procedure of Part A was molded between platens at 180°C under 140,000 lb pressure for 2 minutes and remolded at 150°C under 160,000 lb pressure for 2 minutes to obtain a film about 3 mils thick. A specimen from this film was tested in a typical permeator and showed a W of 113 and a SP of 17% with 3.5% sodium chloride feed.

Example 2: Part A – Flake 66 nylon of the type used for spinning textile fibers, inherent viscosity 1.92 (105 g) was dissolved in 905 g of a mixture of formic acid and acetic acid prepared by reacting 480 g of acetic anhydride with 718 g of 90% formic acid. To this were added 129 g of paraformaldehyde and 798 g of the monomethyl ether of diethylene glycol. The reaction mixture was stirred and heated at 65°C for 30 minutes. The product was poured into water with vigorous agitation. The precipitated polymer was washed with water and dried to obtain 106 grams of N-methoxyethoxyethoxymethyl nylon which on analysis showed 9.5% combined formaldehyde and 8.2% combined –OR ether function.

Part B – A 15% solution of the polymer from Part A in ethanol/water 85/15 was cast into a film. Two layers of this film were passed together between platens at 200°C under 18,000 lb pressure for 10 minutes to yield a film about 3 mils thick. This was tested in a permeator and showed a W of 81 and 25% SP with 3.5% sodium chloride feed.

5-Aminoisophthalic Acid Intermediate

A. Rio; U.S. Patent 3,686,116; August 22, 1972; assigned to Rhone-Poulenc SA, France has discovered that polyamides of diamines with 5-aminoisophthalic acid make semipermeable membranes of good stability and salt rejection. These semipermeable membranes are made from polyamides containing a plurality of units of the formula:

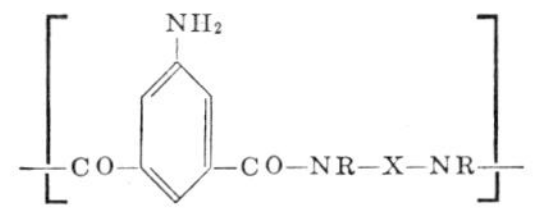

alone or in combination with units of the formula: +CO—Y—CO—NR—X—NR+ in which X and Y are the same or different and each represents a diavalent organic radical, and R is hydrogen, lower alkyl, cycloalkyl of 5 to 6 ring carbons, phenyl radical, lower alkyl-substituted phenyl, or aralkyl, and two R radicals in a unit of formula: —NR—X—NR— may be joined to form a heterocyclic ring containing the 2 indicated nitrogen atoms. A lower alkyl radical is to be understood as an alkyl radical having 1 to 4 carbons. Several different radicals X and Y of different nature can be present in these polymers.

To prepare the polyamides of which the membranes are made, 5-aminoisophthalic acid dichloride in which the amine group has beforehand been blocked can be used as the starting material. A polycondensation of this compound, optionally combined with one or more acid chlorides of formula $Y(COCl)_2$, with one or more polymides of formula $X(NHR)_2$ is carried out. Finally, in a last stage, the polymer is subjected to a treatment to liberate the amine groups.

Example 1: 19.8 grams of bis(4-aminophenyl)methane are dissolved in 150 cm^3 of N-methylpyrrolidone in a 500 cm^3 flask fitted with a stirrer, and cooled to -20°C, and 34.8 grams of 5-phthalimidoisophthalic acid dichloride are added all at once. The mixture is stirred for 2 hours 30 minutes at -10°C and then for 1 hour at 0°C. A sample of the mixture is poured into water. The resulting polymer is filtered, ground, washed and dried. Its specific viscosity is 1.26 (measured at 25°C with a 1% strength solution in N-methylpyrrolidone).

A part of the solution obtained above is used to prepare a membrane. For this, the solution is cast onto a smooth 40 x 22 cm glass plate to a thickness of 0.1 mm. The whole is heated in an oven at 120°C under a reduced pressure of 100 mm of mercury, and kept for 30 minutes under these conditions, then for 30 minutes under a reduced pressure of 30 mm of mercury, and finally for 5 hours under a reduced pressure of 0.5 mm of mercury at 150°C. In this way a 20 μ thick film is obtained, in which the amine group is liberated by treating the film for 20 hours at 25°C with an aqueous solution containing 10% by weight of Na_2CO_3 and 10% by weight of hydrazine hydrate. The resulting film is rinsed with water to remove all traces of hydrazine carbonate.

A disc 9 cm in diameter is cut from this film and is positioned under a grid forming the bottom of a reverse osmosis apparatus consisting of a vertical stainless steel tube 51 cm high and of 8 cm internal diameter, comprising a gas inlet, a pressure release valve, a safety valve and a magnetic stirrer. A sintered metal disc is located under the membrane and the assembly is held fixed by means of a silicone elastomer joint and a metal base equipped with fixing screws and a device for removing the water passing through the membrane.

The efficiency of the membrane in reverse osmosis is measured as follows. 1.5 liters of a 3.5 g/l aqueous solution of sodium chloride are introduced into the apparatus, pressure is applied to the solution, and the flow rate of solution at the apparatus outlet, and the sodium chloride content of the solution, are measured in the usual manner. For the membrane described above, the flow rate of liquid across the membrane is 1 $l/m^2/24$ hr under a pressure of 90 bars and the degree of rejection of salt is 92.25%.

Example 2: A membrane is prepared by casting the polyamide solution of Example 1 on the glass support and then drying at 90°C under a reduced pressure of 1 mm of mercury for 1 hour and immersing the film in water at 25°C for 24 hours. A 40 μ thick film is thus obtained, in which the amine group is liberated as in the preceding example. Under the reverse osmosis conditions of Example 1, under a pressure of 50 bars, this film shows a rate of flow of liquid of 0.21 $l/m^2/24$ hr and a degree of rejection of NaCl of 99.75%.

Phosphorus-Containing Polyamides

A. Blanchard, A. Rio and G. Vivant; U.S. Patent 3,619,424; November 9, 1971; assigned to Rhone-Poulenc SA, France describe semipermeable membranes which have good resistance to hydrolysis and which can be stored dry, made from polyamides or copolyamides containing a plurality of units of the formula:

(1) 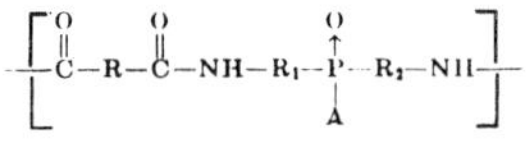

in which R, R_1 and R_2 are the same or different divalent hydrocarbon radicals and A represents a hydroxyl group, a monovalent hydrocarbon radical, a lower alkoxy group or a group of formula:

(2) 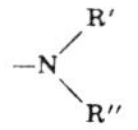

in which R' and R", which may be the same or different, represent hydrogen atoms or monovalent hydrocarbon radicals. More specifically, R may be an alkylene radical of 2 to 10 carbons. R_1 and R_2 may be alkylene radicals of 1 to 10 carbons. A can represent alkyl of 1 to 10 carbons such as methyl, ethyl or butyl, a cycloalkyl group such as cyclopentyl or cyclohexyl, aryl such as phenyl, alkylaryl such as tolyl, or an aralkyl group such as benzyl. Where A is a lower alkoxy group, it can represent a methoxy, ethoxy, propoxy or butoxy group. R' and R" represent alkyl radicals of 1 to 10 carbons, especially methyl, ethyl or propyl. The polyamides containing units of formula (1) in which A is a hydroxyl group or a hydrocarbon group are prepared by polycondensation, in accordance with the usual methods of one or more diacids of the formula:

(3) HOOC–R–COOH

in which R has the significance given above, or of an anhydride or a chloride of such a diacid, with one or more diamines of the formula:

(4) 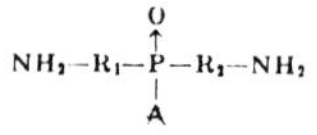

in which R_1 and R_2 have the significance given above and A is a hydroxyl group or a hydrocarbon radical, optionally associated with a diamine which does not contain phosphorus, e.g., of the formula:

(5) $H_2N–R_3–NH_2$

in which R_3 is a divalent aliphatic or aromatic group such as hexamethylene, m- or p-phenylene or a radical of the formula:

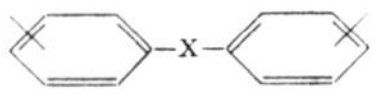

where X is oxygen or methylene. Examples of diacids of formula (3) are succinic, adipic, cyclohexane-1,4-dicarboxylic, orthophthalic, isophthalic and terephthalic acids, these acids may be used by themselves or mixed with one another. Examples of the diamines of formula (4) are bis(m-aminophenyl)methylphosphine oxide, bis(p-aminophenyl)methylphosphine oxide, bis(m-aminophenyl)phenylphosphine oxide, bis(3-aminopropyl)phenylphosphine oxide, 3-aminopropyl-4-aminobutyl-methylphosphine oxide, bis(m-aminophenyl)phosphinic acid and bis(p-aminophenyl)phosphinic acid. Examples of diamines not containing phosphorus of formula (5) which can be associated with the diamines of formula (4) in order to obtain copolyamides possessing units of formula (1), are hexamethylenediamine, m- or p-phenylenediamine, bis(m-aminophenyl)methane, bis(p-aminophenyl)methane and bis(m- or p-aminophenyl) oxide.

Phosphorus-containing polyamides possessing units of formula (1) in which A is a methyl radical, R_1 and R_2 are p-phenylene radicals and R is a p-phenylene, octamethylene or tetramethylene radical have been described by T. Ya. Medved et al, *J. Polym. Sci. USSR,* 5, 386 (1964); phosphorus-containing polyamides in which R_1 and R_2 are trimethylene or tetramethylene groups, R is a tetramethylene or octamethylene group and A is a methyl or phenyl group have been prepared by J. Pellon et al, *J. Polym. Sci.,* A1, 3561 (1963).

The phosphorus-containing polyamides possessing units of formula (1) in which A is a hydroxyl group, and which are called polyamides with phosphinic acid units, are obtained by polycondensation of the diacid or diacids of formula (3) with the diamine of formula:

(6)

$$NH_2-R_1-\overset{\overset{\displaystyle O}{\uparrow}}{\underset{\underset{\displaystyle OH}{|}}{P}}-R_2-NH_2$$

in accordance with the conventional processes for the polycondensation of diacids and diamines.

Example 1: A 15% solution of the polyterephthalamide of bis(m-aminophenyl)methylphosphine oxide in dimethylacetamide, having a viscosity of 120 poises measured at 25°C, is prepared. This solution is cast onto a smooth glass plate 40 x 22 cm, and introduced into an oven and kept there for 1 hour at 85°C under a pressure of 1 mm of mercury and then for 4 hours at 150°C under the same pressure. The whole is then cooled and the film detached from the support. A transparent film 10 μ thick is thus obtained.

A disc 9 cm in diameter is cut from this film and is positioned under a grid forming the bottom of a reverse osmosis apparatus consisting of a vertical stainless steel tube 51 cm high and of 8 cm internal diameter, comprising a gas inlet, a pressure release valve, a safety valve and a mechanical stirrer. A sintered metal disc is placed under the membrane and the whole is held fixed by a silicone elastomer joint and a metal base equipped with fixing screws and a device for removing the water passing through the membrane.

The efficiency of the membrane in reverse osmosis is measured as follows. 1.5 liters of an aqueous 3.5 g/l sodium chloride solution are introduced into the apparatus. A pressure is then applied to the solution. The flow rate of the solution at the apparatus outlet is measured and the sodium chloride content of the solution is measured in the conventional manner. For the membrane described above, the flow rate is 4.52 $l/m^2/24$ hr and the degree of rejection of the salt is 87% under a pressure of 100 bars. Under the same conditions under a pressure of 50 bars, the flow rate of water is 2.76 $l/m^2/24$ hr and the degree of rejection of salt is 86%.

Example 2: The membrane is prepared as in Example 1 but from a solution containing 0.2% by weight of cumyl peroxide based on the polyterephthalamide. Under the same conditions of reverse osmosis as in Example 1 and under a pressure of 50 bars, the membrane obtained shows a flow rate of 2 $l/m^2/24$ hr and a degree of rejection of salt of 98.7%.

Example 3: A membrane is prepared as in Example 1 but the film is first partially dried by heating to 85°C under a reduced pressure of 1 mm of mercury for 30 minutes and then immersed for 5 minutes in water maintained at 85°C. Under the same conditions of reverse osmosis as in Example 2 the membrane obtained shows a flow rate of 75 $l/m^2/24$ hr and a degree of rejection of salt of 89.4%.

Example 4: A phosphorus-containing copolyamide possessing a plurality of phosphonic acid units is prepared by polycondensation of terephthaloyl chloride with a mixture of 25 mol % of bis(m-aminophenyl)phosphinic acid and 75 mol % of bis(p-aminophenyl)-methane, using the following procedure: 1.55 grams (0.62 x 10^{12} mol) of bis(m-aminophenyl)phosphinic acid, 3.7 grams (1.8 x 10^{12} mol) of bis(p-aminophenyl)methane and 5 grams (2.5 x 10^{12} mol) of terephthaloyl chloride in 50 cm^3 of N-methylpyrrolidone are reacted in the presence of 2 cm^3 of triethylamine for 3 hours at −10°C. The reaction mixture is added to water at 20°C to precipitate the copolyterephthalamide. The whole is then acidified to pH 1 by adding hydrochloric acid. The polyamide is filtered off, washed with distilled water on the filter and dried at 100°C under a pressure of 200 mm of mercury. 8.1 grams of a polyamide with phosphinic acid units are obtained, having a specific viscosity of 0.89 measured at 25°C on a 1% by weight solution in a mixture of N-methylpyrrolidone/hexamethylphosphotriamide (25/75% by weight)

A 10 μ thick membrane is prepared by casting a 10% strength solution in the mixture of N-methylpyrrolidone/hexamethylphosphotriamide (25/75) onto the glass plate of Example 1, drying the film for half an hour at 80°C and then immersing it for 5 minutes into water at 80°C. The membrane thus obtained is then used for carrying out a reverse osmosis in the apparatus of Example 1 and under a pressure of 55 bars. The membrane yields a flow rate of 10 $l/m^2/24$ hr and a degree of rejection of salt of 93%.

Polyamides Having Polyoxyethylene Grafts

A process described by *S. Rosenbaum; U.S. Patent 3,472,766; October 14, 1969; assigned to The Dow Chemical Co.* utilizes a permeable polyamide membrane having nonrandomly substituted poly(oxyethylene) grafts where the membrane contains from about 40 to about 65% by weight of combined ethylene oxide. The permeable membranes are prepared from a polymeric material having a group reactive to ethylene oxide such as an >NH group. Of particular interest are the polyamide or nylon type polymers, as well as the polyureas and polyurethanes. The polyamides are generally prepared by reacting a diamine with a dicarboxylic acid or by condensation of an amino carboxylic acid. The polyureas are generally prepared by reaction of a diamine with a diisocyanate and the polyurethanes by reaction of a diisocyanate with a glycol.

The permeable membrane, itself, is prepared by reacting the solid polymeric material with ethylene oxide to produce nonrandomly substituted polyoxyethylene grafts according to the procedure described in Hydroxyethyl Nylon, a New Flexible Super Polyamide Derivative by H.C. Haas et al, *J. Polymer Sci.,* 15, 427 (1955).

While the research described by Hass et al was aimed at producing polymers having flexibility over a wide temperature range it has now been found that such polymers containing from about 10 to about 65% by weight of combined ethylene oxide possess useful permeability properties. The ethoxylated polymeric materials may be fabricated into a hollow fiber, film, etc. depending upon the type of apparatus. While the process is not dependent on the physical shape of the polymeric permeable membrane, a hollow fiber is preferred. The membrane thickness ranges from about 2 to 300 mils with the membrane preferably being as thin as possible.

All manner of osmotic and dialytic processes may be performed with the permeable membrane of this process. As the percent of combined ethylene oxide increases the permeability of the membrane to low molecular weight solutes is enhanced. It is preferred to utilize a membrane containing from about 40 to 65% by weight of combined ethylene oxide for high permeability for hydrophilic solutes, the membrane having particular interest to the field of extracorporeal hemodialysis. For separations such as reverse osmosis or ultrafiltration, it is preferred to utilize a membrane containing from about 10 to about 40% by weight of combined ethylene oxide. For purposes of illustrating the process, the following nonlimiting examples are presented.

Example: A 10 gram sample of a commercial nylon 6-6 (available as Zytel 42) was ground to pass a 20 mesh screen and then heated at 80°C for 40 hours with 60 ml of liquid ethylene oxide in a Parr bomb. After cooling the excess ethylene oxide was allowed to evaporate and the remaining material washed several times with water in a Waring Blendor before drying over molecular sieves at 40°C/0.1 mm. The polymeric product was granular, rubbery, and contained about 55% by weight of combined ethylene oxide by elemental analysis. Product yield was 21.6 grams. A film was fabricated from the polyethoxylated graft nylon 6-6 prepared above by melt pressing 0.4 gram for 5 minutes at 225° to 230°C under 333 psi pressure, followed by cooling to 25°C under pressure. The resulting film was 2.5 to 3 mils in thickness and had an area of 9.6 in^2.

The permeability of the polyethoxylated graft nylon 6-6 to hydrophilic solutes was tested by measuring the rate of dialysis of sodium chloride and urea. A permeability cell having the membrane (film) clamped to the bottom was immersed in a water bath maintained at 37°C. The membrane had an area of 20 cm^2. The solution containing the dissolved solute was placed in the cell and the contents of the cell stirred during the dialysis experiment.

The concentration of the solute was periodically determined and the reciprocal permeability (resistance to permeation) was expressed as the time in minutes required for the concentration of the solute in the cell to fall to one-half its original ($t_{1/2}$). All values are corrected to a 1 mil membrane thickness. Sodium chloride was determined by a conductometric titration for chloride. Urea was determined by the method of Walt and Crisp, *Anal. Chem.*, 26, 452 (1954) which is based on the yellow-green color produced when p-dimethylaminobenzaldehyde is added to urea in HCl solution. Comparative tests were made to an uncoated commercial cellophane membrane with the following results:

Membrane	$t_{1/2}$, minutes NaCl	Urea
Cellophane	16	76
Ethoxylated nylon 6-6	13	36
Nylon 6-6	*	*

*Too slow to measure

Poly(trans-2,5-Dimethylpiperazinfumaramide)

L. Credali and P. Parrini; U.S. Patents 3,743,596; July 3, 1973; 3,743,597; July 3, 1973; and 3,687,842; August 29, 1972; all assigned to Consiglio Nazionale Delle Ricerche and Montecatini Edison SpA, Italy describe an osmosis process which utilizes as the semipermeable membrane, a blend of at least two synthetic polyamides selected from the group consisting of poly(trans-2,5-dimethylpiperazinfumaramide), poly(2-methylpiperazinfumaramide) and poly(trans-2,5-dimethylpiperazinmesaconamide) or a blend of at least one synthetic polyamide selected from the group consisting of poly(trans-2,5-dimethylpiperazinfumaramide), poly(2-methylpiperazinfumaramide) and poly(trans-2,5-dimethylpiperazinmesaconamide), and a member selected from the group consisting of polypyrrolidone, polycaprolactam and polyhexamethylenadipamide.

The polyamide structure is obtained by reacting piperazine or a substituted derivative with a dihalide of fumaric acid or a substituted derivative. The preparation of such polyamides is described, for example, in Italian Patents 793,191 and 852,477 as well as in U.S. Patent 3,143,527. The polyamide structure is characterized by the structural unit of the formula

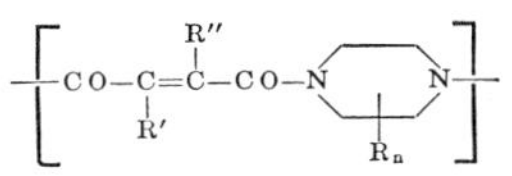

where the two carbonyl groups are bound in a trans position to the two adjacent carbons bound to each other by a double bond; n is either zero or a whole number from 1 to 8; R is a substituent such as alkyl, e.g., methyl or ethyl; cycloalkyl; alkoxy; aryl; aryloxy; arylalkoxy or halogen; and R' and R" are each selected from the group consisting of hydrogen; alkyl, e.g., methyl or ethyl; cycloalkyl; aryl and halogen.

Example 1: This example demonstrates that the polymeric materials used in accordance with the process have a high permeability to water, and that their use in a process for the desalination of a saline solution, according to the principle of reverse osmosis, allows one to considerably reduce the concentration of salt dissolved therein.

(A) Preparation of Completely Dry Films for Reverse Osmosis – The films of polymeric materials were prepared from solutions of the polymer in a suitable solvent. The concentration of the polymer in the solution was between 5 and 10% by weight. Either formic acid or a chloroform/methanol mixture (in a weight ratio of 88/12) was used as the solvent. The deaerated homogeneous solution was spread over a glass plate by means of a film-spreader. The thus formed films were permitted to dry at 30°C for several hours, until complete evaporation of the solvent had occurred. Thereafter, the films were removed from the glass plate, and they were then tough, transparent and homogeneous.

By regulating the thickness of the film spreader and the concentration of the solution, it was possible to obtain films with a final thickness between 6 and 100 microns. Films with a thickness below 6 microns were prepared by immersing a glass plate vertically into the polymer solution. This glass plate remained in the solution for at least 10 minutes. It was vertically extracted from the solution at a speed of about 0.5 cm/sec. The glass plate remained in the solution for at least 10 minutes. It was vertically extracted from the solution at a speed of about 0.5 cm/sec. The glass plate was then placed horizontally to rest for a few hours at 30°C, until complete evaporation of the solvent had occurred. Then the glass plate was immersed into water so as to allow detachment and flotation of the film.

By regulating the speed of extraction of the glass plate from the solution and the concentration of the solution itself, it was possible to obtain films with thickness varying from 0.2 to 6 microns. The films, prepared according to the method described above, had good mechanical resistance. The mechanical properties are recorded in Table 1.

TABLE 1: MECHANICAL PROPERTIES OF COMPLETELY DRY FILMS

Type of film	(a)	(b)	(c)	(d)
Mechanical properties:[1]				
Tensile strength (kg./cm.^{2})	730	810	835	637
Elongation at break (percent)	3	5	7.8	23.7
Elastic modulus (kg./cm.^{2})	29,600	31,000	28,000	35,600

[1] Determined at 23° C., 65% relative humidity, according to ASTM-D 882.

(a) Poly(2-methylpiperazinfumaramide); film obtained from HCOOH solution.
(b) Poly(trans-2,5-dimethylpiperazinfumaramide); film obtained from HCOOH solution.
(c) Poly(trans-2,5-dimethylpiperazinfumaramide); film obtained from $CHCl_3/CH_3OH$ solution.
(d) Poly(trans-2,5-dimethylpiperazinmesaconamide); film obtained from $CHCl_3/CH_3OH$ solution.

(B) Use of the Completely Dry Films in the Desalination of Saline Solutions – The films prepared as described in (A) above were put into a standard type reverse osmosis cell. An aqueous saline solution, the content of which varied from test to test, was used as the feed. The linear flow rate of the feed solution to the surface of the film was 100 cm/sec, and the pressure was between 50 and 150 atm. In Table 2 the data and results are given which were obtained when employing as the feed an aqueous solution containing 5,000 ppm of NaCl. The values of permeability to water, P_{H_2O} (in g/cm sec), were calculated from the flux and saline rejection values according to the method of Lonsdale, Merten & Riley, *J. Appl. Polymer Sci.,* 9, 1341 (1965).

The following data, obtained with completely dry films, show that the permeability to water of the films of this process is very high and is distinctly superior to that of a film of cellulose acetate obtained by casting a 20% solution of Eastman 398-3 cellulose acetate in acetone. In column 9 of Table 2 the flow rate of water is recorded, calculated for a film with a uniform thickness of 1 micron. These values show that the higher permeability to water, P_{H_2O} of the films of this process, with the thickness remaining the same, enables one to obtain a flow rate of at least five times greater than that which is obtained with a film of cellulose acetate. (The values given in column 9 were calculated by multiplying the values in column 3 by the values in column 5.) The saline rejection data given in column 7 show that this property is greater than 80% for sodium chloride, and increases with increasing pressure. At a pressure of 105 atm (tests 2, 3 and 4) the value is 96.5%, very close to that for cellulose acetate.

TABLE 2: OSMOTIC PROPERTIES OF COMPLETELY DRY FILM

(1) Type of membrane	(2) Solvent for preparing the film	(3) Thickness of the film (microns)	(4) Pressure of feed (atm.)	(5) Flow of desalinized water (liter/day/m.²)	(6) Saline concentration in the product water (p.p.m. of NaCl)	(7) Saline rejection (percent)	(8) Permeability to water P_{H_2O} (gr./cm. sec.)	(9) Flow of water determined for film of uniform thickness of 1 micron (liter/day·m.²)
A	HCOOH	25	80	43.0	900	82.0	2.3×10^{-6}	1,075
B	HCOOH	40	50	9.8	440	91.2	1.4×10^{-6}	392
B	HCOOH	40	80	12.0	295	94.1	1.0×10^{-6}	480
B	HCOOH	40	105	13.5	175	96.5	0.87×10^{-6}	540
B	$CHCl_3/CH_3OH$	30	50	10.0	430	91.4	1.1×10^{-6}	300
C	$CHCl_3/CH_3OH$	48	50	8.2	910	81.8	1.4×10^{-6}	393
D	Acetone	40	50	1.55	76	98.4	0.22×10^{-6}	62
D	do	40	80	2.42	70	98.6	0.21×10^{-6}	97

A = poly(2-methyl-piperazinfumaramide);
B = poly(trans-2,5-dimethyl-piperazinfumaramide);
C = poly(trans-2,5-dimethyl-piperazinmesaconamide);
D = cellulose acetate Eastman 398-3.

The results obtained by employing a saline solution containing 10,000 ppm of $MgSO_4$ at a pressure of 50 atm as the feed are reported in Table 3.

TABLE 3: CHARACTERISTICS OF REVERSE OSMOSIS AND SALINE REJECTION IN TEST WITH $MgSO_4$ OF COMPLETELY DRY FILMS

Number of test	Type of membrane	Solvent for preparing the film	Thickness of the film (microns)	Flow of desalinized water (liter/day. m.²)	Saline composition of the product (p.p.m. of $MgSO_4$)	Saline rejection, percent
9	Poly(trans-2,5-dimethylpiperazinfumaramide)	HCOOH	28	14.7	20	99.8
10	Poly(trans-2,5-dimethylpiperazinmesaconamide)	$CHCl_3/CH_3OH$	22	19.8	63	99.3

These data, obtained with completely dry films, show that the saline rejection of $MgSO_4$ for the films of this process is greater than 99%. The results obtained by using poly(trans-2,5-dimethylpiperazinfumaramide) films of different thicknesses, obtained from HCOOH solution, have been recorded in Table 4. An aqueous saline solution containing 5,000 ppm of NaCl under a pressure of 80 atm was used as the feed.

TABLE 4: OSMOTIC PROPERTIES OF COMPLETELY DRY FILM

Number of test	Thickness of the film (microns)	Flow of desalinized water (liter/day.m.²)	Saline rejection (percent)	Permeability to water $P_{(H_2O)}$ (gr./cm. sec.)
11	36	15.5	94.4	1.2×10^{-6}
12	22	24.0	93.1	1.1×10^{-6}
13	13	38.0	93.8	1.0×10^{-6}
14	8.5	64.3	96 1	1.2×10^{-6}

These data, obtained with completely dry films, show that the flow rate of the produced soft water increased with decreasing thickness of the films while the permeability to water remained relatively constant (about 1.1×10^{-6} g/cm sec). Moreover, there was no decrease in the saline rejection as the thickness of the film decreased.

Example 2: A solution containing 7 grams of poly(trans-2,5-dimethylpiperazinfumaramide) and 0.3 gram of $Mg(ClO_4)_2$ in 92.7 grams of formic acid was spread over a glass plate by means of a film spreader. The film thus formed was maintained at a temperature of 42°C for 20 minutes and was then immersed in water for a few hours at room temperature. Thereafter, the film was removed from the glass plate and kept in water for 2 hours at 80°C. The film had a white, dull appearance and displayed good tenacity and mechanical resistance.

The thus obtained film had a thickness of about 70 microns. It was put into a standard type reverse osmosis cell. Water at a pressure of 50 atm was used as the feed. The amount of water which passed through the film was 5,000 l/day/m^2 of film surface. This rate of flow remained constant. This film was conveniently used as a support or carrier of high permeability.

Example 3: Use of Thin Supported Films in the Desalination of Saline Solutions – According to the procedures described in Example 1, poly(trans-2,5-dimethylpiperazinfumaramide) films were prepared, having thicknesses between 1.7 and 8.5 microns. These films were supported on a film, prepared according to the procedure described in Example 2, and having a thickness of 70 microns. The coupled film was put into a standard type reverse osmosis cell. The arrangement inside the cell was such as to have one face of the thin film in contact with the saline feed solution and the other face in contact with the porous support prepared according to the procedure of Example 2. An aqueous solution containing 10,000 ppm of NaCl, at a pressure of 80 atm, was used as the feed.

The results thus obtained are reported in Table 5. Also reported in Table 5 is a test in which the thin film was carried on a cellulose acetate support obtained by casting a solution consisting of: Eastman 398-3 cellulose acetate, 11% $Mg(ClO_4)_2$, 5.5%; H_2O, 10% and acetone, 73.5%. The film formed from this solution had been gelled in water immediately after its formation.

TABLE 5: OSMOTIC PROPERTIES OF COMPLETELY DRY ULTRATHIN FILMS OF POLY(TRANS-2,5-DIMETHYLPIPERAZINFUMARAMIDE) PLACED ON POROUS SUPPORTS

Number of test	Thickness of film (microns)	Solvent for preparing the film	Flow of desalinized water produced (liter/day m.²)	Saline rejection, percent	Kind of the porous support
15	8.5	HCOOH	64.0	97.8	Support of Example n°2.
16	6.2	HCOOH	85.5	98.2	Do.
17	6.0	HCOOH	89.5	98.2	Do.
18	4.0	HCOOH	142	96.0	Do.
19	1.7	$CHCl_3/CH_3OH$	240	94.8	Do.
20	6.0	HCOOH	63.4	97.6	Support of cellulose acetate.[a]

[a] Cellulose acetate Eastman 398-3—Composition of casting solution Eastman 398-3, 11%; $Mg(ClO_4)_2$, 5.5%; H_2O, 10%; acetone, 73.5%.

The data of Table 5 show that by using the films of this process, carried on porous supports, it is possible to obtain saline rejections of sodium chloride greater than 98%, a value comparable with that of the saline rejection of cellulose acetate films.

Piperazine-Dicarboxylic Acid Reaction Products

L. Credali and P. Parrini; U.S. Patent 3,696,031; October 3, 1972; assigned to Consiglio Nazionale Delle Ricerche and Montecatini Edison SpA, Italy describe polyamides which are obtained by reacting piperazine or a substituted derivative with a dicarboxylic aliphatic, cycloaliphatic, or aromatic acid. The polyamide structure is characterized by the structural unit of the following formula where —CO—X—CO— is a radical derived from any dicarboxylic acid capable of yielding a polyamide by reaction with a piperazine, with X being a bivalent radical, for instance, an alkylene, alkenylene, alkadienylene, arylene, or cycloalkylene radical, or being altogether absent (zero) as in the case of oxalic acid; n is either zero or a whole number from 1 to 8; and R is a substituent such as alkyl, e.g., ethyl or methyl; cycloalkyl; alkoxy; aryl; aryloxy; arylalkoxy; or halogen. The preparation of such polyamides is described, for example, in U.S. Patent 2,913,433.

Specific examples of dicarboxylic acids that may be used to form the polypiperazinamides include: oxalic acid, adipic acid, pimelic acid, suberic acid, azelaic acid, sebacic acid, trans-trans-muconic acid, terephthalic acid and cis-hexahydroterephthalic acid. Examples of the resultant polymeric materials include: poly(piperazinisophthalamide), poly(trans-2,5-dimethylpiperazinterephthalamide), poly(trans-2,5-dimethylpiperazinadipamide), poly(trans-2,5-dimethylpiperazinisophthalamide), poly(trans-2,5-dimethylpiperazin-trans-trans-muconamide), etc.

Example: This example demonstrates that the polymeric materials used in accordance with the process have a high permeability to water, and that their use in a process for the desalination of a saline solution, according to the principle of reverse osmosis, allows one to considerably reduce the concentration of salt dissolved therein.

(A) Preparation of Completely Dry Films for Reverse Osmosis — The films of polymeric materials were prepared from solutions of the polymer in a suitable solvent. The concentration of the polymer in the solution was between 5 and 10% by weight. Either formic acid or a chloroform/methanol mixture (in a weight ratio of 88/12) was used as the solvent. The deaerated homogeneous solution was spread over a glass plate by means of a film spreader. The thus formed films were permitted to dry at 30°C for several hours, until complete evaporation of the solvent had occurred. Thereafter, the films were removed from the glass plate, and they were then tough, transparent and hydrogeneous.

By regulating the thickness of the film spreader and the concentration of the solution, it was possible to obtain films with a final thickness between 6 and 100 microns. Films with a thickness below 6 microns were prepared by immersing a glass plate vertically into the polymer solution. This glass plate remained in the solution for at least 10 minutes. It was vertically extracted from the solution at a speed of about 0.5 cm/sec. The glass plate was then placed horizontally to rest for a few hours at 30°C, until complete evaporation of the solvent had occurred. Then the glass plate was immersed into water so as to allow detachment and flotation of the film. By regulating the speed of extraction of the glass plate from the solution and the concentration of the solution itself, it was possible to obtain films with thicknesses varying from 0.2 to 6 microns.

(B) Use of the Completely Dry Films in the Desalination of Saline Solutions — The films prepared as described in (A) above were put into a standard type reverse osmosis cell. An aqueous saline solution containing 5,000 ppm of NaCl was used as the feed. The linear flow rate of the feed solution to the surface of the film was 100 cm/sec, and the pressure was between 50 and 80 atm. In the following table the data and results are given. The values of permeability to water, P_{H_2O} (in g/cm sec), were calculated from the flux and saline rejection values according to the method of Lonsdale, Merten & Riley, *J. Appl. Polymer Sci.*, 9, 1341 (1965).

The following data, obtained with completely dry films, show that the permeability to water of the films of this process is very high and is distinctly superior to that of a film of cellulose acetate obtained by casting a 20% solution of Eastman 398-3 cellulose acetate in acetone.

In column 9 of the following table, the flow rate of water is recorded, calculated for a film with a uniform thickness of 1 micron. These values show that the higher permeability to water, P_{H_2O}, of the films of this process, with the thickness remaining the same, enables one to obtain a flow rate of at least twice or more times greater than that which is obtained with a film of cellulose acetate. (The values given in column 9 were calculated by multiplying the values in column 3 by the values in column 5.)

The saline rejection data given in column 7 show that this property may vary with the degree of substitution of the piperazine and with the nature of the dicarboxylic acid. In general the saline rejection is very high, in some cases being even greater than that for cellulose acetate.

Osmotic Properties of Completely Dry Films

(1)	(2)	(3)	(4)	(5)	(6)	(7)	(8)	(9)
Type of membrane	Solvent for preparing film	Thickness (microns)	Pressure (atmospheres)	Flow of water (liters/day/m.^{2})	Salt content of water (p.p.m. of NaCl)	Saline rejection (percent)	P_{H_2O} Permeability to water (gr./cm. sec.)	Flow of water determined for a film of unitary thickness of 1 micron (liters/day·m.^{2})
Poly(piperazinisophthalamide)	HCOOH	70	80	2.34	40	99.2	3.5×10^{-7}	164
Do	HCOOH	36	80	5.50	70	98.6	4.3×10^{-7}	198
Do	HCOOH	9	80	26.5	80	98.4	5.1×10^{-7}	238
Do	HCOOH	70	50	2.16	85	98.3	5.4×10^{-7}	151
Poly(trans-2,5-dimethylpiperazinterephthalamide).	HCOOH	28	50	5.0	270	94.6	5.0×10^{-7}	140
Poly(trans-2,5-dimethylpiperazinisophthalamide).	$CHCl_3/CH_3OH$	40	50	3.5	1750	65.0	4.9×10^{-7}	140
Poly(trans-2,5-dimethylpiperazinadipamide).	HCOOH	60	80	3.9	950	81.0	5.0×10^{-7}	234
Poly(trans-2,5-dimethylpiperazin,trans-trans muconamide).	HCOOH	24	50	11.5	755	84.9	9.9×10^{-7}	276
Cellulose acetate [a]	Acetone	40	50	1.55	76	98.4	2.2×10^{-7}	62
Do [a]	do	40	80	2.42	70	98.6	2.1×10^{-7}	97

[a] Cellulose Acetate Eastman 398-3 (trademark of Eastman Kodak Chem. Co., U.S.A.).

Interface Condensation Polymers

L.C. Scala, D.F. Ciliberti and D. Berg; U.S. Patent 3,744,642; July 10, 1973; assigned to Westinghouse Electric Corporation have found that unstirred, interface condensation polymerization between a pair of fast reacting intermediates, such as an aliphatic or carbocyclic primary diamine and an aliphatic or carbocyclic diacyl halide or a carbocyclic disulfonyl halide, dissolved in solvents immiscible or relatively immiscible in each other, in situ on and/or in a porous support, will provide reverse osmosis desalination membrane-substrate composites.

These supported interface condensation desalination membranes show excellent resistance to biodegradation and compaction under pressure, resulting in longer useful membrane life and can be made cheaply using mass production techniques. They also have excellent stability in hot water and better strength for high pressure usage than cellulose acetate membranes. The reverse osmosis membrane test system used for testing flat membranes was of the standard type, and used Monel metal and stainless steel as construction materials. Round cells were used for testing 1.0" diameter membranes.

Example 1: A 6 mil (0.006") thick circular filter, made of fibrous cellulosic esters, having a diameter of 47 mm (1.8") and a pore size of 0.054 microns, was dipped for 1.5 minutes in an unstirred, room temperature (25°C) aqueous 0.8 molar solution of 1,6-hexanediamine in aqueous 1.6 molar sodium hydroxide. The saturated filter was air dried for 1.5 minutes, then dipped for 1.5 minutes into an unstirred, room temperature (25°C), 0.4 molar solution of adipyl chloride in trichloroethylene solvent. This treatment was followed by a 0.5 minute dip in distilled water to wash the membrane-substrate composite. This procedure provided a membrane about 0.1 to 0.2 mil thick on both sides of the filter.

Membrane-substrate composites made this way were mounted in a steel reverse osmosis test system heretofore described above. Saline water containing 5,200 ppm impurity salt as NaCl was fed into the cell and against the membrane-substrate composite at 600 psi. After 30 hours exposure, the salt rejection was 80% and the pure water flux was 0.49 gallon per square foot-day. After 90 hours exposure, the rejection was 87% and the pure water flux was 0.44 gal/ft^2-day. The membranes were run for 4 days without apparent compaction difficulties. The use of a very small pore filter as the substrate, in this example, contributed to low pure water flux.

Example 2: Membrane-substrate composites made according to Example 1, but using m-xylylenediamine and adipyl chloride, with the same polyamine and diacyl chloride concentrations and solvents as in Example 1, gave a salt rejection of 65% and a pure water flux of 0.85 gal/ft^2-day, after 90 hours exposure, using a 5,200 ppm saline feed at 600 psi.

Example 3: Membrane-substrate composites made according to Example 1, but using sebacyl chloride, with the same polyamine and diacyl chloride concentrations and solvents as in Example 1, gave a salt rejection of 46% and a pure water flux of 0.85 gal/ft^2-day, after 90 hours exposure, using a 5,200 ppm saline feed at 600 psi.

Example 4: Membrane-substrate composites made according to Example 1, but using a circular filter support substrate with a pore size of 0.3 micron and cyclohexane as the solvent for adipyl chloride, with the same polyamine and diacyl chloride concentrations as in Example 1, gave a salt rejection of 85% and a pure water flux of 2.5 gal/ft^2-day, after 24 hours exposure using a 5,200 ppm saline feed at 600 psi. This example demonstrated the importance of the support pore size.

Polymerized β-Lactam

W. Busch and F. Schwarzer; U.S. Patent 3,703,570; November 21, 1972; assigned to Kalle AG, Germany describe a process for the preparation of a semipermeable membrane which has a better resistance to heat and chemicals than do the known semipermeable membranes. This is achieved by starting from a process for the preparation of a semipermeable membrane, in which the membrane is prepared by pouring a layer of a plastic solution, if desired containing salt, onto a support, a portion of the solvent is evaporated from the plastic layer while the latter is on the support whereby the plastic layer is provided with a surface layer poor in solvent, and the remaining solvent of the plastic layer is then removed by means of a liquid which is miscible with the solvent but does not dissolve the plastic material.

In accordance with the process, a plastic solution is used in the process for pouring the layer which contains, as the plastic material, a polyamide which is a polymer of a β-lactam unsubstituted at the nitrogen or a mixture of such β-lactams. The plastic materials employed for the preparation of the semipermeable membrane are known from the publication of Graf, Lohaus, Börner, Schmidt, and Bestian in the journal *Angewandte Chemie,* vol. 74 (1962), pp. 523–530. They are polymers, the basic substance of which is acetidin-2-one, which has the following structure:

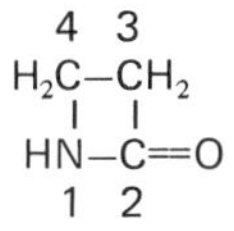

In this formula, substituents replacing hydrogen atoms of the two methylene groups are indicated as being in 3- or 4-position. Useful β-lactams are, for example, 4,4-dimethylacetidin-2-one, 3,4,4-trimethylactidin-2-one, 4,4-diethylactidin-2-one, 4-methyl-4-propylacetidin-2-one or 4-phenylacetidin-2-one. In the preparation of the membrane, the poly-β-lactam is dissolved, for example, in 85 to 100% by weight formic acid, concentrated sulfuric acid, methanolic calcium rhodanide, calcium chloride or zinc chloride solution, trifluoroethanol, dichloroacetic acid, fluoraldihydrate, hexafluoroacetone hydrate or dimethyl formamide containing lithium chloride.

Other solvents dissolving the poly-β-lactam are also useful. Water is generally used as the liquid for precipitating the poly-β-lactam but other precipitating liquids, e.g., methanol or acetone, also are useful; 85 to 100% formic acid and 40% methyl alcoholic calcium rhodanide solution have proved to be the best of all solvents tested.

The poly-β-lactam solution is poured onto a substantially smooth support, e.g., a glass plate, so that the solution forms a layer. The layer is left standing for some time on the support, whereas solvent evaporates from the surface of the plastic layer. After standing, the plastic layer is immersed, together with the support, into a liquid which is a precipitating agent for the poly-β-lactam and is miscible with the solvent of the solution. If desired, the plastic layer is again immersed once or several times into a precipitating liquid. The time during which the poured plastic layer is left standing on the support depends upon the permeability desired of the membrane prepared. Its permeability, of course, also depends on other factors, e.g., the type of solvent, the temperature of the layer, or the degree of saturation with the solvent in the atmosphere above the layer surface.

Example 1: A copolymer containing 70% by weight of 4-methylacetidin-2-one and 30% by weight of 3,4-dimethylacetidin-2-one is dissolved in concentrated formic acid to form a 25.25% by weight solution which, at 20°C, has a viscosity of 23,000 cp.

The solution is briefly subjected to a vacuum produced by means of a water-jet pump in order to remove air. A 0.26 mm thick layer of the solution is then poured onto a glass plate by means of a doctor blade and the layer is left standing in the open on the glass plate for 5 seconds at room temperature. The glass plate is then immersed, together with the layer, several times into distilled water at a temperature of 20°C until the immersion water no longer exhibits an acid reaction after immersion. The membrane thus obtained has a thickness of about 0.10 mm.

For determining the semipermeability of the membrane, water containing 0.05% by weight of sodium chloride and 0.05% by weight of magnesium sulfate is passed over the surface in a turbulent stream under a superatmospheric pressure of 40 atm. At a temperature of 20°C, 200 liters of water per square meter penetrate the membrane in 24 hours while 90% by weight of the magnesium sulfate and 50% by weight of the sodium chloride are retained by the membrane. At a temperature of 80°C, 1,630 liters of water per square meter of the membrane are passed through in 24 hours, the same percentage by weight quantities of the salts being retained.

The membrane obtained is thus suitable, for example, to produce potable water from brackish water. The example also shows that the membrane can be used at an elevated temperature and that the relative permeabilities are substantially unchanged by a temperature change.

Example 2: A polymer of 4-methylacetidin-2-one is dissolved in concentrated formic acid to form a 12.7% by weight solution which, at 20°C, has a viscosity of 12,650 cp. After removing the air, as described in the above Example 1, the solution is poured in a 0.26 mm thick layer onto a glass plate by means of a doctor blade and the layer is left standing in the open on the glass plate for 30 seconds at room temperature. The glass plate is then immersed, together with the layer, in distilled water until the immersion water remains neutral. The membrane obtained has a thickness of about 0.10 mm. At 20°C, it permits 5,000 liters of water per square meter to pass through under a superatmospheric pressure of 40 atm gauge.

4 liters of an aqueous solution were passed through in a circulating process for one hour, on 100 cm^2 of the membrane on one side, under a superatmospheric pressure of 40 atm. The solution originally contained 2% by weight of carboxymethylcellulose and 0.4% by weight of sodium chloride; its viscosity was 20 cp at 20°C. After circulating for one hour, 2 liters of the solution had diffused through the membrane. The remaining solution contained the entire quantity of the carboxymethylcellulose, i.e., 4% by weight, but its content of sodium chloride was 0.4% by weight as before. The membrane is thus suitable for enriching and desalting water-soluble high-molecular weight substances by dialysis.

Crosslinked Polyamideimide-Polyvinylidene Fluoride Interpolymers

A process described by *L.T.C. Lee and K.-J. Liu; U.S. Patent Application (Published) B 276,993; January 28, 1975; assigned to Allied Chemical Corporation* is directed to the preparation of a class of interpolymer membranes based on the hydrophilic polyamide-imide of [1,ω-alkylene(2,6-diketopiperazine)-1,4-dicarboxamide] (referred to as ADD).

The membranes (by employing modifications of composition and procedure) are applicable as selective membranes for dialysis and solvent extraction processes, as membranes in reverse osmosis processes, as ultrafiltration membranes in pressure-drive processes and in hollow fibers structures.

Thus, it has been found that substantially improved interpolymer membranes of excellent strength and controllable degree of swollen character may be formed by preparing a mixture either by way of a solution or a melt blend of two polymers comprising: (a) a highly hydrophilic ADD polyamideimide and (b) a relatively hydrophobic matrix polymer, forming a film from the mixture and upon attaining proper porosity in the film, crosslinking the mixture to form a stable membrane. When the membrane is derived by casting a film from solution, the cast film is dried to an extent sufficient to remove solvent until the desired pore size is attained and then crosslinked to produce a membrane which has essentially a fixed swollen capacity.

Crosslinking may be attained either by incorporating a suitable crosslinking agent or by subjecting the film cast from solution or extruded from a melt to a condition such as radiation, which induces crosslinking. In some applications, the melt extruded membrane may be employed without crosslinking. A layered membrane may also be formed by drying and partially, or wholly, curing a first layer in the foregoing manner and thereafter casting (or extruding) a superimposed layer, or layers, in sequence and drying and crosslinking the entire layered composite.

These membranes provide very low salt-passage in spite of the fact that one component of the membrane system (i.e., ADD polymer) is known to interact strongly with certain ions and exhibit outstanding high water absorption. The following examples illustrate the process. All parts are parts by weight, except where otherwise expressly stated.

Example 1: A solution was prepared by dissolving 68.0 parts of poly[hexamethylene-(2,6-diketopiperazine)-1,4-dicarboxamide] (HDD) and 22.6 parts of polyvinylidine fluoride in 22.0 parts of hexamethylphosphoric triamide (HMP) and 2.50 parts of dimethylformamide (DMF). This clear, homogeneous solution was then filtered to remove dust particles through a sintered glass filter. Shortly before membrane casting, 9.4 parts of crosslinking agent consisting of 1 part BF_3-MEA (a complex of boron trifluoride and monoethanol amine) and 5 parts of epoxide-novolac, was added to the solution. The solution was vigorously agitated for 10 minutes, then transferred to a centrifuge tube which was placed in a centrifuge for 5 minutes to remove the air bubbles. The clear solution was immediately cast on the glass plate with a Gardner Knife set at a suitable gate.

The membrane so cast was placed level into an oven at 120°C for 1 hour and then cured for 3 hours at 150°C. The membranes were cooled to room temperature and then immersed in water to separate them from the glass surface. The membrane was ready for testing after 48 hours immersing in water. The membrane so obtained has a thickness of 22.5 μ, a water absorption of 28.9%. The salt rejection was measured using a 0.01 M sodium chloride solution employing a reverse osmosis apparatus at 25°C and 900 psi pressure. The sample collected was titrated by the standard silver nitrate method. The result indicated 99% salt rejection. The flux rate is 5.8 gal/ft^2/day/μ.

Example 2: The procedure of Example 1 was followed, except that the casting glass surface was pretreated with 1.0% carboxymethylcellulose. The obtained membrane has a thickness of 11.8 μ and a water absorption of 26.5%. The salt rejection of 0.01 M NaCl at 25°C and 600 psi is higher than 98%. The flux rate is 13.8 gal/ft^2/day/μ.

Example 3: The procedure of Example 1 was followed except that 76.5 parts of HDD, 14.5 parts of polyvinylidene fluoride, and 9.0 parts of Epoxy-Novolac was used. The obtained membrane has a thickness of 12.75 μ and a water absorption of 38.5%. The salt rejection of 0.01 M NaCl solution at 25°C and 600 psi is higher than 95.0% with a flux rate of 12.5 gal/ft^2/day/μ.

Ether Treatment

According to a process described by *J.B. Ganci, J.H. Jensen and F.H. Smith; U.S. Patent 3,808,303; April 30, 1974; assigned to E.I. du Pont de Nemours and Company* the solute rejection properties of permselective membranes made of nitrogen-linked synthetic organic polymers in reverse osmosis processes are improved by exposing the membranes to solutions of ethers having carbon to oxygen atomic ratios between 2.5:1 and about 5:1, and having formula weights larger than about 400. Such treatments are even more effective when combined with exposure of the membranes to water at temperatures of 50° to 90°C.

A preferred ether is polymethyl vinyl ether with an average molecular weight of about 50,000. In this structure all the R's are hydrogen and the ratio of carbons to oxygen is 3:1. Also useful are copolymers of vinyl methyl ether and vinyl ethyl ether. Commercially available polymers of methyl vinyl ether include "Gantrez" M-154 and M-093.

Isocyanate Treatment of Acid Grafted Membranes

A process described by *D.G. Pye; U.S. Patent 3,554,379; January 12, 1971; assigned to E.I. du Pont de Nemours & Co.* involves the improvement of acid grafted, substantially linear nitrogenous polymeric membranes for use in the reverse osmosis separation of components of aqueous solutions and mixtures and particularly to the treatment of such membranes to reduce their tendency to swell in water. The process concerns the method of treating the membranes to convert acid groups present in the membrane polymer into less hydrophilic groups thereby reducing the membrane potential for dimensional change in water.

The concept extends to the improvement of permeation separation apparatus employing hollow filament, reverse osmosis membranes and is based in part on the utilization of nylon hollow filaments acid grafted over their entire lengths in continuous filament form. The practical and economic advantages attendant such utilization are based conceptually on the feasibility of converting acid groups to a more desirable form after grafting in contrast to avoiding the formation of acid grafts in given areas during the grafting process.

The graft copolymers which are modified to provide the selectively permeable membranes used in the process are described in detail. These copolymers, in turn, are prepared from substantially linear nitrogenous condensation polymers. Such polymers are characterized by recurring $-N<$ atoms as an integral part of the polymer chain and preferably bear at least about 300 titratable acid groups per 10^6 grams of polymer, at least about 200 of the acid groups being chemically bonded by a carbon-to-carbon linkage to a catenarian carbon atom of the polymer and removed by at least one carbon atom from the catenarian carbon atom. The nitrogenous polymers from which the graft copolymers are made include the polyamides, which are characterized by recurring

$$\begin{array}{c} \mathrm{R} \quad \mathrm{O} \\ | \quad \| \\ -\mathrm{N}-\mathrm{C}- \end{array}$$

links in the polymer chain, where R may be hydrogen or an organic radical. High molecular weight fiber-forming polyamides, now well known as nylons, are preferred in forming these products. In making the starting polymers for the process, hydrophilic side chains are grafted onto such a nitrogenous polymer by known means, such as high energy ionizing radiation or the action of free radical generating catalysts. The material grafted is desirably a polymerizable unsaturated organic acid or a univalent metal salt of such acid which may be applied to the nitrogenous polymer either before, during, or after application of the grafting means. The preferred material grafted is acrylic acid, although numerous other materials are suitable for grafting including methacrylic, ethylacrylic, crotonic, propiolic, styrenecarboxylic, and acids.

By the expression "300 titratable acid groups per 10^6 grams of polymer" is meant the number of equivalents of $-COOH$ per 10^6 grams of polymer, for example as determined by the method of Taylor and Waltz, *Analytical Chemistry,* 19, 448 (1947). If the graft copolymer is insufficiently soluble in hot benzyl alcohol, a satisfactory determination of acid groups may be made by boiling the finely divided copolymer sample gently in 0.1 N NaOH followed by back titrating excess base with 0.1 N HCl using bromocresol green indicator. When the graft copolymer has its acid groups in the form of univalent metal salts, these groups are first converted to acid form by hydrolysis of the salt with strong acid prior to the determination noted above.

When the grafted moiety of the graft copolymer is acrylic acid, a concentration of 300 titratable acid groups per 10^6 grams of polymer is obtained when the graft copolymer shows a grafted weight gain of acrylic acid of about 3%. Thus, the nitrogenous polymers as defined above having at least about 3% by weight of grafted acrylic acid are regarded as containing at least about 300 titratable acid groups per 10^6 grams of polymer.

Selectively permeable membranes are prepared by preforming membranes of the starting nitrogenous polymers from which the graft copolymers are made or by forming into membranes the copolymers resulting from the grafting process. The membranes, whether preformed or postformed, may take any known physical configuration employed in permeation separation processes, such as uniplanar film or a hollow filament, and methods known in the art, such as extrusion and melt or solution spinning, may be employed in their formation.

The technique for reducing the water swelling tendency of selectively permeable membranes of the graft copolymers involves treatment with a chemical reactant which is effective to reduce the hydrophilic nature of the free acid groups present in such graft copolymers. A simple chemical modification involves treatment of acid grafted substantially linear nitrogenous polymers to effect conversion of free acid groups into their polyvalent metal salts. This modification is easily obtained by soaking the graft copolymer in membrane form in a water solution of a polyvalent metal salt.

The preferred polyvalent metal salts are salts of the divalent alkaline earth metals magnesium, calcium, and barium and the trivalent metals iron and aluminum, but salts of other polyvalent metals may also be used, including water-soluble salts of strontium, manganese, nickel, copper, zinc, mercury, and lead. These metals may be used in the acetate, chloride, nitrate or sulfate form.

Another chemical modification which converts the carboxylic acid groups of acid grafted linear nitrogenous polymers into less hydrophilic groups is esterification. A particularly useful reactant for esterifying carboxylic acid groups is diazomethane. The reaction is usually carried out in an ethereal or hydrocarbon solution and converts the acid groups into the methyl ester with evolution of nitrogen.

$$RCOOH + CH_2N_2 \longrightarrow RCOOCH_3 + N_2$$

In carrying out this modification with membranes of acid grafted linear nitrogenous polymers, it is convenient to soak the polymer membrane in a solution of diazomethane at 25° to 50°C for a few hours or until the characteristic color of diazomethane is dissipated. The membrane is then drained and rinsed with a series of solvents ending with a water-soluble or easily evaporated solvent. The chemical reactions involved are known (Eistert, B. *Newer Methods of Preparative Organic Chemistry.* W. Foerest, Ed., New York, Academic Press, Vol. 1, 1948, p. 513). The preferred chemical modification of carboxylic acid grafted linear nitrogenous polymers to reduce their tendency to swell with water involves conversion of the carboxylic acid groups to amide groups by reaction with an isocyanate.

$$RCOOH + R'NCO \longrightarrow RCONHR' + CO_2$$

In this equation —COOH represents any acid group in the graft copolymer and —CONHR′ represents the resulting amide group. R′ in the isocyanate (and the amide) may be a hydrocarbon group of 18 carbons or less and free of aliphatic carbon-to-carbon unsaturation, or such a hydrocarbon group carrying another isocyanate (—NCO) substituent. When

RNCO is a diisocyanate, any isocyanate groups remaining in R after the above reaction are converted to amine groups when the treated portion of the graft copolymer comes in contact with an aqueous medium. Preferred isocyanates for effecting this modification comprise phenylisocyanate, 1-naphthylisocyanate, and toluene diisocyanate. Amide modification is conveniently carried out by soaking a polymer membrane in the technically pure isocyanate or a solution of the isocyanate in a solvent such as xylene or orthodichlorobenzene or a mixture of solvents such as xylene and dimethylacetamide.

A suitable rate of reaction is obtained with thin membranes within about 1 to 6 hours at temperatures in the range of about 100° to about 150°C. Treatments for about 4 hours at about 125°C are preferred for membranes between 10 and 50 microns in thickness, using solutions containing about 25 parts of phenyl isocyanate in a mixture of 5 to 20 parts of dimethylacetamide and 70 to 55 parts of xylene. Such conditions result in an increase in weight of the treated polymer, in essentially complete disappearance of an infrared absorption peak at 5.8 microns characteristic of carboxyl carbonyl compounds, and in destruction of the ability of the treated polymer to form metallic salts.

The products of such treatments exhibit reduced swelling upon soaking in water. In a preferred form of the process, there is provided an improved procedure for assembling permeation separation apparatus adapted for the reverse osmosis separation of components of aqueous solutions and mixtures and comprising at least one cast wall member in sealing relationship with a plurality of selectively permeable, capillary membranes of an acid grafted, substantially linear nitrogenous polymer.

Example 1: A sheet of 1 mil nylon 66 film was placed in an unsealed polyethylene bag. The bag was placed on a tray of crushed solid carbon dioxide and passed five times through the beam of a Van de Graaf apparatus operated at 2 mev and 250 microamperes. The bag was then covered with crushed solid carbon dioxide and taken to a laboratory for treatment with acrylic acid. The cold irradiated film was transferred from the bag to a tray containing 25 parts by volume acrylic acid and 75 parts by volume water and allowed to soak at room temperature for 15 minutes. It was then subjected to a series of 5 minute washings, initially in a first tray with distilled water and serially thereafter in five separate trays with fresh distilled water. After hanging in air to dry, the film showed a weight gain of about 15% of acrylic acid. The above procedure was repeated and the additional weight gain was 10.8%.

A sample of this film was immersed in phenyl isocyanate in a vessel blanketed with nitrogen and heated at 125°C. After 1 hour the film was removed and washed once with dimethylformamide and 3 times with pentane, and dried at room temperature under reduced pressure. In infrared measurements, the absorption peak at 5.8 microns typical of carbonyl compound was substantially reduced in the isocyanate treated film in comparison with the untreated acid grafted film.

Samples of the isocyanate treated and untreated acid grafted film were immersed in aqueous 5% sodium carbonate solution and visually observed. The untreated film became swollen, limp, and soft, but the treated film showed no apparent change in size or texture. Similar reductions in swelling upon immersion in water were obtained when acrylic acid grafted nylon films were treated in similar manner with 1-naphthyl isocyanate and with toluene diisocyanate. This example shows that treatment of acrylic acid grafted nylon films with an isocyanate reduces the carboxyl carbonyl activity introduced by the grafting reaction and greatly reduces the tendency of the grafted film to change in size and texture upon exposure to water.

Example 2: A bundle was prepared from 366 yarns, each containing 17 filaments of hollow nylon 66 fiber with an average inside diameter of about 20 microns and outside diameter of about 50 microns for a void content of 16%. The fiber bundle was protected by a cover of woven nylon mesh and placed in a polyethylene bag. The bag was laid on a bed of crushed solid carbon dioxide and irradiated by passing it through the beam of a Van de Graaf apparatus six times while operating the Van de Graaf apparatus at 2 mev and 250 microamperes. The fiber bundle was then removed from the bag, immersed for 15 minutes in a 25% by volume aqueous solution of acrylic acid, washed five times for 5 minutes each with distilled water, and dried. The weight gain was about 44%.

The ends of the bundle were immersed (under nitrogen) in a mixture of 25 volumes of phenyl isocyanate, 10 volumes of dimethylacetamide, and 65 volumes of xylene for 1 hour at 125°C. The bundle was washed with pentane, dried and immersed in 5% Na_2CO_3 solution. The isocyanate treated portion remained unaffected while the untreated portion swelled extensively. The bundle was then immersed in a dilute aqueous copper sulfate solution. The treated portions remained colorless but the untreated portions turned blue indicating that the carboxyl reactivity of the treated portion had been destroyed by the isocyanate treatment. This example shows the treatment of acrylic acid grafted nylon in hollow filament form with an isocyanate to reduce swelling in water and demonstrates the reduction in carboxylic acid activity obtained.

Example 3: Three hollow nylon fiber bundles containing about 84 yarns each containing 17 filaments of nylon 66 hollow fiber were grafted with acrylic acid by the general procedure of Example 2. An irradiation dosage level of six passes through the Van de Graaf beam was used, the bundles were treated with 25% aqueous acrylic acid, and the weight gains were about 23%. The bundles were dried overnight at 90°C in a vacuum oven and a 10" length of one end of each bundle was immersed under nitrogen in a 10% by volume solution of phenyl isocyanate in dimethylacetamide for 15 minutes at room temperature. The temperature was then raised to 125°C for 4 hours, after which the bundles were rinsed with dimethylacetamide and dried.

The isocyanate treatment was repeated with the other bundle ends, after which the entire bundles were immersed in 5% aqueous sodium carbonate at 50°C for 2 hours, rinsed with distilled water, and dried. The three acrylic acid grafted hollow nylon fiber bundles with isocyanate treated ends were assembled into a desalination apparatus. The openness of the hollow fibers for flow of water through their capillary openings was demonstrated by applying water at a pressure of 200 psig to one end of each permeator and measuring the rate of water flow through the fibers. The flow rates were in essential agreement with calculated flow rates assuming an effective internal diameter of 21 to 24 microns.

The water desalination properties of the three treated hollow fiber bundles were determined by applying to their outer surfaces at a pressure of 210 psig an aqueous solution containing 700 ppm calcium sulfate, 400 ppm magnesium sulfate, and 400 ppm sodium sulfate formulated to simulate many midcontinent groundwaters. The rate of water passage through the walls of the hollow fibers was measured and the sulfate content of the permeated water was determined. The following data were obtained for a considerable time after equilibrium had been established.

Bundle	Permeation Rate, gal./ft.^{2}/day	Salt Passage, percent
A	0.025	6
B	0.025	5
C	0.027	5

In contrast, an acrylic acid grafted nylon hollow fiber bundle the ends of which were not treated in accordance with the process exhibited no through-flow when assembled into a permeator of the type aforedescribed and subjected to the same testing conditions, as the hollow interiors of the epoxy resin encapsulated fiber ends had become constricted as a result of excessive swelling in water. This example illustrates an improved process for assembling a hollow filament reverse osmosis water desalination apparatus which includes treating a portion of a bundle of hollow fibers with a reactant which converts free carboxylic acid groups into less hydrophilic groups and encapsulating the treated portion of the bundle in a setting resin.

In related work *H.H. Hoehn and D.G. Pye; U.S. Patent 3,524,546; August 18, 1970; assigned to E.I. du Pont de Nemours & Company* describe a process for separating the components of fluid mixtures, particularly the desalination of seawater, by the use of membranes formed from copolymers of nylon grafted with titratable acid groups. *J.W. Richter and H.H. Hoehn; U.S. Patent 3,567,632; March 2, 1971; assigned to E.I. du Pont de Nemours & Company* describe permselective barriers or membranes which are prepared from synthetic, organic, nitrogen-linked aromatic polymers of the formula —LR— where L is a selected nitrogen-containing functional linkage such as an amide linkage, and R is a selected aromatic-containing linkage such as phenylene. The membranes are useful in separating components of fluid mixtures or solutions such as water containing dissolved salts.

Acid Treatment

L.A. Cescon and H.H. Hoehn; U.S. Patent 3,551,331; December 29, 1970; assigned to E.I. du Pont de Nemours & Company describe a process for the reverse osmosis separation of a liquid mixture containing at least about 25% by weight water which comprises

(A) passing the mixture in contact with one surface of a substantially linear aliphatic polyamide resin membrane which is about 2 to 75 microns thick and is characterized, when dry, by

- (1) a wide angle x-ray diffraction pattern indicating a high degree of crystallinity, the crystal perfection index being at least about 90, and
- (2) a small angle x-ray diffraction pattern indicating the presence of scattering centers which, when determined by the Dismore small angle soft x-ray method, have an extrapolated intercept scattering intensity at zero scattering angle of about 50 to 220 calculated by the method of Guinier, and is characterized, when wet, by
- (3) a water permeability of about 50 to 50,000;

(B) and recovering from the other side of the membrane liquid mixture which has passed through the membrane and which contains a reduced amount of one component of the mixture.

When the membrane is in the form of a hollow fiber, it should have an outside diameter of 10 to 250 microns, a wall thickness of 2 to 75 microns, and a ratio of the cross-sectional area of the internal bore of the fiber to the total cross-sectional area within the outer perimeter of the fiber of 0.12 to 0.60, and be further characterized, when dry, by a wide angle x-ray diffraction pattern in which the (100) and (010, 110) diffraction arcs have orientation angles of less than about 100°. The process is based upon the discovery that the rate of water permeation through a thin polyamide membrane can be materially increased without an excessive decrease in the physical strength of the membrane or its salt rejection by treating the membrane with a particular treating agent to a specified degree under controlled conditions. This process comprises

(A) treating a substantially linear aliphatic polyamide resin membrane about 2 to 75 microns thick with a liquid treating composition containing, by weight,

- (1) about 1 to 100% of a treating agent selected from the group consisting of
  - (a) protonic acids having a pKa in water not greater than about 10.3 and a pH not in excess of about 6.3 for a 0.01 molar aqueous solution at 25°C,
  - (b) lyotropic salts containing a cation and an anion in which the anion is higher in the list than the cation, and
  - (c) Lewis acids selected from the group consisting of aluminum halides of the formula AlX_3 in which X is chlorine or bromine, and boron halides of the formula BX'_3 in which X' is fluorine, chlorine or bromine, and
- (2) about 0 to 99% of a solvent for the treating agent which is essentially

chemically inert toward the treating agent and the membrane and is essentially a nonsolvent for the membrane;

at a temperature at least high enough to maintain the composition as a single liquid phase but not in excess of the boiling point of the composition for at least one second, the conditions of temperature, time, concentration of treating agent and choice of solvent being such that, when the membrane is dried to constant weight, treated under the conditions rinsed to remove the treating agent and redried to constant weight, a weight loss of about 1 to 35% is obtained, and

(B) rinsing the membrane to remove the treating agent with a rinse medium which contains at least 25% by weight water, is a solvent for the treating agent, and is essentially inert toward the membrane under the rinse conditions.

Commercial polyamide resins which are suitable for making the treated membranes of this process may contain small amounts of unreacted monomers, delustering agents, coloring materials, and other components with no significant effect on their membrane properties. The procedure used to treat polyamide membranes to render them suitable for use in the reverse osmosis process described is relatively simple. A suitable polyamide membrane is treated with a treating agent selected from the group consisting of certain protonic acids, selected lyotropic salts and specified Lewis acids.

The treating composition may be heated or cooled to the desired treating temperature and brought into contact with the membrane by immersing the membrane in the composition or by any other convenient technique. After the desired treating time, the treating composition is removed from contact with the membrane and the membrane is washed substantially free of the treating composition with a suitable rinse medium. The resulting treated membrane is then ready for use in the reverse osmosis separation of aqeuous mixtures. Typical treating agent-solvent combinations which are useful are given below.

Treating Agent	Solvent
Formic acid	Water
Formic acid	Chloroform
Acetic acid	Water
Chloroacetic acid	Water
Chloroacetic acid	Chloroform
Dichloroacetic acid	Water
Dichloroacetic acid	Chloroform
Trichloroacetic acid	Water
Phosphoric acid	Water
Phosphoric acid	2,2′-diethoxydiethyl ether-water
o-Chorophenol	Ethanol
Chloral hydrate	Water
Phenol	Water
Calcium chloride	Methanol-water
Potassium thiocyanate	Methanol
Zinc nitrate	Methanol
Cobaltous nitrate	Methanol
Manganous thiocyanate	Methanol
Manganous bromide	Methanol
Manganous nitrate	Water
Ferric chloride	Methanol
Boron trifluoride	Methanol

Preferred treating agent concentrations depend on the activity and the solubility of the treating agent in the solvent and on the treating temperature. For instance, in the case of aqueous formic acid, the preferred treating compositions contain between about 45 and about 70% by weight formic acid, and produce useful membranes by treatments of a few minutes to a few hours at temperature between about 80°C and ambient. On the other hand, when the treating composition is formic acid in chloroform, the formic acid concentration preferably is in the range of about 1 to 4%. Concentrations of formic acid in chloroform as low as 4% produce drastic overtreatments in treating times as short as 7 minutes at ambient temperature. Another preferred treating composition contains 15 to 25% calcium chloride, 50 to 70% methanol, and 10 to 25% water. Preferred concentrations of lyotropic salts such as calcium chloride and zinc chloride are higher in water than in methanol or methanol-water mixtures. Other preferred compositions include boron trifluoride at concentrations up to about 20%, but below its maximum solubility, in nonreactive oxygenated solvents such as methanol; and aluminum trichloride at concentrations up to about 35%, but below its maximum solubility, in nonreactive hydrocarbon solvents.

Tannin Treatment

According to a process described by *C. Chen and J.B. Ganci; U.S. Patent 3,886,066; May 27, 1975; assigned to E.I. du Pont de Nemours & Co.* semipermeable membranes having permanently reduced solute passage under reverse osmosis conditions are obtained by contacting nonporous semipermeable membranes with solutions of hydrolyzable tannins. The polymers suitable for use in the semipermeable membranes of this process are nonporous semipermeable polymers such as polyarylene ether sulfones, cellulose acetates, and the substantially linear synthetic organic aromatic polymers.

Example 1: Substantially as described in U.S. Patent 3,184,436, a mixture of 70 parts of isophthaloyl chloride and 30 parts of terephthaloyl chloride was added gradually to a dimethylacetamide solution of a substantially stoichiometric amount of a mixture of 80 parts of metaphenylene diamine and 20 parts of the calcium salt of metaphenylene diamine-4-sulfonic acid. The resulting polymer nominally contained 1.5% sulfur, equivalent to 0.46 sulfonic acid moieties per 1,000 grams.

Acetamide containing 20 parts of lithium nitrate and 6 parts of lithium chloride (based on the polymer present) was spun through a 150 hole spinneret to obtain hollow fibers substantially as described in U.S. Patent 3,567,632. These hollow fibers were extracted with water to obtain asymmetric hollow fiber membranes with an outside diameter of 87 microns and an inside diameter of 40 microns. These membranes were kept wet with water until used.

Four lengths of the 150 filament skeins of hollow fiber membranes, each about 1 meter long, were placed into separate 100 ml bottles. Two of the bottles were filled with deionized water and the other two were filled with a solution made by dissolving 0.5% tannic acid (analytical reagent grade) in deionized water and adjusting to pH 4 with sodium hydroxide. The bottles were held in a water bath controlled at 55°C for 16 hours. The hollow fiber membranes were removed from the bottles, rinsed with deionized water, and stored in deionized water until assembled into reverse osmosis separation units.

In assembling such units each 150 filament skein of hollow fibers was doubled to obtain 300 parallel fibers and, while wet with water, was inserted into a plastic tube fitted with a side tube near each end. The hollow fiber membranes were sealed into the plastic tube with epoxy resin at each end and then were carefully cut outside the end seals to open the hollow fibers for fluid flow. An end fitting and a pressure gauge were attached at one end for measurement of dead end pressure and the other end was attached to a buret for measurement of water flow from the open ends of the hollow fibers under reverse osmosis conditions. The resulting units contained hollow fiber membranes with an effective length of 24" and a nominal effective area of 0.25 ft^2 of membrane surface.

The permselective properties of these hollow fiber membranes were determined under reverse osmosis conditions at an applied pressure of 600 psig with an aqueous solution containing 0.5% sodium chloride in a system containing Monel metal parts. After about 24 hours under these conditions one of the units containing membranes held at 55°C for 16 hours in deionized water had a flux rate of 1.36 gfd and a salt passage of 2.84% under conditions such that about 2.8% of the feed solution passed through the unit was passed through the membrane (2.8% conversion) and the dead end pressure was 90 psig. After about 72 hours one of the units containing membranes held at 55°C for 16 hours in deionized water had a water flux rate of 1.92 gfd and a salt passage of 2.88% under conditions such that about 7.8% of the feed solution passed through the membrane and the dead end pressure was 112 psig. After about 48 hours the units containing membranes held at 55°C for 16 hours in solutions containing 0.5% tannic acid had water fluxes of 2.07 and 1.67 gfd and salt passages of 0.33% and 0.53% under conditions such that about 6% of the feed solutions passed through the membranes and the dead end pressures were 251 and 237 psig.

After 27 days exposure to the sodium chloride solution under reverse osmosis conditions the tannin-treated membranes had water fluxes of 1.79 and 1.74 gfd and salt passages of 0.29 and 0.26% under conditions such that 4.4% and 4.7% of the feed solutions passed through the membranes and the dead end pressures were 250 and 246 psig. During this period there was no significant decline in the water flux rate of the membranes contacted with solutions containing tannic acid.

Example 2: Hollow fiber membranes with an o.d. of 91 microns and an i.d. of 41 microns were prepared as described in Example 1 and three reverse osmosis units (A, B, and C) were assembled as described therein. The hollow fiber membranes from units B and C were removed from their pressure tubes and placed into bottles containing 10 ml of aqueous 0.1% tannic acid. After about 5 minutes for equilibration, a 2 ml portion of the tannic acid solution was removed from each bottle for analysis.

The membranes from unit B were removed from the tannic acid solution after 4 hours, rinsed thoroughly with water, and reinstalled in their pressure shell. Membranes from unit C were removed from the tannic acid solution after 21 hours and similarly rinsed and installed in their shell. The amounts of tannic acid absorbed by these membranes were determined by analysis of the residual solutions. The properties of the membranes in units A, B and C were determined under conditions such that about 3.7% of the feed solution passed through the membranes and the dead end pressure was 141 psig. After about 24 hours the membranes had the separation properties described in the following table.

Reverse Osmosis Unit	A	B	C
Soaking time (hours)	None	4	21
Tannic acid absorbed (micrograms per square centimeter)	—	2.3	9.3
Water flux (g.s.f.d.)	2.7	2.9	2.6
Salt passage (%)	6.5	1.6	1.3
Conversion (%)	6.2	8.6	9.7
"Dead end" pressure (p.s.i.g.)	139	160	158

Example 3: To a solution of approximately 80 grams of pelletized sulfonated polyphenylene ether sulfone (Bakelite polysulfone P-1700) dissolved in 800 ml of chloroform was added 40 g of chlorosulfonic acid over about 5 min as the mixture warmed slightly. A gummy second phase separated during about 30 minutes as sulfonation proceeded. Hexane was added, the supernatant liquid was decanted, and the second phase was dissolved in about 200 ml of dimethylacetamide.

The resulting solution was neutralized with approximately 20 grams of lithium hydroxide monohydrate with stirring. The solution was warmed with an inert gas sweep to remove low boiling solvents and then was added to acetone in a high speed blender to obtain the sulfonated polyphenylene ether sulfone in granular form. The separated polymer was washed with acetone, rinsed with water until the rinsings were no longer alkaline and then dried.

Asymmetric semipermeable membranes were prepared from this polymer by procedures analogous to those of U.S. Patent 3,567,632. Films with a nominal thickness of 0.015" were cast with a doctor knife onto smooth plates using a solution containing 20% polymer and 30% lithium nitrate (based on the polymer present) in dimethylacetamide. These films and the supporting plates were heated on a hot plate at 100°C in a ventilated hood for 2 minutes and were then immersed in water for about 1 hour to obtain permselective membranes, which were kept wet with water until use.

One of these membranes was shown to contain 0.64 sulfonic acid groups per 1,000 grams of polymer by the procedure of converting the polymer to the acid form by soaking for 30 minutes each in three different portions of 1/10 molar hydrochloric acid; washing free of excess acid with deionized water; and titrating to pH 7 with 0.1 N sodium hydroxide in 50 ml of water containing 25 mg of sodium chloride. One of the titrated membranes contained 8.7% sulfur, equivalent to 0.46 sulfonic acid groups per 1,000 grams of dry polymer. Another of these membranes contained 0.55% lithium, equivalent to 0.80 sulfonic acid groups per 1,000 grams of polymer.

Another of these membranes was installed in a reverse osmosis test cell. After 4 days exposure to a sodium chloride solution under the reverse osmosis conditions of Example 1, this membrane had a water flux of 10.4 gfd and a salt passage of 7.1% under conditions such that about 5% of the feed solution passed through the membrane. The test cell was then filled with an aqueous solution containing 0.005% (50 parts per million) of tannic acid and 2% citric acid adjusted to pH 4 with ammonium hydroxide.

The solution was maintained in contact with the film membrane for 2 hours at a pressure of 75 psig and the cell was then rinsed substantially free of tannic and citric acids. After an additional 20 hours exposure to the solution containing 0.5% sodium chloride at 600 psig, the membrane had a water flux of 9.1 gfd and a salt passage of 3.0% under conditions such that less than 5% of the feed to the test cell passed through the membrane.

In related work, *J.B. Ganci; U.S. Patent 3,853,755; December 10, 1974; assigned to E.I. du Pont de Nemours and Co.* has found that the passage of salts through a nonporous semipermeable membrane with a coating layer of a hydrous heavy metal composition under reverse osmosis conditions is reduced after the membrane and the coating layer have been contacted with a solution of a hydrolyzable tannin.

In earlier work, *F.W. Tober; U.S. Patent 3,660,281; May 2, 1972; assigned to E.I. du Pont de Nemours and Co.* describes a process which comprises treating a membrane for the selective permeation separation of dissolved and colloidally dispersed materials in aqueous solutions or mixtures consisting essentially of substantially linear aliphatic polyamide resin, with an aqueous solution of a hydrolyzable tannin until the membrane contains sufficient tannin to inhibit degradation of the resin from deterioration. The presence of an effective amount of hydrolyzable tannin in the membranes has been found to significantly lengthen the useful life of such membranes.

Replacement of Membrane-Associated Water

According to a process described by *P.F. Dismore; U.S. Patent 3,710,945; January 16, 1973; assigned to E.I. du Pont de Nemours and Co.* selectively permeable, polymeric membranes for use in the reverse osmosis separation of components of aqueous mixtures are chemically treated to preserve their permeability characteristics and to protect against loss thereof during the obtention of a water-tight seal with the potting compounds generally employed in permeation separation devices. The desired results are achieved by effecting a substantial replacement of membrane-associated water with an inert organic medium exhibiting a lower volatility than water.

Depending on the degree of hydrophilicity and water solubility of the organic medium, replacement of the membrane-associated water is accomplished by direct contact including successive contacts of the water-wet membrane with aqueous solutions of increasing organic medium concentration and/or indirect contact including successive contacts of the membrane with organic media of decreasing hydrophilicity. Reversal of the replacement process to reintroduce water to the membrane structure restores the operability of the members and renders them effective for their intended desalination and purification functions.

Example 1: Selectively permeable membranes were prepared by treating polymer films, formed from the condensation product of hexamethylene diamine and adipic acid, with an aqueous solution of 62% formic acid in the manner described in U.S. Patent 3,551,331. Specifically, the polymer films were soaked in the formic acid solution for one hour and then thoroughly washed with deionized water. A first portion of the membrane, designated as Sample 1A, was allowed to dry at room temperature for several days. A second portion of the membrane, designated as Sample 2A, was preserved in water until tested for permeation properties. A third portion of the membrane, designated as Sample 3A, was allowed to dry at room temperature for several days after being treated in accordance with the following schedule:

	Bath Composition (vol %)	
Soaking Period (hr)	**Glycerin**	**Water**
4	10	90
64	20	80
5	40	60
3.5	60	40
16	80	20
4	100	-

The three membrane samples were then tested for permeation properties at an applied pressure, using brackish sulfate solution containing 206 ppm Ca^{++}, 137 ppm Na^{+}, 80 ppm Mg^{++}, 1,084 ppm $SO_4^{=}$ at 500 psi as test liquid. Test results obtained are set forth in the table below.

Sample	W_m	% SR†
1A	115	93
2A	660	82
3A*	660	92

*Washed with water prior to testing. †Salt Rejection

Example 2: A copolymer of metaphenylene diamine and a mixture of isophthaloyl (70 mol %) and terephthaloyl (30 mol %) chloride was prepared in the manner described in U.S. Patent 3,063,966, infra, by simultaneously adding to a reaction vessel fitted with a stirrer a solution of 43.9 parts of isophthaloyl chloride and 10.98 parts of terephthaloyl chloride dissolved in 1,600 parts of methylene chloride and a solution of 48.87 parts of metaphenylene diamine hydrochloride and 109.3 parts of triethylamine dissolved in 1,600 parts of methylene chloride. An additional 400 parts of methylene chloride was used for rinse purposes and added to the reaction mass. After 10 minutes, polymer having an inherent viscosity of 1.44 and a melt temperature of 370°C was formed.

A casting solution was then prepared by mixing 30 grams of the prepared copolymer with 3 grams lithium chloride and 167 grams dimethylacetamide and used to cast a membrane on a glass plate. The thickness of the membrane was determined by using a doctor knife set at 15 mils. Curing of the membrane was effected by placing the glass plate with the cast film thereon in a vacuum oven at 140°C for 140 minutes. Thereafter, the film and plate were immersed in ice water and film was removed and sectioned for treatment in accordance with the process.

One part of the resultant membrane, designated as Sample 1B, was allowed to dry out at room temperature. A second part of the membrane, designated as Sample 2B, was kept in water. A third part of the membrane, designated as Sample 3B, was soaked in a bath of 100% glycerin for 19 hours and allowed to dry at room temperature for several days.

The three samples of membrane were then tested as in Example 1 except that a 0.3% chloride solution was employed as test liquid and a pressure of 460 psig was applied. The results are given in the following table.

Sample	W_m	% SR
1B	1.4	94.0
2B	630	87.0
3B*	340	90.0

*Washed with water prior to testing.

OTHER NITROGEN-CONTAINING MATERIALS

Poly(N-Amido)Imides

According to a process described by *W.J. Wrasidlo; U.S. Patent 3,816,303; June 11, 1974; assigned to The U.S. Secretary of the Interior* semipermeable membranes are prepared from poly(N-amido)imides, i.e., polymers characterized by the recurring structural group

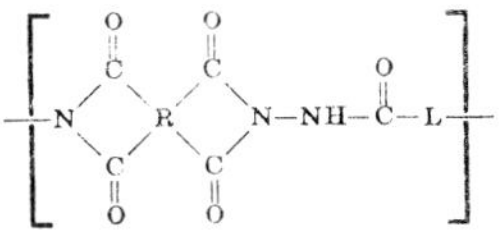

where R is a tetravalent organic radical and L is a divalent organic radical.

The poly(N-amido)imides of the process are prepared by reactions of the type conventionally employed in preparation of imides. This involves initial reaction of a tetracarboxylic acid dianhydride with an amidohydrazide to form a polyamic acid, and subsequent dehydration to form the corresponding poly(N-amido)imide. The overall reaction is represented by the following general equation:

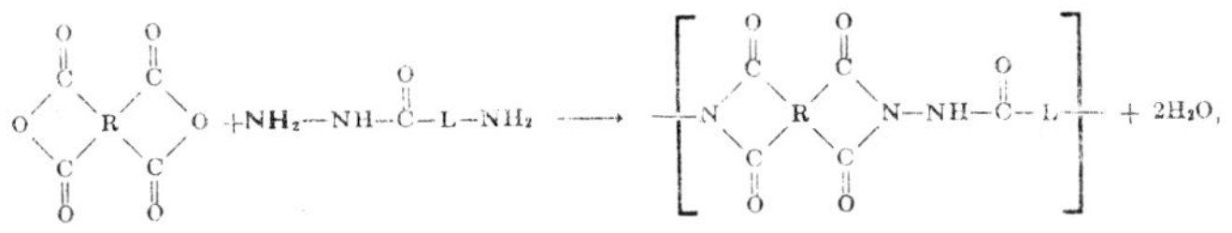

where R and L are as defined above. The reactions are carried out in conventional solvents such as dimethylacetamide (DMAc), dimethylformamide (DMF), dimethylsulfoxide (DMSO), ethyl iodide, methylpyrrolidone, pyridine or 2-pyrrolidone. Suitable reaction temperatures for preparation of the polyamic acids will generally vary from about 0° to 25°C, with the optimum temperature depending on the particular reactants, molecular weight of the product, etc. Generally, ambient conditions of temperature, as well as pressure, are satisfactory, although a lower temperature, e.g., about 0° to 15°C may improve efficiency of the reaction in some cases.

Stoichiometric amounts of the reactants are usually satisfactory, with concentrations of reactants in the solvent being about 10 to 25%. The order of mixing and means for mixing the reactants and solvent are not critical; however, a slow addition of the hydrazide to a solution of the anhydride in the solvent has generally been found to give polyamic acids of desirably high molecular weight. The polyamic acid generally has a weight average MW of about 20,000 to 100,000.

Conversion of the polyamic acid to the corresponding polyimide may be achieved by heating the solution of the polyamic acid at a temperature of about 100° to 200°C for a period of about 5 minutes to 24 hours. This is most readily accomplished by heating a cast film of the solution to a temperature of about 135° to 160°C for about 1 to 2 hours. Alternatively, the conversion may be accomplished by means of chemical dehydration, e.g., by treatment of the solution with a mixture of 50 vol % pyridine in acetic anhydride. A percent conversion to polyimide of about 60 to 70 will normally be desirable, since a relatively high degree of imidization is usually required to provide membranes having high salt rejections and high mechanical stability under reverse osmosis conditions. Molecular weights of the resulting imide polymers will generally range from about 10,000 to 80,000.

Examples 1 through 8: These examples illustrate preparation of the N-amido imide polymers. In each example, the procedure consisted of slow addition of a stoichiometric amount of a hydrazide to a solution of a dianhydride in a suitable solvent at 0°C. The solution was stirred for a period of 2 hours to complete the reaction of the hydrazide and the dianhydride to form the amic acid. Conversion of the amic acid to the imide form was achieved by adding to the amic acid reaction mixture 50 vol % mixture of pyridine and acetic anhydride, and allowing the resulting mixture to react, with stirring, for a period of about 30 minutes.

The following table gives the hydrazide, the dianhydride, the solvent, the concentration of the dianhydride in the solvent, inherent viscosity of the N-amido imide polymer (determined at 25°C at 0.5% polymer concentration in DMAc), and the structural formula of the characteristic group of the polymer.

Ex.	Monomers: Dianhydride	Monomers: Hydrazide	Conc. (wt. percent)	Reaction solvent	Inherent viscosity (dl./g.)	Structure of polymer
(1)	3,3',4',4'-benzophenone	Isophthalic dihydrazide	18	DMAc	1.58	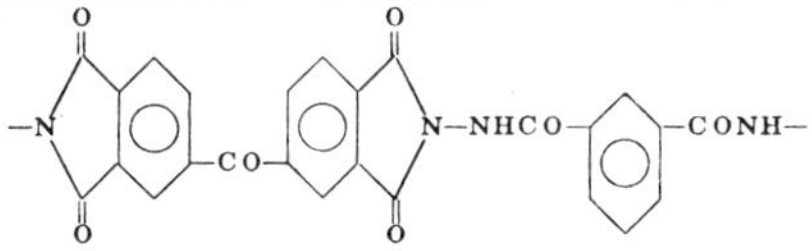
(2)	do	p-Aminobenzoic hydrazide.	20	DMAc	1.26	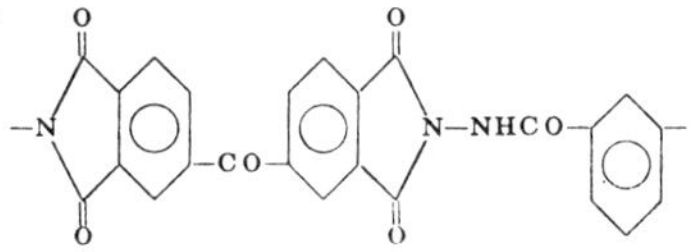
(3)	do	Oxalic dihydrazide	18	DMF	1.02	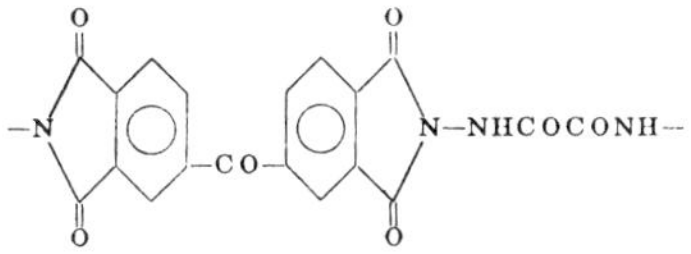
(4)	do	Succinic dihydrazide	10	DMSO	0.85	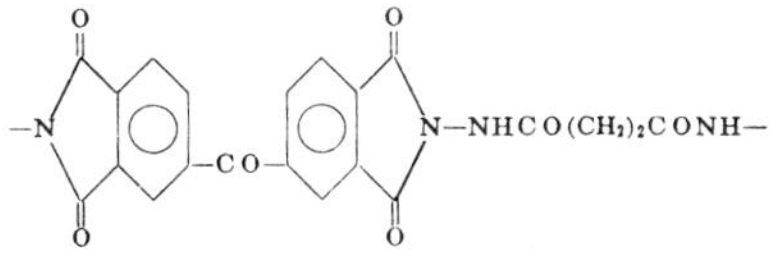
(5)	do	Glutaric dihydrazide	15	DMF	1.41	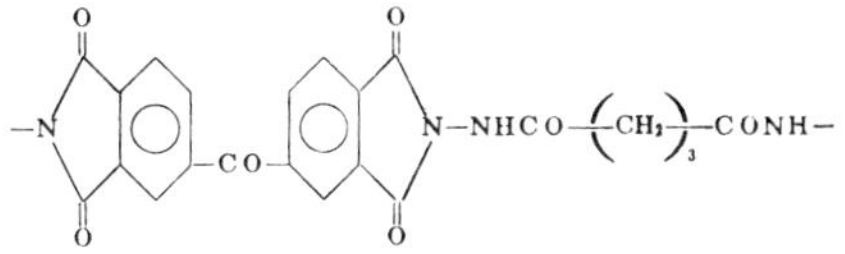
(6)	do	Adipic dihydrazide	18	DMAc	1.32	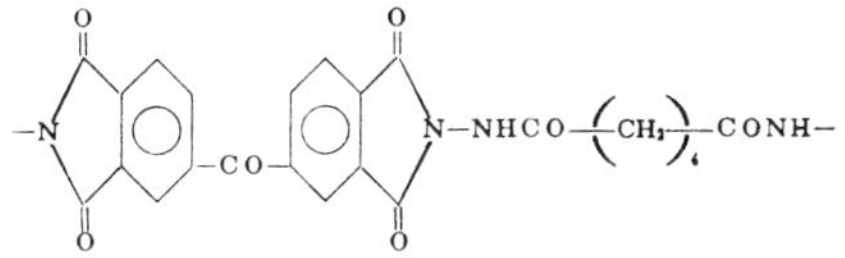

(continued)

Ex.	Monomers: Dianhydride	Monomers: Hydrazide	Conc. (wt. percent)	Reaction solvent	Inherent viscosity (dl./g.)	Structure of polymer
(7)....	Pyromellitic..........	Isophthalic dihydrazide.	18	DMAc	0.95	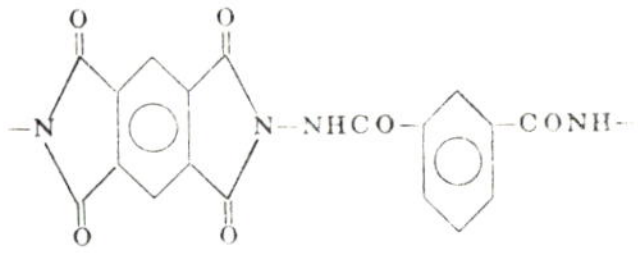
(8)....	1,2,3,4-cyclopentane........	do...............	10	DMAc	0.68	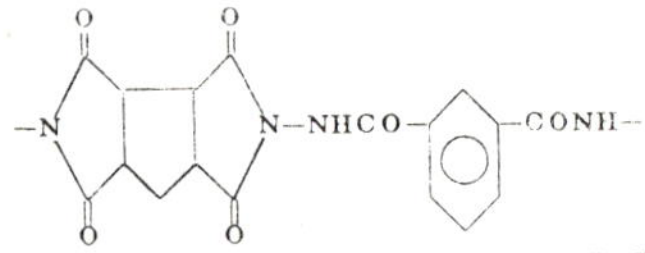

Examples 9 through 17: In these examples, films, essentially pinhole and particle free, were prepared by filtering the imide polymer solutions obtained in Examples 1 through 8, respectively, through 5 μ Teflon (polytetrafluorethylene) filters at 450 psi pressure, and casting 50 ml of the solution onto a glass plate using a casting knife with a 10 mil knife gap, to give about 4 ft^2 of wet film. Prior to casting, the glass plate was cleaned several times with acetone and coated with a thin layer of DMAc.

After casting, the glass plate was covered with a 1" high chamber to prevent deposition of dust, and the film was dried by means of heating lamps at 73°C for 1 hour at a humidity level of less than 3% (RM). After cooling to ambient temperature, the film was stripped from the glass plate, rolled into a Dacron (polyethylene terephthalate) cloth, and dried for an additional hour at 135°C. The resulting films had thicknesses as shown in the following table.

Example	Film Thickness (μ)	Water Flux (gfd-2000 A)	Rejection (%)
9	28	28	99.95
10	30	18	99.29
11	28	31	98.60
12	20	19	99.65
13	32	22	99.20
14	35	17	98.80
15	24	21	98.05
16	26	23	97.30
17	16	2.9	99.35

Reverse osmosis properties of the thus-prepared films were tested in an Amicon pressure cell consisting of a stainless steel porous plate with a 3" diameter membrane area, a Teflon-lined chamber, a magnetic stirring assembly, an inlet port for gas pressurization and a product water outlet located beneath the porous plate. The films, on a support consisting of No. 40 filter paper, were pressurized to 1,500 psi using a 3.5% NaCl feed solution. Water flux and NaCl rejection were measured periodically over a period of 2 days.

Results, including those for a 39.8% acetyl cellulose acetate control (Example 17), are given in the above table. The films of Examples 9 through 16 correspond to the polymers of Examples 1 through 8, respectively. Water fluxes in the table are expressed in gfd normalized to a thickness of 2000 A (viz, the approximate thickness of the barrier in an asymmetric cellulose acetate membrane). As seen from the data of the above table, the poly(N-amido)imides of Examples 1 through 8 all exhibit superior water flux and salt rejection, particularly as compared to cellulose triacetate.

Poly-2,2'-(m-Phenylene)-5,5'-Bibenzimidazole

According to a process described by *W.C. Brinegar; U.S. Patents 3,841,492; October 15,*

1974; and 3,720,607; March 13, 1973; both assigned to Celanese Corporation the membrane is formed of poly-2,2'-(m-phenylene)-5,5'-bibenzimidazole. The membrane is formed by (a) providing a solution of the polybenzimidazole polymer of the formula indicated, (b) depositing a film of the solution upon a support, (c) evaporating an amount of solvent from the film sufficient to allow the formation of a thin solid layer on the exposed surface of the film having a density which is substantially greater than that of the remaining portion of the film on which the solid layer of increased density is formed, and (d) washing the resulting film with a nonsolvent for the polymer to remove residual solvent.

The reverse osmosis membranes exhibit performance characteristics which are more highly resistant to deterioration upon the passage of time, are capable of operation at high pressures (e.g., 1,000 psi and up), are capable of operation at elevated temperatures (e.g., 100°C), and exhibit outstanding chemical stability.

Example 1 through 3: A solution having a viscosity of 400 poises at 30°C was prepared employing N,N-dimethylacetamide as solvent containing 15% by weight poly-2,2'-(m-phenylene)-5,5'-bibenzimidazole based upon the total weight of the solution, and 2% by weight lithium chloride based upon the total weight of the solution. The dissolution of the polymer was accomplished by agitating the same while in particulate form with the N,N-dimethylacetamide solvent (in which the lithium chloride was previously dissolved) while in a closed zone at a temperature of about 230°C. The resulting solution was next filtered to remove any residual solids.

Quantities of the solution while at ambient temperature (i.e., 25°C) were poured onto level smooth glass supports having upright edges extending above the surface to a uniform height of 4 mils (Examples 2 and 3) or 8 mils (Example 1) as indicated in the table. A doctor blade was passed over the surface of the solutions at a rate of about 20 in/min while resting upon the upright edges of the glass supports to insure that wet films of uniform thickness were positioned upon the supports.

Next a stream of air at ambient temperature (i.e., 25°C) was caused to impinge upon the exposed surfaces of the wet films at a velocity of about 2 ft/sec. As indicated in the following table, the wet films were exposed to the stream of air for either 5 minutes (Example 2) or 15 minutes (Examples 1 and 3). During this period N,N-dimethylacetamide solvent was evaporated from the exposed surfaces of the films thereby forming a relatively thin dense solid layer on the surfaces thereof. The remaining portions of wet films were of substantially lesser density than the solid layers formed upon the exposed surfaces.

The resulting films while still present upon the smooth glass supports were immersed in a vessel of cool water having a temperature of about 12°C. While immersed in water, residual quantities of N,N-dimethylacetamide were essentially completely removed from the films and the remaining polybenzimidazole polymer situated beneath the thin surface layers was coagulated to a solid porous consistency. The reverse osmosis membranes formed upon the removal of residual solvent had thicknesses of approximately 8 mils in Example 1 and of approximately 4 mils in Examples 2 and 3.

The resulting membranes were next utilized in the desalination of 0.5% by weight aqueous sodium chloride solutions provided at 25°C by reverse osmosis. The membranes were kept moist from the time of their formation to the time of their utilization in the reverse osmosis separation. Each membrane was stripped from the smooth glass support and positioned upon a filter paper and a porous sintered metal plate which served to support the same during desalination.

Each membrane, together with its porous support, was positioned within a conventional reverse osmosis apparatus of the flat plate type with the sodium chloride solution circulating on one side of the membrane and pure water being withdrawn on the opposite side of the membrane. The more dense surface of the polybenzimidazole membrane was exposed and faced the aqueous sodium chloride solution. A pressure of 600 psi was applied to the sodium chloride solution.

The rejection value is a relative measure of the ability of the membrane to retard passage of the component being separated from the solution usually expressed as a weight percentage of the total. Flux refers to the amount of fluid passing through the membrane per unit time and is generally expressed as gfd. As indicated in the following table, flux rates of 1.8 to 14.5 gfd and rejection rates of 53 to 95% were observed for the examples.

Example No.	Wet Film Thickness	Solvent Evaporation Time	Flux	Rejection (%)
1	8 mils	15 min	1.8	93
2	4 mils	5 min	14.5	53
3	4 mils	15 min	1.9	95

Examples 4 through 9: The polybenzimidazole polymer solution described in Example 1 was placed in a spinning jet at ambient temperatures. The jet which was used had an o.d. of about 0.036" and an i.d. of about 0.029". The polybenzimidazole polymer was spun through the outer ring of the jet nozzle while nitrogen was forced through the inner ring of the nozzle. As the polymer was spun through the outer ring, it was fed into a coagulation bath formed of water.

During exposure to the bath which lasted about five seconds the hollow filament was drawn to a drawdown ratio of about 5:1. After exposure to the bath the hollow fil fiber was fed into an ethylene glycol annealing bath maintained at about 175°C. After about 10 minutes' exposure to the bath, the resulting hollow fil fiber exhibited a wall thickness of about 30 to 40 microns, an overall diameter of about 150 microns, and a tensile strength of about 1.25×10^3 psi.

The above fiber was then cut into 2 foot lengths and about 100 of these filaments were glued together near each end and placed inside a sealed pipe. The glued ends were attached to each end of the pipe to form a liquid tight seal between the inside of the pipe and the outside of the pipe. The pipe itself was equipped with two valves located near each end of the pipe which valves communicated with the inside of the pipe and thus the outside surface of each of the fibers in the bundle. Using the above apparatus desalinized water was prepared by pumping salt-containing water through the interior of the above apparatus under the desired pressure and collecting salt-reduced water at either end of the pipe.

Six desalination apparatus were constructed and operated at the following water pressures with the following results. Test pressure refers to the pressure at which water was introduced into the interior of the desalination apparatus. Flux refers to the flow rate of desalinized water in gal/ft^2 of hollow fil fiber outside surface area per day. Rejection refers to the percent of salt in the introduced salt water that is removed. In all instances 0.5 weight percent salt water solutions were used, with the following results:

Example No.	Test Pressure (psi)	Flux (gfd)	Rejection (%)
4	200	1.25	77.0
5	200	0.66	94.4
6	400	1.6	95.2
7	500	1.7	98.7
8	400	1.2	98.4
9	600	1.4	99.1

According to a process described by *A.A. Boom; U.S. Patents 3,737,042; June 5, 1973; and 3,699,038; October 17, 1972; both assigned to Celanese Corporation* a member cast in the form of either a hollow filament or a flat film formed of a polybenzimidazole polymer and possessing reverse osmosis properties is subject to an annealing process by contacting the membrane with an organic liquid under conditions found capable of substantially enhancing the properties of the same. These polybenzimidazole membranes find particular utility in reverse osmosis desalination procedures. The annealing step of the process substantially improves the performance of the membranes in such a separation.

M.J. Ram; U.S. Patent 3,851,025; November 26, 1974; assigned to Celanese Corporation describes a process for the production of a semipermeable polybenzimidazole membrane in the form of a hollow filament. A solution of the polybenzimidazole polymer in N,N-dimethylacetamide solvent is extruded through a concentric hollow jet spinneret while continuously introducing a liquid alkane into the center portion of the extruded solution. Following coagulation, washing, and annealing a highly uniform polybenzimidazole hollow filament results which is particularly suited for use in reverse osmosis desalination procedures.

Pyrrone Polymer Precursors

A process described by *L.W. Frost and P.-K. Lee; U.S. Patent 3,817,386; June 18, 1974; assigned to Westinghouse Electric Corp.* relates to polyamino-acid-amide polymers containing pendant carboxyl and amino groups, and supported permselective membranes made from such polymers and used in reverse osmosis systems. In particular it concerns polymeric intermediates in the synthesis of imidazopyrrolone polymers. These imidazopyrrolone polymers are formed, in one case, by heat curing the product of an aromatic dianhydride and an aromatic tetraamine reacted in a polar aprotic solvent, as described by Bell and Jewell, *J. Poly. Science.,* (A-1), vol 5, pp 3043-3060 (1967) and U.S. Patent 3,414,543, and shown in simplified form below:

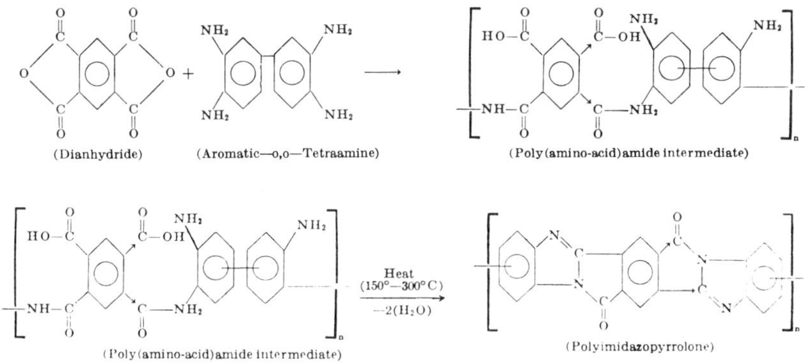

The arrows indicate isomerism between the two positions shown. The polymers will generally contain from about one carboxyl radical and one amino radical per 150 to 1,000 units of polymer molecular weight. The membrane will have a relatively uniform cross section of polymer chains having outstanding thermal stability and superior chemical resistance and anticompacting properties, as compared to standard cellulose acetate reverse osmosis membranes. They may be specifically applicable to treatment of certain hot waters or industrial wastewaters where the present cellulose acetate membranes cannot be applied. Because they are substantially uniform throughout, they are not as easily damaged, as cellulose acetate membranes containing thin osmotic skins, and either side can be used against the high pressure in the reverse osmosis system.

Example: A solution of 4.286 grams (0.02 mol) of 3,3'-diaminobenzidine (DAB) in 129.4 of N,N-dimethylacetamide (DMAC) was stirred rapidly while adding a solution of 6.444 grams (0.02 mol) of 3,3',4,4'-benzophenonetetracarboxylic dianhydride (BTDA) in 80.0 grams of DMAC at room temperature. Approximately 80% of the solution was added within about 15 minutes. The remaining 20% was added at the rate of about one drop every 10 seconds. The resulting solution was filtered to give a clear red-brown polyamic acid-amine (polyamino-acid amide) casting solution of moderate viscosity. This polymer solution, diluted as necessary, was cast on glass plates 10.5 x 10.5 cm in size with a doctor blade. As shown in the table below, the solvent was evaporated at various temperatures and times on a hot plate. Membranes were formed which were separated from the glass plates by soaking the plates and membranes in hot water. The membranes were between 0.05 to 0.8 mil thick and had a substantially uniform cross section. They were kept immersed in water until put into the reverse osmosis test unit. Reverse osmosis performance of the prepared membranes was tested in a standard stainless steel reverse osmosis membrane test unit, using round 5.08 cm diameter sections cut from the cast membranes. The test membranes were used against sodium chloride solution at a feed pressure of 600 psi and the brine circulation rate was 150 ml/min. The test results are summarized in the table below for membranes cast from 1 to 5% polymer solutions:

Performance of DAB-BTDA polyamino-acid-amide membranes.
Feed concentration: 0.5% NaCl sol. (5,000 ppm) at room temp.
Feed pressure: 600 p.s.i.

Casting Solution	Casting Conditions	Hours On Test — Membrane Side Facing Pressure	Salt Rejection % Initial - Final	Water Flux gal./sq. ft./day Initial - Final
No. 1 5% solids in DMAC	dried at 50°C for 20 hours; lifted in 50°C water; 0.8 mil. thick	168 air evaporation side	98.0–97.7	0.7–0.4
No. 2 5% solids in DMAC	dried at 50°C for 3 hours; lifted in 80°C water; 0.4 mil. thick	100	93.8–98.1	0.9–1.0
No. 3 5% solids in DMAC	dried at 50°C for 1 hour; lifted in 60°C water; 0.2 mil. thick	90 glass cast side	99.0–99.2	0.47–0.51
No. 4 5% solids in DMAC	dried at 50°C for 1 hour; lifted in 60°C water; 0.2 mil. thick	90 air evaporation side	98.0–99.0	0.98–1.1
No. 5 2% solids in DMAC	dried at 50°C for 1 hour; lifted in 60°C water; 0.1 mil. thick	95 air evaporation side	98.3–98.7	0.67–0.68
No. 6 2% solids in DMAC	dried at 25°C for 15 hours; lifted in 25°C water; 0.05 mil. thick	72	93.2–98.5	1.44–1.89
No. 7 1% solids in DMAC	dried at 25°C for 15 hours; & 95°C for 3 min.; 0.1 mil. thick; lifted in 89°C water	78	80.5–96.1	3.10–2.81
No. 8 1% solids in DMAC	dried at 45°C for 1 hour; lifted in 78°C water; 0.05 mil. thick	47	80.4–98.3	3.8–3.0
Feed concentration: 3.57% NaCl sol. (35,700 ppm) at room temp. Feed pressure: 600 p.s.i.				
No. 9 1% solids in DMAC	dried at 45°C for 1 hour; lifted in 78°C water; 0.5 mil. thick; previously tested for 386 hours against 0.5–1.96% NaCl sol.	120	80.4–69.3	0.9–0.92
No. 10 5% solids in DMAC	dried at 55°C for 1 hour; lifted in 25°C water; 0.1 mil. thick; previously tested for 24 hours against 1.96% NaCl sol.	120	70.5–68.0	1.2–1.22

As can be seen, membrane No. 3 exhibited salt rejection efficiencies up to 99.2% for brackish water, which is comparable to that of cellulose acetate reverse osmosis membranes. While pure water flux rates obtained were below those of cellulose acetate, significant increases do appear possible, and the outstanding thermal stability, superior chemical resistance and anticompaction properties of the membranes of this process would make them more widely applicable than current cellulose acetate membranes.

The pure water flux rate of the prepared membrane was dependent upon membrane thickness, preferred membranes being 0.01 to 10 mils thick. Unlike cellulose acetate membranes, the polyamino-acid-amine membranes of this process appear to be symmetrical, as both the air evaporation side and the glass cast side of the membrane showed almost the same desalting behavior. For seawater, a multiple stage treatment, where the membrane would be cast inside fine pored resin bonded sand or other type tubular modules, with seawater passing inside the series of tubes, may be required.

Vinyl Pyrrolidone Polymers and Polyisocyanates

A process described by *R.L. Riley and C.R. Lyons; U.S. Patent 3,661,634; May 9, 1972; assigned to The U.S. Secretary of the Interior* provides semipermeable membranes prepared from a mixture of a vinylpyrrolidone polymer and a polyisocyanate, the mixture being cured with water under conditions of controlled humidity. Particularly useful polyisocyanates are "blocked" polyisocyanates which have decreased reactivity toward water at ambient temperatures, but which provide reactive polyisocyanate functionality toward water at elevated temperatures.

Example 1: Anhydrous polyvinylpyrrolidone (PVP) with an average molecular weight of 360,000 is dissolved in an anhydrous solvent consisting of equal volumes of pyridine and N,N'-dimethylformamide. Methylene bis(4-phenylisocyanate) (MDI) is dissolved in an anhydrous solvent also consisting of equal volumes of pyridine and N,N'-dimethylformamide to form a series of solutions having different concentrations. The polyvinylpyrrolidone solution is mixed with the series of MDI solutions to provide a series of 3 casting solutions each containing 13.5 wt % PVP, and varying amounts of MDI. These 3 casting solutions are each immediately cast on clean glass plates in a chamber in which air containing an amount of water vapor such that it would have a RH of 10% at 25°C and 1 atm of pressure, is passed over the surface of the film-coated plates.

After 8 hours of drying in this chamber, the films are heated in an oven at 120°C for 4 hours. The cured semipermeable membranes are removed from the glass plates by soaking in water, and are transparent, strong and imperfection-free. The water-sorption of each membrane is measured by soaking the membrane in distilled water for 48 hours at room temperature, blotting the membrane and then weighing it. The water is then removed from the membrane at room temperature under vacuum over calcium sulfate and the membrane is weighed again. The water content of the membrane when swollen with water at room temperature at equilibrium is then obtained by difference, and the water-sorption is then determined as a weight-percent, based on the weight of the water-swollen membrane.

Reverse osmosis experiments are performed on each membrane using a test separation device generally similar to that described in U.S. Patent 3,133,132. A 0.8 wt % aqueous sodium chloride solution is fed to the separation device adjacent to the semipermeable membrane at a pressure of 34 atm, and a sufficient flow of this feed solution is maintained past the membrane to keep the effect of the boundary layer at a low level. The membrane constant and the percent of salt rejection are calculated for each membrane from the flow rate through the membrane and the salt concentration in the output fluid. The following table presents the data for these membranes. The "isocyanate equivalent" heading represents the ratio, present in each of the casting solutions used to prepare the membranes, of mols of isocyanate groups of the MDI, to mols of pyrrolidone groups.

Semipermeable Membrane Number	Isocyanate Equivalent	Thickness (microns)	Water Sorption (percent)	Membrane Constant Assuming 0.2μ Membrane Thickness ($g/cm^2/sec\text{-}atm$)	Salt Rejection (percent)	Water Permeability ($10^{-9}/cm^2/sec$)	Sodium Chloride Permeability ($10^{-10}\ cm^2/sec$)
1	0.448	35	31	9.95×10^{-5}	35.8	27.00	1,400.00
2	0.620	80	26	3.65×10^{-5}	58.0	10.00	209.00
3	1.100	57	16	6.26×10^{-4}	90.05	1.70	0.47

Example 2: A series of three semipermeable membranes is prepared and tested exactly as in Example 1 except that polymethylene polyphenylisocyanate (PMPI) is used as the polyisocyanate instead of methylene bis(4-phenylisocyanate). The isocyanate equivalent heading represents the ratio, present in each of the casting solutions used to prepare the membranes, of mols of isocyanate groups of the PMPI, to mols of pyrrolidone groups. The cured membranes are transparent, strong and imperfection-free. The table below presents the data for these membranes.

Semipermeable Membrane Number	PMPI Equivalent	Water-Sorption (percent)	Membrane Constant Assuming 0.2μ Membrane Thickness (g/cm²/sec-atm)	Salt Rejection (percent)	Water Permeability (10^{-9} g/cm²/sec)	Sodium Chloride Permeability (10^{-10} cm²/sec)
1	0.45	34	9.21×10^{-5}	52.4	25.0	655.0
2	0.70	28	3.13×10^{-5}	72.3	8.5	94.0
3	1.20	17	8.11×10^{-6}	97.25	2.2	1.8

Thermoosmotic Membrane from Vinylpyrrolidone

L. Lazare; U.S. Patent 3,276,996; October 4, 1966 describes membranes for a thermoosmosis process. The membrane, which constitutes the diaphragm when the unit is in operation, is one which is absorptive of water without itself dissolving; chemically inert to water and electrolytes such as salt or others to be processed; electrolyte excluding, i.e., only slightly or totally impermeable to electrolyte solutes such as are found in typical solutions, in the absence of applied electrical potential; and the membrane is thermoosmotic with respect to water, that is, transmits water through under the influence of a temperature gradient through the diaphragm.

Examples of materials which meet these requirements and can be used as the chief component of the thermoosmotic membranes are polymers and, for best results, crosslinked polymers of the following monomers: the N-vinyl lactams, examples of which are N-vinylpyrrolidone and N-vinylimidazole; C_2–C_4 alkylene oxides such as ethylene, propylene and butylene oxides; vinyl alcohol, made as the polyvinyl alcohol by hydrolysis of polyvinyl acetate; cellophane; cellulose acetate; and nitrocellulose, e.g., pyroxylin of degree of nitration about 11 to 12% N.

The crosslinking is effected by any known procedure including exposure of the selected polymer in film form, to ionizing radiation, or chemical crosslinking of monomers, during polymerizing thereof, by polyfunctional agents. Examples of such crosslinking agents are polycarboxylic acids such as fumaric, oxalic, adipic, maleic and phthalic acids; any water-soluble polyhydroxy compound as plasticizer, e.g., dimethylhydroxyethylene urea; organic diisocyanates, e.g., hexamethylene, 2,4-tolylene, and 4,4'-diphenylmethane diisocyanate; and divinyl monomers, e.g., monomers having two ethanoid bonds (or groups) as in methylene bisacrylamide or the divinyl monomers.

The chemical crosslinking is effected by direct copolymerization in situ of the monomer and crosslinking agent. They are mixed together in the absence of oxygen but with an initiator of ethenoid bond polymerization and then warmed to 60° to 90°C. Suitable initiators are benzoyl, lauroyl, and t-butyl peroxide. Proportions of the crosslinking agent are approximately 0.1 to 2% by weight of the mixed monomers.

When the membrane selected is not in itself sufficiently impermeable to the saline ions, that is, insufficient in electrolyte exclusion, the membrane may be and suitably is combined with a surface coating of extreme thinness which is less permeable to the ions of the saline than is the membrane itself. Because of the exceptional thinness of the coating, its decreased water permeability is acceptable. Examples of such superficial coatings that may be applied to the membranes of kind described are either in the form of the free acid or its sodium or other alkali metal salt; quaternary and tertiary amine derivatives of polyvinyl aromatic C_{6-9} hydrocarbons such as polystyrene, polyvinyl toluene, and polyvinyl xylene, in which the group $-CH_2-N(CH_3)_3Cl$, for instance, is attached to a carbon of the aryl rings; melamine formaldehyde resins (e.g., 50 to 70%) mixed with guanidine (30 to 50%); and polyacrylic or polymethacrylic acid.

L. Lazare; U.S. Patent 3,386,912; June 4, 1968; assigned to The Puraq Company describes the utilization of separatory membranes which are highly permeable to water while being impermeable to electrolytes, together with selected solutions of partially water-soluble polymers in an extraction cell where a stream of seawater passes on one side of and in contact with the separatory membrane and the steam of polymer solution passes on the other side of and in contact with the separatory membrane. The differential osmotic pressures of the seawater and the polymer solution, together with a counter current relation of the stream, provides a driving force for extracting saline-free water from the stream of seawater, passing the same through the membrane to enrich the polymer solution which at elevated temperatures will dissolve substantial amounts of water. The water-enriched polymer solution is then reduced as to its temperature at which phase separation will take place to provide a saline and polymer-free water phase and a polymer solution of reduced water content for recycling to repeat the extractive procedure.

The membrane is formed of a hydrophilic monomer such as vinylpyrrolidone which may be copolymerized with a strength imparting, hydrophobic monomer such as vinyl acetate, methyl methacrylate, butadiene, styrene, chloroprene, acrylonitrile and mixtures, together with a crosslinking agent such as triallyl cyanurate, diallyl maleate, diallyl sebacate and divinylbenzene.

By way of example the membrane may be formed by copolymerizing 60% vinylpyrrolidone, 20% styrene, 19.5% butadiene and 0.5% triallyl cyanurate, all by weight, in Teflon coated, spaced glass plate molds, to form the membrane of a thickness of the order of about 1⁄32" and which will swell in water to a thickness of about 1⁄20" to 1⁄16". Also, the membrane may be formed by copolymerizing 75% vinyl alcohol, 24.5% butadiene and 0.5% diallyl maleate, all by weight, to form a thin water swellable sheet which is effective to pass water while being impermeable to salt, and which will have reasonable strength and resistance to normal pressure conditions in the cell.

N-3-Oxohydrocarbon-Substituted Acrylamide Polymer

In a process described by *D.I. Hoke; U.S. Patent 3,520,804; July 21, 1970; assigned to The Lubrizol Corp.* films of polymers (both homopolymers and interpolymers) of N-3-oxohydrocarbon-substituted acrylamides, preferably N-(1,1-dimethyl-3-oxobutyl)acrylamide or diacetone acrylamide, are prepared from the appropriate polymers in bulk, solution, suspension, or emulsion; the latter includes latexes which are particularly useful. The films may be crosslinked by reacting with a wide variety of reagents. They are particularly useful as membranes for the desalination of water by hyperfiltration. The following are examples of typical monomers suitable for preparing the polymeric compositions of this process.

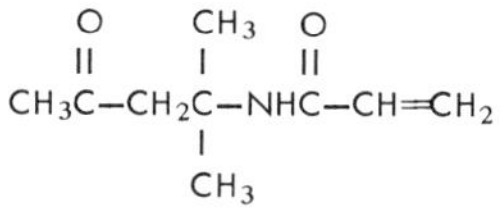

N-(1,1-dimethyl-3-oxobutyl)acrylamide

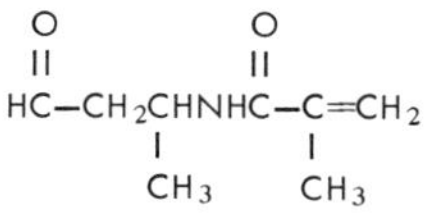

N-(1-methyl-3-oxopropyl)methacrylamide

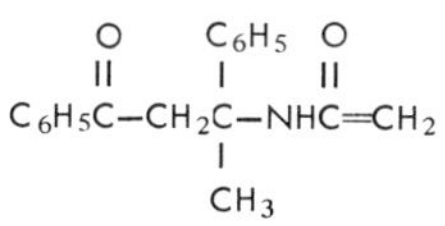

N-(1,3-diphenyl-1-methyl-3-oxopropyl)acrylamide

Example 1: A reaction vessel is flushed with a dilute aqueous solution of potassium persulfate at 80°C for two hours, and then rinsed with distilled water. The flask is charged with 400 grams of distilled water and 90 grams of diacetone acrylamide. Vinyl acetate, 10 grams, is added and the mixture is purged with nitrogen for one-half hour and heated to 50°C. Finally, 1 gram of sodium lauryl sulfate and 0.1 gram of ammonium persulfate are added and the polymerization reaction is allowed to proceed for one hour at 50° to 70°C. The mixture is then filtered through a fine-mesh conical paint filler. There is obtained 495 grams of milky white 20% aqueous emulsion of diacetone acrylamide-vinyl acetate (9:1) copolymer latex.

Example 2: The procedure of Example 1 is repeated, except that the monomer mixture comprises 50 grams of diacetone acrylamide and 50 grams of vinyl acetate. The product is a latex of a 1:1 copolymer.

Example 3: The procedure of Example 1 is repeated, except that the monomer mixture comprises 5 grams of diacetone acrylamide and 95 grams of vinyl acetate. A copolymer latex with a 5:95 weight ratio of diacetone acrylamide to vinyl acetate is obtained.

Example 4: A solution of 5,600 grams of diacetone acrylamide, 2,400 grams of 2-ethylhexyl acrylate, 200 grams of polyvinyl alcohol, and 120 grams of dioctyl sodium sulfosuccinate (75% solids) in 12,800 grams of water is purged with nitrogen and a solution of 8 grams of ammonium persulfate in 100 grams of water is added. The solution is heated to 60°C over 1½ hours as polymerization occurs, and is then heated to 65° to 70°C and held there for one hour. Finally, it is cooled to 30°C and filtered through cheesecloth. The product is a 41.4% aqueous latex of a diacetone acrylamide-2-ethylhexyl acrylate (70:30) copolymer.

These films are useful wherever a breathable or semipermeable film is desirable. Typical applications include formation of membranes for use in equipment designed to supplement bodily functions, such as artificial kidneys; in purification of chemical products, both aqueous and nonaqueous; and in removal of dissolved impurities, especially salt, from water by hyperfiltration.

Electrodeless Glow Discharge Polymerization

H. Yasuda; U.S. Patent 3,775,308; November 27, 1973; assigned to The U.S. Secretary of the Interior has found that a composite membrane characterized by high chemical stability and mechanical strength, as well as desirable reverse osmosis characteristics, may be prepared by electrodeless glow discharge polymerization of an organic monomer in the presence of a porous support.

Polymerization is achieved by subjecting the monomer, in vapor form, to the influence of electromagnetic energy, such as radio frequency or microwave energy, of a frequency and power sufficient to cause an electrodeless glow discharge in the monomer vapor. The growing polymer species deposit onto the surface of the porous support due to the decrease of vapor pressure of the growing polymer molecules. This deposition of growing species results in formation of a highly crosslinked film on the surface of the support, the thickness of which is nearly linearly dependent on the polymerization rate and reaction time and will usually range from about 300 to about 2,000 A.

The state of the monomer vapor at which polymerization occurs is that of a low temperature plasma consisting of various species of ionized and excited gases and vapors. One of the characteristic features of such a plasma is its liquid-like property, which enables deposition of polymer on surfaces of any shape. The porous support may, therefore, be in various forms, e.g., sheet, tubular membrane, or hollow fibers.

Example 1: In this example, 4-vinylpyridine was polymerized by means of a radio frequency of 13.56 MHz. The porous support consisted of 0.025 μ pore size Millipore filter, type VSWP, having the dimensions 90 mm diameter and a thickness of 140 microns. This support was placed in the polymerization vessel with the top side of the filter toward the inside of the tube and with the opposite side fitted tightly against the glass wall of the polymerization vessel. The polymerization vessel was evacuated to a vacuum of less than 1 μ Hg and the support was degassed for a period of 10 minutes at this vacuum. The monomer, 4-vinylpyridine, which was previously degassed, was introduced into the vessel up to 200 μ Hg and polymerized at a discharge wattage of 30 watts for approximately 8 minutes. The resulting composite membrane, consisting of the porous support coated with a thin layer (about 0.05 μ in thickness) of the polymer, was then tested in a reverse osmosis apparatus consisting of four high-flow cells in a test loop pressurized by a 5.1 gpm capacity pump. The gap between membrane surface and the top surface of the cells was adjusted to 1 mm to insure a sufficient flow rate of feed solution at the membrane surface. The membrane was found to give 99.9% salt rejection with a water flux of 30 gfd during a two week test period with a feed of 1.2% NaCl and a pressure of 1,200 psi.

Example 2: Using the apparatus described above, a membrane was prepared from 4-picoline on porous polysulfone substrate. The polysulfone was prepared by coagulation in water of the cast polysulfone in DMF solution, and had a pore size comparable to that of the Millipore filter of Example 1. The substrate was mounted and degassed as in Example 1. The 4-picoline vapor was then admitted to a pressure of 540 μ Hg. The sample was polymerized at 30 watts for 8 minutes. In reverse osmosis testing this sample gave 98% salt rejection at a flux of 6.4 gfd. Changing the run conditions of 1.2% NaCl at 1,200 psi to 3.5% NaCl at 1,500 psi did not substantially alter the results.

Example 3: Using the apparatus described above, a membrane was prepared from 3,5-lutidine on a porous glass tube which had a pore size of 50 A. The porous glass tube was mounted in the center of the reaction chamber by sliding onto a smaller tube extending from the monomer inlet. The monomer was then admitted from a 4 liter reservoir containing 1,700 μ Hg monomer vapor and the plasma was immediately initiated. The sample was polymerized at 150 watts for 5 minutes. In reverse osmosis testing with 3.5% NaCl at 1,500 psi this sample gave 99% rejection at a flux of 0.40 gfd (approximately 70% of water flux for nontreated porous glass tube).

Example 4: Using the apparatus described above, a membrane was prepared from N-vinylpyrrolidone on porous polysulfone substrate (prepared as described in Example 2). The substrate was mounted and degassed as in Example 1. The monomer was admitted into the pumping system to give a steady-state pressure in this flow system of 40 μ Hg. The plasma was immediately initiated and the polymerization of the monomer continued for 30 minutes at 30 watts. In reverse osmosis this sample gave 90% salt rejection at 11.0 gfd at 1.2% NaCl and 1,200 psi.

Plasma Discharge Using Allylamine

According to a process described by *J.C. Fletcher, J.R. Hollahan and T. Wydeven; U.S. Patent 3,847,652; November 12, 1974* allylamine and chemically related compounds are polymerized as thin films in the presence of a plasma discharge. The monomer compound can be polymerized by itself or in the presence of an additive gas to promote polymerization and act as a carrier. The polymerized films thus produced show outstanding advantages when used as reverse osmosis membranes.

ACRYLIC, POLYOLEFIN AND OTHER ORGANIC AND INORGANIC MEMBRANES

ACRYLIC

Isotactic and Syndiotactic Polymethacrylate

H. Tanzawa, Y. Sakai, T. Matsushita, Y. Fujii, K. Sugita and H. Maekawa; U.S. Patent 3,896,061; July 22, 1975; assigned to Toray Industries, Inc., Japan have found that a semipermeable membrane comprising isotactic polymethyl methacrylate and syndiotactic polymethyl methacrylate has utility as a transparent hydrogel. The resulting semipermeable membrane may be utilized in the form of thin sheet or hollow fiber to separate components of a solution, for example, to remove undesirable microsolutes from blood. A semipermeable membrane of the process is produced by the steps of:

(a) dissolving isotactic polymethyl methacrylate and syndiotactic polymethyl methacrylate in a solvent capable of dissolving the polymers to prepare a sol;
(b) lowering the temperature of the sol to the temperature sufficient for the system to be transformed to the gel phase; and
(c) replacing the solvent in the gel with water, whereby a hydrogel type semipermeable membrane having high water content is obtained.

Example: A mixture of 5 parts of syn-PMMA-1 and 1 part of iso-PMMA-1 was dissolved into 34 parts of dimethyl sulfoxide (DMSO) at 120°C, where syn-PMMA-1 is syndiotactic polymethyl methacrylate obtained by an ordinary radical polymerization process (MW 660,000) and iso-PMMA-1 is isotactic polymethyl methacrylate (MW 1,450,000) obtained by a Grignard process. The tacticities of polymers may be determined from a triad, i.e., the structural configuration of any 3 adjacent monomer units along the molecular chains of the polymer, using NRM spectrum. I is the percent of triad isotactic structures, S the percent of triad syndiotactic structures and H the percent of triad heterotactic structures. The tacticity of this mixture by triads was I = 27%, S = 45% and H = 28%. The viscosity of the DMSO solution of PMMA was 10 poise. When the solution was cooled, it formed a gel. The sol-gel transition temperature was 70°C and the sol-gel transition was reversible.

The solution was poured onto a glass plate at 120°C, thereby forming a thin film having a thickness of 125 microns, and was cooled at room temperature until the gelation of the film was brought to completion. This thin film of gel was immersed in an ice water bath, (0°C), whereby the DMSO in the gel film was replaced with water by mutual diffusion, and the hydrogel structure of PMMA was formed. The hydrogel film obtained was washed with water several times and stored in water, or in air after glycerin treatment.

The hydrogel film had a water content of 73%, it was completely transparent, and it provided excellent properties for use as a permselective membrane. The following table shows the observed values of the permeability of water and solute for a 0.5% aqueous solution of sodium chloride at 30°C. In the table, hydrogel PMMA is the membrane of this process, and "Cuprophane", whose values are shown for reference purposes, is a conventional commercial membrane made with regenerated cellulose.

PERMEABILITY COEFFICIENT	HYDROGEL PMMA	"CUPROPHANE"
P1	4.7×10^{-12}	6.4×10^{-14}
P2	3.2×10^{-6}	8.8×10^{-7}

P1, the water permeability coefficient, and P2, the solute permeability coefficient, are defined by the following equations:

$$Jv = P1 \frac{\Delta P}{\lambda}$$

$$Js = P2 \frac{\Delta C}{\lambda}$$

where Jv is the total volume flux, Js is solute flux, ΔP is the trans-membrane pressure, ΔC is the difference in trans-membrane concentration and λ is the thickness of the membrane.

It is clearly seen from the table that the permeability of the hydrogel PMMA membrane made by the process is excellent in comparison with "Cuprophane". The molecular weight cut-off level of the membrane of the process was 40,000 and the membrane rejected bovine serum albumin completely. The tensile strength at the breaking point and the ultimate elongation of the membrane of the process were 50 kg/cm^2 and 40%, respectively. The membrane was useful for hemodialysis, blood filtration, and the ultrafiltration and separation of water-soluble macromolecules.

Galactose Methacrylate-Methyl Methacrylate Copolymer

In a process described by *A. Sharples and G. Thomson; U.S. Patent 3,542,908; November 24, 1970; assigned to the U.S. Secretary of Interior* reverse osmosis membranes are prepared by copolymerizing a hydrophobic and a hydrophilic monomer after rendering the hydrophilic monomer temporarily hydrophobic. After polymerization the hydrophilic portion of the copolymer is reestablished. Casting dopes containing the copolymer, a solvent, and water may be formed into asymmetric membranes by casting on a smooth surface followed by immersing in water. Conventional swelling agents may be added to the dope if desired. A reverse osmosis membrane of this type is formed from a galactose methacrylate-methyl methacrylate copolymer.

The water uptake for a galactose methacrylate-methyl methacrylate copolymer membrane is proportional to the amount of galactose methacrylate in the copolymer. For example, it has been found that in homogeneous membranes, a 10 mol percent of galactose methacrylate will give a 10 weight percent uptake of water upon immersion in water at 20°C. Correspondingly, 14.2 mol percent galactose methacrylate will give a 14% water uptake and 20% galactose methacrylate will give a 19% water uptake.

Homogeneous films were prepared from copolymers having different amounts of hydrophilic polymer to determine its salt rejection characteristics. These films were prepared by casting a 5 weight percent solution of the copolymer in equal volumes (1:1 v/v) of ethanol and chloroform onto a smooth glass surface. The particular organic solvent used is not critical. The films were dried in vacuo to give a final thickness of about 20 microns. Three films were made having a galactose methacrylate mol percent of 10, 14.2 and 20% and were tested in a standard reverse osmosis cell at 1,500 psi. The rejection characteristics of these films are shown on the following page.

Sample Percent Galactose Methacrylate in Copolymer	Thickness, microns	Percent Rejection Water Flux, gal/ft^2/day	Percent Rejection NaCl from 1% Solution	Percent Rejection $MgSO_4$ from 1% Solution
10	14	0.08	97	98.7
14.2	10	0.36	91	99.5
20	20	0.62	73	98.9

The rejection of sodium chloride decreases and the throughput of water increases as the membrane is made more hydrophilic. Also, like prior membrane materials, these membranes reject divalent ions more easily than sodium ions, however they maintain very good divalent ion salt rejection even with high hydrophilic content. The table above shows that by increasing the galactose methacrylate content from 10 to 20 mol percent the flux is increased from 0.08 to 0.62 gal/ft^2/day while the rejection of $MgSO_4$ remains essentially unchanged. Copolymers having as much as 33 mol percent galactose methacrylate may be formed. Beyond that point the copolymer becomes water-soluble. The useful range of mol percent of galactose methacrylate in the copolymer for film formation is from 1 to 33 percent. The following examples are illustrative of asymmetric membrane casting techniques.

Example 1: A casting dope containing 20% by weight copolymer comprising 10 mol percent galactose methacrylate and 90 mol percent methyl methacrylate; 68% by weight acetone; 10% by weight water; and 2% by weight zinc chloride was prepared. This mixture was cast for 1 minute at 20°C on a smooth glass surface to an initial thickness of 0.4 mm. It was then immersed into water at 15°C and removed from the glass plate. When tested at 1,500 psi this membrane has a product water flux of 44 gal/ft^2/day, a rejection of a 0.7% magnesium sulfate solution of 90% and of a 1% sodium chloride solution of 33%.

Example 2: A casting dope containing 20% by weight copolymer comprising 10 mol percent galactose methacrylate, 90 mol percent methyl methacrylate; 69% by weight acetone; 10% by weight water; and 1% by weight magnesium perchlorate was prepared. The same casting procedure was used as in Example 1. When tested at 1,500 psi this membrane had a product flux of 4.1 gal/ft^2/day and a rejection for a 1.8% sodium chloride solution of 54%.

Hydrous Oxide-Polyacrylate Dual Layer Membrane

The process of hyperfiltration, or reverse osmosis, as well as methods for the formation of dynamically formed hyperfiltration membranes, is well documented in the technical and patent literature; see, for example, chapter 8 of *Principles of Desalination,* Academic Press, 1966, and U.S. Patents 3,413,219 and 3,449,245. Dynamically formed membranes are created by circulating, under pressure, a feed containing various well-known membrane-forming additives through a porous substrate. Among the more useful additives for this purpose are colloidal hydrous metal oxides formed by contacting a porous face of a substrate containing micron-size pores with an aqueous solution containing colloidal particles formed from a salt of a polyvalent metal capable of forming a hydrous metal oxide selected from the group consisting of zirconium(IV), thorium(IV), iron(III), tin(IV), aluminum(III) and hafnium(IV), or from a solution of naturally occurring and synthetically produced polycarboxylic acids.

Hydrous oxide membranes, when operated at optimum pH values, i.e., in acid solution, have been found to form rather easily on a rather wide variety of porous supports and have usefully high rejection and fluxes. With hydrous oxide membranes, however, best salt rejections are found at high anion exchange capacities which occur at acidities higher than are found in most natural saline waters. In addition, hydrous oxide membranes are prone to be poisoned by polyvalent anions such as sulfate.

In recent years, organic polyelectrolytes, particularly polycarboxylic acids such as polyacrylic acids [S.B. Sachs, W.A. Baldwin, and J.S. Johnson, *Desalination* 6, 215 (1966)] have found favor for use as a hyperfiltration membrane because the pH of their optimum performance frequently matches acidities of natural brackish waters.

However, there have been certain disadvantages. Product water fluxes are usually lower than other members of the dynamic membrane-forming class, and the presence of polyvalent counterions such as Ca^{++} and Mg^{++} in feeds reduces their effectiveness. A more serious disability is concerned with obtaining reproducible performance. Difficulty has been experienced in forming polyacrylate membranes directly on porous supports and even when they are successfully deposited, they show unstable and irreproducible performance in terms of salt rejection.

J.S. Johnson, Jr.; U.S. Patent 3,743,595; July 3, 1973; assigned to the U.S. Atomic Energy Commission has found that the advantages of hydrous oxide and polycarboxylic acid membranes can be combined by forming a hydrous oxide sublayer, exposing it at acidic pH to a feed containing a polycarboxylic acid, and then adding base to bring the feed to a neutral range, e.g., to a pH in the range 5 to 10. The resulting dual-layer membrane eliminates the disadvantages of the component single layers while retaining the best features of both.

Example 1: In this and succeeding examples, experimental apparatus used to form the membrane and test its salt rejection ability consisted of a closed loop. In many cases the membrane was formed on nylon filter fabric impregnated with a copolymer of vinyl chloride and acrylonitrile, the average pore size being 0.45 micron (Acropor AN450).

This fabric was wrapped around a porous stainless steel section, having an average pore size in the range of 5 to 10 microns, which was enclosed in a transparent pressure jacket. Aqueous feed was circulated under pressure through the annulus between the wrapped stainless steel tube and the jacket, with product water being collected from the inside of the porous tube. The aqueous feed contained membrane-forming materials. Rejections were determined from conductivity measurements on the feed and transpirated product. Membranes were formed and tested at a pressure of 1,000 psig and a temperature of 25°C.

In this first example a comparison was made between a hydrous zirconium oxide single-layered membrane and the dual-layer hydrous (zirconium) oxide-polyacrylate membrane. The hydrous zirconium(IV) oxide membrane formed on 0.45 μ Acropor at 37 ft/sec from a colloidal dispersion filtered out 62% of the salt from a 0.05 M sodium chloride solution. At pH 6, rejection of the zirconium(IV) membrane dropped. Rejection remained low after addition of neutralized polyacrylic acid (sodium polyacrylate). However, on adding acid to reach a solution pH of about 2.7 and then base to bring the pH back to 5.6, rejection rose dramatically to a much higher value, 82%, than the single (zirconium oxide) layer.

In addition, the following table summarizes the results of carrying out hyperfiltration through a wide range of acidity. In the table, the first row gives the rejection and flux obtainable with this membrane. The second row gives the properties of the single-layer hydrous oxide membrane after rinsing out the membrane-forming solution with 0.05 M sodium chloride to a pH of 6, and it will be seen that the higher pH is reflected by a marked decrease in salt rejection.

	0.025μ Millipore,[a] 37 ft./sec.		0.45μ Acropor AN, 14 ft./sec.		0.45μ Acropor AN, 28 ft./sec.		0.45μ Acropor AN, 37 ft./sec.	
pH	Robs,[b] percent	Flux, g.p.d./ft.²	Robs, percent	Flux, g.p.d./ft.²	Robs, percent	Flux, g.p.d./ft.²	Robs, percent	Flux, g.p.d./ft.²
			0.001 M Zr(IV)					
3.1	73	35	59	360	59	380	62	630
6.0	43	40	31	290	30	420	32	690
			20 p.p.m. PAA					
5.6	29	45	9	440	11	440	23	700
2.7	49	17	52	280	51	300	53	370
5.6	85	17	68	175	69	175	82	210
8.3	75	24	66	150	74	145	85	180
			After operation overnight					
7.6	73	30	65	155	69	170	80	240
10.8	73	30	67	145	69	160	81	210
7.7	69	35	58	185	61	210	74	300

[a] A commercially available filter aid material.

[b] $Robs = 1 - \frac{Cw}{Cf}$ where Cw = concentration of salt in product and Cf = concentration of salt in feed.

The data in the third row was taken after the addition of 20 ppm of polyacrylic acid having a molecular weight of about 300,000. The fourth and fifth rows indicate the effect of a pH cycle into the acid range with return to a more neutral pH, showing the heightened rejections which result in the relatively neutral pH region. High rejections were recorded even at a pH as high as 10.8, although some degeneration in rejection can be seen when the pH was lowered to a 7.7 operating condition. In any event, the data dramatize the high rejections and transpiration fluxes obtainable with a dual-layer membrane at neutral and near-neutral pH conditions.

Example 2: Of even more practical importance is the performance of the dual membrane with feed solutions of compositions typical of natural waters, containing sulfate, magnesium, and calcium in addition to sodium and chloride ions. The table below summarizes the salt rejection quality of a single-layer polyacrylate membrane in comparison to that obtained with a dual-layer hydrous oxide-polyacrylate membrane using dynamically formed membranes, tubular configuration at 1,000 psig.

	Without Zr(IV) oxide sublayer (5 ft/sec), $\underset{\sim}{R}_{obs}$, percent			With Zr(IV) oxide sublayer (35 ft/sec), 0.45 μ Acropor An, $\underset{\sim}{R}_{obs}$, percent		
Feed	Chloride	Total anion	$Mg^{2+}+Ca^{2+}$	Cl	Total anion	$Mg^{2+}+Ca^{2+}$
Coalinga:						
1X	53	60	66	81	91	99
3X	47	56	63	77	92	98
Foss reservoir, 2X	33	39	40	70	86	90
Dalpra farm:						
1X		58				
2X				78	92	98
Sea water	35	38	40	72	63	78
0.05 M NaCl	80			92		88

NOTE.—Coalinga (1X: 0.0028 M $NaHCO_3$, 0.0096 M Na_2SO_4, 0.0018 M $CaCl_2$, 0.0020 M $MgCl_2$); Foss reservoir (1X: 0.005 M $MgSO_4$, 0.005 M $CaSO_4$, 0.0014 M NaCl, 0.0026 M $NaHCO_3$); Dalpra farm (0.008 M $NaHCO_3$, 0.019 M Na_2SO_4, 0.003 M $MgSO_4$, 0.003 M $CaCl_2$, 0.00004 M $FeCl_3$, 0.00002 M $MnCl_2$); Sea water (0.47 M NaCl, 0.03 M $MgSO_4$, 0.04 M $MgCl_2$, 0.002 M $NaHCO_3$).

The single-layer membrane was supported on a 0.3 μ ceramic tube; it was mentioned earlier that it is difficult to form single-layer membranes on many supports, and 0.45 μ Acropor is one of them. Not only were rejections poorer with the single layer than with the dual layer, but production rates were only 5 to 8 gpd/ft^2 for the single-layer, while they were from 23 to 35 gpd/ft^2 for the dual layer. At the low fluxes through the single-layer, concentration polarization should not seriously affect $\underset{\sim}{R}_{obs}$.

It is seen that the performance of the dual-layer membrane surpasses the single-layer hydrous oxide membrane both in terms of anion and polyvalent counterion (Mg^{2+} and Ca^{2+}) rejection for treatment of brackish water types such as Coalinga, Foss reservoir, and Dalpra farm, as well as seawater.

It is not necessary to introduce the polyacrylate in the salt form, but only to expose the hydrous oxide sublayer to polyacrylic acid under acid conditions below a pH of about 4. The mechanism appears to be that a fraction of the carboxyl ions, ionized even at acid pH, become attached to anion exchange sites of the hydrous oxide and the resulting attachment is not broken on adding base unless high pH values, of the order of 11 for hydrous zirconium oxide, are reached. On raising the pH, the attached polyacrylic acid is converted to the salt form, and ion exchange capacity and salt rejection increase. At high alkalinity, i.e., at a pH of ∿11, salt rejection decreases by apparent loosening of the polyelectrolyte from the hydrous oxide, but high rejection levels can be restored by a cycle through an acidic pH.

In the preceding description, the dual hydrous oxide-polyacrylate membrane was formed by first depositing the hydrous oxide on a microporous surface at an acid pH of about 3 after which a solution of polyacrylic acid was passed over the hydrous oxide membrane at an acid pH to form the desired dual-layer membrane, whereupon the pH was increased to 8 by successive base (usually sodium hydroxide) additions. It has also been found that a mixed membrane of favorable properties can also be formed by depositing both hydrous oxide and polyacrylic acid on porous supports from a single stock solution.

Polymerization of Frozen Monomer Solution

C.B. Chapman and R.A. Haldon; U.S. Patent 3,880,763; April 29, 1975; assigned to Imperial Chemical Industries; England describe an asymmetric membrane which comprises a laminar vinyl polymer film in which one layer comprises a semipermeable membrane and the other layer comprises a porous structure. The asymmetric membrane is made by nonthermal polymerization of a frozen monomer solution, while one surface is maintained in a liquid state, and subsequent removal of the solvent.

Polymers suitable as asymmetric membranous articles for use in separations include poly-2-hydroxyethyl methacrylate, poly-2-hydroxypropyl methacrylate, polyvinylpyrrolidone, polyethyl acrylate, polymethyl methacrylate, polyvinylene carbonate, polyvinyl alcohol, polyvinyl acetate, polymethyl methacrylate/galactose methacrylate, polyvinyl methyl ether/maleic anhydride, poly-2-hydroxyethyl methacrylate/methacrylic acid and copolymers containing any of the above components.

Example 1: Hydroxyethyl methacrylate (4.5 parts), ethylene glycol dimethacrylate (0.5 part) and uranyl nitrate (0.1 part) were mixed with 3.33 parts of an inert liquid comprising an aqueous ethylene glycol solution (4:1 by volume water:ethylene glycol) to give a homogeneous solution. A film of the solution was cooled between two thin glass plates with the lower plate maintained at -40°C. When the inert liquid in the film had solidified by freezing, a glass cell containing circulating water at approximately 20°C was placed on the top glass plate such that the top surface of the frozen film melted.

The film was irradiated for 10 minutes by a 125 watt medium pressure mercury lamp located at a distance of approximately 5 cm above the top surface of the film. The inert liquid was then removed by melting to yield an asymmetric membrane. A photomicrograph showed the membrane to be 0.5 mm thick and to consist of a porous structure of 0.45 mm thickness having adjacent one face a semipermeable membrane of 50 microns thickness.

Example 2: An aqueous solution which was 5.8 molar in 2-hydroxyethyl methacrylate, 2.2×10^{-4} molar in methylene blue and 7.1×10^{-3} molar in sodium p-toluene sulfinate was unidirectionally frozen by spreading on a glass plate maintained at -63°C. The frozen film was partially polymerized by irradiating for 10 minutes with a 750 watt tungsten-iodine lamp located at a distance of 7 cm above the film. The frozen film was then warmed to allow the water to melt. A layer at the bottom of the film which had not polymerized in the frozen film was polymerized at room temperature.

The resultant membrane was asymmetric. A photomicrograph showed the membrane to be 0.75 mm thick and to consist of a porous structure of 0.67 mm thickness having adjacent one face a semipermeable membrane of approximately 80 microns thickness.

Example 3: A rotating steel drum was cooled to a surface temperature of approximately -60°C in methanol/solid CO_2. The drum was then rotated at about 60 rpm such that it touched the surface of a solution consisting of, 125 parts acrylamide, 60 parts ethylene glycol, 240 parts water and 1.4 parts uranyl nitrate.

A film of approximately 0.1 cm thickness was picked up on the drum. The solution was removed and the film was heated by a 350 watt rod heater situated about 0.2 cm from the film surface. After two minutes the surface of the film was just wet after passing the heater. The film was then irradiated with a 750 w tungsten-iodine lamp situated approximately 6.5 centimeters from the drum surface such that the light fell on the film immediately after passing the heater.

After 5 minutes irradiation the lamp was switched off and the frozen aqueous ethylene glycol removed by washing in cold methanol. The membrane film was next removed from the drum, allowed to warm up to room temperature in methanol and allowed to dry in air.

Microscopic examination revealed a membrane film 0.80 mm thick, with a porous layer 0.76 mm thick and a nonporous layer of approximately 40 microns thickness.

Vinyl Copolymers Subjected to Water Conditioning

W.M. Muir and R.M. Livingston; U.S. Patent 3,822,330; July 2, 1974; assigned to Babcock & Wilcox Limited, England describe a process for the manufacture of a semipermeable membrane which comprises (A) preparing a casting solution containing a vinyl copolymer and an organic polar solvent; (B) casting the casting solution as a film; (C) evaporating the solvent from the film from step (B); (D) contacting the film from step (C) with water for at least 10 minutes. The semipermeable membranes are used for reverse osmosis or hyperfiltration treatment of various feed solutions.

Example 1: A copolymer of styrene with methacrylic acid containing 50% acid was prepared as a casting solution as follows: 32 g of copolymer, 245 g of dimethylformamide and 70 g of acetone.

The casting solution was cast at room temperature as a film using a doctor blade with a blade opening of 0.005 inch, after ten minutes of solvent evaporation at room temperature, the plate was immersed in water at room temperature for 10 minutes to produce a membrane 0.0035 inch in thickness. This flat-plate membrane was tested in a test cell with whey feeds of differing concentrations for fractionation properties.

Feed – Whey, 6% solids. Applied pressure: 620 pound per square inch gauge.

Test Period Elapsed (hours)	Permeate Flux Rate ($gal/ft^2/day$)*
1	12.8
2	12.4
3	12.8
4	11.9
5	11.9
6	12.9
7	11.9
8	11.9

*Product in all instances showed 99% protein rejection and clear eluate.

Feed – Whey, 30% solids. Applied pressure: 650 pounds per square inch gauge.

Test Period Elapsed (hours)	Permeate Flux Rate ($gal/ft^2/day$)*
1	8.9
2	7.1
3	7.1
4	7.0
5	7.0

*Product showed over 98% protein rejection and clear eluate.

Example 2: A copolymer of styrene with methacrylic acid containing 50% acid was prepared as a casting solution as follows: 30 g of copolymer, 270 g of dimethylformamide and 25 g of water.

The casting solution was cast at room temperature as a film using a doctor blade opening of 0.005 inch. After 15 minutes drying at room temperature the plate was immersed in water at room temperature for 10 minutes to produce a membrane of approximately 0.003 inch in thickness. This flat plate membrane was tested in a test cell with 6% whey solution at 630 pounds per square inch gauge, as shown on the following page.

Test Period Elapsed (hours)	Permeate Flux Rate (gal/ft²/day)
1	11.4
2	11.8
3	11.7

Protein rejection was over 99% throughout the test period but lactose rejection was not measured.

Example 3: A copolymer of methyl methacrylate with methacrylic acid containing 40% acid was prepared in dope form as follows: 10 g of copolymer, 10 g of water, 71 g of acetone and 19 g of dimethylformamide.

The dope was cast at room temperature into a membrane with a doctor blade with a blade opening of 0.012 inch. After five minutes of drying at room temperature, the plate was immersed in water at room temperature for 10 minutes to produce a membrane of about 0.004 inch in thickness. This flat-plate membrane was tested in a cell with a 1% sodium chloride solution at 1,000 pounds per square inch gauge.

Test Period (hours)	Salt Rejection, %	Flux (gal/ft²/day)
1	17	16
2	17	16
3	17	15
4	18	14

Example 4: Cellulose acetate polymer powder was made into a casting dope according to the formulation: 22 g of cellulose acetate polymer and 85 g of dimethylformamide. The dope which was at a temperature in the range of 40° to 50°C, was cast into a membrane with a doctor blade opening of 0.015 inch. The plate and film were placed in a vacuum oven at 45°C for 5 hours at a vacuum of 26 inches of mercury after which the plate was immersed in water at a temperature of about 25°C for 10 minutes to produce a membrane of 0.0032 inch in thickness. This flat-plate film was tested at 1,000 pounds per square inch gauge with a 1% sodium chloride solution.

Test Period (hours)	Salt Rejection, %	Flux (gal/ft²/day)
6	73	3.6
16	68	5.2
18	72	5.6

POLYOLEFIN

Vinyl Acetate-Maleic Acid Copolymers

S.S. Kremen, G.E. Foreman and J.M. Chirrick; U.S. Patent 3,877,978; April 15, 1975; assigned to the U.S. Secretary of the Interior describe a method of improving the rejection performance of semipermeable membranes by incorporating an effective amount of a supplemental polymer containing a substantial amount of acetyl groups. A suitable supplemental polymer is a copolymer of vinyl acetate and one or more reactants, such as maleic acid or a half-ester thereof, crotonic acid, and lower alkyl esters of acrylic acid. The copolymer may be stabilized by subsequent treatment with a multivalent cation.

Example 1: Sheetlike cellulose acetate asymmetric reverse osmosis membranes are initially fabricated from E-398 cellulose acetate resin and are used in the treatment of saline water to produce potable water for a period of about 24 months. Initially, the asymmetric membranes exhibit a salt rejection of about 97.5% (based on sodium chloride) and a water flux of 10 gallons per square foot per day (gfd). However, after operation for the indicated period, the salt rejection of the membrane decreases to about 94% and the water flux measures 16 gfd.

A supplemental polymer is prepared by reacting vinyl acetate with the ethyl ester of maleic acid in equimolar amounts. The reaction is conducted at 60°C and atmospheric pressure, using continuous agitation and using ammonium ion as a catalyst. The copolymer which is formed has the following general repeating unit:

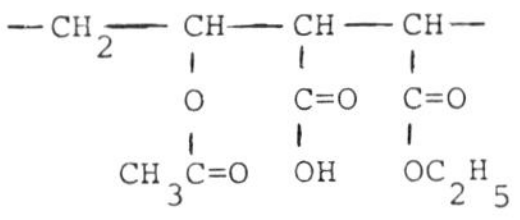

The reaction under these conditions produces a polymer having a molecular weight between about 10,000 and 50,000. The polymer exhibits good solubility in water having a pH of 9, and a 5% by weight solution is formulated at this pH, using ammonia to provide mild alkalinity.

The cellulose acetate asymmetric semipermeable membrane is immersed in this dilute solution of the supplemental polymer for 5 minutes to absorb the polymer onto the active surface layer of the semipermeable membrane. Thereafter, the supplemental polymer is insolubilized by immersion of the membrane for about 5 minutes in a dilute aqueous solution of hydrochloric acid containing about one weight percent of zinc chloride, which solution has a pH of 4.

Following this treatment, insolubilization of the supplemental polymer is complete, and the treated semipermeable membrane is tested using essentially the same feed solution as it has previously been operating on for the 24-month period. The treated membrane exhibits a salt rejection of 97.5% and a water flux of 10 gfd. Substantially continuous operation under these conditions for a period of greater than 180 days shows that the salt rejection remains about 97.5% without further treatment and no noticeable change in flux occurs. Thus, the treatment process using the supplemental polymer is considered to be most effective and relatively permanent.

Example 2: Additional samples of the cellulose acetate asymmetric semipermeable membrane tested in Example 1 are treated using a supplemental polymer made from vinyl acetate and maleic anhydride where the mixture reacted had a ratio of 3 mols of vinyl acetate to 1 mol of maleic anhydride. The copolymerization reaction is carried out under substantially the same conditions and produces polymers having a molecular weight between 10,000 and 50,000.

Treatment of the membrane in one weight percent solution and insolubilization of the absorbed polymer are carried out in the same manner as in Example 1. Testing of the treated membrane shows that the salt rejection increases to 98% and exhibits water flux of 10 gfd. Testing for 180 days thereafter shows the salt rejection capability remains above 97.5%.

Crystalline Polymers of Vinyl Methyl Ether

C.A. Lukach and H.M. Spurlin; U.S. Patent 3,645,890; February 29, 1972; assigned to Hercules Incorporated describe reverse osmosis membranes useful for desalination of water which comprise thin films of crystalline homopolymers or vinyl methyl ether which are either supported on a microporous substrate or in the form of thin walled hollow fibers. (For the sake of brevity, vinyl methyl ether is referred to as VME, and polyvinyl methyl ether as PVME.

Example 1: A 4 x 5 inch sheet of filter material composed of mixed esters of cellulose of 0.01 μ pore size (VF filter material) was floated on the surface of deionized water, with the dull side down (glossy side up), for 5 minutes. The sheet was lifted with forceps, tilted to drain off excess water, and placed, wet side down, upon a glass plate. A piece of adhesive tape was placed across the top edge to hold the sheet to the glass.

A 1.0 g sample of crystalline (32%) PVME of RSV 9.6 was dissolved in 50 ml of 1,1,2,2-tetrachloroethane. The 2% (w/v) solution was filtered under pressure through a 1.2 μ pore size filter material. The polymer solution was placed near the top of the above prewet VF support sheet, beneath the leading edge of a 10-mil casting knife (2½ inches wide). After drawing the solution across the entire length of the VF support sheet with the casting knife, the glass plate was placed in a level position in a cabinet through which filtered air was circulated. After 16 hours, a dry PVME film, of calculated thickness 0.20 mil, was obtained over the VF support sheet.

A 47 mm diameter disc was cut from the above membrane (including the VF support sheet). Soaking in water helped to remove the cut disc from the glass plate. The membrane was then evaluated as follows.

Test Apparatus and Method – Each test cell consisted of a 6-bolt Millipore 47 mm high-pressure filter holder (Cat. No. XX45 047 00) which holds a 47 mm diameter membrane on a support screen between two stainless steel plates. An additional hole was drilled into the top plate, so that the brine solution could enter the cell, circulate over the membrane, and leave the cell. The permeate was collected from the bottom port of the cell and analyzed.

Eight such cells were connected in parallel, through a series of suitable valves, to three back pressure regulators, a pump and a 100-gallon reservoir, to provide a recirculating assembly capable of evaluating eight membranes simultaneously at two different pressures (four cells at each pressure). Pumping rates were up to 0.5 gallon per minute (30 gallons per hour).

The brine solution in the reservoir contained dissolved NaCl and Na_2SO_4 in amounts providing the ion concentration noted in the examples. It was analyzed each day for Na^+ and Cl^- concentration, using a Beckman Expandomatic pH meter and appropriate electrodes. The sulfate ion was calculated from these two values, using the expression

$$\text{ppm } SO_4^{=} = 2.09 \text{ ppm } Na^+ - 1.35 \text{ ppm } Cl^-$$

Sulfate ion concentrations were also determined independently by a titration method. These values usually agree well with the calculated value. The brine solution also contained a small amount of both ethyl violet and Rhodamine B. When a membrane contained a small pinhole, a pink-purple dot from these dyes was visible on the GS filter below the membrane after testing.

The membrane to be tested was cut to a 47 mm diameter size and placed in the test cell atop two 47 mm Millipore GS filter discs (0.22 μ pore size; 45,000 gfd at 1,500 psi). Depending upon the number of cells in operation, the brine solution was circulated through the cell, and over the membrane at the rate of 200 to 400 ml/min. The membranes were first kept at 500 psi and then at 1,000 psi long enough to collect enough permeate for analysis. The membranes were then kept at 1,500 psi for extended lengths of time, while permeate samples were measured and analyzed periodically. Percent rejection of any ion was calculated from the ppm of the ion in the permeate and the feed solution as follows:

$$\%\text{ rejection of x} = \frac{\text{ppm x in feed} - \text{ppm x in permeate}}{\text{in ppm x in feed}} \times 100$$

where x is any ion, (e.g., Na^+, Cl^- or $SO_4^{=}$).

Analysis of the permeate collected, compared to the feed solution gave the salt rejections as shown on the following page.

Pressure, p.s.i.	Colorless permeate				
		Sodium		Chloride	
	G.f.d.	P.p.m.	Percent rejection [1]	P.p.m.	Percent rejection [1]
500	-------	-------	-------	-------	-------
1,000	-------	-------	-------	-------	-------
1,500	0.14	-------	-------	-------	-------
1,500	0.16	[2] 33.5	99.1	[2] 53	98.6

[1] Feed solution analysis, 3,500 p.p.m. Na^+, 3,750 p.p.m. Cl^-.
[2] Samples at 1,500 p.s.i. combined.

Examples 2 through 5: Salt rejections of membranes prepared in a similar manner to Example 1 (from different PVME samples with supports wet with different solvents) and evaluated at 1,500 psi are given below in the following table.

	PVME			Dry film thickness (mils)	Percent rejection			Flux rate (g.f.d.)
	RSV	Percent crystallinity	Support prewet with—		Na^+	Cl^-	SO_4^{--}	
Example:								
2	2.2	37	TCE [1]	0.7	98.8	98.5	93	0.21
3	9.6	32	TCE [1]	0.5	92.9	92.3	-------	0.05
4	9.6	32	TCE [1]	0.2	97.4	93.4	96.7	0.12
5	9.6	32	H_2O	0.1	98.0	97.6	96.9	0.16–0.23

[1] Tetrachloroethane.

Reticulation of Polyvinyl Alcohol

L. Nicolas and R. Dick; U.S. Patent 3,837,500; September 24, 1974; assigned to Anvar-Agence National de Valorisation, France describe diaphragms for inverse osmosis essentially which consist of a film-producing polymer having a sufficient number of mobile hydrogen groups to give, after reticulation, compounds which are insoluble in water. The method covers the preparation of ultra-thin layers which can be employed for hyper-filtration purposes, and also includes the superficial reticulation of a dry film of polymers which include the polyvinyl alcohols and polyvinylpyrrolidone.

The reticulation agent is preferably chosen from the diisocyanates and may comprise a solvent for a swelling agent for the polymer and a catalyst for the reticulation reaction. An important feature of the process is that the degree of reticulation of such diaphragms varies progressively throughout the thickness of the film, the most highly reticulated layer being a very thin superficial layer of the diaphragm.

The following examples illustrate the basic characteristics of the method of preparation of diaphragms in the vapor phase. The flow-rate measurements in inverse osmosis given in these examples were taken with an apparatus constructed according to the model described by Sourirajan (*Ind. Eng. Chem. Proc. Design and Development Fundamentals* 3, 206, 1964).

The pump employed to circulate the brine under a pressure of 100 bars above the diaphragm delivers 13.8 liters per hour. For reasons of operational safety of the equipment, the circulation of the brine was stopped during the night and at the week-end, so that the results are lower than those which are obtained with a continuous circulation of the brine. The useful surface of the diaphragm is 7.5 cm^2 and the thickness of the layer of brine above the diaphragm is 2.8 cm.

Example 1: A film of polyvinyl alcohol of 6 microns in thickness (dry film) was prepared from a 5% aqueous solution by weight of Elvenol 73-125 (polyvinyl alcohol containing at least 1% of polyvinyl acetate) by pouring the solution over a plate of nickel-plated brass, and by evaporation of the water in a stove with a partial vacuum, brought up to 60°C.

The film, previously stretched on a glass support and dried at 60°C for one hour was exposed for 14 hours by its nickel face (face remaining in contact with the plate of nickel-plated brass during the evaporation of the water) to vapors of reactants obtained by evaporation at 60°C of a 20% mixture of dimethyl sulfoxide and 80% of diisocyanate of toluylene, and then rinsed with benzene and dried at 60°C for 15 hours.

The exposure to the reactant vapors and also the drying before and after exposure to the reactant vapors were carried out in a desiccator containing P_2O_5. The desiccator employed had a volume of 1.6 liters and the crystallizer containing the powdered P_2O_5 had a diameter of 7.6 cm. The rinsing with benzene and the transfers of the film were carried out under protection against humidity in a glove box. Under these conditions, the thickness of the reticulated layer, determined after dissolving the nonreticulated portion in hot water, was 1.1 microns.

The film of 6 microns in thickness, deposited on a porous support (Millipore AAWP) having a pore diameter of about 0.8 micron so that the reticulated face is in contact with the support, was tested directly without thermal treatment. By hyper-filtration, under a pressure of 100 bars, of brine containing 3.5% of NaCl prepared from distilled water, there were obtained the results indicated in the table below. At the end of the operation, the thickness of the film had practically not varied.

Inverse osmosis: pressure ~ 100 bars
3.5% solution of NaCl
Thickness of the reticulated layer: 1.1 μ
Porous support: Millipore AA (Pore diameter 0.8 micron)
Thermal treatment: None.

Duration of test Hours	Flow-rate of water l/sq.m. per day	Rejection of NaCl %
5	270	79.7
6.5	266	82.0
31	250	84.9
55	250	86.9
75	225	88.0

A film of polyvinyl alcohol of 5 microns in thickness (dry film) was prepared from an 8% aqueous solution by weight of Rhodoviol 25/100 M (copolymer of polyvinyl alcohol and vinyl acetate containing 8.5% of acetate) following the same method as for the film of Example 1. The film, stretched on a glass support and dried for 1 hour at 80°C, was exposed for 20 hours by its nickel face to vapors of reactants obtained by evaporation at 60°C of a mixture containing 50% of dimethyl sulfoxide and 50% of diisocyanate of toluylene, and then rinsed with benzene and dried at 60°C for 15 hours.

The exposure to the reactant vapors, and also the drying before and after exposure to the reactant vapors were carried out in a desiccator containing P_2O_5. The rinsing with benzene and the transfers of the film were carried out in the ambient air. The thickness of the reticulated layer was 4.9 microns. The film was tested directly without thermal treatment. The test was carried out under the same conditions as for Example 1. The results are given in the table below.

Inverse osmosis: Pressure ~ 100 bars
3.5% solution of NaCl
Thickness of the reticulated layer: 4.9 μ
Porous support: Millipore AA (diameter of pores 0.8 micron)
Thermal treatment: None.

Duration of test Hours	Flow-rate of water l/sq.m. per day	Rejection of NaCl %
8	22	80.6
80	21.5	83.7
104	19	89.7
130	18	93.0
150	18	94.7
175	18	96.2
210	18.5	97.3
234	18.5	97.5
256	18	97.5
330	18	98.0

The coefficients of permeability of this film, calculated following the method put forward by Lomsdale, Merton and Riley (*J. Appl. Polymer Sci.* 9, 1341, 1965) were compared with those of homogeneous diaphragms of cellulose acetate thermally treated in water at 80°C and not treated. These coefficients of permeability to water (PW) and to NaCl (PS) were determined after 300 hours for a 3.5% solution of NaCl and under a pressure of 100 bars.

Polymer	PW g/cm/sec	PS sq cm/sec	PW/PS g/cu cm
Cellulose acetate E - 398 - 3 Not thermally treated	2.5×10^{-7}	2.8×10^{-9}	90
Cellulose acetate E - 398 - 3 Thermally treated	1.2×10^{-7}	1.2×10^{-10}	1000
Reticulated polyvinyl alcohol Rhodoviol 25/100 M	1.7×10^{-7}	1.9×10^{-10}	900

Polyvinyl Chloride

A process described by *G. Bourat and A. Fabre; U.S. Patent 3,751,536; August 7, 1973; assigned to Rhone-Poulene SA, France* relates to microporous membranes based on polyvinyl chloride (PVC). The membranes are microporous opaque PVC membranes having a porosity between 40 and 85%, preferably between 50 and 80%, the pores having a diameter less than 0.1 μ. (The diameters of the pores are measured by observing the surface of the membranes under the electron microscope.) The porosity as used here is defined as the number

$$100\left(1-\frac{\text{weight of the porous membrane}}{\text{weight of a membrane of the same size but free of pores}}\right)$$

The membranes may be prepared by leaching a film consisting of an essentially homogeneous mixture of PVC and of poly-p-dialkylaminostyrene (referred to as PPDMAS). The latter plays the role of a pore-forming agent. The alkyl radicals generally have 1 to 4 carbon atoms. The following examples illustrate the process.

Example 1: The following were successively dissolved in 70 g of cyclohexanone, with stirring: at 60°C, 6 g of PVC (AFNOR viscosity index = 100), and at 23°C, 24 g of PPDMeAS (molecular weight about 1,900) (PPDMeAS is used to denote poly-p-dimethylaminostyrene).

This solution was cast on a glass plate so as to form a liquid film 0.2 mm thick. The cyclohexanone was evaporated in an oven at 70°C and the film obtained was immersed for 24 hours in a bath of 0.5 N aqueous hydrochloric acid provided with a liquid circulator. After washing with distilled water until the chloride ions had disappeared in the wash waters, an opaque membrane having a pore diameter of less than 0.1 μ and a substitution resistance of 0.2 $\Omega\text{-cm}^2$ was obtained. It contained no elementary nitrogen. Water covering one face of the membrane under a relative pressure of 0.05 bar passed through this membrane at a flow rate of 48 $\text{mm}^3/\text{min-cm}^2$.

The membrane was furthermore used in an ultrafiltration operation on an aqueous solution containing 0.5 g/l of albumin and 9 g/l of NaCl, the pressure being 2 bars; with albumin of average molecular weight 70,000, no albumin was noted in the ultrafiltrate. With albumin of average molecular weight 45,000, it was found that at the end of the operation the ultrafiltrate contained 25% by weight of the albumin employed.

Example 2: The following were successively dissolved in 1,000 g of cyclohexanone, with stirring: at 60°C, 125 g of PVC (AFNOR viscosity index = 100) at 23°C 350 g of PPDMeAS (molecular weight about 2,000). 12.5 g of octyl phthalate were then added and the whole was heated at 70°C for 1 hour. The solution was left to stand for 15 hours so as to allow any trapped bubbles of air to escape.

A grid-reinforced membrane was then prepared as follows. A 75 μ thick polyamide woven fabric consisting of 35 μ diameter fibers, and weighing 26 g/m^2, was immersed in the solution prepared above, dried, again immersed in the same solution and again dried. The octyl phthalate was extracted by immersion in methanol for 2 hours at 23°C. The film obtained was still transparent. The PPDMeAS was extracted by immersion for 24 hours in a leaching bath consisting of an 0.5 N aqueous solution of HCl, the bath being provided with a liquid circulator.

After washing with distilled water until chloride ions had disappeared in the washwaters, an opaque membrane was obtained, the pores of which had a diameter of less than 0.01 μ; its resistance was 0.2 $\Omega\text{-cm}^2$. The elementary nitrogen content corresponded exactly to the amount of nitrogen present in the grid.

Water covering one face of the membrane under a relative pressure of 0.05 bar passed through this membrane at a flow rate of 0.012 $cm^3/min\text{-}cm^2$. The membrane was furthermore used in an ultrafiltration operation on an aqueous solution containing 10 g/l of albumin of average molecular weight 60,000 and 9 g/l of sodium chloride. Under a relative pressure of 2 bars, the flow rate of ultrafiltrate was 0.06 $cm^3/min\text{-}cm^2$, and the albumin content of the material which has permeated was 1 g/l.

Porous Polyethylene

D.B. Pall and S. Krakauer; U.S. Patent 3,048,537; August 7, 1962; assigned to Pall Corporation describe a process of preparing porous polyethylene articles which are nonporous in selected areas. It has been found that porous articles can be prepared from low density polyethylene, by sintering a layer of particles under moderate pressure sufficient to take up the natural slight shrinkage of the particles during sintering and at the same time compress the layer just a little to minimize shifts of positions of the particles while they are soft. A layer of polyethylene particles is confined between inert nonadhering surfaces and brought to a temperature above the softening point of the particles while applying to the layer through a surface a pressure within the range from 10 to 50 psi.

Examples 1 through 7: Seven bottle caps were prepared, four using powdered polyethylene (density 0.92) which passed a 50 mesh sieve and three using 100 mesh polyethylene. In preparing each cap the powder was distributed to a depth of 0.174 inch in a flat-bottomed female bottle-cap mold of stainless steel, 0.174 inch deep. The mold was circular and the face of the male portion of the mold had a raised ridge portion extending all the way around the perimeter of the base, so that the perimeter of the layer would be compressed to a greater extent than the central portion, and give a rim or flange for gripping the bottle neck.

The top of the mold was brought down upon the layer of polyethylene at a pressure of either 30 or 50 psi, and the mold brought to a temperature of 280 to 285°F. As the polyethylene softened, the top portion of the mold sank into its limiting position, and the layer was reduced to two-fifths of the original layer's thickness. The layer was sintered while confined to the mold at this temperature for 60 minutes. The mold was then quenched and the cap removed. The caps obtained had a flanged rim of sintered nonporous material, a skin of sintered but porous material, and an interior of unsintered material. These caps could be fitted to a bottle filled with nonviscous fluid, such as after-shave lotion. When inverted, the cap filled with the fluid, which could be expressed from the cap by pressing the cap against a surface, such as the skin, even after the bottle had been tipped so that the body of fluid drained from the cap.

The caps were tested for tensile strength, pore size and fluid permeability. Maximum pore size was determined by bubble point measurement. The cap was immersed in ethyl alcohol until the pores were filled with alcohol. The cap then was placed in a jig and air applied to one side while a 1 mm depth of alcohol was maintained on the other side. The bubble point was the air pressure in millimeters necessary to cause the first bubble to appear on the alcohol side of the filter.

The fluid permeability was determined by pressure drop of air through the filter. The flow of air was measured as standard cubic feet per hour per 0.8 square inch area, and converted to standard cubic feet per minute per square foot of filter area at 1 psi differential pressure. The caps were found to have a satisfactory bubble point and fluid permeability. The tensile strength of the cap was such that it was impossible to tear it by hand.

	Example Number						
	1	2	3	4	5	6	7
Process data:							
Material size (mesh*)	50	50	50	50	100	100	100
Pressure during sintering (psi)	30	50	30	50	30	30	50
Sintering temperature (°F)	280-285	280-285	350	350	280-285	350	280-285
Sintering time (hr)	1	1	1	1	1	1	1
Filter data:							
Depth of die and thickness of initial layer (inches)	0.174	0.174	0.174	0.174	0.174	0.174	0.174
Approximate thickness of finished filter (inches)	0.105	0.105	0.105	0.105	0.105	0.105	0.105

*Material passes through this mesh.

Graft Polymerization of Acid and Basic Monomers on Perhalogenated Polymers

A. Chapiro and A.-M. Jendrychowska-Bonamour; U.S. Patent 3,666,693; May 30, 1972; assigned to Centre National de la Recherche Scientifique, France have found that perhalogenated olefin polymers containing grafted acid and basic functionality, where the acid and the basic functionality are both substantially neutralized, have vastly improved water swellability as compared with common prior art graft copolymers having the same proportion of grafted comonomer.

It has been discovered that, if acid organic groups and basic organic groups are incorporated together into the perhalogenated polyolefin membrane, the rate of swelling of this mixed membrane is superior to the rates observed in the case of the simple membranes containing naught but acid organic groups. The acid organic or carboxylic groups are each derived from a monomeric acid that is polymerizable alone (acrylic acid) or from an acid (maleic acid) that is polymerizable in mixture with another monomer. The basic organic groups are derived each from a polymerizable vinylic monomer of the pyridine type (vinyl-4-pyridine) or of the amine type (p-aminostyrene).

Example 1: A polytetrafluoroethylene (PTFE) film of 0.1 mm thickness and weighing 0.9852 g was sealed under vacuum in a glass flask containing an aqueous solution of 50% acrylic acid, containing 0.25% of Mohr's salt (ferrous ammonium sulfate). This flask was irradiated with the gamma rays from a cobalt-60 source at an intensity of 240 rads per minute. After 38 hours of irradiation, the film was removed from the flask, washed in water and dried. Its weight was 1.349 g. This film was washed for 24 hours with a 0.1 normal potassium hydroxide solution. After drying, its weight was 1.532 g. Two fragments of this film were sealed under a vacuum in flasks containing an 80% aqueous solution of vinyl-4-pyridine. Each flask was subjected to different doses of gamma rays at an intensity of 29 rads per minute. Then the films were washed in water, and then in a solution of 10 normal hydrochloric acid solution, and finally dried. The results are summarized in the following table.

Wt. of film after 1st graft and neutralization	Dose in second graft (in rads)	Wt. of film after 2nd graft and neutralization	% swelling of film in water (i.e., g. of water per 100 g. of dry film
0.0760	38,900	0.0735	25.9
0.0705	86,400	0.0790	39.8

Example 2: A PTFE film of 0.05 mm thickness and weighing 1.0070 g was sealed in vacuum in a glass flask containing an 80% aqueous solution of vinyl-4-pyridine. This flask was subjected to gamma rays at 36 rads per minute. After a total dose of 94,000 rads, the film weighed, after drying, 1.1475 g, which corresponds to an amount of grafting of about 14%. Several fragments of this film were treated for 24 hours with a 0.1 N solution of hydrochloric acid, and then irradiated in vacuo with gamma rays at 27.5 rads per minute in a 50% aqueous solution of acrylic acid containing 0.25% by weight of Mohr's salt. The films were then washed in water and then in a 0.1 N solution of potassium hydroxide. The results are summarized in the following table.

Weight of film after 1st graft and neutralization	Dose in second graft (in rads)	Weight of film after 2nd graft and neutralization	% Swelling of film in water
0.0365	26,600	0.0325	6.15
0.0385	68,500	0.0345	14.5
0.0395	149,000	0.0375	21.3
0.0445	192,000	0.0465	24.7

Example 3: A PTFE film 0.05 mm thick, weighing 1.1030 g, was grafted with vinyl-4-pyridine as in Example 2, but with a total dose of 119,000 rads instead of 94,000. It weighed 1.5965 g after drying, which corresponds to an amount of grafting of about 45% (instead of 14). Several fragments of this film were neutralized with hydrochloric acid and then treated with acrylic acid as in Example 2. The results are summarized in the following table.

Weight of film after 1st graft and neutralization	Dose in 2nd graft (in rads)	Weight of film after 2nd graft and neutralization	Swelling of film in water in percent
0.0650	41,000	0.0490	53.1
0.0565	78,000	0.0445	47.1
0.0535	108,000	0.0420	43.0
0.0615	183,000	0.0500	58.0

It is to be noted that, in each of the three examples above, cases are found in which the weight of the film diminished after the second grafting. This effect presumably corresponds to some extractions of the polymer. Nevertheless, the swelling of this film in water is very high, higher than that of membranes containing only acrylic acid or vinyl-4-pyridine. Its permeability to water is likewise higher, whereas its electrical resistance is lower. All the membranes thus prepared have very good resistance to strong bases and strong acids.

Polyvinyl Carbonate

A process described by *P.A. Cantor and R.E. Kesting; U.S. Patent 3,332,894; July 25, 1957; assigned to U.S. Secretary of the Interior* involves the production of membranes useful in reverse osmosis operation. The membranes are cast from a solution containing polyvinylene carbonate and may be formed into dried films of gelled membranes in single or multiple layers. Vinylene carbonate may be synthesized according to the following reaction

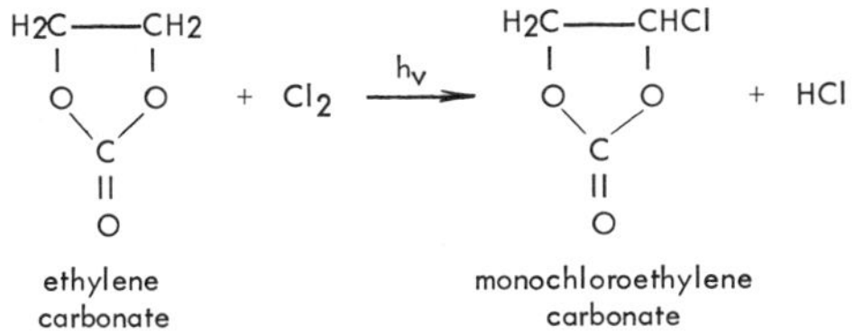

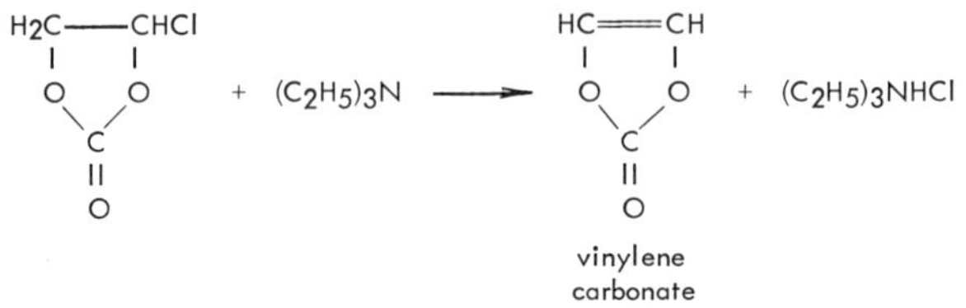

Alternatively, the chlorination may be conducted in a carbon tetrachloride solution. In practice, it has been found that the reactions cited above do not result in a product of desired purity for forming high molecular weight polyvinylene carbonate. The chief impurities are chloroethylene carbonate and dichloroethylene carbonate. To remove traces of these halogen compounds the vinylene carbonate prepared by the above methods should be purified by heating with sodium borohydride, and then redistilled. Such a treatment removes traces of halogen compounds which interfere with a subsequent polymerization.

When using commercially prepared vinylene carbonate monomer, which contains approximately 3% chlorine, care should be taken to provide for a careful distillation to remove these chlorine compounds prior to polymerization. It is preferable to distill such monomers at reflux ratios in the range of from 5/1 to 20/1 rather than to rely upon a sodium borohydride treatment. However, if the monomer sample has much less chlorine present, a simpler distillation or a sodium borohydride treatment will suffice.

The polymerization of the purified monomer may be carried out in bulk using azobisisobutyronitrile in the range of 0.075 to 0.50 g/100 ml of monomer as the initiator. The molecular weight of the resulting polymers can be varied by varying the catalyst concentrations. For example, at high catalyst concentrations the molecular weight diminishes. Initiation of the polymerization at temperatures of from about 60° to 75°C in vacuo give excellent results. The monomer and initiator are normally charged to a vessel and subjected to a plurality of freeze-thaw cycles on a vacuum system to remove oxygen, after which the vessel is sealed and placed into an oven at about 65°C for about 25 to 100 hours.

The resulting polymer is dissolved in dimethylformamide to the extent of 10% and precipitated by methanol, washed in methanol, and dried in vacuo at 40°C. Using this polymerization procedure, molecular weights approaching 1 million have been obtained. Fully dried films of polyvinylene carbonate may be prepared from casting solutions of from 18 to 20% polyvinylene carbonate in dimethylformamide. The films are cast on glass or other smooth surface, partially dried in air at from 20° to 40°C and as an optional step, further dried in a convection oven at from 40° to 60°C.

Immersion of the dried films in water for a few minutes causes them to float free of the glass plates. The thickness of the films can easily be controlled by varying the thickness of the casting solution which is spread on the smooth surface. Uniformity is achieved through the use of a doctor blade. For use in reverse osmosis, the desired thicknesses are in the range of from 0.4 to 4.0 mils. For the casting of very thin films i.e., less than 0.1 millimeter acetone is the preferred solvent. The advantage of thin films is attainment of high product-water fluxes.

Example 1: Films of various thicknesses were prepared from casting solutions of 18.2 to 20% polyvinylene carbonate in dimethylformamide. The films were cast on glass, partially dried in air at room temperature, and further dried overnight at 55°C in a conventional oven. As a comparison membrane, a dried film of cellulose acetate was cast in a similar manner from a solution of acetone. Immersion of the dried films in water for a few minutes caused them to float free of the glass plates. The free films were stored in water before use; during this storage they absorbed about 17% of water. Evaluation by reverse osmosis was conducted at 1,500 psi pressure with a 3.5% sodium chloride solution in a 3 inch test cell. From the flux and salt retention, the water permeability, P_w (g/cm-sec), was calculated using the method of Lonsdale, Merten and Riley, *J. Appl. Polymer Sci.,* 9, 1341, (1965).

Data was obtained on fully dried films of polyvinylene carbonate and values of water permeability were calculated. This data shows that the water permeability for polyvinylene carbonate films is fairly constant, ranging from 3.0 x 10^{-7} to 5.8 x 10^{-7} g/cm-sec with membrane thicknesses ranging from 0.5 to 3.75 mils. The best salt retention, 94%, was obtained with the thickest film. With some of the thinner films, folds or creases appeared in the membrane after testing and salt may have leaked through these or other imperfections, thus giving the lower salt retentions obtained with the thinner films.

The best polyvinylene carbonate film (94% retention) was compared with a fully dried cellulose acetate film cast from a 20% solution of Eastman 398-3 cellulose acetate in acetone, which had approximately the same water flux. From the comparative film thicknesses and water permeabilities, it was seen that polyvinylene carbonate has a water flux about 2.5 times greater than cellulose acetate. The water permeability of the film containing 5% polyacrylic acid showed a value 2.9 times greater than that for cellulose acetate at a salt retention of 98.2%.

Example 2: A very thin air-dried film of polyvinylene carbonate was prepared by casting from an 11% polyvinylene carbonate (32,000 molecular weight) in acetone solution using a 5 mil knife gap setting on glass. The fast-drying casting solution air dried in one minute before immersion in water. The film containing 16.8% H_2O and was 6.3 microns thick. On testing at 1,500 psi using 3.5% sodium chloride solution, a flux of 4.8 gfd was obtained with 70% salt retention.

OTHER ORGANIC MEMBRANES

Polyethers Containing Pendant Metal Thiosulfate Groups

E.J. Vandenberg; U.S. Patent 3,706,706; December 19, 1972; assigned to Hercules Incorporated describe modified polyethers containing pendant metal thiosulfate groups. The modified polyethers are prepared by reacting a polymer of an epihalohydrin or 1,4-dihalo-2,3-epoxybutane with a metal thiosulfate.

The following examples illustrate the preparation of the modified polyethers. All parts and percentages are by weight unless otherwise indicated. The dimethylsulfoxide used as solvent was dried prior to use by passage through a molecular sieve column. Unless otherwise indicated, the stabilizer used in the examples was the reaction product of 1 mol of crotonaldehyde and 3 mols of 3-methyl-6-tert-butyl-phenol, the product being essentially 1,1,3-tris(2-methyl-4-hydroxy-5-tert-butylphenyl) butane. The molecular weights given in the examples are weight average molecular weights of the polymer.

When molecular weight is indicated by reduced specific viscosity (RSV), the term denotes the specific viscosity divided by concentration as determined on a solution of the polymer in a suitable solvent. The reduced specific viscosities of the modified polyethers were run on the dialyzed aqueous solution at 25°C by taking a portion of the dialyzed solution and diluting to 0.1% polymer concentration using sufficient added potassium chloride to give 0.1 molar potassium chloride, unless otherwise indicated.

Example 1: Five parts of an epichlorohydrin-ethylene oxide copolymer (70% by weight epichlorohydrin molecular weight of 2.7 x 10^6; RSV of 3.8 as determined on a 0.1% solution in 97% α-chloronaphthalene-3% acetylacetone at 100°C; and containing 26.3% chlorine) containing 0.3% of stabilizer was dissolved in 138 parts of dry dimethyl sulfoxide at 65°C under a nitrogen atmosphere, and 9.3 parts of sodium thiosulfate

$$(Na_2S_2O_3 \cdot 5H_2O)$$

was added to the solution. The reaction mixture was then tumbled at 65°C for 2 hours under a nitrogen atmosphere, after which time the insoluble material (unreacted sodium thiosulfate and sodium chloride by-product) was separated from the solution and from 3 to 4 volumes of water was added to the solution.

A dimethyl sulfoxide-soluble, water-insoluble precipitate was formed (6.4% of total isolated product) and was separated from the solution, after which the remaining solution was dialyzed against distilled water to remove unreacted, dissolved sodium thiosulfate and dimethyl sulfoxide. The water-soluble polymer had an RSV of 0.10 and a molecular weight greater than 1 x 10^6. A portion of the dialyzed solution was then stripped of solvent and dried for 16 hours at 80°C in vacuo (0.4 mm), giving a rubbery film (4.36 parts solids based on the total solution) which was insoluble in and unswollen by water and thus was highly cross-linked. Analysis for sulfur and chlorine showed that the water-soluble polymer contained 13% of

$—CH—CH_2—O—$
|
$CH_2—S—SO_3Na$

units.

Example 2: The procedure of Example 1 was repeated except that the reaction mixture was tumbled for 9 hours. No water-insoluble, dimethyl sulfoxide-soluble precipitate formed on addition of water. The water-soluble polymer had an RSV of 0.07 and a molecular weight greater than 1 x 10^6. A portion of the dialyzed solution was stripped of solvent, freeze dried under vacuum and then dried for 3 hours at 50°C under vacuum (0.4 mm). This product (4.52 parts based on total solution) was a soft, white rubber which was insoluble in and very slightly swollen in water. Analysis for sulfur and chloride showed that the water-soluble polymer contained 21% of

$—CH—CH_2—O—$
|
$CH_2—S—SO_3Na$

units.

Example 3: Example 1 was repeated except that the reaction mixture was tumbled for 24 hours. No water-insoluble, dimethyl sulfoxide-soluble precipitate was formed. The water-soluble polymer had an RSV of 0.14 and a molecular weight greater than 1 x 10^6. The product recovered from a portion of the dialyzed solution (5.0 parts based on the total solution) gave a tough film of low extensibility which was insoluble in and unswollen by water. Analysis for sulfur and chlorine showed that the water-soluble polymer contained 28% of

$—CH—CH_2—O—$
|
$CH_2—S—SO_3Na$

units

Example 4: Example 1 was repeated except that the reaction mixture was tumbled for 72 hours. No water-insoluble, dimethyl sulfoxide-soluble precipitate was formed. The water-soluble polymer had an RSV of 1.2 and a molecular weight greater than 0.5 x 10^6. The product which was recovered from a portion of the dialyzed solution (5.5 parts based on total solution) gave a hard, brittle film and was insoluble in water, and unswollen by water and slightly swollen by dimethyl sulfoxide. Sulfur and chlorine analyses (13.5% sulfur and 13.2% chlorine) showed that the water-soluble polymer contained 41% of

$—CH—CH_2—O—$
|
$CH_2—S—SO_3Na$

units.

To a portion of the dimethylsulfoxide solution (prior to dialysis) equivalent to 1 part of polymer was added 2 parts of water and then 1 part of a 3% aqueous solution of hydrochloric acid. The resulting solution was stable at room temperature, but after heating on a steam bath for about 6 hours, gave a polymeric precipitate, indicating that hydrolysis of the sodium thiosulfate groups to the nonionic polymercaptan groups had taken place. The precipitate was insoluble in dimethyl sulfoxide, indicating that crosslinking had occurred.

Example 5: Example 1 was repeated except that the reaction mixture was tumbled for 72 hours at 95°C. No water-insoluble, dimethyl sulfoxide-soluble precipitate was formed on addition of water, and the solution was stable on heating on a steam bath for 20 hours. The water-soluble polymer had an RSV of 3.1 and a molecular weight greater than 0.5×10^6.

The product which was recovered from a portion of the dialyzed solution (7.6 parts based on the total solution) gave a hard, brittle film which was insoluble in and unswollen by water. Analysis for sulfur and chlorine showed that the water-soluble polymer contained 70% of

$$\begin{array}{l} -CH-CH_2-O- \\ \ \ | \\ \ \ CH_2-S-SO_3Na \end{array}$$

units. To each of two separate portions of the dialyzed solution was added an aqueous solution of either zinc chloride or calcium chloride in an amount sufficient to provide 10 mols of zinc or calcium chloride per sodium ion.

Each portion was permitted to stand at room temperature for 72 hours, after which time the portion to which the calcium chloride was added was a solution containing a trace of insoluble matter and the portion to which the zinc chloride was added was a solution containing a slight amount of precipitate. Each portion was dialyzed 4 times against 1 molar zinc chloride or calcium chloride and then against distilled water until all of the soluble chloride ions were removed and precipitates formed. The precipitates were separated from the water-soluble fractions and then were washed and dried.

The water-soluble fractions were recovered by stripping off the solvent, and the products dried under vacuum. The calcium-containing product was a colorless brittle solid and the zinc-containing product was a brittle brown solid. Sodium and zinc or calcium analyses on the water-soluble fractions indicated that 77.5% of the sodium was replaced by zinc and that 57.1% of the sodium was replaced by calcium, after allowance for the small amount of precipitate (1% of total product in the case of calcium and 5.5% of the total product in the case of zinc).

Another portion of the dialyzed solution was treated at room temperature with 10% aqueous sodium sulfide in an amount sufficient to provide 1 mol of sodium sulfide for each thiosulfate group. At the end of 10 minutes the solution had turned to a solid yellow gel which gradually shrank after a total of 25 minutes to form a small solid phase (polymer) and a liquid phase. At the end of one hour the polymer was in the form of a small hard cylinder. The polymer, after air drying, was hard and tough.

Still another portion of the dialyzed solution was treated with sodium sulfide as above, except that a 10-fold molar excess of sodium sulfide was used. The gel-like precipitate was formed in 3 minutes and readily separated from the liquid phase as a compact discrete solid phase. The utility of the polymers of this process and particularly the polymers containing

$$\begin{array}{l} -CH-CH_2-O- \\ \ \ | \\ \ \ CH_2-S-SO_3Na \end{array}$$

units as membranes for water desalination is demonstrated by the following examples.

Examples 6 through 9: Films were prepared by casting a portion of the dialyzed solution prepared in Examples 2, 4 or 5 over a water-wet VF Millipore support sheet (4" by 5" sheet of filter paper composed of esters of cellulose of 0.01 μ pore size) supported on a glass plate and drying first at room temperature for 16 hours and then at 80°C for 16 hours.

Discs (47 mm diameter) were then cut from the film and sheet and evaluated as membranes by supporting the disc in tests cells assembled from high pressure filter holders and through which a brine solution was circulated at a rate of 200 to 400 ml/min. The membranes were tested at 1,500 psi for extended lengths of time and permeate was collected from the bottom port of the cell and analyzed for Na^+, Cl^- and $SO_4^=$ ions. The brine solution contained approximately 7,000 ppm, NaCl (2,730 ppm Na^+, 4,200 ppm (Cl^- and 3,000 ppm Na_2SO_4 (930 ppm Na^+, 2,000 ppm $SO_4^=$). Rejections at 1,500 psi are given below.

	-------- Film --------		--- Percent Rejection ----			
Example No.	Example No.	Thickness (mils)	Na^+	Cl^-	$SO_4^=$	Flux Rate (gfd)*
6	2	1.0	99.6	98.6	94.1	0.12
7	4	1.5	65	46	71	3.5
8	5	1.0	89	85	78	3.9
9	5	0.5	61	52	80	16.1

*Measure of the number of gallons per square foot per day which can be forced through the membrane.

Polymeric Epoxides

A process described by *C.L. Lukach, H.M. Spurlin, E.J. Vandenberg and W.L. Young III; U.S. Patent 3,567,631; March 2, 1971; assigned to Hercules Incorporated* involves the use of reverse osmosis membranes characterized by a salt-rejecting layer of an amorphous polymer derived from an epihalohydrin and an alkylene oxide. The polymeric epoxide from which the reverse osmosis membranes are prepared is selected from one of the following classes:

(A) amorphous copolymers of at least one alkylene oxide having from 2 to 4 carbon atoms and an epihalohydrin in which the alkylene oxide comprises from 5 to 99 mol percent;
(B) Bunte salts of the copolymers of (A); and
(C) amorphous terpolymers of an epihalohydrin, an alkylene oxide having from 2 to 4 carbon atoms and an amine having the formula:

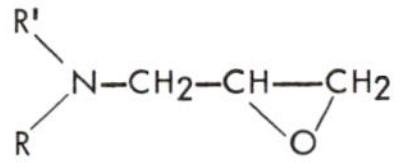

or

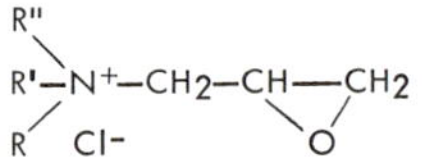

where R, R' and R'' are alkyl groups, in which the epihalohydrin and amino epoxide each comprise at least one mol percent and the alkylene oxide comprises at least 40 mol percent.

The polymeric epoxide has a reduced specific viscosity (RSV) of at least 0.5 as measured on a 0.1% solution of the copolymer in alpha-chloronaphthalene (or other appropriate solvent) at 100°C, and a weight average molecular weight of at least 50,000. The following examples illustrate the process.

Example 1: A 4 x 5 inch sheet of filter material composed of mixed esters of cellulose of 0.01 μ pore size (Millipore Corporation VF filter material) was floated on the surface of deionized water, with the dull side down for 5 minutes. The sheet was lifted with forceps, tilted to drain off excess water, and placed, wet side down, upon a glass plate. A piece of adhesive tape was placed across the top edge to hold the sheet to the glass.

A 10 g sample of a copolymer of ethylene oxide and epichlorohydrin, containing 75 mol percent EO and of RSV 4.9 was dissolved in 200 ml of toluene. The solution was filtered under pressure through a 1.2 μ pore size filter. To 20 ml of this 5% (w/v) solution was added 2.0 ml of a filtered solution of a basic polyamide (crosslinking agent) in toluene (20%). This amount of polyamide corresponds to 40 weight percent, based upon the weight of polymeric epoxide present. After mixing well by stirring, the polymer solution was placed near the top of the prewet VF substrate beneath the leading edge of a one mil casting knife 2½ inches wide.

After drawing the solution across the entire length of the substrate with the casting knife, the glass plate was placed in a level position in a cabinet through which filtered air was circulated. After 4 hours, a dry film of calculated thickness 0.5 mil was obtained deposited on the substrate. The glass plate containing the film and substrate was then heated at 80°C for 16 hours.

The basic polyamide employed in this example was prepared by reacting polymerized soybean oil acids with triethylene tetramine to produce an intermediate resin having an amine number of 225 and a Brookfield viscosity of 500 poises at 40°C, and then reacting 9 parts of the resin with 1 part of tetraethylene pentamine for 2 hours at 200°C, giving a final polyamide having an amine number of 300 and a viscosity at 25°C of about 500 poises. A 47 millimeter diameter disc was cut from the above membrane. Soaking in water helped remove the cut disc from the glass plate. The membrane was then evaluated as follows.

Test Apparatus and Method – Each test cell consisted of a 6 bolt 47 mm high pressure filter holder which holds a 47 mm diameter membrane on a support screen between two stainless steel plates. An additional hole was drilled into the top plate, so that the brine solution could enter the cell, circulate over the membrane, and leave the cell. The permeate was collected from the bottom part of the cell and analyzed. Eight such cells were connected in parallel, through a series of suitable valves, to three back pressure regulators, a pump and a 100 gallon reservoir, to provide a recirculating assembly capable of evaluating eight membranes simultaneously at two different pressures (four cells at each pressure). Pumping rates were up to 0.5 gallon per minute (30 gallons per hour).

The brine solution in the reservoir, unless otherwise indicated, contained approximately 7,000 ppm NaCl (2,730 ppm Na^+, 4,200 ppm Cl^-) and 3,000 ppm Na_2SO_4 (930 ppm Na^+, 2,000 ppm $SO_4^=$. It was analyzed each day for Na^+ and Cl^- concentration, using a Beckman Expandomatic pH meter and appropriate electrodes. The sulfate ion was calculated from these two values, using the expression ppm $SO_4^=$ equals (2.09 ppm Na^+ – 1.35 ppm Cl^-).

Sulfate ion concentrations were also determined independently by a titration method. These values usually agreed well with the calculated value. The brine solution also contained a small amount of both ethyl violet and Rhodamine B. When a membrane contained a small pinhole, a pink-purple dot from these dyes was visible on the GS filter below the membrane after testing. The membrane to be tested was cut to a 47 mm diameter size and placed in the test cell atop two 47 mm Millipore GS filter discs (0.22 μ pore size; 45,000 gfd at 1,500 psig).

Depending upon the number of cells in operation, the brine solution was circulated through the cell and over the membrane at the rate of 200 to 400 ml/min. The membranes were kept first at 500 psi and then at 1,000 psi for long enough times to collect enough permeate for analysis. The membranes were then kept at 1,500 psi for extended lengths of time while permeate samples were measured and analyzed periodically. Percent rejection of any ion was calculated from the ppm of the ion in the permeate and the feed solution as follows. If X is any ion, (e.g., Na^+, Cl^- or $SO_4^=$), then

$$\text{percent rejection of X} = \frac{\text{(ppm X in feed minus ppm X in permeate)}}{\text{ppm X in feed}} \times 100$$

Permeate Rate and Analysis – Analysis of the permeate collected, compared to the feed solution, gave the following salt rejections.

Pressure	Flux, gfd	Colorless Permeate — Percent Rejection* Sodium	Chloride	Sulfate
500	1.0	86.4	82.4	–
1,000	1.74	92.6	99.9	–
1,500	1.83	96.7	95.5	98.5
1,500**	1.74	97.5	96.4	99.6

*Feed solution analysis 3,300 ppm Na^+; 3,750 ppm Cl^-; 2,230 ppm $SO_4^=$.
**After an additional 580 minutes.

Examples 2 through 11: Salt rejections (at 1,500 psi) of membranes prepared in the same manner as Example 1 with different amounts of polyamide crosslinking agent are given in the table below. Example 11 demonstrates the presence of a nonsolvent for the polymer.

Example	Percent Cross-linking Agent	Percent Rejection Na^+	Cl^-	$SO_4^=$	Flux, gfd
2	0	17	9	–	6.1
3	2	41	25	–	3.5
4	5	69	61	–	1.5
5	10	75	65	–	3.0
6	20	96.5	95	98.6	1.4
7	30	92.6	92	–	1.4
8	75	95.9	95.6	–	1.1
9	50	96.9	95.6	–	0.77
10	75	95.9	94.5	–	0.8
11	45*	87	82	95.9	1.8

*Coating solution contained 10 volume percent nonsolvent (heptane).
NOTE: Examples 2 through 8, film was 0.5 mil thick; Examples 9 through 11, film was 1.25 mils thick.

C.A. Lukach; U.S. Patent 3,567,630; March 2, 1971; assigned to Hercules Incorporated also describes reverse osmosis membranes which are characterized by a salt-rejecting layer of an amorphous copolymer of an alkylene oxide and allyl glycidyl ether.

The amorphous copolymer from which the reverse osmosis membranes are prepared is comprised of from 1 to 60 mol percent of allyl glycidyl ether and from 40 to 99 mol percent of an alkylene oxide having from 2 to 6 carbon atoms and has a reduced specific viscosity (RSV) of at least 0.5 as measured on a 0.1% solution of the copolymer in alpha-chloronaphthalene (or other appropriate solvent) at 100°C, and a weight average molecular weight of at least 50,000.

Polyoxetanes with Functional Groups

In U.S. Patent 3,459,687, there are described ion exchange resins in the form of a powder or of granules, obtained from poly-3,3-bis(chloromethyl)oxetane (PBCMO) by reaction with amine.

R. Margraff; U.S. Patent 3,839,235; October 1, 1974; assigned to Rhone-Poulenc SA, France describes a process which provides homogeneous membranes and in particular, ion exchange membranes from polyoxetanes such as PBCMO. It has been found that this can be achieved with polyoxetanes possessing functional groups of a new type, these polyoxetanes can be nonquaternized or quaternized. Nonquaternized polyoxetanes comprise recurring units of the formula as shown on the following page.

(1) 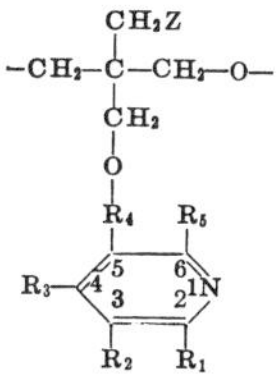 and/or (2) 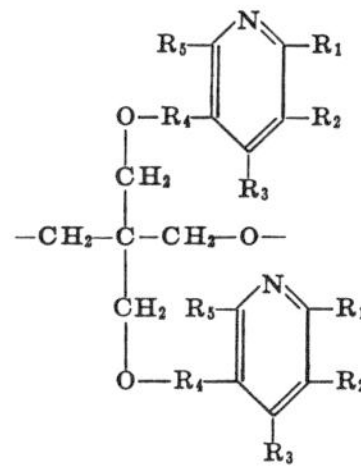

and optionally, units of the formula:

(3) 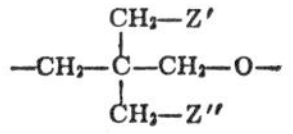

in which each of R_1, R_2 and R_3, which may be identical or different, represents a hydrogen atom or a lower alkyl radical, it being also possible for R_1 and R_2 together to form a buteno bridge (thus forming a quinoline nucleus with the pyridine ring), each of Z, Z' and Z'', which may be identical or different, represents one of various monovalent atoms or radicals such as halogen atoms, especially chlorine, or hydroxyl, alkoxy, aryloxy, acyloxy or nitrile groups, it being also possible for Z' and Z'' together to be a single atom of oxygen i.e., a free oxetane ring is formed, R_5 represents a hydrogen atom or a lower alkyl radical, and —O—R_4 represents the divalent oxygen atom —O—, or, together with R_5, forms the trivalent radical of the formula:

(4) 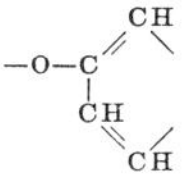

in which the free valencies of the end carbon atoms are respectively connected as follows: one to carbon atom 5 of the pyridine nucleus and the other to carbon atom 6 of the nucleus.

The radicals Z, Z' and Z'' preferably have at most 6 carbon atoms if they do not contain an aromatic ring and at most 12 if they do contain an aromatic ring. In the nonquaternized polyoxetanes, the proportion of nitrogen-containing heterocyclic structures per recurring ether unit (that is to say, per oxygen atom of the basic skeleton) is on average between 10 and 200%, and preferably between 20 and 100%.

The process provides quaternized polyoxetanes, derived from the polyoxetanes defined above and which comprise recurring units of the formula:

(5) 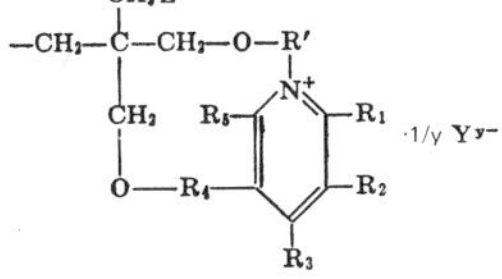 and/or (6) 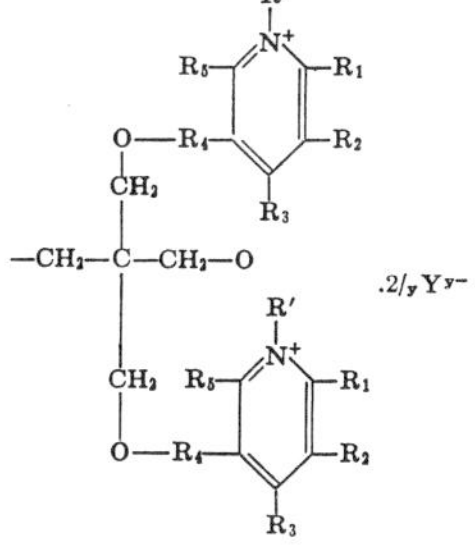

optionally combined with units of formulas 1 and/or 2 and/or 3 in which, R_1, R_2, R_3, R_4 and R_5 are as defined above, R' represents a monovalent or polyvalent organic radical, Y^{y-} denotes an anion such that H_yY represents water or an inorganic or organic acid, and y is a positive integer representing the charge on the ion Y.

The nature of the anion Y^{y-} is not critical to this process. Furthermore, the anion Y^{y-} can be changed by the usual methods of ion exchange, especially during the use of the polyoxetanes according to the process, without in any way fundamentally changing the nature and the properties of the quaternized polyoxetanes. Usually Y^{y-} represents a halide, hydroxyl, nitrate, sulfate, phosphate or arylsulfonate, carbonate, alkylcarboxylate or aralkylcarboxylate anion.

The polyoxetanes can be used as anion exchangers of the weak base type (nonquaternized polyoxetanes) or of the strong base type (quaternized polyoxetanes). They can be formed into membranes, the crosslinked membranes being produced by crosslinking a film obtained by casting a noncrosslinked form.

These membranes can be reinforced with a support or screen-reinforcement such as a woven fabric, a grid or a net, this support generally forming less than 70% by weight, and preferably less than 40%, of the supported membrane. These membranes can be used in numerous fractionation processes, and in particular, in electrodialysis, direct or inverse osmosis and dialysis; they can also be used as separators in electrical cells, in particular in electrolytic cells.

The membranes show an excellent uniformity of properties and of distribution of ionic sites; this uniformity arises essentially from the possibility of dissolving the noncrosslinked, weak base (nonquaternized) and strong base (quaternized) forms of the polyoxetanes (and then of casting this solution). The membranes also show good mechanical properties; the membranes based on quaternized polyoxetanes, in particular, also show a low electrical resistance. The following examples illustrate the process.

Example 1: 4.5 g of 3-hydroxy-pyridine-N-oxide are added to 20 cm^3 of a 2 M solution of potassium methylate in methanol; the solution is evaporated to dryness, 50 cm^3 of anhydrous hexamethylphosphotriamide (HMPT) are added and the mixture is heated to 140°C, in an atmosphere of dry nitrogen; a solution prepared from 200 cm^3 of HMPT and 15.5 g of poly-3,3-bis(chloromethyl)oxetane (PBCMO) of average molecular weight equal to 190,000 is then added. The mixture is heated, whilst stirring, in an atmosphere of dry nitrogen, for 4 hours at 140°C.

The reaction mixture is then run into 4 liters of water, the precipitate is washed with water and then with methanol and redissolved in 3 liters of tetrahydrofuran (THF). This new solution is filtered, concentrated and run into 4 liters of 0.1 N aqueous HCl; the new precipitate obtained is washed with water and then with methanol and dried.

16 g of polyoxetane N-oxide containing 2.3% by weight of nitrogen are thus obtained, which corresponds to a replacement of about 14.5% (by number) of the chlorine atoms by hydroxy-pyridine-N-oxide. It shows IR absorption bands characteristic of N-oxides (545 cm^{-1} and 670 cm^{-1}).

This polyoxetane is dissolved in 200 cm^3 of dioxane and is then reduced by passing in a gaseous stream of SO_2 at the rate of 2 l/hr over a period of 4 hours. The reduced polymer precipitates during the reduction; it is filtered off, washed with boiling water and then with methanol, and dried. 16 g of polyoxetane possessing pyridine groups are thus obtained; the compound no longer shows the IR absorption bands of the N-oxides.

1.6 g of this polyoxetane are dissolved at 100°C in 12 cm^3 of formic acid, the solution is cast on a 100 cm^2 glass plate and then dried for 4 days at 20°C; the film obtained is detached by immersing the film-plate assembly for 8 hours at 90°C in 10 N potassium hydroxide solution; the film is then washed with methanol and crosslinked by immersion for 48 hours at 20°C in a solution of 0.4 g of p-bis(chloromethyl)benzene in 50 cm^3 of methanol.

After washing with methanol, the quaternization is terminated by immersing the crosslinked membrane for 24 hours at 20°C in a solution of 2 cm³ of CH_3I in 50 cm³ of methanol. The membrane produced in this way has a substitution resistance of 3 Ω-cm² and a permeation selectivity of 73% (measurements carried out as described in French Patent 1,584,187).

Example 2: 9 g of 3-hydroxy-pyridine-N-oxide are added to 40 cm³ of a 2 M solution of potassium methylate in methanol; the solution is evaporated to dryness, 50 cm³ of anhydrous hexamethylphosphotriamide (HMPT) are added and the mixture is heated to 135°C in an atmosphere of dry nitrogen; a solution prepared from 200 cm³ of HMPT and 15.6 g of poly-3,3-bis(chloromethyl)oxetane (PBCMO) of average molecular weight equal to 190,000 is then added. The mixture is heated, whilst stirring, in an atmosphere of dry nitrogen, for 7 hours at 140°C.

The reaction mixture is then run into 4 liters of water; the precipitate is washed with water and then with methanol and redissolved in 2 liters of boiling methylene chloride containing active charcoal powder. This new solution is filtered when hot, concentrated, and run into 1 liter of ethyl ether. The new precipitate obtained is washed with water and then with methanol, and dried.

19.5 g of polyoxetane N-oxide containing 4.4% by weight of nitrogen are obtained, which corresponds to a replacement of about 31.8% of the original chlorine atoms. It shows the IR absorption bands characteristic of N-oxides.

5 g of this polyoxetane N-oxide are dissolved in 250 cm³ of boiling acetic acid. 3 g of iron powder are added and the solution is kept at the boil for 8 hours, filtered when hot, and precipitated by adding water; the precipitate is washed with water, dried and redissolved in methylene chlorine; the solution is filtered and the solid reprecipitated in petroleum ether and dried.

The polyoxetane with pyridine groups produced in this way is insoluble in methanol and does not show any IR absorption bands corresponding to N-oxides. When immersed in a solution of 2 g of methyl iodide in 50 cm³ of methanol, a limpid solution is obtained. By adding ethyl ether to this solution, a light yellow powder is precipitated which is soluble in water and consists of polyoxetane containing methyl-pyridine groups.

Anisotropic High Flow Polycarbonate Membranes

A.S. Michaels; U.S. Patent 3,615,024; October 26, 1971; assigned to Amicon Corporation has prepared an anisotropic high flux low pressure polymeric membrane capable of being dried without loss of beneficial mechanical and processing characteristics. It has in a continuous polymer phase a barrier layer containing pores from 1 to 1,000 millimicrons in diameter and an open porous support layer. The polymer absorbs less than 10% moisture at 100% relative humidity at 25°C.

The submicroscopically porous anisotropic membranes consist of an integral macroscopically thick film of porous polymer, usually more than about 0.002" and less than 0.050" in thickness. One surface of this film is an exceedingly thin, but relatively dense barrier layer of skin of from 0.1 to 5.0 microns thickness of microporous polymer in which the average pore diameter is in the millimicron range, for example, from 1.0 to 1,000 millimicrons, i.e., 1/10 to 1/100 thickness of the skin. The balance of the integral film structure is a support layer comprised of a much more coarsely porous polymer structure through which fluid can pass with little hydraulic resistance.

The membranes are prepared by: forming a casting dope of a polymer in an organic solvent; casting a film of the casting dope; preferentially contacting one side of the film with a diluent characterized by a high degree of miscibility with the organic solvent and a sufficiently low degree of compatibility with the dope to effect rapid precipitation of the polymer; and maintaining the diluent in contact with the membrane until substantially all the solvent has been replaced with the diluent.

Example: A 15% solution, i.e., a casting dope, was prepared by dissolving powdered poly-(bisphenol A carbonate) in N,N'-dimethylformamide (DMF). The solution was prepared at a temperature of from 60° to 70°C. Subsequently the solution having a viscosity of 20 cp (measured at 23°C with a Brookfield spindle at 60 rpm) was drawn down in a 0.014" film on a glass substrate. Prior to the drawdown, tape was placed on the glass along the intended edges of the membrane to assure its continued adherence to the glass plate during washing. This adherence is necessary in order to avoid the precipitation of a barrier layer on the backside of the membrane.

Moreover, the tape aids in minimizing membrane shrinkage in subsequent processing steps. The drawn film was covered with a shimmed-up glass plate to protect it from dust and the like as it levelled for a period of 1 minute. Then the drawn film was washed by submersion in a water bath, i.e., the diluent, for 15 minutes at 20°C, during which time a porous poly-carbonate membrane was formed. The membrane has an anisotropic structure, i.e., there was a relatively tight thin skin portion on the upper surface thereof, but a much looser structure below this skin.

In testing the membrane, 400 cc of an aqueous solution containing 1% by weight of Dextran 110 (a trade name for a dextran of 110,000 molecular weight) was placed in a pressurized batch cell. The first 10 cc of filtrate through the membrane are discarded. The next 10 cc of filtrate are used to judge initial performance characteristics of the membrane. At the applied pressure of 50 psig, a water flow rate of 1.54 cm^3/min-cm^2 (i.e., 541 gal/ft^2/day) was measured, and all of the Dextran passed through the membrane. An albumin solution of 0.03% weight concentration was also subjected to ultrafiltration with the membrane of the example: 90% of the albumin was rejected at a flow rate through the membrane of 71 gallons per square feet per day.

Composite Anisotropic Polysulfone Membrane

J. Shorr; U.S. Patent 3,556,305; January 19, 1971; assigned to Amicon Corporation describes a composite membrane. For some time, investigators active in the membrane art have attempted to solve the problem of forming reasonably strong membranes obtaining maximum diffusion activated separation by forming bifunctional membranes usually of an ultrathin diffusion barrier supported on a thicker, and more rugged, substrate which provides less resistance to flow than does the diffusion barrier.

Among such attempts are those described in U.S. Patent 2,960,462 where a very thin film of polymer is coated on a slightly more permeable polymer to form a diffusion membrane; in U.S. Patent 3,022,187, where a metallic oxide film at or near one surface of a metallic substrate forms a relatively tight barrier; in U.S. Patent 3,335,545 which describes the use of a thin liquid film over a porous substrate to obtain gas separation; and in U.S. Patent 3,225,107 which describes grafting a thin polymeric coating on at least one side of a polymeric membrane.

Moreover, a considerable amount of work has been done on so-called "dynamically formed" membranes, i.e., membranes formed by taking a relatively porous substrate and continuously coating the substrate with a polymeric film material which is maintained in dilute solution in the liquid mass being subjected to separation.

This continuous deposition is required because the polymeric film is subject to being carried into the pores and through the porous substrate. The films which are sufficiently soluble in the liquid mass being separated are usually subject to lateral erosion in continuous process devices.

The problems associated with the formation of dynamically formed membranes are indicative of the problems generally encountered by those trying to form an ultrathin diffusive polymer on a hydraulically porous substrate. By hydraulically porous substrate is meant one whose minimum average pore size is 10 angstroms, thus making it permeable to fluid flow without regard to material of construction. The term is meant to exclude substrates which are substantially pore-free and which permit the selective passage of molecules largely by activated diffusion.

This process involves forming a tripartite membrane comprising a selected anisotropic porous substrate, an ultrathin adhesive layer over the porous substrate, and a thin diffusive membrane formed over the adhesive layer and bound to the substrate by the adhesive layer. The adhesive polymer is conveniently coated on the substrate by dipping the selected porous substrate into a dilute solution of the adhesive polymer or otherwise forming a thin coat under low pressure conditions whereby the adhesive polymer is largely retained on the substrate surface, not pushed into the pores.

For this reason, adhesive polymers which are sparingly soluble in water are most easily utilized. Since many separation processes involve the diffusion of water through a membrane, such water-soluble adhesive polymers also provide in most cases the least resistance to the desired flow properties of the finished membrane. However, when separations which do not involve the transmission of water are to be carried out, the adhesive polymer may be selected for its solubility in some other vehicle for deposition upon the porous substrate.

By adhesive layer is meant a very thin layer, not more than 1500 angstroms, preferably less than 500 angstroms, in thickness, of a substance capable of holding materials together by surface attachment. More particularly, the adhesive must be such that (1) it is capable of acting as a holding adhesive, i.e., capable of attaching one adherent to another and holding them in place and (2) it is capable of acting as a sealing adhesive where it provides a joint between the two adhesives which, in effect, aids in sealing the adjacent surface of the anisotropic porous substrate adherent. Such an adhesive coat is believed to function not only as a means for bonding the diffusive membrane formed thereover to the substrate, but also is believed to form a protective coating over the substrate which prevents the diffusive membrane material from itself getting into the pores of the substrate.

Poly(methyl vinyl ether/maleic anhydrides) known as Gantrez AM are useful adhesive polymers. Conveniently used is Gantrez AN-139 of specific viscosity of 1 to 2. Solutions of this material are compatible with solutions of other useful adhesive polymers such as sodium polyacrylate known as Acrysol; such naturally derived products as albumin, sodium alginate, sodium carboxymethyl cellulose, starch, oxidized starch, dextrin, casein; carboxyvinyl polymers such as those known as Carbopol 934.

The porous substrate should be a highly anisotropic and porous membrane of the type described in U.S. Patent 3,549,016. Such membranes are advantageous because they combine, in an integral substrate member (1) a barrier skin on one side which has sufficiently small micropores to form a highly desirable surface on which to form, and retain, the adhesive polymer film and (2) a macroporous substructure which offers no significant additional resistance to liquids which pass the aforementioned barrier layer, but does provide mechanical support for the integral barrier layer and subsequently coated polymer layers superimposed upon it.

The anisotropic porous substrates which are most advantageously utilized have distinctive barrier skins of from 0.1 to 5 microns in thickness. The average pore size is in the millimicron range, i.e., from 10 to 1000 angstroms. Surfaces having pore sizes in this range provide the most suitable method of maintaining the membrane's operating characteristics for prolonged periods of operation.

These anisotropic porous substrates are highly anisotropic, submicroscopically porous membranes formed of polymers having good mechanical integrity, most advantageously those crystalline and/or glassy thermoplastic polymers known to the art. By crystalline and glassy polymers are meant those materials which possess from 5 to 50% by weight crystallinity as measured by x-ray diffraction techniques known to the art and/or a glass transition temperature (Tg) of at least 20°C.

Particularly advantageous are polymers of inherently low water sorptivity, which unlike the cellulose acetate materials known to the membrane art may be allowed to dry during storage without losing their beneficial mechanical and processing characteristics. These polymers are those having water absorptivities of less than 10% by weight of moisture at 25°C, and 100% relative humidity. The procedure used in forming the porous substrate used in the illustrative working example follows:

> 30 grams of a Polymer 360 polysulfone polymer was dissolved at 50° or 60°C in 100 cc of dimethylsulfoxide (DMSO). The resultant solution was cooled to 25°C and centrifuged down to remove any sedimentary matter therefrom. Next, the solution was drawn down into a film 12 mils thick. This casting was done on a clean glass surface with the edges of the casting overlying a taped perimeter which is to provide means to assist in removing the casting from the glass without tearing it.
>
> The casting was allowed to level for 1 minute and then submerged into a gently circulating water bath for a period of 20 minutes. This water bath is maintained at about 25°C. Next, the resulting membrane was stripped off the glass plate and cut into the desired membrane shape.
>
> The membrane was found to reject completely a polysaccharide of 100,000 average molecular weight, to reject 75% of a polysaccharide having a 20,000 molecular weight. These rejections were measured with 1% aqueous solutions under 50 psig operating pressure and at 30°C. At 100 psig applied pressure and 25°C the membrane exhibited a flux of 50 gallons per square foot per day to distilled water.

The material Polymer 360 is particularly advantageous for use because, it has been found, the rheological properties of this polymer allow the formation of a porous network having an unusually high hydraulic permeability and yet being highly retentive of macromolecular solutes. The polymer is a thermoplastic having a chain formed of diphenyl and phenyl groups which groups are linked by sulfone groups and oxygen atoms.

The inherent viscosity of the material is 0.42 to 0.46 as measured in a 1% solution of the material in dimethylformamide. The working examples which are set forth below were carried out in a so-called thin channel cell of essentially rectangular dimensions having ⅜" entrance and exit ports at either end thereof. Exposed membrane surface was 4.75" by 1.5". Depth of the channel was 0.010". Flow through the cell was maintained at 100 cc/min or a flow velocity of 16 cm/sec.

Example 1: An anisotropic membrane, formed of Polymer 360 as described above, was soaked for 4 hours in an aqueous solution containing 1,000 ppm of polyvinyl methyl ether (Gantrez M-155). The membrane was then mounted in a stainless steel membrane separation cell with the barrier skin layer facing in what may be called the upstream direction. An aqueous solution comprising 2,000 ppm of polyvinyl alcohol (Elvanol 50-42) was prepared.

A pressure of 100 psig was exerted on the solution as it was filtered through the adhesive coated anisotropic membrane for 30 minutes. During this period a thin diffusive membrane of polyvinyl alcohol formed over the adhesive. After a 30 minute period, the system was flushed and a 1% aqueous sugar solution was passed through the resultant membrane at 1,500 psig. The object of this experiment was to evaluate the effect of the membrane in retaining sugar while passing water through. The test was run over a period of 15 hours during which retentions of 85 to 92% sugar were achieved with an average water flux of 50 gallons per square foot per day.

Example 2: An anisotropic membrane of the type described in Example 1 was soaked for 2 hours in a solution containing 1,000 ppm polyvinyl methyl ether (Gantrez 155). The membrane was mounted as before and an aqueous polymer solution comprising 1,000 ppm of poly(methyl vinyl ether/maleic anhydride) (Gantrez AN-139) and 200 ppm of polyvinyl methyl ether was passed through for 30 minutes. Afterwards the system was flushed with water, and a 2% sugar solution was passed through at 1,500 psig. About 85% of the sugar was retained. Flux rates were maintained at 94 gallons per square foot per day over the 15 hour test period.

In related work *A.A. Massucco; U.S. Patent 3,556,992; January 19, 1971; assigned to Amicon Corporation* describes an anisotropic ultrafiltration membrane comprising a support layer, a barrier layer, and an adhering coating of irreversibly compressed gel, this coating adhering to the membrane when in contact with flowing liquid at a liquid shear rate of 16,000/sec. The membrane is prepared by placing an irreversible gel forming material in contact with such a membrane in sufficient concentration to form a gel and applying liquid pressure at least as high as the desired operating pressure to the membrane sufficient to irreversibly compress the gel into a coating adhering to the membrane even when in contact with flowing liquid at a liquid shear rate of 16,000/sec.

The membrane will be less permeable (e.g., to gel-forming material) than the untreated membrane, when both are used at the same operating pressure. The process is therefore useful as a method of carrying out an ultrafiltration of a polymeric gel-forming material to form a membrane having improved and stable permeability characteristics.

The pressure of the liquid feed stream on the upstream side of the anisotropic membrane is increased to a pressure above the desired working pressure and maintained at the elevated level until the flow rate of the gel-forming material through the membrane decreases to a stable or constant value. The pressure of the feed stream may thereafter be dropped without substantially affecting the permeability characteristics created at the higher pressure.

Collagen

According to a process described by *W.S. Higley; U.S. Patent 3,644,202; February 22, 1972* a collagen membrane is employed as a reverse osmosis membrane in desalination of water. The process is particularly applicable to desalination of brackish waters containing substantial amounts of divalent ions.

The membranes are preferably prepared from a microcrystalline form of collagen, i.e., collagen in which the average crystal size is less than about 1 μ. Colloidal casting dispersions of the collagen, preferably in concentrations of about 0.25 to 1.0% by weight, are prepared by dispersing the collagen in a mixture of water and an organic solvent such as methanol, ethanol, isopropanol or acetone.

Since the cast collagen films are adhesive in nature it may be desirable to employ casting surfaces that permit ready release of the dried films. Suitable surfaces of this type are wax-extended polyethylene sheeting and silicone release paper. The membrane may, however, be cast directly on a suitable support such as a filter material.

Crosslinking of the collagen films is usually desirable in order to provide sufficient wet strength. This may be achieved by either thermal or chemical means or a combination of the two. Drying the wet membrane overnight at about 35° to 50°C, followed by heat treatment at a temperature of about 110° to 150°C for about 2 to 6 hours has been found to give good results. Addition of a crosslinking agent such as hexamethoxymethylmelamine to the casting solution in an amount of about 0.06%, and subsequent heat treatment of about 110° to 150°C, also gives a film having good strength and desalination properties. Examples of other suitable crosslinking agents are glyoxal, quinone and chrome alum. Optimum treatment conditions for a given crosslinking agent are best determined experimentally.

The collagen membranes have the additional advantage that they may be deposited on porous substrates having an average pore size less than the average particle size of the dispersed colloidal collagen particle. Examples of suitable substrates are Metrical filters of about 2,000 angstrom average pore size. These filters consist of cellulose triacetate.

Example 1: A casting dispersion was prepared by dispersing 0.64 wt % of 113 H, a microcrystalline collagen (crystal size $>1\ \mu$) in a 1/1 (wt/wt) solution of water-methanol in a

Waring Blender. Avitene H is prepared from bovine hide collagen by a process described in *Journal of Applied Polymer Science,* 11, 481 (1967). Collagen and water-methanol mixtures were charged to the blender, allowed to soak for 5 minutes and blended for 5 minutes at a transformer setting of 35 v, followed by 10 to 15 minutes at a setting of 90 v. The resulting dispersion was filtered through a 70 x 100 fiber count per inch nylon parchment and cast in the form of a film 100 mils thick on a wax-extended polyethylene sheeting taped to glass.

The film was dried overnight at 47°C and then heat-treated at 150°C for 6 hours to effect crosslinking. Three circles of the film were water-wet and mounted on No. 50 Whatman filter paper in a reverse-osmosis unit and tested at 800 psi using 10,000 ppm magnesium sulfate feed solution. The 1.45 mil wet-thick film held up well with actual fluxes of 5.0 to 5.2 gfd and 74.4 to 75.4% magnesium sulfate rejection.

A limited life study was continued by allowing the film to stand in contact with the feed solution for 24 hours after the initial shutdown. It was then restarted, run for an additional 1.6 hr and shut down for 72 hr and then restarted again. Over this period, the actual flux increased slightly to 5.8 gfd and the magnesium sulfate retention fell to 69.3% indicating only a slight deterioration of reverse-osmosis properties.

The calculated fluxes of the collagen membranes based on wet thickness of 2500 A ranged from 734 to 822 gfd with magnesium sulfate rejections of 69.3 to 75.4% over the same time range. Cellulose acetate membranes of the same wet thickness would give only 10 to 12 gfd under similar conditions. Thus, the high fluxes demonstrated for collagen membranes make them unique for magnesium sulfate desalination.

Example 2: Collagen films were cast on solvent-wet Metricel filters of 2000 and 75 A average pore size to give a membrane consisting of a collagen film bonded to the surface of the supporting filter. Collagen casting dispersions (0.64% microcrystalline collagen with 0.056% hexamethoxymethylmelamine crosslinking agent) in a 1:1 water-methanol mixture were cast on the solvent (1:1 water-methanol) filled Metricel filters followed by air drying and heat treatment at 150°C for 6 hours. The composite membranes were tested in a reverse osmosis cell at 800 psi pressure with 1.0% sodium chloride feed solution. The details are presented in the following table. The above composite membranes would be expected to give much higher retentions using a brackish water feed solution containing a preponderance of divalent ions such as Ca^{++} and Mg^{++}.

Cast wet thickness, mils	Dried film thickness, mils	Filter pore size, A.	Operation hours	Actual flux, gf.	Calculated flux (2,500 A. thick) Dry	Calculated flux (2,500 A. thick) Wet	Salt retention T percent
50	6.1	2,000	0.50	9.3-9.7	228, 236	290, 299	27.7, 24.8
			0.95	9.6-9.8	234, 238	297, 302	24.7, 22.5
			1.30	9.7-10.0	237, 244	301, 308	28.5, 26.2
			3.10	9.3-9.7	228, 236	290, 299	25.4, 23.9
30	3.65	75	3.10	0.27	4.0	5.1	79.4

Particulate Filled Polymers

C. Berger; U.S. Patent 3,567,666; March 2, 1971 describes a composite product which consists of a polymer and particulates with the latter serving as a selective barrier to the passage of certain molecules and ions. This is possible because of the special uniform pore structure of the particulates and a continuous channel structure in the polymer. The complementary properties of the granular particulates and the channel-containing film gives rise to unexpected properties in the separation of ions from solvents, gas separation and applications in galvanic cells.

Typical among such applications is the use of a film or tubular form in reverse osmosis. Cellulose acetate is usually employed in conventional systems and seriously deteriorates over a period of time due to the high pressures employed and when applied to high saline water such as seawater. The films of the process are not affected by the high pressure which cause film collapse in cellulose acetate and can readily be employed in highly saline solutions.

The channels which envelop the particulates must be equal to or less in diameter than the pores in the particulates in order to prevent the separable species from finding an alternate migration route through the separating means. The particles in question may be inorganic, polymeric or metallic. The preferred materials are those consisting of predominantly silicon dioxide such as these constituted of at least 95% silica. The pore structure of some of these materials is produced by leaching out certain molecular phases of specific glasses.

One example of such a material is Corning Vycor 7935. The pore diameter of this material is 35 to 45 angstroms but suitable materials such as Porasil exist which extend the pore diameter range from 20 to 1500 angstroms. The chemical composition of Vycor 7935 is 96% silicon dioxide and about 4% B_2O_3, Al_2O_3, Cr_2O_3. The composition of Porasil is essentially 100% silicon dioxide. In both of these materials the pore diameters are uniform with little variance.

Other materials that may be employed are inorganic materials such as molecular sieves, microporous Teflon (polytetrafluoroethylene) and polyvinyl chloride as well as porous aluminum and stainless steel. The pore diameter range will fall approximately in the range described for Porasil and must be reasonably uniform. For all materials the gross diameter of the particles themselves must be five to ten times the diameter of the pores contained within the particle itself. A suitable mesh size for the gross particle is less than 325 mesh. The following examples illustrate the process.

Example 1: To 5.0 grams of a 20% polyvinylidene chloride solution in dimethylacetamide are added about 1.0 cm^3 of dimethylacetamide. The additional solvent is required so that uniform mixing can occur. 3 grams of powdered 96% silica Vycor glass is added to the solution and after uniformly dispersing it, the mixture is poured into a Petri dish and placed in an oven at 100°C for 0.6 hour.

At the end of this period the film is immersed in water for 2.0 hours and then removed for testing. The disc is 8.7 cm in diameter and 0.5 mm in thickness. The film is strong and slightly flexible. After trimming the film is placed in a reverse osmosis cell as depicted in *Membranes, Technology and Economics* by R.N. Rickles, p 106, Noyes Data, 1967.

At a pressure of 40 atmospheres and a feed solution of NaCl of 5,000 ppm a salt rejection of 75% was obtained in the product effluent. Two experiments identical to the preceding were performed except that Georgia kaolin and olivene were employed instead of the Vycor glass powder of the example. In several experiments a salt rejection of 0 to 10% were recorded. Natural materials such as kaolin particulates have nonuniform pores, cracks and fissures thereby having an inefficient ability to screen or hinder ion transport while uniform molecular channels can indeed screen efficiently.

Example 2: The experiment was similar to Example 1 except that 9.0 grams of powdered glass was used and 3.0 cm^3 of dimethylacetamide. The salt rejection was about 80%.

Example 3: An experiment similar to Example 1 was performed except that no heat was applied prior to immersion. The film was not as strong and was more flexible than that of Example 1. Moreover, the salt rejection was 65%. Such films may be reinforced by pouring the original polymer solution onto a synthetic or natural fiber mat which serves as a backing after film formation. Salt rejection remains about the same when a 0.125 mm mat of Webril (polypropylene) is used.

Example 4: The experiment was similar to Example 1 except that Porasil Type A glass (100 A pore diameter) was used. The salt rejection was about 70%.

Example 4: An experiment similar to Example 4 was performed except that polyphenylene oxide was employed as the polymer and chloroform as the solvent. Salt rejection rates were about 70%. If the oven temperature for the preceding is maintained at 125°C for two hours the salt rejection rate averaged about 75%. Moreover, if only 3.0 grams of the polymer solution is employed (or approximately a 50% by weight loading of inorganic particulates) then the rejection ratio is diminished to about 10 to 40%.

Example 6: Using ingredients described in Example 2 a tube about one inch in diameter and nine inches long was formed and then processed as in Example 1. Employing one tube of the reverse osmosis apparatus as described in *Membranes Technology and Economics,* p 110, a salt rejection in one stage of about 78% was achieved.

Red Blood Cells

In a process described by *L.S. Meriwether; U.S. Patents 3,593,852; July 20, 1971 and 3,736,204; May 29, 1973; both assigned to American Cyanamid Co.* brackish or saline water may be purified by forcing the water under comparatively high pressure through a reverse osmotic membrane consisting of the reconstituted membranes derived from living cells such as red blood cells. The permeability characteristics of the external membrane of the red blood cell have been studied at length (see Passow, Herman "Ion and Water Permeability of the Red Blood Cell." In *The Red Blood Cell,* edited by C. Bishop and D.M. Surgenor, page 71, New York: Academic Press, 1964.)

The intact red blood cell is so small that the membrane surrounding it cannot be used as such as a useful dialysis membrane, having an area of only about 150 square microns. It has been found that the ghosts or cell membranes of red blood cells can be mechanically subdivided by the use of ultrasonic energy so that the cell membranes or ghosts are fragmented mechanically without unduly altering their physical or chemical characteristics. The resulting fragments are believed to be lipoprotein subunits of the parent cell membranes.

The fragmented cell membranes are then reconstituted into a fragile membrane with a thickness of the order of from about 50 to 300 angstroms. These reconstituted membranes possess the same uniquely useful permeability characteristics as the membrane of the intact blood cell. These reconstituted membranes are supported on an intermediate support layer which has a great permeability to water, but sufficient mechanical strength to support the reconstituted membrane formed from the cell walls; and the reconstituted membrane on the intermediate support layer is mechanically supported by a strong grid which may be silica, porcelain, sintered glass, porous sintered metal or a fine woven metallic screen of about 20 to 200 mesh, or a perforated metallic plate.

The remarkable selective permeability to ions of the plasma membrane of living cells, which membranes have a thickness of the order of 100 angstroms, is utilized by reconstituting such membranes in useful sizes, and supporting the reconstituted biological membrane on a very permeable but mechanically stronger support layer which may be porous cellulose acetate or similar material, which in turn is supported by a mechanical grid.

The plasma membranes of living cells have remarkable permselectivity properties. Although freely permeable to water and small anions, they have a very low permeability to small cations such as Na^+ and K^+. Typically, cell plasma membranes consist mainly of protein and phospholipids and may be classified as lipoprotein in nature. A representative type of cell plasma membrane is that of the red blood cell, which consists of 45% protein, 45% phospholipid plus cholesterol and 10% polysaccharide. The protein and lipid are tightly complexed together since mechanical disruption of the membrane produces lipoprotein subunits which are not readily degraded further except by chemical treatment as, e.g., with detergents.

The mechanically disrupted subunits produced by the ultrasonic disruption of the cell membranes reaggregate at an interface in the presence of certain salts into large continuous membranes having a thickness and composition similar to that of the intact cell membrane, but which have such size and shape as to be useful for many purposes. For use in the fields of biology and medicine because of the biological natures of the materials, it may be important that the membrane have originated from a particular species. For the purposes of forming osmotic membranes, the red blood cells of mammals in general are satisfactory. For instance, in medical applications or tests human blood cells would be preferable in order that the characteristics be that of the human cell membrane, but for osmotic purposes either the human blood cell or more conveniently the blood cells of cattle, sheep, pigs, or other animals from slaughterhouses may be used and are an economical source of such red blood cells.

Figure 4.1 is a representation of the membrane of this process. The drawing is not to scale because no single drawing could show both the permselective membrane from the living cell wall and the mechanically strong grid with both to scale. The permselective membrane **11** is of the order of 50 to 1000 angstroms, as indicated, and as reconstituted from living cell walls is diagrammatically represented. Representations of such cell membranes are idealized at best because the thickness is such that it could only be examined with an electron microscope, and any such electron microscope would either be working with a replicate or a dehydrated cell membrane which would undoubtedly change in characteristics during preparation for study by the electron microscope.

The intermediate support layer **12** is again illustrated, but to a different scale as the thickness of this layer may run from less than 1/100 of a millimeter to around a millimeter or more. This layer is approximately 1,000 times as thick as the permselective membrane itself. The third layer, or support layer **13**, is also shown but again, here, while shown as a perforated plate, the size and thickness can vary with the size of the cell being used. It is this layer which furnishes the primary gross mechanical support and for large dialytic areas may have any of the configurations currently used in dialytic cells or filter cells in which the reinforcement is adequate for the type of pressure involved. This may be a mesh screen of 200 or 300 mesh supported by a perforated steel plate or a sintered plate or a thin, perforated plate, again supported by a solid plate having run off grooves or as may be chosen for the particular dialytic cell in question.

FIGURE 4.1: MEMBRANE STRUCTURE CONTAINING RED BLOOD CELLS

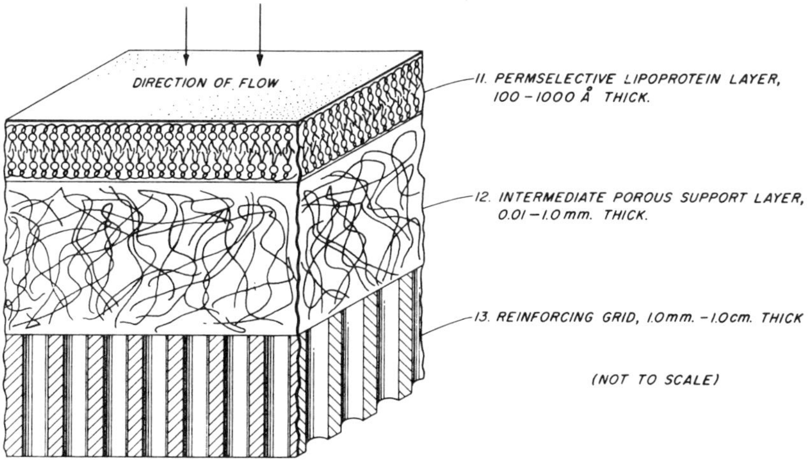

Source: L.S. Meriwether; U.S. Patent 3,736,204; May 29, 1973

For large installations, balanced type plates are used in which by analogy to conventional filter plate techniques, the gross mechanical pressure is counterbalanced by having a membrane on each side of a saline water feed plate and a reinforcing plate having grooves for the expression of purified water to a purified water-collecting trunk.

The balanced pressures from adjacent cells permit the plate to be primarily subject to balanced pressures with only the end plates of a stack having sufficient rigid massive reinforcement to withstand the osmotic pressures over the large cell areas which are employed for major saline water purification operations.

Example 1: Isolation of Membranes – Human blood (440 ml) from a patient with polycythemia vera was collected into 110 ml of ACD solution (each 100 ml of ACD contained 1.32 g sodium citrate, 0.44 g citric acid, and 1.47 g dextrose). Red cells were freed of plasma by being washed twice in 1:10 dilutions with solution A [0.14 M NaCl, 0.001 M EDTA, 0.005 M tris(hydroxymethyl)aminomethane (Tris) adjusted to pH 7.4 with hydrochloric acid] and by being harvested after each wash by centrifugation at 4,000 gravities for 5 minutes.

The ghosts were prepared by a modification of the hemolysis procedure of J.T. Dodge, C. Mitchell and D.J. Hanahan, *Arch. Biochem. Biophys,* 100, 119 (1963), in which the phosphate buffer was replaced by Tris and EDTA. Cells were hemolyzed by a 1:20 dilution in solution B (0.007 M NaCl, 0.001 M EDTA, 0.005 M Tris, pH 7.4). After hemolysis and each wash, the ghosts were harvested by centrifugation at 14,000 gravities for 10 minutes. The entire isolation procedure was carried out at 4°C.

Freshly packed ghosts were diluted with an equal volume of solution C and 40 ml aliquots were disrupted in 50 ml beakers in an ice bath using a model S75 Bronson Sonifier with a precooled ⅜ inch probe. The membranes were disrupted by two successive 30 sec ultrasonic bursts separated by a 3 min cooling period, using tap 6 of the Sonifier. During the sonication procedure, the sample temperature remained between 2° and 10°C. Examination of the sonicated sample on a Leitz Laborlux microscope with phase contrast optics at 400X demonstrated that no intact ghosts remained.

A 100 ml portion of the above sonicated ghost sample was brought to 10^{12} M in $CaCl_2$ by the addition of a 1 M solution of $CaCl_2$ and then shaken vigorously in air for 15 minutes at 25°C. The resulting reaggregated membrane suspension was then layered on an Eastman HT-00 cellulose acetate film by suction filtration of the suspension through the film while the latter was mounted in a Pyrex Millipore filter holder. The coated film was washed thoroughly with water and then mounted and tested in a reverse osmosis test cell. The membrane displayed a salt rejection of greater than 90% when 3,000 ppm NaCl solution was passed through it at 1,000 psig.

Example 2: Another portion of freshly packed ghosts, prepared as in Example 1, was sonicated as in Example 1 except that greater membrane disruption was induced by treatment with 20 successive 30 seconds ultrasonic bursts, each separated by a 3 minute cooling period, using tap 8 of the Sonifier. Examination of the sonicated sample in the Leitz microscope under phase contrast at 2,000X and in an RCA 3F Electron Microscope at 20,000X demonstrated that no intact ghosts or vesicles were present in the sample. The sample was then centrifuged at 90,000 x *g* for 2 hours at 0° to 5°C in a Spinco Model L Ultracentrifuge. The yield of soluble membrane lipoprotein present in the clear supernatant solution was 80%.

A small aliquot of the above soluble lipoprotein solution was injected at the surface of a 10^{12} M $MgCl_2$ solution contained in a Langmuir trough. The resulting reconstituted lipoprotein membrane formed at the air-water interface was compressed to 20 dynes/cm and a series of ten layers of membrane were coated onto an Eastman RO-89 cellulose acetate film by the Langmuir-Blodgett dipping technique. The coated film was then tested for desalination of seawater at 1,500 psig in a reverse osmosis test unit and was found to produce potable water.

INORGANIC MEMBRANES

Radio Frequency Sputtering Technique Using Platinum and Silicon Nitride

G.L. Booman and J.R. Delmastro; U.S. Patent 3,794,174; February 26, 1974; assigned to the U.S. Atomic Energy Commission describe a membrane for use in reverse osmosis and electrodialysis separation processes comprising a thin metal layer with a thin layer of insulator material on each side thereof, the membrane being porous with the pores having uniform radii from 5 angstroms and lower to approximately 100 angstroms, and means associated with the metal layer for controlling the surface charge density by external electrical circuitry whereby ionic transfer through the pores is controlled.

The membrane is made by depositing sequentially by radio frequency sputtering technique 5 to 500 angstrom layers of platinum and silicon nitride on a 50 to 100 micron thick glass substrate to form a thin sandwich. Then there are chemically etched 300 micron diameter holes in the glass substrate by using a mask of sputter-etched molybdenum metal with a thin molybdenum layer functioning as a stopping layer for the etching step.

Next the molybdenum masking layer is removed and, using the 300 micron holes in the substrate as a pattern, the molybdenum stopping layer is chemically etched through. Next the resulting sandwich structure is irradiated with fission fragments to produce damage tracks in the Si_3N_4 layers, after which these layers are chemically etched along the fission fragment damage tracks. Finally the metal membrane layer is chemically etched through, using the small holes in the surrounding Si_3N_4 dielectric layers as a mask.

Presintered Metal Oxide Membranes for Fuel Cells

C. Berger and F.C. Arrance; U.S. Patents 3,490,953; January 20, 1970 and 3,462,314; August 19, 1969; both assigned to McDonnell Douglas Corporation describe the construction of an ion exchange membrane of high transverse strength and high conductivity provided, and especially suited for application in fuel cells, by presintering a substance selected from the group consisting of water-insoluble hydrous metal oxides and water-insoluble acid salts, and an inorganic acid when the substance is the hydrous oxide, and mixing the presintered material with an inorganic acid, e.g., phosphoric acid, and a material referred to as a water balancing agent.

It has also been found that by further treating the resulting mixture including presintered material, with an inorganic acid followed by additional sintering, ion exchange membranes of substantially higher transverse strength can be obtained without any sacrifice of ionic conductivity.

Thus, inorganic ion exchange membranes are produced which have a transverse strength of at least 4,000 psi and which can range from 4,000 to 10,000 psi, and which have a transverse strength of the order of 8,000 to 10,000 psi where the presintering operation described is followed by a subsequent sintering operation. The high strength ion exchange membranes thus formed have good to excellent ionic conductivity.

Thus, for example, the ion exchange membranes produced according to the process can have a resistance of less than 30 ohm/cm^2 at 25°C, e.g., ranging from 3 to 20 ohm/cm^2, and often less than 10 ohm/cm^2, at 35°C. On the other hand inorganic ion exchange membranes produced by prior art procedure have a transverse strength ranging generally from 500 to 2,500 psi and a conductivity corresponding to a resistance of 50 to 150 ohm/cm^2.

Fuel cell membranes produced according to the process and of high strength and high conductivity have been found to be particularly advantageous for use in fuel cells particularly hydrogen-oxygen and hydrocarbon fuel cells at varying temperatures. Thus, for example, such ion exchange membranes have functioned efficiently in hydrogen-oxygen fuel cells at temperatures of the order of 65°C, and also function satisfactorily in hydrocarbon fuel cells operating at temperatures of the order of 100° to 125°C.

One of the starting materials employed according to the procedure for producing the ion exchange membrane is a water-insoluble hydrous metal oxide. Examples of such hydrous oxide ion exchange or ion conducting materials which can be employed preferably include the insoluble hydrous oxides of zirconium, titanium, antimony, tungsten, silicon, scandium, bismuth, vanadium, molybdenum, chromium and aluminum.

The term "insoluble hydrous metal oxides" includes those water-insoluble solids containing one or more metal cations, oxide ions, hydroxide ions, and an indeterminate quantity of water, and includes hydrous hydroxides. Such hydrous metal oxides do not necessarily have a definite stoichiometric combination or a definite crystal structure and they may contain ionic impurities as well. Examples of additional hydrous metal oxides are given in U.S. Patent 3,437,580.

The preferred starting materials are the insoluble hydrous metal oxides. When such starting materials are employed, such material is mixed with an inorganic acid prior to the initial sintering operation. Such inorganic acids can be any of those used to form the above noted acid salts, and including phosphoric acid, boric acid, molybdic acid, phosphomolybdic acid, tungstic acid, and sulfuric acid. The acid, preferably phosphoric acid, can be employed in stoichiometric amount with respect to the hydrous metal oxide to form the corresponding acid salt, but an excess of oxide or acid can be used. According to one mode of operation, approximately equal parts by weight of the metal oxide and acid can be used.

The presintering operation is carried out by subjecting a granulated or compacted mixture of the hydrous metal oxide and inorganic acid or of the acid salt, to temperature ranging from 200° to 1000°C, preferably between 300° to 600°C. After presintering, the resulting presintered material composed essentially of an acid salt, e.g., zirconium phosphate, is treated with an inorganic acid of the above type, preferably phosphoric acid, and there is also added a so-called water balancing agent. These water balancing agents are inorganic additives of controlled water vapor characteristics capable of retaining water and providing suitable water vapor pressure especially at temperatures above 100°C when such ion exchange membranes are incorporated in a fuel cell.

Any water balancing agent can be incorporated in the ion exchange or ion conducting material which will balance the amount of water in the membrane at a given temperature, and particularly at temperatures above 100°C to provide maximum conductivity of such membrane at such temperature. Thus, water balancing agents can be employed which when present in an inorganic membrane incorporated in a fuel cell provide a water vapor pressure of from 10 to 200 mm at 100°C and atmospheric pressure. Examples of suitable water balancing agents which can be employed together with the ion conducting material include aluminosilicates, activated alumina, aluminum sulfate, silicic acid, colloidal silica, silica gel, phosphorus pentoxide, copper sulfate, ammonium acid phosphate and calcium chloride.

The preferred water balancing agents for the process are the aluminosilicates such as those marketed as Zeolite, Zeolon and the like, and including, for example, sodium and potassium aluminosilicates, and magnesium, calcium, barium and strontium aluminosilicates. These materials can be used separately, but often mixtures of these aluminosilicates are used, for example, complex mixtures of both the alkali metal and alkaline earth metal aluminosilicates. Ion exchange membranes containing a water balancing agent of the type described above, are described in U.S. Patent 3,497,389.

Referring to Figure 4.2, the fuel cell **11** comprises a pair of back plates **10** which when assembled hold together a pair of adjacent neoprene gaskets **12** and **13** with the ion exchange membrane of enhanced strength and conductivity, designated **14** sandwiched between the gaskets **12** and **13**. In this embodiment the ion conducting membrane **14** is composed of zirconium phosphate containing an aluminosilicate as a water balancing agent. The assembly of members **10, 12, 13** and **14** can be accomplished by use of any suitable adhesive or glue.

The central portion of the ion conducting membrane **14** is covered or coated with a platinum black catalyst on both sides of the membrane, indicated at **20** and **21**. Prior to assembly of members **10, 12, 13** and **14**, tantalum screens **18** and **19** impregnated with platinum black, and preferably also Teflon for waterproofing purposes, are placed in the central portion of gaskets **12** and **13**, respectively, with the peripheral edges of the screen positioned between membrane **14** and the respective gaskets **12** and **13**. Following assembly of the above noted parts, it will be seen that chambers **16** and **17** are formed on opposite sides of the ion conducting membrane **14**, chamber **16** containing the screen **18** and the catalyst electrode **20**, and chamber **17** containing screen **19** and the catalyst electrode **21**. The screens **18** and **19** are of a corrugated or mesh material.

FIGURE 4.2: FUEL CELL EMPLOYING INORGANIC ION EXCHANGE MEMBRANE

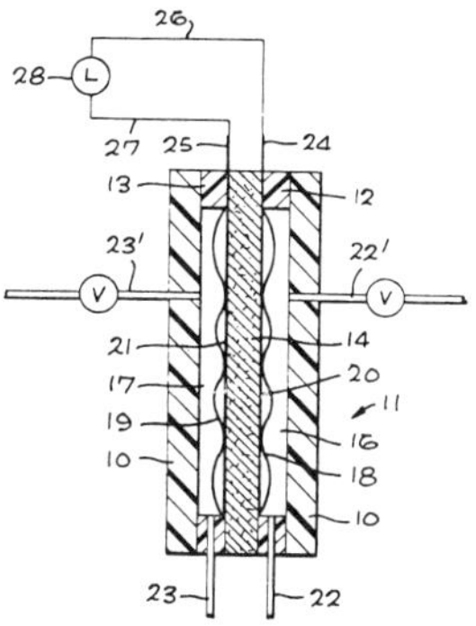

Source: C. Berger and F.C. Arrance; U.S. Patent 3,490,953; January 20, 1970

The fuel cell **11** is provided with a valved inlet **22'** to chamber **36** for passage of an oxidizer, e.g., oxygen gas, into such chamber, and a valved inlet **23'** to chamber **17**, for passage of hydrogen in the case of a hydrogen-oxygen fuel cell, or a hydrocarbon, e.g., ethane, in the case of a hydrocarbon fuel cell, into such chamber, via a conduit **29**.

A first conduit **22** which passes through gasket **12** and communicates with chamber **16**, serves for removal of excess water and excess oxygen gas from chamber **16**, and a second conduit **23** which passes through the gasket **13** and communicates with the opposite gas chamber **17**, serves as an outlet for excess hydrogen, or for excess hydrocarbon and carbon dioxide in the case of a hydrocarbon fuel cell, from chamber **17**. Terminals **24** and **25** are connected respectively to the tantalum screens **18** and **19**, such terminals extending exteriorly of the fuel cell.

Terminals **24** and **25** are connected in an external circuit including the electrical wires **26** and **27** and a load indicated at **28**. Where the fuel cell described above is employed as a hydrogen-oxygen fuel cell, hydrogen in chamber **17** reacts at the catalyst electrode or anode **21** and is oxidized to form hydrogen ion which migrates through the ion conducting membrane **14** and reacts with hydroxyl ion adjacent the catalyst electrode or cathode **20**, which hydroxyl ion is formed by reduction of the oxygen in chamber **16** at such catalyst electrode or cathode, forming water.

Example 1: A mixture of equal parts of hydrous zirconium dioxide and phosphoric acid is prepared, dried and granulated. Such mixture is then presintered at a temperature of about 500°C for 20 hours. Then the presintered mixture is mixed with phosphoric acid and Zeolon H (an aluminosilicate) in a 1:1 ratio by weight of these three materials. The resulting mixture is compacted at about 5,000 psi pressure and formed into membranes.

These membranes formed from the presintered materials have a transverse strength of about 5,000 psi and have low electrical resistance of less than 10 ohm/cm^2 at room temperature. When incorporated in a fuel cell as described above and illustrated in Figure 4.2 and the fuel cell is operated as a hydrogen-oxygen fuel cell at 65°C such fuel cell operates effectively for more than 500 hours at 0.5 volt and a current density of 20 ma/cm^2.

In comparison with the above procedure, membranes are prepared by mixing hydrous zirconium dioxide, phosphoric acid and Zeolon H in a 1:1:1 weight ratio of such components. This mixture is compacted into membranes and such membranes sintered at 500°C. The resulting membranes have a transverse strength of only 2,500 to 2,700 psi, as compared to twice that value or 5,000 psi for the membranes prepared by the presintering procedure.

Example 2: The membranes formed in Example 1 by the presintering procedure are saturated with H_3PO_4, and subjected to an additional sintering at about 500°C for a period of 60 minutes. The resulting membranes have a transverse strength of the order of 10,000 pounds per square inch as compared to 5,000 psi for the presintered membrane. These high strength membranes produced by the second sintering operation, however, also have a low electrical resistance of the order of the electrical resistance of the presintered membranes of Example 1, i.e., an electrical resistance less than 10 ohm/cm^2 at ambient or room temperature.

An ion exchange membrane produced by this dual sintering procedure is sprinkled on opposite sides with platinum black and employed as the ion exchange membrane in a fuel cell as described above with tantalum screens impregnated with platinum black pressed against opposite sides of the membrane as described above. The fuel cell operates effectively employing hydrocarbon fuels such as ethane, propane and butane and oxygen as oxidizer, at temperatures of 100° to 125°C.

In related work, *C. Berger and A.D. Kelmers; U.S. Patent 3,497,389; February 24, 1970; assigned to McDonnell Douglas Corp.* describe inorganic ion exchange membranes containing a water balancing agent such as the aluminosilicates.

Metal Oxide Membranes for Fuel Cells

C. Berger; U.S. Patents 3,392,103; July 9, 1968 and 3,497,394; February 24, 1970; both assigned to McDonnell Douglas Corporation has found that certain strong ceramic membranes formed from inorganic base materials of insoluble hydrous metal oxides and acid salts, upon treatment with acid or alkaline solution, at high pressures and temperatures may be converted to ion exchange membranes having substantially the transverse strength of the original ceramic membrane.

The ion exchange membranes thus formed, are particularly adapted for use in fuel cells and batteries where extremely strong, ion selective, membranes are required to maintain ion separation between the electrodes of the battery or fuel cell and wherein operating temperatures may approach and exceed 125°C. Moreover, the method has the distinct advantage of allowing membranes to be stored in an inert form for indefinite periods of time without change and to be converted into high strength permselective membranes as needed by simple procedures.

Specific examples for hydrous metal oxides follow. In each of the following examples, the metal oxide starting materials forming the ceramic membranes were produced in accord with the methods described in U.S. Patent 3,437,580. The method of measuring resistivity can vary depending upon the particular system in which the membrane is placed. The resistivity of all the membranes in the following examples is given in terms of 90°C and 60% relative humidity (RH) for the sake of uniformity. Extrapolations have been made to 90°C and 60% relative humidity in many instances. The resistivity figures are therefore approximate.

Example 1: Group III Hydrous Metal Oxides — A flame sprayed aluminum oxide membrane having a resistivity of 3×10^6 ohm-cm, a modulus of rupture of 6,500 psi, and no ion exchange capacity was supported in a 10 liter autoclave containing 1 liter of a 50% hydrochloric acid solution. The alumina membrane was then exposed to steam at 1,000 psi and approximately 300°C for 325 hours. After exposure, the membrane had a resistivity of 35 ohm-cm, an ion exchange capacity of 1.0 meq/g and a modulus of rupture of 5,200 psi.

Example 2: Group III Hydrous Metal Oxides — A porous ceramic membrane 2" in diameter and 0.02" thick was prepared from scandium oxide (Sc_2O_3) by compacting at 20 tons total load and sintering at 1800°C. As fired, the membrane had a resistivity of 2.7×10^6 ohm-cm, a modulus of rupture of 5,000 psi and no measurable ion exchange capacity. The ceramic membrane was supported in a 10 liter autoclave containing 1 liter of a 35% phosphoric acid solution and exposed to superheated steam at 2,000 psi and approximately 340°C for 250 hours. After exposure to the steam, the membrane had a resistivity of 50 ohm-cm, a modulus of rupture of 4,700 psi, and an ion exchange capacity of 1.05 meq/g.

F.C. Arrance and C. Berger; U.S. Patent 3,437,580; April 8, 1969; assigned to McDonnel Douglas Corporation describe the preparation of hydrous metal oxide membranes.

Thorium Hydrous Oxides

K.S. Rajan and A.J. Casolo; U.S. Patent 3,479,266; November 18, 1969; assigned to U.S. Secretary of the Interior describe a process for separating electrolytes from their solutions by employing inorganic ion exchange membranes which have greater stability to high temperatures, chemical attack and fouling than do organic membranes customarily used in such processes. Membranes useful in this process comprise a mixture of thorium hydrous oxide with a hydrous oxide of a second metal and those membranes are produced by coprecipitating a mixture of two metal salts together with a binder material followed by hydrolysis to form the hydrous oxide of the metals.

Due to the ability of the membranes to remain stable at relatively high temperatures, to resist degradation and fouling and to remain stable under corrosive and oxidizing conditions, they have great utility in electrical membrane separatory processes; particularly in electrodialysis and transport depletion processes. Preparation of the coprecipitated hydrous oxide membranes involves first dissolving a hydrolyzable salt of thorium and hydrolyzable salt of the coprecipitating cation in a solvent together with a matrix-forming or binding material. Any soluble hydrolyzable salt may be used but the nitrate is preferred because of its solubility and its general ease of hydrolysis.

The resulting mixture is then spread on a smooth surface, such as a glass plate, to form a uniform film of predetermined thickness. The film as cast conveniently is approximately 5 mils in thickness. After casting the film, it is heated in order to drive off substantially all of the solvent. Heating temperature is not critical so long as it is not sufficiently high to damage the matrix or binder material.

A temperature range of about 70° to 110°C is preferred. After driving off the solvent, the resulting film is exposed to a hydrolyzing agent, preferably aqueous ammonium hydroxide, to form the hydrous oxides of the elements present.

Matrix-forming or binder materials which may be used include polytetrafluoroethylene, polyvinylidene fluoride, polyvinyl butyral and silicon based polymers. Polyvinylidene fluoride is particularly preferred because of its excellent thermal and chemical stability and its good mechanical properties. The binder material comprises from about 30 to 50% by weight of the finished membrane.

The weight ratio of thorium hydrous oxide to the coprecipitated metal hydrous oxide may range from 20 to 1 to 1 to 1. Some variation of electrical properties occurs as the ratio of the two hydrous oxides change. A ratio of about 5 to 1 has been found to be particularly useful. The following example illustrates the process.

Example: Inorganic anion exchange membranes were prepared by coprecipitating the hydrous oxide of thorium with each of the hydrous oxides of magnesium, calcium, uranium, aluminum, chromium, zirconium, cerium, and tin. In each case, the procedure was as follows.

Ten parts by weight of thorium nitrate and 2 parts by weight of the coprecipitating cation as the nitrate salt were dissolved in 10 parts by weight of dimethyl formamide by heating to 90°C. To this solution was added 15 parts by weight of a 20% solution of polyvinylidene fluoride in dimethyl acetamide. The mixture was cast while hot into a film of 25 mils thickness on a glass plate. The film was dried in an oven at 90°C for 2 hours in order to drive off substantially all of the solvent. After drying, the film was hydrolyzed in a 10% aqueous solution of ammonium hydroxide in order to form the hydrous oxides of the coprecipitated metals. Electrical properties of the membranes are set out in the following table.

Electrical Properties

Membrane	Resistance ohm-cm.^{2}	Transference No. (0.05 to 0.10 M KCl)
Thorium-calcium hydrous oxide	3.0	0.88
Thorium-magnesium hydrous oxide	2.5	0.90
Thorium-uranium hydrous oxide	2.5	0.87
Thorium-aluminum hydrous oxide	3.0	0.82
Thorium-chromium hydrous oxide	3.5	0.86
Thorium-zirconium hydrous oxide	3.0	0.86
Thorium-cerium hydrous oxide	2.0	0.93
Thorium-tin hydrous oxide	2.5	0.90

Electrical resistances of the membranes were determined by measurement using a conductance bridge and a standard cell having a cross-section area of 5 cm^2. The procedure consisted of placing the equilibrium membrane in the cell, filling the cell with 0.5 M NaCl and measuring the resistance with the conductance bridge. Membrane resistance was taken as the difference in cell resistance with and without the membrane.

Membrane transference numbers reported are the transference numbers of chloride ions through each membrane. They were determined by the procedure described in Office of Saline Water Test Manual for Permselective Membranes, *OSW Research and Development Progress Report No. 77,* 1964. The membranes were found to have good stability when used in the electrodialysis of a sodium chloride brine at 60°C and were nonfouling when exposed to a low concentration of dodecyl benzene sulfonate.

In related work, *J.I. Bergman, D.E. Anthes and R.S. Braman; U.S. Patent 3,463,713; August 26, 1969; assigned to the U.S. Secretary of the Interior* describe the demineralization of solution by electrodialysis using a zirconium phosphate cation exchange membrane and a thorium oxide anion exchange membrane.

Aluminum Vanadate and High Temperature

J.A. Bittles; U.S. Patent 3,499,537; March 10, 1970; assigned to McDonnell Douglas Corporation describes a process for preparing an aluminum vanadate ion exchanger which comprises dissolving a soluble aluminum salt and vanadium pentoxide in a solution selected from an aqueous acid or base solution and precipitating an aluminum vanadate ion exchanger from the acid solution by addition of base, or from the base solution by addition of acid. The precipitate is removed from the solution and preferably treated or digested to remove any soluble salt.

The resulting aluminum vanadate in particulate form is dried, and the dried powder is pressed and sintered at 2,000 to 20,000 pounds per square inch and 500° to 800°C for 2 to 8 hours to produce an aluminum vanadate ion exchange membrane. Alternatively, the dried aluminum vanadate powder can be mixed with a fusible ceramic, followed by pressing and sintering to produce an aluminum vanadate ion exchange membrane.

The aluminum vanadate ion exchanger which is precipitated from acid solution, that is, by addition of a base, as described above, functions as an aluminum vanadate cation exchanger, and the aluminum vanadate which has been precipitated from a base solution, that is, by addition of acid as noted above, functions as an anion exchanger. The procedure provides an efficient sintered aluminum vanadate ion exchange membrane, which after sintering still retains the high ion exchange activity and conductivity characteristics of the particulate or powdered aluminum vanadate, is highly resistant to deterioration and fouling, and which remains stable and operable, retaining its ion exchange and permselective properties at elevated temperatures above 125°C, e.g., up to about 900°C and above.

The aluminum vanadate ion exchange membranes produced according to this process have varying thicknesses, e.g., ranging from 0.005 to 0.050 inch. These membranes have high strength, e.g., a transverse strength of 1,000 to 5,000 psi, have high anion or cation transport numbers for the anion or cation exchange membranes, respectively, e.g., ranging from 0.8 to 0.99, high current efficiency ranging from 85 to 95% and high electrode efficiencies of from 80 to 99%, and operate efficiently as ion exchange membranes at elevated temperatures up to 900°C for prolonged periods of operation without deterioration or fouling as a result of contact with corrosive chemical solutions or subjection to such high temperatures.

Such membranes are not only effective as electrodialysis membranes, but also can be used in fuel cells and batteries. Also, aluminum vanadate ion exchange membranes produced according to the procedure have the additional advantage, that they can be designed either as a cation or anion exchanger. The following are examples of the process.

Example 1: Aluminum oxide and vanadium pentoxide were dissolved in 1.0 molar HCl solution, in an amount to provide concentrations of aluminum (Al^{+3}) and vandium (V^{+5}) ions each in 0.1 molar concentration. Aluminum vanadate was then precipitated from such solution by adding a solution of 2.0 molar sodium hydroxide dropwise until no further precipitate was obtained. The gelatinous precipitate was then filtered and dried to produce particulate aluminum vanadate which functions as a cation exchanger.

Example 2: Aluminum chloride and vanadium pentoxide were each dissolved in 0.1 molar potassium hydroxide solution, in a concentration such as to provide aluminum and vanadium ions each in a concentration 0.1 molar. Aluminum vanadate was precipitated by adding 1.0 molar hydrochloric acid to the alkaline solution containing aluminum and vanadium ions, until no further precipitate formed. It was observed that a precipitate first commenced to form at a pH of about 10 and that precipitation was complete at a pH of about 7. The precipitate was filtered and dried. The particulate aluminum vanadate product thus produced functions as an anion exchanger.

Example 3: 70.6 millimols vanadium pentoxide and 141.2 millimols aluminum chloride were dissolved in 1 molar HCl solution. Aluminum vanadate was precipitated from such solution by adding 20% potassium hydroxide solution, with precipitation ceasing at a final pH of 4. The precipitate was removed from the solution and was first digested or treated with 0.01 molar, then with 0.1 molar hydrochloric acid solution and finally with water. The digested product was dried at 110°C, providing 20 grams of aluminum vanadate. Such aluminum vanadate product functions as a cation exchanger.

Examples 4 through 7: Procedure similar to that noted in Example 3 was carried out in each of Examples 4 through 7 as shown on the following page, employing the solutions and treatments outlined in the table on the following page for each of these examples.

It will be noted that in Example 4 as shown in the table on the following page, the digestive treatment produced a potassium salt form of the ion exchanger, in Example 5, a hydrogen salt form, in Example 6 a lithium salt form, and in Example 7, a sodium salt form. Each of the dried aluminum vanadate products of Examples 4 through 7 function as cation exchangers.

Ex.	Product Composition	Procedure	Digestions	Yields, grams
4	Aluminum vanadate precipitated from acid solution, potassium salt.	35.3 mmoles V_2O_5 in 1 molar HCl and 70.6 mmoles $AlCl_3$ in 1 molar HCl, precipitated with 1.98 mmoles KOH—final pH 4.	0.01 M HCl, 0.1 M HCl, 1 M KCl, and H_2O.	10
5	Aluminum vanadate precipitated from acid solution, hydrogen salt.	do	0.01 M HCl, 0.1 M HCl, and H_2O.	10
6	Aluminum vanadate precipitated from acid solution, lithium salt.	do	0.1 M HCl, 0.1 M HCl, 1 M Li Cl and H_2O.	10
7	Aluminum vanadate precipitated from acid solution, sodium salt.	do	0.01 M HCl, 0.1 M HCl, 1 M NaCl and H_2O.	10

Example 8: The aluminum vanadate (base precipitated) dried and powdered ion exchanger of Example 2 was placed in a die, and the powder pressed at pressures of 6,000 to 20,000 pounds per square inch to form membranes of a size 1.5 inch in diameter and having a thickness of 0.032 inch. The resulting pressed membranes were set upon ceramic tile, placed in a furnace and sintered at temperature of 600°C for approximately 4 hours.

Treatment of Silver Filter with Metal Salts, Polyelectrolytes and Humic Acid

J.S. Johnson, K.A. Kraus, A.E. Marcinkowsky, H.O. Phillips and A.J. Shor; U.S. Patent 3,449,245; June 10, 1969; assigned to the U.S. Atomic Energy Commission describe a method of separating an aqueous solution of metal salts into a first solution depleted in the metal salts and a second solution enriched in the metal salts comprising the steps of providing a permeable substrate having pores with diameters in the range of 30 A to 5 microns, contacting the substrate with an aqueous phase containing a dispersible material, thereby forming a finely pored permeable membrane on the substrate, and forcing a portion of the aqueous solution through the permeable membrane under pressure. In carrying out the process an aqueous phase containing a dispersible material is forced through a porous substrate.

The substrate may be treated with the dispersible material by contact before its contact with the aqueous salt solution or it may be treated by maintaining a low concentration of the additive in the salt solution. Even when a separate step is used to pretreat the substrate with the additive, some additive may be provided in the salt solution, and this procedure is preferred since it provides a source of material to heal defects in the membrane which may develop during processing.

The permeable substrate may be of any material capable of maintaining its integrity under the pressures involved and in the presence of the water and its dissolved salts. The chemical nature of the substrate is not significant, and materials of such widely diverse natures as metal filters, porcelain frits, porous carbon, glass frits, and organic materials such as highly permeable cellophane may be used. The diameter of the pores in the substrate may suitably range from 30 A to 5 microns. When the additive is a water-insoluble substance it is preferred that the pore diameter be about the order of the diameter of the additive particles.

If the additive is water-soluble it is desirable to use a substrate with as large a pore size as can be used that will restrain permeation by the additive. The additives used in the process do not pass through the pores of the substrate even though the molecular dimensions of these substances are small compared to the diameters of the pores in the substrate, and are thus able to form a salt-rejecting membrane on the substrate.

It has been found that many different substances having a wide variety of physical and chemical properties will function as a salt-rejecting membrane when an aqueous phase containing them is forced through a porous substrate. The additive may be of a nature to form a true solution, an emulsion or a suspension in water; it may be an electrolyte or a neutral material; and it may be either an organic or an inorganic substance. These additives may be more specifically characterized as falling into one of the following classes: neutral organic polymers, polyelectrolytes, organic ion exchangers, inorganic ion exchangers, and polyvalent metal salts.

Example 1: A permeable substrate was made by cutting disks 2 centimeters in diameter from a 40 micron thick silver filter having a nominal pore size 0.2 micron in diameter and mounting the disk supported by a porous metal frit in a hyperfiltration apparatus. This disk was then treated by passing through it an aqueous solution 0.02 molar in NaCl and 0.002 molar in $ThCl_4$. A solution of the same composition was circulated past the membrane at a pressure of 2,200 psi. The transmission rate through the barrier was over 330 gpd/ft^2 and the salt rejection was 90%. This demonstrates a high rejection coupled with a high permeation rate. The procedure in this example was typical of the procedure of Examples 1 through 13. Example 2 illustrates a modification in which a substrate is pretreated with a solution different from that whose rejection is being investigated.

Example 2: A porous silver disk similar to that of Example 1 was conditioned by forcing through it a pretreating solution 0.02 molar in NaCl and 0.001 molar in $ThCl_4$ (pH 2 to 3). The pressure used to force the liquid through the membrane ranged from 100 psi at the beginning to 500 psi at the end of the conditioning step. The permeation rate decreased from 15 centimeters per minute to 1 centimeter per minute during this pretreatment. A feed solution 0.025 molar in NaCl, 0.012 molar in $MgCl_2$, and 0.0001 molar in $ThCl_4$ was circulated past the conditioned membrane at a pressure of 400 psi.

The transmission rate through the membrane was 60 gpd/ft^2, and the salt rejection was 68%, based on total chloride analysis; separate analyses gave for $MgCl_2$ a rejection of 82% and for NaCl, 55%. Other experiments were carried out using other additives and other salts. The data for these experiments, together with the data for Experiments 1 and 2, are given in the table below.

Substrate – Silver: 0.2 Micron Nominal Pore Diameter

	Feed Solution						
Example Number	Salt		Additive		Pressure (psi)	Flow Rate (gpd/ft.^{2})	Salt Rejection (Percent)
1	0.02M	NaCl	0.002M	$ThCl_4$	2,200	330	90
2	[0.025M	NaCl]					
	[0.012M	$MgCl_2$]	0.001M	$ThCl_4$	400	60	68
3	0.02M	NaCl	0.002M	$ZrOCl_2$	2,100	320	85
4	0.02M	NaCl	0.002M	$FeCl_3$	2,200	180	30
5	0.03M	NaCl	0.002M	$La(NO_3)_3$[1]	600	530	14
6	0.02M	NaCl	0.002N	polyelectrolyte[2]	2,200	1,000	17
7	0.02M	NaCl	0.002N	polyelectrolyte[2]	650	150	65
8	0.05M	NaCl	0.01%	polyelectrolyte[3]	1,900	30	73
9	0.05M	NaCl	0.001%	polyelectrolyte[3]	1,900	65	55
10	0.025M	$MgCl_2$	0.001%	polyelectrolyte[3]	1,950	50	12
11	0.025M	Na_2SO_4	0.001%	polyelectrolyte[3]	2,000	40	86
12	0.05M	NaCl	0.01	wt. % polyvinyl pyrrolidone	1,850	30	18
13	0.05M	NaCl	0.0001	wt. % polyvinyl alcohol	2,000	80	17
14a	0.05M	NaCl	0.001	wt. % hydroxyethyl cellulose	1,900	45	20
14b	0.05M	NaCl	0.0001	wt. % hydroxyethyl cellulose	2,000	45	20
14c	0.02M	NaCl	0.001	wt. % hydroxyethyl cellulose	1,200	30	22
15	0.02M	NaCl	Bentonite[4]		1,850	180	35
16[5]	0.02M	NaCl	Bentonite[4]		1,700	120	40
17[5]	0.02M	NaCl	Cation exchange resin[6]		1,600	180	45

[1] The pH of this solution was adjusted to 7.5.
[2] Polyvinylbenzyl trimethyl ammonium chloride.
[3] Methyl vinyl ether-maleic anhydride copolymer which hydrolyzes in water to produce carboxylic acid groups.
[4] The NaCl solution was shaken with bentonite and then permitted to settle for 3 hours.
[5] The frit was a 1.2 micron silver frit.
[6] Dowex-50 0.5% cross-linked.

Examples 3 through 5: These examples in addition to Examples 1 and 2 demonstrate the use of hydrolyzable polyvalent ions in the preparation of salt-rejecting membranes from porous bodies. Except for Example 5, where the pH of the additive was first adjusted, the experiments were carried out at the natural pH of the additive-water mixture. The pH of the additive was, however, adjusted in a number of other cases. Thus, with Th(IV), experiments were carried out in the presence of 0.02 molar HCl and after addition of one and two mols of NaOH per mol of Th(IV). Acidification tended to decrease rejection of NaCl.

Addition of base in this case did not seriously affect rejection but had the advantage of permitting estabilshment of the salt-rejecting layer rapidly with substrates of larger pore diameter (1.2 microns compared with 0.2 micron). In these examples the additive was still in solution; precipitation of a hydrous oxide and its deposition on a porous substrate, however, should also lead to high flux rejection membranes. As far as the rejection with these materials is determined by their ion exchange characteristics, it should be recognized that variation of pH is not always beneficial since ion exchange characteristics vary with pH (for details see "Ion Exchange Properties of Hydrous Oxides," K.A. Kraus, H.O. Phillips, T.A. Carlson and J.S. Johnson, *Proceedings of Second United Nations International Conference on the Peaceful Uses of Atomic Energy,* 28, 3 to 16, 1958.)

In related work *K.A. Kraus, J.S. Johnson and A.J. Shor; U.S. Patent 3,334,928; October 3, 1967; assigned to the U.S. Atomic Energy Commission* have found that a membrane having excellent separative qualities is formed by providing humic acid as the dispersible material in an aqueous phase forced through a permeable substrate. The use of humic acid as the dispersible agent in the water to be treated has the advantages that it is present as a contaminant in most natural waters, and can be prepared easily. Once the membrane has been established by passing through the porous substrate water containing a relatively high concentration of humic acid or inert constituents whose rejection properties can be enhanced by humic acid, it can be maintained by the addition of little or no material to the water to be treated.

In carrying out the process a dynamic membrane is established by passing an aqueous phase containing humic acid through a porous substrate. As is described in U.S. Patent 3,449,245, the permeable substrate may be of any material capable of maintaining its integrity under the pressures involved and in the presence of water and its dissolved salts. The chemical nature of the substrate is not significant, and materials of such widely diverse natures as metal filters, porcelain frits, glass frits, and organic materials such as permeable papers may be used. Small-diameter porous carbon tubes will be especially useful.

The lower limit of humic acid concentration necessary to establish and maintain a rejection layer is not known. However, rejecting films have been established with as little as 2.5 ppm humic acid and experiments indicate that films can be maintained with as little as 0.1 ppm humic acid. While humic acid by itself forms a separative membrane, it has been found that superior results can be achieved with the use of an ion exchanger, either organic or inorganic, with the humic acid. The concentration of ion exchanger used to establish the membrane may be as low as 10 ppm. Once this membrane has been established its superior properties continue even in the absence of an ion exchanger in the feed.

Example: A permeable substrate was made by cutting disks from a 40 micron thick silver filter having a nominal pore diameter of 0.2 microns. These disks were supported by a porous metal frit and mounted in apparatus adapted to circulate a high pressure aqueous solution over the disks. Humic acid was obtained by soaking oak leaves in water for 5 to 7 days and decanting the resulting liquid. This liquid in acid-base titration was found to contain 3 to 4 meq of titratable material per ml. Evaporation of the liquid produced 3 mg of dry material per ml of liquid. Enough liquid was added to an aqueous solution 0.1 M in NaCl to give an additive concentration of 25 ppm. This feed solution was circulated past the disks at a pressure of 1,500 psi. The flow rate through the disks was 60 gpd/ft^2 and salt rejection 48%.

In a continuation of this work *J.S. Johnson and K.A. Kraus; U.S. Patent 3,503,789; Mar. 31, 1970; assigned to the U.S. Atomic Energy Commission* describe the use of weakly ionizable amines, such as polyvinyl pyridine.

Graphitic Oxide

In a process described by *L.C. Flowers and D. Berg; U.S. Patent 3,457,171; July 22, 1969; assigned to Westinghouse Electric Corporation* graphitic oxide is made by oxidizing graphite flakes. A suspension of the graphitic oxide is deposited on a porous supporting substrate to form a thin continuous film of graphitic oxide. A resinous polymer is employed as a binder for the graphitic oxide film. The supported film is inserted into a reverse osmosis pressure cell. A demineralized water product is obtained.

Colloidal Hydrous Oxide on Silver Frit

K.A. Kraus and J.S. Johnson; U.S. Patent 3,413,219; November 26, 1968; assigned to the U.S. Atomic Energy Commission describe a method of making a solute-rejecting permeable membrane which comprises passing an aqueous phase containing hydrous metal oxide colloidal particles through a permeable substrate having pores with diameters of 30 A to 20 microns.

In carrying out the process a dynamic membrane is established by passing an aqueous phase containing a colloidal hydrous oxide through a porous substrate. As is described in U.S. Patent 3,449,245, the permeable substrate may be of any material capable of maintaining its integrity under the pressures involved and in the presence of water and its dissolved salts. The chemical nature of the substrate is not significant, and materials of such widely diverse natures as metal filters, porcelain filters, sintered glass, porous carbon, and organic materials such as porous paper may be used.

The diameter of the pores in the substrate may range from 30 A to 20 microns and the preferred pore diameter range is from 0.1 to 5 microns. Any polyvalent metal salt capable of forming a colloidal hydrous metal oxide may be used to form a dynamic membrane in the process. Examples of typically useful salts are those of tri- and tetravalent metals such as iron, zirconium, tin(IV), and thorium. In view of earlier observations on ion exchange properties of hydrous oxides, it is expected that colloidal dispersions of the hydrous oxides or pentavalent metals [e.g., Nb(V) and Ta(V)] and of hexavalent metals [e.g., U(VI)] will also be effective.

In the process colloidal particles must be formed from the polyvalent metal salt, and methods of forming colloidal aggregates are known in the art. It has been found that excellent results may be achieved by heating an aqueous solution of the hydrous metal oxide or by adjusting the pH until a turbid solution is obtained. The concentration of colloidal hydrous oxides in the solution used to form the dynamic membrane varies somewhat with the oxide used, but usually ranges upward from 10 ppm.

After formation, the membrane will continue to have rejecting properties for a day or more, though rejection gradually decreases. If, however, a small concentration of additive, sometimes as little as 0.1 ppm, is kept in the solution, rejection properties are maintained. Continued presence of additive also repairs defects which may appear in the membrane.

Example 1: A permeable silver frit having a nominal pore diameter of 0.8 micron was mounted in hyperfiltration apparatus adapted to pass an aqueous solution under pressure across one face. A 0.02 molar NaCl feed was circulated past the frit at a pressure of 25 psi. The flow through the frit was over 35,400 gpd/ft^2 with no NaCl rejection. To this solution was added a 0.1 molar $FeCl_3$ solution, which had been boiled to form a colloidal dispersion, to bring the feed solution to 0.0005 molar Fe(III). At the end of two hours, the permeation rate was 70 gpd/ft^2 at 35 atmospheres and the rejection of NaCl was 60%. In a similar run using $FeCl_3$ the rejection of NaCl was 30%. As can be seen from Example 1 colloidal ferric oxide forms a salt rejection barrier having a high flow rate even at low pressure. Example 2 below shows the salt rejection and flow rates achieved using another hydrous metal oxide.

Example 2: A dynamic hydrous Zr(IV) oxide membrane was formed on a silver frit identical to that of Example 1 by contacting it with a 0.02 molar NaCl solution containing a colloidal dispersion of 0.001 molar Zr(IV). The colloidal hydrous Zr(IV) was formed by boiling a 0.25 molar $ZrOCl_2$ solution for several hours at which time precipitation became barely apparent. An aqueous solution 0.025 molar in $BaCl_2$ and 0.05 molar in HCl was then passed over the membrane at 500 psi. The rejection of $BaCl_2$ was 68% and the flow rate through the membrane was 130 gpd/ft^2. A result different from those in Example 1 was the observation that the concentration of HCl in the effluent was 11% higher than its concentration in the feed. This shows that separations of solutes from one another, as well as separations of solutes from solvent, are possible with this membrane.

Application to Discrete Particles

E.R. Brownscombe and L.R. Kern; U.S. Patent 3,331,772; July 18, 1967; assigned to the Atlantic Refining Company describe semipermeable membranes which are composed of numerous discrete particles rather than formed as integral films or sheets. The diameter of at least one dimension of these particles should be less than 100 μ and preferably less than 10 μ. In other words, there is a maximum cross-sectional thickness of the size noted in at least one direction. Also, a minimum size restriction is imposed. The finely divided particles should have an average diameter greater than 0.10 μ and preferably greater than 1.0 μ. Membranes prepared from particles much larger than 100 μ in overall size have reduced efficiency for demineralizing water while membranes comprised of particles smaller than about 0.10 μ have reduced flow rates (due to low permeability).

The membrane material may be prepared in the finely divided form by any of a number of size-reducing techniques or may be formed from solution by precipitation. However prepared, the finely divided osmotic material is suspended in a liquid which acts as a nonsolvent. Water is preferred for this purpose; however, organic liquids can also be used. After the resulting suspension is agitated over an interval sufficient to ensure that the dispersion is substantially homogeneous, the slurry is filtered on a selected porous support.

The liquid passes through the support medium as filtrate and the particulate membrane material is deposited as a filter cake or layer on the surface of the support. The support medium should be readily permeable to water and must be capable of filtering the finely divided osmotic material. Suitable support materials, for instance, are paper, cloth, coarse clays, sandstone, ceramics, finely woven metal mesh, perforated metal, etc.

The deposited membrane material should be compressed to reduce pore space and increase its strength. A preferred technique involves subjecting the membrane to an applied pressure of from 5 to 10 times the pressures which are encountered during the reverse osmotic process. Normally, pressures of 1 to 5 thousand psi suffice for this purpose and significantly increase the effectiveness of the membrane for desalting operations. After the membrane has been formed, it should be preserved by continuing to maintain the membrane under at least a small applied pressure, e.g., one hundred psi. This ensures that the membrane stays firmly pressured against the support member. Also, the membrane should always be kept moist with a liquid to prevent drying out and cracking.

This same technique may be used for repairing conventional (cast) semipermeable membranes. A suspension of finely divided osmotic material is added to the feed solution input to the membrane cell and the particulate material is deposited on the surface of the semipermeable membrane. If necessary, some large, inert particles can be added to bridge and facilitate the plugging of large holes. Repair can be effected while desalting operations are being carried out so that production need not be interrupted.

Example 1: Preparing the membrane is as follows. Finely divided bentonite having the particle size set forth above is used as the membrane material. The bentonite is treated with sodium polyacrylate, suspended in water, and agitated until well-mixed. The bentonite is then deposited as a sheet of discrete particles on a porous support which is permeable to water. Pressure filtration accomplishes this end by interposing the porous medium in such a way that the bentonite is retained while the liquid filtrate (water) passes through the pores of the support.

The bentonite membrane is compressed by a fluid under an applied pressure of several thousand psi. By a similar procedure semipermeable membranes are prepared from sulfonated polystyrene, cellulose acetate, and cellulose acetate propionate. Materials used as porous supports are perforated metal plates and tubes, filter paper, sandstone, and wire mesh screens.

Example 2: Repairing semipermeable membranes is as follows. When leaks or holes develop in the membranes prepared according to Example 1, the membranes are repaired by pressuring a slurry of finely divided membrane material through the damaged membrane.

Preferably, the material used for patching or coating the membrane is the same as the original membrane material; however, dissimilar materials may be employed where desired. Repair may involve forming a second layer on the original membrane but all that is necessary is that the hole or flaws be filled or covered. This repair procedure is not restricted to membranes comprised of numerous discrete particles but is equally applicable to conventional semipermeable membranes. Thus, a cast cellulose acetate membrane may be repaired by depositing a layer of particulate osmotic material on its surface.

Exfoliated Vermiculite

According to a process described by *A.S. Patil, R.G. Rice and J.W. Kraus; U.S. Patent 3,791,969; February 12, 1974; assigned to W.R. Grace & Co.* a permselective membrane is prepared by providing an aqueous suspension of sodium exfoliated vermiculite, replacing at least part of the exchange sites with ions of a metal of Groups I-a and II-a, flocculating the resulting suspension, and dewatering the flocculated mass whereby the vermiculite platelets are aligned into a thin cohesive membrane.

The membrane is useful in a variety of selective separation processes, such as water purification, and to process other solvents having soluble ionic contaminants. The membrane may be further modified to provide desiccating properties by incorporating various solid adsorbents, such as silica gel and activated carbon.

Example 1: An aqueous suspension of sodium exfoliated vermiculite was steeped for 24 hours in a 20% aqueous solution of lithium chloride. Approximately 50% of the sodium was replaced by the lithium. The thus-modified vermiculite was flocculated by the addition of approximately 1 to 5 ml of dilute (20%) hydrochloric acid and the flocculated mass was centrifuged to rapidly remove the water. The individual vermiculite platelets were aligned into a continuous thin film having a thickness of 20 mils and a pore diameter below 1000 A.

Example 2: The process of Example 1 was repeated except that finely divided silica gel in an amount up to 50% by weight of the finished membrane was added to the suspension before it was flocculated.

Example 3: The process of Example 1 was repeated except that 20% by weight of the crystalline zeolite (type 4A, Linde) was added to the suspension before it was flocculated.

Example 4: The permselectivity of the membrane prepared according to Example 1 was tested by wetting the membrane with a NaCl solution and placing in a cell arranged as follows:

		Vermiculite Membrane		
Calomel Electrode	NaCl Solution I (A_1)		NaCl Solution II (A_2)	Calomel Electrode

By varying the concentration (activities) of the NaCl solution on either side of the membrane while maintaining a constant A_1:A_2 ratio, the transport number of sodium ion was determined from electromotive measurements. These are reported below:

Molal activities		Molal concentration			
A_1	A_2	NaCl I	NaCl II	E.m.f. (mv.)	Transport number for Na^+
0.00557	0.00186	0.00605	0.00195	27.9	0.99
0.05	0.0167	0.062	0.0192	26.0	0.92
0.15	0.05	0.205	0.062	23.2	0.82
0.45	0.15	0.673	0.205	16.3	0.58

It can be shown that potential developed in the cell is described by:

$$\Delta E = 59.15 ti \log \frac{a_1}{a_2} \text{ (mv.)}$$

since the ratio of a_1/a_2 is 3 in this case if the membrane is ideal and permselective, i.e., ti = 1, then the potential developed is the maximum one and this ΔE_{max} would be 28.15 mv. For the case where the membrane is less than ideal i.e., ti < 1, the potential will be less than ΔE_{max}. It can be easily shown that for the above cell for constant ratio of solution activities on either side of the membrane

$$\frac{\Delta E}{\Delta E_{max}} = ti$$

As an approximation, the transport number determined can be considered to be representative of the average concentration of solution i.e.,

$$\frac{a_1 + a_2}{2}$$

As the solution concentration increases the transport number of most of the membranes decrease due to leakage of other than the main ion. The table shows that vermiculite membranes are permselective (ti > 0.5) in the region of brackish water.

ION EXCHANGE MEMBRANES

HALOCARBONS

Crosslinked Polyvinylidene Fluoride

H.P. Gregor; U.S. Patent 3,808,305; April 30, 1974 has found that substantially improved fixed-charge membranes of excellent strength and controllable pore size may be prepared by casting a solution including a matrix polymer, polyelectrolyte and a crosslinking agent to form a film that when dried to an extent sufficient to remove solvent until the desired pore size is attained and then crosslinked, has essentially a fixed pore size.

A layered membrane may also be formed by drying and partially, or wholly, curing a first layer in the foregoing manner and thereafter casting a superimposed layer, or layers, in sequence and drying and crosslinking the entire composite. The superimposed layers may comprise a membrane of the same or opposite charge or ultimately a neutral, i.e., a nonionic, membrane layer may be disposed between the sandwiching layers of the same or oppositely charged membranes, i.e., between an anionic and a cationic membrane.

The porosity of these ionic membranes can be controlled in a number of ways. One of these is to employ polyelectrolytes and film-forming matrix polymers which are reasonably uniform in cross-sectional diameter, i.e., they do not have large rings or other substituents in side groups, thus making them capable of coming closer together in the final film and thereby making the pores in the film smaller.

For example, for a very fine pore anion-permeable structure, one would employ as the ionic polymer a polymer of quaternized or partially quaternized polyethyleneimine together with a film-forming polymer such as polyvinyl chloride-acrylonitrile copolymer, e.g., Dynel, or polyvinylidine fluoride, e.g., Kynar, employing an epoxide as crosslinking agent.

Similarly, a very tight cation-permeable structure can be achieved by the employment of vinyl sulfonic acid together with similar film-forming matrix polymers. In systems of this kind, also, avoidance of the bulky benzene ring is achieved and there is less tendency to fouling in such membranes because of the absence of the aromatic structure.

Among the various polyelectrolytes that may be employed, it has been found that sulfonic acid polyelectrolytes such as polystyrene sulfonic acid (PSSA), the sodium salt of polystyrene sulfonic acid (PSSNa), sulfonated polymethyl styrene, and copolymers thereof with other vinyl monomers, and polyvinyl sulfonic acid produce especially good results.

The following examples illustrate the process where all parts are parts by weight, except where otherwise expressly stated. The following general procedure was employed in casting from nonaqueous solvents, all materials were dry as could be obtained by drying at 40°C over P_2O_5 to reasonably constant weight; polymer solutions were filtered prior to use.

Films were cast on clean surfaces such as glass, glass coated with a dilute carboxymethylcellulose solution and dried prior to use, stainless steel belts, thin plastic sheets attached to a glass surface, on paper substrates or on hollow fibers or tubes by drawing them through the casting solution. Other surfaces may also be used. The casting technique usually employs a knife blade, although the membranes may be prepared by dipping or spray techniques.

After thermal or other treatment, the films may be removed from the casting surface by soaking in water, except, of course, when films are bonded to porous supports or to (treated) papers and are intended as an integral part of such composites. Since these formulations employ crosslinking agents prior to casting, appropriate precautions are taken to combine reactive ingredients shortly before the casting procedure itself.

Characterization procedures: Membranes were characterized by a number of conventional evaluation procedures. These included:

(1) Thickness (L) in microns measured using a thickness gauge with the water-wet membrane;

(2) Water content (W) in percent as grams of water per gram of water-wet membrane;

(3) Concentration potential (CP) in millivolts (mv) for the membrane separating 0.1 M and 0.2 M potassium chloride at 25°C;

(4) Specific conductivity (K) in ohm^{-1} cm^{-1} measured in 0.1 M potassium chloride at 25°C;

(5) Hydraulic permeability (HP) in cm/sec atm was determined in water or a dilute salt solution under the imposition of a pressure of 60 to 100 psi;

(6) Salt rejection (SR) in percent was measured using a 0.01 M potassium chloride solution and magnetic stirring above the membrane measured at 25°C and employing a designated driving pressure (usually 60 or 100 psi) and reported as the concentration of the feed solution minus that of the product, the whole divided by the feed concentration and multiplied by 100;

(7) Dye rejection (DR) in percent was measured using a 15 ppm solution of methylene blue, a cationic dye of molecular weight 300 with membranes containing fixed cationic groups, or a 15 ppm solution of erythrosin, an anionic dye of MW 879 was used with membranes containing fixed anionic groups, both measurements being made under the general conditions employed for salt rejection and reported in the same manner;

(8) Protein rejection (PR) in percent was measured using a 60 ppm solution of bovine serum albumin in a pH 7.4 phosphate buffer employing the general conditions of other rejection measurements.

Example 1: A membrane was prepared by dissolving 6.9 parts of Kynar with 2.8 parts of PSSA acid of molecular weight 70,000 and 0.3 part of crosslinking agent A, consisting of a solution made up of 1 part BF_3-MEA (a complex of boron trifluoride and monoethanolamine) and 5 parts of DEN-438 (epoxide-novolac resin), all of these being dissolved in 90 parts of a mixture of 9 parts hexamethylphosphorictriamide (HMP) and 1 part of dimethylformamide (DMF).

All substances and solvents were dry. To prevent premature crosslinking, the matrix (Kynar) and polyelectrolyte were dissolved together and filtered to remove dust particles and then combined with the crosslinking agent in solution shortly before membrane casting. Membranes were cast on a clean glass surface with a Gardner knife set at a suitable gate.

In this run the knife was set so that the cast film had a wet thickness of about half of that, or 150 microns, and being 10% in solids, dried to a final thickness of 15 microns.

The freshly cast films were dried for 1 hr at 100°C and then cured for 3 hr at 150°C. They were then allowed to cool, immersed in water to separate them from the glass surface, washed with water, and then tested. The data and properties of this membrane are presented in the following table.

Example 2: The procedure of Example 1 was followed, except that 6.1 parts of Kynar, 2.4 parts of PSSA and 1.5 parts of crosslinking agent A were used. Properties of this membrane are set forth in the following table.

Example 3: The general procedure of Example 1 is followed, except that 4.7 parts of a polyvinyl chloride-acrylonitrile copolymer (Dynel NYGL) is employed as the matrix polymer, 417 parts of PSSA and 0.6 part of crosslinking agent A were used, all being dissolved in 90 parts of DMF. The properties of this membrane are presented in the following table.

Example 4: The general procedure of Example 1 is followed, except that 4.6 parts of Kynar, 4.6 parts of PSSA and 0.8 part of crosslinking agent B, consisting of a solution made up of 1 part BF_3-MEA and 40 parts bisphenol-A epoxide, available as ERL-2774, were used, all being dissolved in 90 parts of dimethyl sulfoxide (DMSO). The pertinent properties are presented in the table.

Example 5: The general procedure of Example 1 is followed, except that 2.8 parts of Dynel NYGL, 2.8 parts of PSSA and 4.4 parts of crosslinking agent A were used, all being dissolved in 90 parts of DMF. The properties of the membrane are presented in the following table.

Properties of Membranes, Examples 1 through 5

Example	L(μ)	W (percent)	C.P. (mv.)	$K \times 10^3$ (ohm^{-1} $cm.^{-1}$)	$HP \times 10^6$ (cm./sec. atm.)	Percent		
						S.R.[a]	D.R.[a]	P.R.[a]
1	15	16	9.5	0.8	5.0	57	92	100
2	18	10	16.0	1.0	0.06	70	99	100
3	20	58	14.1	3.0	20.0	46	90	100
4	50	45	13.2	5.0	12.0	35	85	100
5	65	27	15.6	5.0	5.0	67	98	100

[a] All measurements of salt, dye and protein rejections were carried out using driving pressures of 60 p.s.i.

Sulfonamide Substituted Fluorinated Polymer Films

A process described by *W.G. Grot; U.S. Patent 3,784,399; January 8, 1974; assigned to E.I. du Pont de Nemours and Company* relates to a film of the fluorinated polymer where one surface of the film has a majority of the sulfonyl groups of the polymer in the $(SO_2NH)_mQ$ form where Q is H, NH_4, cation of an alkali metal, and/or cation of an alkaline earth metal and m is the valence of Q while the other surface is in the nonsulfonamide/salt form.

All parts and percentages are by weight unless specified otherwise. In the example, perselectivity is measured according to the description in the third paragraph on page 636 of the *Encyclopedia of Polymer Science and Technology;* Vol. 8, John Wiley & Sons, Inc., 1968.

Example: A blown film in the form of tubing made from a copolymer of tetrafluoroethylene and

$$CF_2{=}CFOCF_2\underset{\displaystyle CF_3}{\underset{|}{C}}FOCF_2CF_2SO_2F$$

(mol ratio of the former to the latter is 7:1) was inflated with a mixture of air and approximately 1% by volume ammonia gas at room temperature. The tubing had a circumference of 18 inches and a wall thickness of 0.004 inch. The ends were clamped to prevent the escape of the inflating gas mixture. After 15 hr approximately 0.001 inch out of the total thickness of 0.004 inch was converted to the SO_2NH_2 form. This amount of conversion was indicated by staining of the sample and observing the variation in the stain. The tube was deflated and immersed in a 10% solution of sodium hydroxide at 90°C for 48 hr.

This converted the SO_2NH_2 of one surface to the $SO_2NH^-Na^+$ form and the other surface which was approximately 0.003 inch of the total of 0.004 inch to the $(SO_3^-)Na^+$ form. The permselectivity of this sample when exposed to 3N KCl solution on the $(SO_2NH^-Na^+)$ surface and 1N KCl solution on the other surface was 73% as compared to 58% for a sample that had not been treated with ammonia.

An electrolysis cell was constructed and consisted of a graphite chamber which served as the anode and endplate, two silicon rubber spacers which served as gaskets to seal a stainless steel wire screen cathode between them and a tetrafluoroethylene polymer endplate. The membrane sample to be tested was placed between the graphite chamber and the first spacer with the sulfonamide salt $(SO_2NH^-Na^+)$ surface facing the cathode.

Suitable outlets were provided in both the endplates for the electrolyte. A saturated brine of NaCl (26% by weight) was circulated through both compartments of the cell separately while passing an electric current through the cell. The amount of sodium hydroxide produced was determined by titration of the samples taken periodically from the effluent of the cathode department.

The current efficiency was expressed as the percentage of sodium hydroxide produced compared to the theoretical amount that the current passed through the cell should have produced. The current efficiency of the membrane with the sulfonamide/salt surface was 98% while the current efficiency for a sample without such a surface was 83%.

Polyvinyl Fluoride-Acrylonitrile

A process described by *J.G. McKelvey, Jr., and K.S. Spiegler; U.S. Patent 3,132,094; May 5, 1964; assigned to Gulf Research and Development Company* involves a method for deionization of electrolyte solutions by ultrafiltration through an ion exchange membrane, whereby improved, efficient separation of ionic materials from solvent is obtained, even in instances of solutions containing a relatively large proportions of electrolytes.

It has been found that such improved deionization can be obtained by contacting a solution containing electrolyte with one surface of an ion exchange membrane having an exchange capacity (milliequivalents per gram wet membrane): specific hydraulic permeability ($cm^2\ sec^{-1}\ dekabar^{-1}$) ratio of at least 3×10^8, to 1×10^{10}, preferably 5×10^8 to 8×10^8, and establishing a pressure differential between the surface of the membrane contacted by the solution and another surface of the membrane sufficient to force solvent through the membrane. Normally the pressure differential will be in the range of 75 to 15,000 psi, and usually it will be between 500 to 5,000 psi, preferably 500 to 2,000 psi, but greater pressure differentials can be used.

Although the principles of this process are applicable to either cation or anion exchange type membrane, regardless of the type of electrolyte involved, it has been found that especially good results are obtained in filtering dilute solutions by the use of anion exchange membranes, where the electrolyte in the solution to be filtered involves a polyvalent cation and an anion of lower valence, and that especially good results are obtained with cation exchange membranes where the electrolyte involves a polyvalent anion and cation of lower valence.

It is preferred to employ homogeneous ion exchange membranes having exchange capacities greater than 0.3 milliequivalents per gram wet weight, preferably between 0.5 and 2.0 milliequivalents per gram, and having specific hydraulic permeabilities in the range of 5×10^{-11} to 5×10^{-8} cm^2/sec/dekabar, preferably 1×10^{-10} to 2×10^{-8} cm^2/sec/dekabar.

Membranes which have other exchange capacities and other specific hydraulic permeabilities, including heterogeneous membranes, can be used, provided the ratio of exchange capacity to specific hydraulic permeability is as defined above.

Specific examples of membranes having the properties described are membranes cast from a solution of a mixture of a major proportion of a polyvinyl chloride-acrylonitrile copolymer and a minor proportion of a polymer of 1-vinyl-3-methylimidazolium iodide (anion exchange), and membranes cast from a common solution of a major proportion of a polyvinyl chloride-acrylonitrile copolymer and a minor proportion of polystyrene sulfonic acid (cation exchange), but other membranes can be used. The process will produce superior results with solutions containing low concentrations of electrolytes, but it is especially valuable for filtering electrolyte solutions of a strength between 0.1 normal and 2.0 normal, or even higher, for example, approaching the solubility limit of the electrolyte.

Patching Technique for Sulfonated Fluoropolymers

E.H. Cook, Jr., A.T. Emery and B.O. Schoepfle; U.S. Patent 3,852,135; December 3, 1974; assigned to Hooker Chemical Corporation describe a method for patching cation-active permselective diaphragm material consisting essentially of a hydrolyzed copolymer of a perfluorinated hydrocarbon and a fluorosulfonated perfluorovinyl ether.

This is accomplished by forming a solution of a hydrolyzed copolymer of a perfluorinated hydrocarbon and a fluorosulfonated perfluorovinyl ether but having an equivalent weight somewhat below that of the damaged hydrolyzed copolymer of the same chemical type in a lower hydrocarbon alkanol applying the solution to the area adjacent to the damaged section of hydrolyzed copolymer diaphragm, covering the damaged section with a patch of hydrolyzed copolymer of the same equivalent weight as the diaphragm material and removing the lower hydrocarbon alkanol solvent from the covered area.

The patch is thereby firmly bonded to the diaphragm material and by covering the damaged section enables the diaphragm to once again function as a barrier to liquids and gases in the manner of the undamaged sections of the diaphragm. Parts and percentages are by weight and temperatures are given in degrees centigrade, unless otherwise specified.

Example 1: 15 parts of the hydrolyzed copolymer of tetrafluoroethylene and perfluoro (2-fluorosulfonylethoxy)-propyl vinyl ether, having an equivalent weight of about 1,000, which had been cut into small pieces were suspended in 100 parts of ethanol and the suspension was agitated and warmed to about 55° and 60°C until solution occurred, about 1 hr. The resulting solution was filtered, warm to remove a small amount of insoluble material and the clear solution was cooled to ambient temperature.

A section of a 10 mil thick film of a hydrolyzed copolymer of tetrafluoroethylene and perfluoro(2-fluorosulfonylethoxy) propylene vinyl ether having an equivalent weight of about 1,100 on a woven Teflon support which had been used as a cation-active permselective membrane to separate the anode and cathode of a brine electrolysis cell was damaged by cutting the membrane with a knife to form a tear measuring about 1 inch long in the film.

An area of about 2 inches square around the tear in the membrane was thoroughly wetted with the copolymer solution prepared above by brushing. A patch of about 2 inches square of the higher equivalent weight copolymer was likewise wetted with the adhesive solution and then applied to the wetted area around the tear in the membrane. The patch was pressed upon the membrane and held firmly in place for about 30 min to permit the solvent to evaporate. The patch was firmly adhered to the membrane which could then be replaced in service in the cell to provide a diaphragm which was impervious to liquids and gases.

Example 2: 10 parts of a hydrolyzed copolymer of tetrafluoroethylene and perfluoro-(2-fluorosulfonylethoxy)propylene vinyl ether having an equivalent weight of about 900 were suspended in about 100 parts of isopropanol. The suspension was agitated and heated at about 55° to 60°C until solution of the copolymer was essentially complete, about 1 hr. The solution was filtered to remove a small amount of insoluble material and the clarified filtrate was cooled to ambient temperature.

A section of the hydrolyzed copolymer of the character as described in Example 1 above which had been used as the diaphragm in a two compartment brine electrolysis cell was pierced with the point of an ice pick to produce a simulated pin hole. Three such pin holes were made in an area of about 2 square inches of the diaphragm film. The damaged area was brushed with the copolymer solution prepared above, covering an area of about 4 square inches encompassing the pin holes.

This wetted area was covered with a patch of about 4 square inches of the same hydrolyzed copolymer material as the damaged film material, and which had been brushed with the copolymer solution on one side. The patch was pressed wetted side down onto the damaged surface and firmly held thereon while a current of warm air was blown onto the patch to assist in evaporating the solvent from the patched area. When dry, about 30 min, the patch was securely bonded to the film and the diaphragm could be replaced in service and was impervious to liquids and gases to the same degree as the original undamaged membrane material.

Perhalogenated Fluorocarbons

A process described by *W.K.-W. Chen, H.M. Halter, E.B. Hotelling and C.A. Wetmore; U.S. Patent 3,257,334; June 21, 1966; assigned to American Machine & Foundry Co.,* involves electrodialysis membranes comprising in combination a fluoropolymer and a polymer derived from a vinyl monomer.

The fluoropolymer compositions are prepared from at least a two-component system. The first component is a fluoropolymer which may be derived from one or more monomer systems. Therefore, the first component may be a polymer, copolymer or terpolymer, for example. The first component is preferably a long chain carbon compound, either straight or branched, substituted with fluorine atoms in at least half the sites available for substitution. The chain is further substituted with hydrogen or halogen atoms in the remaining sites available for substitution. Such long chain carbon compounds are characterized by the following general formula:

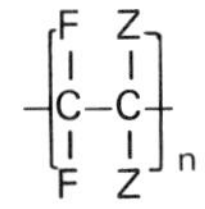

in which Z represents hydrogen or a halogen atom and n is an integer greater than unity. Among such long chain carbon compounds which may be employed as the first component of this composition of matter are: polytetrafluoroethylene, polychlorotrifluoroethylene, polyperfluoropropylene, polyvinylidene fluoride, poly-1,1-difluoro-2,2-dichloroethylene and copolymers of these materials with one another or with other materials having the general formula shown above. The second component of the fluoropolymer combinations and compositions of this process is ethylenically unsaturated and derived from the polymerization or copolymerization of a vinyl type monomer.

The combination resulting from the interaction of the first polymer component and the monomer may have various forms, depending upon the exact method of preparation and the chemical nature of the starting materials. Commonly, the product is a block polymer, in which the first component and the second component form alternating units of varying length in a chain. However, in some cases the monomer second component is grafted onto the first component to form side chains of varying size attached to the main stem or trunk of the first component.

In some cases, the second component forms a cyclized polymer or crosslinked network around and about the fluoropolymer first component. In other cases, the monomer polymerizes to form the second component which is in some cases in the form of a linear homopolymer interlacing the first component without recognizable chemical bonds, but forming an intimate and inseparable plastic alloy. In general, the compositions comprise an intimate molecular combination of two or more polymer systems which are not leached apart in any appreciable degree when treated with solvents.

The compositions are useful as ion exchange materials and particularly as membranes in electrodialysis apparatus. Those compositions which have inherent acidic or basic groups may have ion exchange or chelating or selective absorption properties per se. The following examples illustrate the process.

Example 1: Polytetrafluoroethylene film with a thickness of 0.25 mil was immersed in styrene monomer containing 0.01% by weight of benzoyl peroxide at 65° to 70°C. After 17.5 hr the film was found to contain 19.4% polystyrene which could not be extracted by boiling in benzene. Sulfonation of this film produced a cation exchange membrane with an electrical resistance of 2.9 ohm-cm^2 after equilibration of 0.6N KCl solution.

Example 2: The procedure of Example 1 was repeated except that the styrene (containing 0.01% of benzoyl peroxide by weight) was held at 90° to 95°C for 60 min. The amount of polystyrene incorporated into the film was 10.9% by infrared analysis. Chloromethylation of this film, followed by treatment with N-methylpiperidine for 12 hr at room temperature and washing with distilled water, produced anion exchange membrane with an electrical resistance of 8.1 ohm-cm^2 after equilibration of 0.6N KCl solution.

Example 3: The procedure of Example 1 was repeated except that the film of shaved polytetrafluoroethylene had a thickness of 2.5 mils. After immersion in styrene (containing 0.01% benzoyl peroxide by weight) for 17.5 hr the amount of polystyrene incorporated into the film was only 3.2% by infrared analysis. This experiment again illustrates the advantage of employing thin films of the fluoropolymer starting material.

Polyhalide Resins

J. Place, U. Weissenberg and R.D. Goodenough; U.S. Patent 3,554,905; January 12, 1971; assigned to The Dow Chemical Company describe a process for treating water with halogen released by an anion exchange resin in polyhalide form which is obtained by using a membrane permeable to halogen but essentially impermeable to water and water-soluble anions to separate the polyhalide resin and the water being treated. Such a membrane reduces the polyhalide resin instability arising from the influence of water-soluble anions and alkalinity in the water supply. Particularly effective in this process is a polyethylene film packet of a polybromide resin. The following examples illustrate the process.

Example 1: Treatment of Water with Polybromide Resin — (A) About 100 ml of a moist strong base quaternary ammonium anion exchange resin (Dowex 21K resin, 20 to 50 mesh) in polybromide form containing 25 weight percent bromine was sealed in a 1 mil polyethylene film packet and placed on a perforated polytetrafluoroethylene support in a 5 cm i.d. ion exchange resin column. Tap water at line temperature was passed through the column at different flowrates and the bromine concentration in the eluent water determined. Typical results are given below.

Release of Bromine from Polybromide Resin (25% Br_2) in 1 Mil Polyethylene Packet

Flowrate, ml/min	20	100	600
Eluent Br_2, ppm	0.94	0.09	0.04

(B) In contrast a similar resin column with 100 ml of polybromide resin containing 25.1 weight percent bromine in a water porous cartridge gave eluent bromine concentrations ranging from 0.6 ppm at 100 ml/min to about 2.0 to 2.4 ppm at 590 ml/min during initial operation. However, higher elution rates were observed during later stages of operation apparently related to the formation of the less stable Br_2Cl^- anion from interaction of the polybromide resin with the chloride present in the tap water.

Example 2: In another experiment, a 1 mil polyethylene packet loaded with a moist polybromide resin containing about 50 weight percent bromine was placed in the resin column. Typical results from elution with tap water at line temperature and different flowrates are given below.

Release of Bromine from Polybromide Resin (50% Br_2) in 1 Mil Polyethylene Packet

Flowrate, ml/min	20	50	100	300	500	900	1,700
Eluent Br_2, ppm	158	18	9	5	3.5	2.0	1.0

Although for a given flowrate and resin halogen content, the use of a semipermeable membrane reduces the concentration of halogen in the eluent liquid, concentrations effective for water treatment can be obtained without the instabilities arising from the influence of anions and alkalinity in the water supply. By control of the halogen content of the resin and the diffusion characteristics of the semipermeable membrane effective operation can be achieved for a wide variety of applications and needs.

Vinyl Chloride Polymers

J.H. van der Neut, H.G. Roebersen, A. Schors and O.P. van der Werff; U.S. Patent 2,972,586; February 21, 1961; assigned to Nederlandse Organisatie voor Toegepast-Natuurwetenschappelijk Onderzoek ten Behoeve van Nijverheid, Netherlands have found that by introducing ionic groups with a dissocation constant of at least 10^{-5} into thin films of hydrophobic material with a length and breadth of at least 100 x the thickness in a quantity of at least 0.20 meq/g of dry foil, or film, excellent membranes are obtained.

By introducing these ionic groups in a quantity of 0.20 meq/g of dry film the hydrophobic starting material is converted into a hydrophilic material, owing to the hydrophilic nature of these groups, the hydrophilic material absorbs sufficient water in a salt solution to become sufficiently electrically conductive. If the dissociation constant of the ionic groups is smaller than 10^{-5} or the quantity of ionic groups is less than 0.20 meq/g, the selectivity is not sufficient and/or the resistance of the membrane is too high.

An advantage of this process is that by choosing a foil or film thickness as the starting material and making the incorporation of ionic groups proceed in such a way that the form and the mechanical properties of the film are substantially maintained, it is possible to produce very thin membranes of a thickness of, e.g., 0.02 to 0.3 mm, which is not the case with many processes, according to which monomeric substances are condensed in membrane form or polymeric compounds containing ionic groups, mixed with a nonactive, plastic substance and are rolled into foils.

Films of hydrophobic substances used as starting material may be polyvinyl chloride, copolymers of vinyl chloride such as vinylidene chloride, chlorinated rubber, rubber hydrochloride, polystyrene and copolymers of styrene and polyethylene. The ionic groups which are particularly suitable are sulfonic groups, quaternary ammonium groups, tertiary sulfonium groups, guanidinium groups and pyridinium groups. The membranes obtained according to the process excel by their high selectivity. The cause of this presumably is that these membranes exclusively allow ions to pass on those spots where ionic groups have been introduced, so that practically no transport of oppositely charged ions through the membrane can take place. The following examples illustrate the process.

Example 1: A film of chlorinated rubber with approximately 60% by weight of chlorine containing 40 parts by weight of dioctyl phthalate as plasticizer per 100 parts by weight of chlorinated rubber, which film was 0.09 mm thick, the resistance of which is 0.1N NaCl solution is very high, was treated with a solution of 25% trimethylamine in water at

20°C for 22 ½ hr. The film then had a dark-brown color, its capacity being 0.27 meq ionic groups per gram. The selectivity of the film thus treated between salt solutions of 0.1 and 0.2N KCl is over 80%.

It was found that at the same temperature a solution of 25% trimethylamine in ethyl alcohol (in which chlorinated rubber slightly swells) acts much more quickly and gives a more even and better reproducible result. A film of chlorinated rubber with a thickness of approximately 120μ (in the same composition) was treated with a solution of 25% trimethylamine in ethyl alcohol at 20°C. After 24 hr the resistance had fallen to 0.6 Ω-cm^2 in 0.1N NaCl solution; the capacity was 2.3 meq ionic groups per gram dry substance.

Also with this low resistance and high capacity the membrane produced allows only few ions to pass through in an undesired direction; the diffusion amounted to only 18 x 10^{-6} grams equivalent cm^2 per hour if the membrane was used as partition between a solution of 0.1 N NaCl and distilled water. The selectivity measured from the membrane potential between solutions of 0.1 and 0.2 N KCl was approximately 85%. The shorter the duration of action of the solution, the higher are both the resistance and the selectivity.

Example 2: Commercial Pliofilm (rubber hydrochloride) (chlorine content 31%) without plasticizer, with a thickness of 0.064 mm was treated with sulfuric acid of about 98.3% strength at 20°C for various periods. The properties of the tea-colored films were as follows:

Reaction Time	Resistance in 0.1N NaCl Solution, Ω-cm.^{2}	Selectivity Between 0.1 and 0.2N KCl, percent	Capacity
1 hour	63	94	0.68
1 1/2 hours	25	86 1/2	0.70
2 hours	16	97	0.74
3 1/2 hours	8	98 1/2	0.97

The resistance of a film of the same composition with a thickness of 45 μ, treated at 20°C with the same both was only 2.5 Ω-cm^2 in 0.1N NaCl solution in water after 3 ½ hr. The capacity after this treatment was 1.06 meq/g of dry substance; the selectivity between 0.1 and 0.2 N KCl was practically 100%, between 0.5 and 1N KCl approximately 85% and between 1 and 2 N KCl approximately 70%.

SODIUM STYRENE SULFONATE AND VINYL BENZYL AMMONIUM CHLORIDE POLYMERS

Polyelectrolyte Membrane Structures

A.S. Michaels and R.G. Miekka; U.S. Patent 3,546,142; December 8, 1970; assigned to Amicon Corporation describe solid polymeric materials formed by coprecipitation of water soluble polyelectrolytes having dissociable ionic groups of opposite charge, and to methods of forming solids. The materials are useful as ion sorbing resins, selectively permeable and semipermeable membranes and barriers, protective coatings, artificial biological tissues, microporous sponges and filters, semiconductor elements, and humidity sensitive conductive elements. They may be formed to any desired shape by casting, molding, fiber drawing, or by fabricating methods, such as machining and grinding.

The polyelectrolytes are organic polymers having dissociable ionic groups which impart electrolytic characteristics in forming salts and acids (in the case of anionic polyelectrolytes) and bases (in the case of cationic polyelectrolytes). Typical of these are the polymers of sodium styrene sulfonate and of vinyl benzyl trimethyl ammonium chloride, and other materials of the same general type having a synthetic organic polymeric structure, which, without the ionic groups (sulfonate or quaternary ammonium), would be a water-insoluble film forming material. They are accordingly characterized by a sufficiently high molecular weight to be solid and capable of film formation (typically greater than 50,000) while having sufficient dissociable ionic groups chemically bonded to the polymeric structure to be water-soluble. As a very general rule there should be at least one ionic group for every six repeating monomer (mer) units, or for each average chain interval of twelve carbon atoms.

When solutions of oppositely charged polyelectrolytes are mixed together under proper conditions of concentration and temperature, the oppositely charged macroions coreact ionically and precipitate from solution. Because of their high charge density, polyelectrolytes have a much greater tendency to associate with each other than do their corresponding monomers. Most polyelectrolyte precipitates can be dissolved, or their formation prevented, by the addition of sufficient amounts of "indifferent" soluble ionic salts such as magnesium sulfate or sodium chloride.

Subsequent removal or dilution of these shielding salts results in reprecipitation of the polyelectrolyte complexes. A preferred method of carrying out this process is described in detail below with reference to the formation of a coprecipitate of poly (sodium styrene sulfonate) and poly-(vinyl benzyl trimethyl ammonium chloride) from a solvent mixture of water, acetone, and sodium bromide. Reference is made to Figure 5.1, which is a ternary diagram showing the solubility of the polyelectrolytes in the water-acetone-sodium bromide system.

Example: Poly(sodium styrene sulfonate), (NaSS) an anionic polyelectrolyte having an average molecular weight of 760,000 and poly (vinyl benzyl trimethyl ammonium chloride), (VBTAC) a cationic polyelectrolyte having molecular weight of 300,000 were reacted together to yield transparent solid resins of controlled shape and size. The polyelectrolytes were purified by treating aqueous solutions containing 3 to 5% by weight of the polymers with a mixture of a conventional cation exchange resin in the hydrogen form and an anionic exchange resin in the hydroxide form for 12 hr, and the solutions were then filtered. The anionic polyelectrolyte solution was then passed over a cation exchange resin in the sodium form while the other was passed over an anionic resin in the chloride form to assure that the desired counter-ions were present; the solutions were then dried.

Solutions containing 5 to 10% by weight of each polyelectrolyte dissolved in a solvent consisting of 60% water, 20% NaBr, and 20% acetone were made up with polymer equivalence ratios of 2 VBTAC/1 NaSS, 1 VBTAC/1 NaSS, and 1 VBTAC/2 NaSS. The solutions were poured into paraffin coated glass petri dishes and allowed to gel by evaporation of the acetone. The samples were next placed in an oven at 50°C to evaporate water until, or shortly before, a slight opacity developed at the periphery of the gels due to salt out of NaSS, and were then washed briefly with water to extract some of the NaBr, and replaced in the oven to again allow water evaporation to the point of insipient NaSS precipitation.

This alternate washing and drying procedure was continued until the polymer content was 40 to 50% of the total weight of the samples, after which the polysalts were washed with demineralized water until more salt could be extracted from them, cut into samples of the desired size, and dried at 50% humidity over a sulfuric acid-water mixture to yield amber, transparent solids. The purpose of this rather elaborate wash-dry procedure was to remove as much salt, water and acetone as possible from the polysalts while keeping the overall solvent composition near the region of polysalt solubility. If too much salt was extracted from the dilute gels by prolonged contact with water, internal porosity developed due to contraction of the polymer volume, and the final polysalts were microporous sponges rather than continuous solids. Where a porous structure is desired, drying is carried out without the intermediate washing steps.

After final washing of the polysalts, the interpolymer reactions were found to be surprisingly complete. No sodium could be detected by flame tests made on samples cut from the quaternary-rich polysalts, and no visible silver halide precipitate was formed when silver nitrate solution was added to one tenth molar sodium nitrate solution which had been equilibrated with a sample of NaSS-rich polysalt. The neutral polysalts gave negative tests for both sodium and bromide ions, with the exception that small opaque regions (if present) at the periphery of the samples gave weak positive flame tests for sodium.

FIGURE 5.1: SOLUBILITY OF POLYELECTROLYTES IN ACETONE-WATER-SODIUM BROMIDE SYSTEM

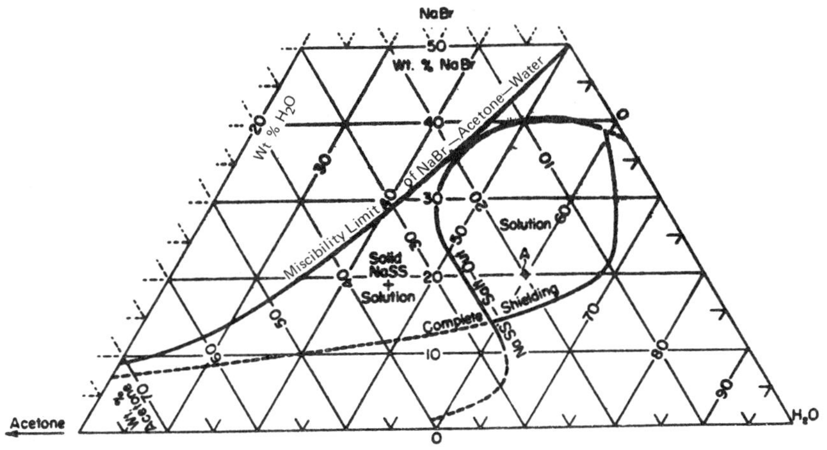

Source: A.S. Michaels and R.G. Miekka; U.S. Patent 3,546,142; December 8, 1970

The mechanism of gel formation demonstrated in the preceding example is readily understood by referring to the ternary diagram given in Figure 5.1. The presence of sodium bromide in sufficient concentration has a shielding effect which prevents the two polyelectrolytes from reacting, while the presence of acetone reduces the amount of sodium bromide required, as shown by curve designated "complete shielding."

Above and to the left of this curve, the polyelectrolytes remain in solution, while at compositions below and to the right of this curve, precipitation occurs. It will also be noted that at relatively high acetone concentrations and/or sodium bromide concentrations a second limiting effect arises from the anionic polyelectrolyte being salted out separately from the cationic polyelectrolyte.

In the above example an initial solution having a composition at the point designated "A" is brought by the elimination of acetone to a composition lying generally to the right at or near the "complete shielding" line. At this point gel formation occurs, and further processing is carried out to maintain the gel near the limiting composition for complete shielding, by removal of both water and sodium bromide until a substantially dry nonporous solid is produced.

Ionically Associated Gel

P.N. Rigopulos; U.S. Patent 3,549,016; December 22, 1970; assigned to Amicon Corp. describes the formation of membrane of a gel structure having a lattice formed of two ionically associated polyelectrolytes under controlled conditions on a solvent impervious casting surface, then exposing the membrane while still on the casting surface to a solvent under controlled conditions. It has also been found that a reinforcing member in the form of a fibrous web, preferably an unwoven fibrous web, may be incorporated in the membrane while it is being formed to provide additional strength.

The ionically associated gel structure may be formed as previously described in U.S. Patent 3,546,142 by forming an aqueous solution containing reactive quantities of a polyanionic organic polymer and of a polycationic organic polymer capable of reacting with each other to form a water-insoluble precipitate, and also containing sufficient electrolyte, which acts as an ion shield, and sufficient organic solvent, miscible with the aqueous solution, to maintain the polymers in solution without precipitation. The gel may then be formed by removing the organic solvent or by reducing the activity of the electrolyte or both until a solid gel structure forms. The membrane may be formed directly from a solution so prepared from two separate polymers by casting the solution into the desired shape on a casting surface, preferably a water impervious casting surface, then causing it to gel by reducing the electrolyte activity or removing the solvent, or both, or if desired such a gel previously formed, e.g., as a finely divided solid, may be dissolved in water containing sufficient ion-shielding electrolyte together with water miscible organic solvent to maintain the solid in solution and this solution may be cast, then caused to gel or solidify in the form of the desired membrane by reducing the activity of the electrolyte or removing solvent or both.

The membranes are useful for a wide variety of ultrafiltration operations such as are involved in biological or pharmaceutical separations, protein concentration, sewage treatment, water clarification, sugar purification, bacteria or virus concentration or filtration, cold sterilization of wine and beer. The membranes are characterized by a unique zone of lower permeability adjacent the face which is exposed to water during the washing step, as pointed out above.

This characteristic is of great practical importance in that when the membrane is used for ultrafiltration, any plugging of the membrane can readily be removed simply by reversing the flow for a short period of time. This anisotropic of nonuniform characteristic of the membrane is also evident by the dissimilar visual appearance of opposite sides of the membrane, the side of lower permeability having a higher sheen or gloss than the other. It is also evidenced by the strikingly dissimilar dye-absorption characteristics of opposite sides of the membrane. For example, when such a membrane is immersed in a solution of red acid fuchsin, almost no dry absorption occurs on the side of low permeability, whereas the opposite side becomes brightly colored.

The membranes are also characterized by high water transmissivity properties. By proper control of the composition of the casting solution and of the washing and annealing conditions, for example, membranes may be produced having flow rates of the order of 50 to 175 gallons of distilled water per 24 hours per square foot of membrane area at a pressure differential of 100 pounds per square inch across the membrane.

The permeability of the membranes is such that one having a water transmissivity of the order of 175 gallons per day per square foot at 100 psi retains completely dextran molecules having an average molecular weight of 10,000 while permitting raffinose, sucrose, sodium chloride, etc. to pass completely through; such a membrane will also retain 80% of polyethylene glycol molecules having an average molecule of 6,000. On the other hand, the permeability of a membrane having a water transmissivity of the order of 60 gallons per day per square foot at 100 psi is such that it retains from 80 to 100% of raffinose molecules and retains 50% of sucrose molecules, while permitting passage of sodium chloride. Both membranes of course retain completely higher molecular weight solutes such as bovine serum albumin.

Example 1: An ionically associated solid gel formed by reacting together equivalent weights of poly(vinyl benzyl trimethyl ammonium chloride) and poly(sodium styrene sulfonate) was prepared as described in Michaels et al, *Journal of Physical Chemistry,* Vol. 65, pages 1, 765 et seq. (1961). One hundred parts by weight of the resulting dry, solid, neutral ionically associated polymer in finely divided form was dissolved with stirring in a solution containing 284 parts by weight of dioxane, 226 parts by weight of concentrated aqueous hydrochloric acid (37% HCl), and 54 parts by weight of distilled water.

A filter paper having a gauge of 4 mils and consisting of polyester fibers (Dacron) was secured with adhesive tape to a casting surface consisting of a smooth glass plate. The solution was then poured over the sheet of filter paper and drawn down by a conventional drawdown bar set at a clearance of 8 mils from the casting surface. Immediately after drawdown, the casting while still on the surface of the glass plate was immersed for 2 to 3 seconds in a bath of water maintained at 60°C.

The water bath contained 0.1% by weight of an anionic surface active agent, the dioctyl ester of sodium sulfosuccinate, in order to minimize the streaking of the casting which might otherwise occur. The casting, while still on the glass casting surface, was then immediately immersed in a second water bath maintained at room temperature. After immersion in the room temperature water bath for 30 seconds, the casting, while still on the glass casting surface, was withdrawn and immersed for 5 sec in a water bath maintained at 90°C, after which it was cooled to room temperature without drying by immersion in a second room temperature water bath for 5 minutes.

The resulting membrane having a total thickness of 8 mils was readily stripped from the casting surface while still wet and stored by immersion in a sealed container filled with water. It was found that little strike-through of the paper by the polymer had occurred, but that adhesion of the polymer to the paper was sufficient to prevent delamination even after wet storage for an extended period of time. The membrane exhibited a water transmissivity rate, using distilled water, of 175 ± 25 gallons per 24 hours per square foot of membrane at a pressure drop of 100 pounds per square inch across the membrane, while at a pressure drop of 500 pounds per square inch the rate was 880 gallons.

In each case the membrane was supported by a sintered stainless steel support member. The membrane exhibited essentially no retention when used for filtration of an aqueous solution containing 0.25% by weight of raffinose at 100 psi, but exhibited 80 to 100% retention of the solute in each case when used for ultrafiltration of aqueous solutions in the same concentration of polyethylene glycol (MW 6,000) and dextran (MW 10,000) at 100 psi. The membrane was found capable of 100% retention of solute when used with an aqueous solution of 0.25% by weight of bovine serum albumin (molecule 69,000) at 100 psi. In the case of the last three solutions, somewhat lower water transmissivity rates (of the order of 60 to 90 gallons per 24 hours per square foot) were observed at 100 psi than were observed with distilled water. The membrane exhibited a tensile strength approximately equal to that of the paper used as a reinforcing member and a burst strength of approximately 25 psi as determined by Mullen burst tests.

Example 2: One hundred parts by weight of the ionically associated dry, solid particulate polymer of Example 1 together with five parts by weight of a nonionic fluorochemical surface active agent (FC170) were dissolved in a solution containing 252 parts by weight of ethanol, 252 parts by weight of aqueous hydrochloric acid (37% HCl) and 63 parts by weight of distilled water. The surface active agent was employed merely to improve the casting characteristics of the solution and minimize streaking. A member was prepared from this solution by the same procedure as described in Example 1.

It was found that the substitution of ethanol for dioxane in the casting solution led to the production of a membrane having lower permeability and also lower water transmissivity. This membrane when tested with aqueous solutions containing 0.25% by weight of sucrose or raffinose, exhibited a water transmissivity of 50 to 75 gallons per 24 hours per square foot at a pressure drop of 100 psi, and a transmissivity of 200 gallons per 24 hours per square foot at 500 psi. The membrane exhibited nearly 100% retention of raffinose and 40 to 60% retention of sucrose from such solutions at 100 psi. The physical strength of this membrane was the same as that of Example 1.

R.W. Baker; U.S. Patent 3,567,810; March 2, 1971; assigned to Amicon Corporation also describes a process for forming improved polymeric anisotropic membranes comprising the additional step of subjecting a cast polymer film suitable for formation of such an anisotropic membrane to an evaporation step prior to precipitation of the membrane, the evaporative step is controlled to selectively evaporate solvent from the surface of the film and thereby selectively increase the polymer concentration at the surface of the film.

The polymeric anisotropic membranes which are formed and described by the process are characterized by improved rejection characteristics. Furthermore, they retain liquid flux characteristics which are surprisingly high in view of the rejection characteristics.

Coprecipitation of Water Soluble Electrolytes

A.S. Michaels and R.G. Miekka; U.S. Patent 3,276,598; October 4, 1966; assigned to The Dow Chemical Company describe solid polymeric films formed by coprecipitation of water soluble polyelectrolytes having dissociable ionic groups of opposite charge. The polyelectrolytes useful in this process are organic polymers having dissociable ionic groups which impart electrolytic characteristics in forming salts and acids (in the case of anionic polyelectrolytes) and bases (in the case of cationic polyelectrolytes). Typical of these are the polymers of sodium styrene sulfonate and of vinyl benzyl trimethyl ammonium chloride, and other materials of the same general type having a synthetic organic polymeric structure, which, without the ionic groups (sulfonate or quaternary ammonium), would be a water insoluble film forming material.

They are accordingly characterized by a sufficiently high molecular weight to be solid and capable of film formation (typically greater than 50,000) while having sufficient dissociable ionic groups chemically bonded to the polymeric structure to be water soluble. As a very general rule there should be at least one ionic group for every six repeating monomer (mer) units, or for each average chain interval of twelve carbon atoms.

When solutions of oppositely charged polyelectrolytes are mixed together under proper conditions of concentration and temperature, the oppositely charged macroions coreact ionically and precipitate from solution. Because of their high charge density, polyelectrolytes have a much greater tendency to associate with each other than do their corresponding monomers. Methods for isolating these exceedingly thin interfacial complex polyelectrolyte films from their original polyelectrolyte environments have been found. These methods are described below as applied to films formed between solutions of poly(vinylbenzyltrimethyl ammonium chloride), and poly(sodium styrene sulfonate) homopolymers.

In the process, two separate solutions (usually aqueous, but not necessarily so), one containing an anionic polyelectrolyte (or polyelectrolytes), the other containing a cationic polyelectrolyte (or polyelectrolytes) are made up with concentrations adjusted such that the polyelectrolytes will react to form an interfacial film when the two solutions are brought into mutual contact. These concentrations may be readily determined by routine experimentation. The two polyelectrolyte solutions are contacted in one of the following manners to allow the formation of an interfacial film.

(A) A washer, screen, or other suitable porous object is dipped into or otherwise coated by one of the polyelectrolyte solutions so that a film of the solution is formed across the hole(s) in the object. This assemblage is then coated with the second polyelectrolyte solution in such a way that it comes into intimate contact with the first liquid coating across the holes in the object. The complex polyelectrolyte film which forms rapidly at the two-solution interface will be anchored peripherally to the supporting structure.

(B) Opposite sides of a porous membrane (filter paper, bacteriological filter membrane, etc.) are contacted more-or-less simultaneously with the solutions containing the oppositely-charged polyelectrolytes. Sufficient time is allowed for the solutions to diffuse together and react to form a complex film within, or at the surface of, the supporting membrane. If the film is to be formed within the supporting membrane structure, it is desirable to wet the membrane before contacting it with the polyelectrolyte solutions. The complex film is next washed free of the excess polyelectrolyte solutions by water or other suitable solvents. If the complex film is supported only at infrequent intervals (e.g., when the supporting substrate is a coarse screen, ring, washer, etc.), great care must be taken to avoid breaking the delicate membrane during the washing process.

By way of illustration, complex films formed by interaction between poly(vinylbenzyltrimethyl ammonium chloride), VBTAC, a cationic polyelectrolyte having a molecular weight of about 300,000, and poly (sodium styrene sulfonate), NaSS, an ionic polyelectrolyte having a molecular weight of about 760,000, have been prepared and treated in the following ways.

Example 1: One side of a clean Plexiglas ring, 3/4" o.d., 5/8" i.d., and approximately 1/16" thick was dipped into a 3% aqueous solution of NaSS in such a way that a film of the solution was formed across the hole in the ring. The ring was then turned over and dipped into a 3% aqueous VBTAC solution so that a film of the latter was formed in contact with the NaSS solution film. A coprecipitated polyelectrolyte membrane formed rapidly at the interface between the two liquid films.

The ring was then suspended edgewise in a beaker of water for 1 hr to wash the excess unreacted VBTAC and NaSS solution from the surfaces of the precipitated film. The plastic ring was then removed from the wash water and placed in a desiccator at 50% relative humidity to dry the film. The final dried film was transparent, but in reflected light resembled the "black" films formed by oil on water, indicating its extreme thinness. By careful manipulation, films formed in this manner could be successfully transferred, wet or dry, to supporting surfaces such as glass, or fine mesh metal screens. The dried films adhered tenaciously to any solid surface with which they came in contact.

Example 2: A piece of filter paper (S & S No. 576) was wetted with water and clamped vertically in a diffusion cell in such a way as to form a barrier between the two compartments of the cell. The diffusion rate of NaCl through the filter paper was determined conductimetrically to be 0.31 $meq/hr/in^2$ for a 0.1 molar concentration gradient across the paper (0.1 M NaCl). After being washed, the filter paper was contacted on one side by a 3% NaSS solution, and on the other side by a 3% VBTAC solution by pouring the solutions simultaneously into the two cell compartments.

The heights of the two solutions were adjusted so that no hydraulic gradient existed across the filter paper. After sufficient time had been allowed for the polymer solutions to diffuse into the filter paper and react to form an interfacial film, the solutions were poured out, and the paper was thoroughly washed with distilled water to remove the excess polyelectrolyte solutions. The diffusion rate of 0.1 M NaCl through the treated paper was remeasured and again found to be 0.31 $meq/hr/in^2$, indicating that the diffusional resistance of the interfacial membrane was negligible compared to that of the filter paper. In a test of the continuity of the imbedded interfacial film it was found that less than 0.01 ml of water flowed through the treated filter paper when a 3 inch hydraulic gradient was applied across it for 24 hr.

OTHER MEMBRANES

Carbodiimide Treatment of Cellulose Acetate

J. Steigman and F.S. Chiccarelli; U.S. Patent 3,457,256; July 22, 1969; assigned to American Cyanamid Company describe a method for treating cellulose membranes which comprises contacting the membrane with a carbodiimide followed by treatment of the intermediate with ammonia, and amine or various acids. It is known that carboxy groups are formed during the alkaline aging step in the manufacture of cellulose membranes.

The ability of the membranes to bind divalent cations from a solution is due to the action of the carboxyl groups on the membrane. It has been found that membranes of this type can be altered so as to eliminate their tendency to adsorb such cations. This result is accomplished by treating the cellulose membrane with a carbodiimide, followed by treatment with aqueous ammonia, amines or acids. This reaction converts the reactive carboxyl groups to various amides or anhydrides and the adsorption of polyvalent cations is prevented.

Furthermore, it is found that when aqueous ammonia is utilized, the effect on the membrane is twofold. First, the adsorption of divalent cations from a 0.00001 molar solution is decreased from about 50% to less than about 5%. Secondly, the time required to reach equilibrium in dialysis is increased from about 4 hr for the untreated membrane to 160 hr for the treated membrane. When mixtures of polyvalent and monovalent cations are placed in solution, a partial separation of the two by dialysis can therefore be effected utilizing the membranes. The following examples illustrate the process.

Examples 1 through 14: Dialysis Procedure for Membranes — A portion of an appropriate inorganic working solution [0.00001 M $Ca(NO_3)_2$] is transferred into a bag produced from a carboxy group containing cellulose membrane which is sealed off at one end with a double knot. The other end of the bag is twisted to reduce air space and is sealed off with a double knot. The excess tubing is cut away. The bag is lowered into a glass tube containing the solution and a portion of an inorganic (radioactive tracer solution).

The tube is tightly stoppered with a polyethylene wrapped stopper and sealed with tape. Shaking is performed using a horizontal shaker maintained at 30°C until equilibrium is reached. The bags are removed and quickly rinsed with distilled water. After removal of the excess water, the bag is placed in a clean tube and punctured. The empty bag is removed, quickly rinsed with water and placed overnight in a glass stoppered centrifuge tube containing 1.0 molar HCl to remove the inorganic material adsorbed during dialysis. Equal portions of the inside and outside solutions are transferred to counting vials containing the scintillation solution described below. Activity measurements are performed with a Packard Tri-Carb Liquid Scintillation Spectrometer.

Scintillation solution:

Naphthalene	60 g
Methanol	100 ml
Ethylene glycol	20 ml
p-Dioxane	880 ml
2,5-diphenyloxazol	4 g
1,4-bis-2-(phenyloxazolyl)benzene	200 mg

With the experiments using Na^{22}, 1.0 ml portions of these dialyzed solutions are counted directly on a Baird-Atomic Gamma Spectrometer. In order to count the 1.0 molar HCl used to remove the adsorbed calcium from the membrane, a determination of the quenching effect of the 1.0 molar HCl on the fluorescent is made. The total activity of a 1.0 milliliter portion of 1.0 molar HCl in the presence of a scintillator is determined. A redetermination of the total activity after the addition of a 0.2 ml spike of radioactive tracer is then made. The percent quenching is calculated and a correction is applied to the sample. Following this procedure, various other experiments are conducted to determine the dialysis of untreated cellulose membranes employing a working solution of 0.00001 M Ca^{++} with Ca^{45} as a tracer. The results are set forth below.

Example	Percent Activity of Ca^{++} in Solution	Percent Activity of Ca^{++} in Membrane	Total Recovery
1	53.0	46.5	99.5
2	49.0	45.0	94.0
3	40.0	55.3	95.3
4	60.6	43.0	103.6
5	46.0	50.6	96.6
6	40.8	47.6	88.5
7	58.1	37.2	95.3
8	46.2	52.8	99.0
9	44.4	45.1	99.5
10	53.7	49.5	103.2
11	58.3	39.9	98.2
12	60.0	38.0	98.0
13	68.0	35.0	93.0
14	55.0	42.0	97.5
Average	52.4	44.8	97.2

The results indicate that an average of 44.8% of Ca^{++} is adsorbed by the untreated cellulose membrane, from a 0.00001 molar solution. A series of membranes is prepared as described in Example 1 and dialyzed with 10^{-3}, 10^{-4} and 10^{-5} molar Na^{+}. A 2 ml portion of radioactive tracer Na^{22} is added. There is no indication of Na^{22} adsorption on the membrane.

Treatment of Cellulose Acetate with Polyol and Sulfolane Solutions

A process described by *W.E. Skiens; U.S. Patent 3,494,780; February 10, 1970;and U.S. Patent 3,532,527; October 6, 1970; both assigned to The Dow Chemical Company* provides a membrane of a film-forming cellulose ester by treating the membrane with a solution of a placticizing compound for the cellulose ester and a secondary additive, different from the plasticizing compound, and which is not a solvent for the cellulose ester and is a poor plasticizer for the cellulose ester, for example, a polyol, ethanol or water.

The important feature of the process is the treating of the membrane with the solution of the combined secondary additive and the plasticizing compound. For example, treating the membrane with a combined solution of a polyol and sulfolane compound will usually provide membranes with a water permeability on the order of 15 to 20 times higher than that obtained with membranes treated with an equal amount of the sulfolane compound alone. On the other hand, the secondary additives contemplated in the process, polyols, are normally very poor or nonexistent plasticizers or swelling agents for the cellulose esters and when used alone, they are not able to provide the permeation properties desired.

The plasticizers or plasticizing compounds used in the treatment can be any of those that first of all, of course, plasticize the cellulose ester, and are compatible with the treating solution. For example, such materials include dimethyl sulfoxide, dimethyl formamide, butyrolactone, N-methyl formamide, dimethyl acetamide, caprolactam, 2-pyrrolidone, malonitrile, triacetin, tetramethylene sulfone (frequently referred to as sulfolane) and ring substituted derivatives such as 2,4-dimethylsulfolane, 3-sulfolanyl acetate, etc. Beneficially and preferably, the plasticizer for the cellulose ester that is employed in the treating solution is a sulfolane compound and particularly tetramethylene sulfone (which is frequently referred to as "sulfolane") and ring-substituted derivatives such as 3-ol esters and ethers are discussed in U.S. Patents 2,219,006 and 2,451,299. Preferably, those sulfolane compounds that are employed in the process are represented by the structural formula:

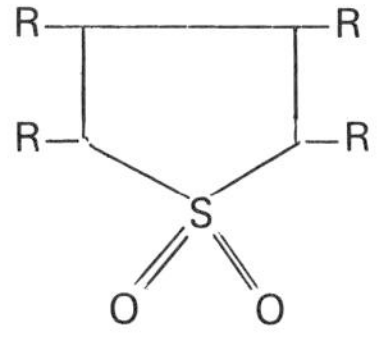

where R represents hydrogen or a methyl radical. Advantageously and preferably, as mentioned, sulfolane is employed, i.e., with reference to the above formula, where each R is hydrogen. The secondary additives that are added with the plasticizing compound in the membrane treating solutions are generally those having hydroxyl groups and having a molecular weight up to about 4,000. The particular one chosen will depend on the acetyl content of the polymer and should be one that will form a solution with the plasticizing compound.

Exemplary of some of the secondary additives that can be employed are water, methanol, ethanol and such alcohols when high salt rejection membranes are desired. When dialysis, type membranes are desired polyols are preferably employed and include such materials as ethylene glycol, diethylene glycol, triethylene glycol, tetraethylene glycol, etc., and including propylene glycol, dipropylene glycol, etc., and mixtures of ethylene and propylene glycol units and such other polyols as glycerin. Advantageously and beneficially, the secondary additives which are polyols having a molecular weight up to about 2,000 are used in the process. Exceptionally excellent results are obtained when the cellulose ester is plasticized and extruded according to the procedure described in U.S. Patent 3,532,527. The following examples illustrate the process.

Example 1: Cellulose triacetate (43.6% acetyl) was blended with a 15:1 mixture (by weight) of tetramethylene sulfone and polyethylene glycol (having MW about 200). The weight ratio of the combined tetramethylene sulfone and polyethylene glycol to cellulose triacetate was about 0.4:1. The plasticized mass was molded into a membrane film at a temperature of about 250°C. The film was removed from the mold and immersed in a solution of a 1:1 mixture (by weight) of tetramethylene sulfone:polyethylene glycol (having a molecular weight of about 200) and maintained therein at about 25°C for about 4 hr. The membrane was removed from this bath and washed with water at 50°C for about ½ hr.

The membrane was then tested to determine its water permeability and salt permeability properties. This was accomplished by mounting the membrane in a test cell using an aqueous sodium chloride solution. It was found that the water permeability of the membrane was about 6 to 7 times greater than that of a commercial dialysis tubing (regenerated cellulose). Additionally, no salt was restrained by the membrane, the salt freely flowing through the membrane. Water permeabilities 100-fold less result if membranes prepared in the identical manner of Example 1 are not post treated according to this process.

Example 2: A membrane film was cast from a polymer solution containing 10% cellulose triacetate (acetyl content 43.6%) 2.8% tetramethylene sulfone and the remainder methylene dichloride. The membrane after casting and air drying was treated by soaking in a solution containing 50% tetramethylene sulfone in water at 25°C for 10 min. The membrane was then removed from the treating bath and washed in 60°C water for ½ hr. The water permeability of this membrane was 5 to 7 times better than a membrane prepared from the above polymer solution but without the sulfolane-water bath treatment. The salt rejection of both the treated and untreated membranes was about 97.8%.

Example 3: A membrane film was cast from a polymer solution containing 10% cellulose diacetate (38.3% acetyl, 40 second ASTM falling ball viscosity), 2.4% tetramethylene sulfone and 0.2% polyethylene glycol (average molecular weight 1,450) and the remainder acetone. The film after casting was air dried a sufficient time to allow the acetone to evaporate. The membrane was then treated in a bath containing 50% tetramethylene sulfolane and 50% polyethylene glycol (average molecular weight 1,450) at 25°C for 1 hr. The membrane was then leached for ½ hr in water at 60°C. The water permeability was found to be equivalent to that of commercial dialysis membranes (i.e., regenerated cellulose) under similar test conditions. No salt was restrained by the membrane, the salt flowing freely through the membrane.

Casting from Butyrolactone Solutions, Nylon and Sulfonic Acid

A process described by *M. Mindick and H.I. Patzelt; U.S. Patent 2,957,206; October 25, 1960; and U.S. Patent 3,297,595; January 10, 1967; both assigned to Nalco Chemical Company* involves electronegative and electropositive ion selective barriers having high permselectivities and low ohmic resistances. It has been discovered that the production of the membranes from a butyrolactone solution of the polymers results in improved electrochemical properties. The permselectivities are high and the resistances are low. The strength of the membranes is also increased. The butyrolactone is stable in the presence of the polyelectrolyte, especially an acidic polyelectrolyte such as polystyrene sulfonic acid.

Butyrolactone is a good solvent for both of the polymers employed in the production of the membranes, and it enables the use of higher concentrations of the polymers. Disadvantages encountered in employing a mixture of two dissimilar solvents are overcome. The volatility of butyrolactone is also relatively low, enabling the use of heat for producing the solution of the polymers, and the viscosity of the solution is favorable. These characteristics are especially significant in the commercial production of the membranes. The following example illustrates the process.

Example 1: In the production of the ion exchange membranes, it is generally preferred to dissolve the film-forming polymer and the polyelectrolyte in separate portions of the butyrolactone, and then mix the solutions. Dynel is dissolved in gamma-butyrolactone, in the proportion of 1 part by weight of Dynel to 4 parts of the lactone, at 200°F with vigorous agitation. The solution is filtered to remove any polymer gel particles. Polystyrene sulfonic acid, 1 part is dissolved in 4 parts of gamma-butyrolactone at 100°F, with agitation.

In manufacturing the membranes, 23 parts of the polystyrene sulfonic acid solution are mixed with 77 parts of the Dynel solution, heated to 160°F, filtered, and vacuum deaerated. The solution is then ready for casting. It may be cast on a glass or metal plate. For producing large quantities of the membranes, it is preferably deposited in a layer, by passing through an elongated slot onto a rotating drum or revolving belt. The layer is then dried and removed by soaking with a polar solvent or peeling the film from the casting surface. The above casting solution contains 4.6% of polystyrene sulfonic acid, 15.4% of Dynel, and 80% of gamma-butyrolactone. The resulting membrane contains 23% of the polyelectrolyte and the balance Dynel, on a completely dry basis.

Example 2: Anion permeable membranes are prepared in like manner to Example 1, which contain, for example, 27.5% of poly(vinyl benzyl trimethyl ammonium iodide) and the balance Dynel, on a dry basis. The same Dynel solution as in Example 1 may be employed. The polyelectrolyte solution is produced by dissolving 1 part of the electrolyte in 4 parts of gamma-butyrolactone, and heating to 120°F with agitation.

In production of the membrane, 27.5 parts of the electrolyte solution are mixed with 72.5 parts of the Dynel solution, heated, filtered, and vacuum deaerated as in Example 1. The casting solution contains 5.5% of the polyelectrolyte, 14.5% Dynel, and 80% of gamma-butyrolactone. The membranes are then produced in the same manner, pouring a layer of sufficient thickness to produce the membrane thickness desired, or employing a doctor blade suitably adjusted for the desired thickness.

H.P. Gregor and H.I. Patzelt; U.S. Patent 3,004,909; October 17, 1961; assigned to Nalco Chemical Company also describe the preparation of electropositive selective permeable membranes.

Electrolyte Surface Treatment

Y. Mizutani, R. Yamane, T. Sata and R. Izuo; U.S. Patents 3,510,418 and 3,510,417; May 5, 1970; assigned to Tokuyama Soda Kabushiki Kaisha, Japan describe an ion permeable membrane which comprises principally an ion exchange membrane composed of an insoluble infusible synthetic organic polymer having an ionic group chemically bonded, the surface of which membrane having been treated with an electrolyte containing an ion component having a molecular weight of at least 100, the dissociation constant of the electrolyte being at least 0.001, and the amount of the electrolyte substance being present in an amount of at least 0.1 mg for each square decimeter of the ion exchange membrane.

It was found that by treating the surface of the ion exchange membrane with a specific electrolytic substance an ion permeable membrane could be obtained which could affect the permeation selectivity of ions of smaller valence for among those having the same charge. It was moreover found that this ion permeable membrane, after having been used continuously for a 1-month period, still maintained its selectivity to specific ions effectively. The electrolytic substance which can be used for this purpose (referred to also as the electrolytic treating agent) are those water-soluble substances which have an ion component having a molecular weight of at least 100 and moreover whose dissociation constant is at least 0.001.

The following examples illustrate the process. In the examples the permselectivity between the ion of smaller valence M_1 and the ion of greater valence M_2 having the same charge has been indicated by the relative transport number.

$$P^{M_2}_{M_1}$$

$$(1) \qquad P^{M_2}_{M_1} = \frac{(t_{M_2}/t_{M_1}) \text{ intramembrane}}{(C_{M_2}/C_{M_1}) \text{ intrasolution}}$$

where t_{M_1} and t_{M_2} are the transport numbers of ions M_1 and M_2, respectively, and C_{M_1} and C_{M_2} are the concentrations of ions M_1 and M_2, respectively.

The electrodialysis apparatus employed was that of two compartments separated by either an anion or a cation selective membrane. A prescribed quantity of the solution of known concentration was introduced into the two compartments, following which an electric current was passed between silver chloride electrodes. The intramembrane cation transport numbers were computed from the changes that occur in the amount of ions in each compartment before and after passage of the electric current, and then the relative transport number was obtained by substitution of these values in Equation (1). The measurement was made at a temperature of 25°C, and both compartments were vigorously stirred.

Separately, a multicompartment type of electrodialysis apparatus constituted by disposing a plurality of pairs of anion and cation selective permeable membranes was used, into each of which compartments was introduced seawater of the following composition at the rate of 6 cm/sec. A current was then passed through the apparatus via electrodes provided at both ends thereof at a current density of 2 amp/dm^2 of the membrane area. The effective area of the membrane was 1 dm^2 and the temperature of the seawater was 30°C. Composition equivalent per liter: Cl, 0.53; SO_4, 0.05; Ca, 0.02; Mg, 0.11; K, 0.01; Na, 0.44.

The solutions concentrated by electrodialysis separately in the alternate compartments were analyzed after their compositions reached equilibrium, and the relative transport number was obtained using the approximate Equation (2).

$$(2) \qquad P^{M_2}_{M_1} = \frac{(C_{M_2}/C_{M_1}) \text{ concentrated solution}}{(C_{M_2}/C_{M_1}) \text{ seawater}}$$

Example 1: (A) One part of finely divided powder of polyvinyl chloride, 0.90 part of styrene, 0.10 part of 50% divinyl benzene, 0.3 part of dioctyl phthalate and 0.01 part of benzoyl peroxide were homogeneously mixed, and the resulting mixture was applied to a 1.6 mesh polyethylene net. The so-treated net was covered with cellophane on both surfaces, and polymerized by heating for 3 hr at a temperature of 110°C. The obtained film was sulfonated for 24 hr with a 98% sulfuric acid of 50°C, thereby to give a cationic exchange membrane having a sulfonic acid group as an exchange group. With the use of this cationic exchange membrane, an electrolytic solution of a mixture of 0.2 N NaCl and 0.2 N $CaCl_2$ was subjected to electrodialysis. The cationic transport number, direct current resistance, and relative transport number, P^{Ca}_{Na}, as measured by the two-compartment electrodialysis method, were 0.98, 7 Ω cm^2, and 2.5, respectively.

(B) In carrying out the abovementioned electrodialysis, poly-2-vinyl pyridine hydrochloride with a molecular weight of 30,000 was added to an electrolytic solution in the anodic compartment to a concentration of 20 ppm. It was found that the cationic transport number, direct current resistance and relative transport number, P^{Ca}_{Na}, are 0.98, 10 Ω cm^2, and 0.4. The presence of about 1 mg/dm^2 of poly-2-vinyl pyridine hydrochloride on the surface of the membrane was observed. Subsequently, the electrolytic solution containing poly-2-vinyl pyridine hydrochloride was discharged from the dialysis vessel. The same electrolytic solution as noted above containing no poly-2-vinyl pyridine hydrochloride was added, and the same experiment was carried out, with the same results. The presence of 0.8 mg/dm^2 of poly-2-vinylpyridine hydrochloride on the membrane surface was observed.

(C) Instead of the polyvinyl pyridine, lauryl pyridinium chloride was added to the anodic compartment to a concentration of 300 ppm. It was found that the cation transport number, direct current resistance and relative transport number, P_{Na}^{Ca}, are 0.95, 11 Ω cm^2, and 1.2, respectively.

Example 2: A polymeric latex consisting of 30 parts of styrene and 70 parts of butadiene was applied to a glass fabric. A membranous substance obtained after drying was sulfonated for 5 hr with a 95% sulfuric acid of 30°C to thereby give a cationic exchange membrane having a sulfonic acid group as an exchange group. With the use of this cationic membrane, an electrolytic solution of a mixture of 0.4 N KCl and 0.1 N $MgCl_2$ was subjected to electrodialysis. The cation transport number, direct current resistance and relative transport number P_K^{Mg}, as measured by the two-compartment electrodialysis method were 0.99, 6 Ω cm^2, and 0.9, respectively.

In the above mentioned dialysis, a polymeric substance obtained by treating poly-2-vinylpyridine with methyl iodide to thereby convert about one-half of the pyridyl group to a methyl pyridinium group was added to the electrolytic solution on both sides of the membrane to a concentration of 5 ppm. As a result, the cation transport number, direct current resistance and relative transport number, P_K^{Mg}, were 0.98, 7 Ω cm^2, and 0.3, respectively.

The presence of about 0.4 mg/dm^2 of poly-2-vinylpyridine hydrochloride on the surface of the membrane was observed. When instead of the methyl iodide treated poly-2-vinylpyridine, polyacrylic acid with a molecular weight of 3,000 was added to both sides of the membrane to a concentration of 300 ppm, the cation transport number, direct current resistance and relative transport number, P_K^{Mg}, were 0.99, 7 Ω cm^2, and 0.8 respectively.

Subsequently, a polymeric substance obtained by treating poly-2-vinylpyridine with methyl iodide to thereby convert about one-half of the pyridyl group into a methyl pyridinium group was added to the anodic side to a concentration of 5 ppm, and polyacrylic acid with a molecular weight of 500 was added to the cathodic side to a concentration of 300 ppm. It was found that the cation transport number, direct current resistance and relative transport number, P_K^{Mg}, are 0.99, 7 Ω cm^2 and 0.2, respectively.

Example 3: Finely divided powder of polyvinyl chloride (1.0 part), 1.5 parts of 4-vinylpyridine, 0.1 part of 50% divinylbenzene, 0.3 part of dioctyl phthalate and 0.02 part of benzoyl peroxide were homogeneously mixed to form a paste. The resulting paste was applied to a fabric of polyvinyl chloride. The treated fabric was covered with cellophane and polymerized by heating for 3 hr at 90°C. The obtained film was immersed for 24 hr at 25°C in a solution composed of 2 parts of methyl iodide and 8 parts of methanol, and then thoroughly washed with hydrochloric acid. There was obtained an anionic exchange membrane having a pyridinium group as an exchange group.

With the use of a multicompartment electrodialysis apparatus in which the anionic exchange membrane and the cationic exchange membrane of Example 1 (A) were provided, an experiment was conducted on the concentrating of sea brine. First, electrodialysis was performed for 3 days by introducing a sea brine to which had been added polyethylene imine with a molecular weight of 30,000 to a concentration of 10 ppm. Subsequently, a sea brine to which the compound was not added was introduced, and the electrodialysis was continued.

The results obtained are shown in the following table in comparison with the results obtained by subjecting sea brine to dialysis with the use of a membrane which was not treated with polyethylene imine. It was observed that 0.6 mg/dm^2 of polyethylene imine was deposited on the surface of the membrane so treated. It is clear from this example that selective permeation of Na ions is markedly improved by treating a cationic exchange membrane with polyethylene imine, and that the ionic exchange membrane of this process has a prolonged selectivity.

Period After Start of Experiment	Before Treatment	2	Days 5	7	10	15	20	30
Sea Brine To Be Treated	Sea Brine	Sea Brine Containing Polyethyleneimine	- - - - - - - - - -	- - -	Sea Brine	- - -	- - -	- - - -
Electric current effect of entire ions	0.91	0.91	0.92	0.91	0.90	0.92	0.91	0.91
P^{Ca}_{Na}	1.6	0.6	0.6	0.6	0.5	0.6	0.7	0.8
P^{Mg}_{Na}	1.0	0.4	0.3	0.3	0.3	0.3	0.5	0.6

Prestressed Reinforced Membrane

P.D. Foote and M.R.J. Wyllie; U.S. Patent 3,009,578; November 21, 1961; assigned to Gulf Research & Development Company have found that the conductivity of heterogeneous ion exchange membranes can be improved without sacrifice in mechanical strength, and that the permselectivity of heterogeneous membranes can be improved without sacrifice in specific conductivity.

Briefly, a plurality of elongated reinforcing members are subjected to tension along their longer axes, and while such tension is maintained, a continuous, coherent resinous structure comprising an electrolytically inert organic plastic matrix having a plurality of finely divided, water-insoluble, electrolytically conductive ion exchange particles substantially uniformly distributed in a proportion such that the opposing surfaces of such structure are conductively connected by a plurality of concatenated groups of such electrolytically conductive particles, is formed about the reinforcing members and then caused to harden, so that the reinforcing members become bonded to the continuous resinous structure. Although the continuous resinous structure is advantageously formed by molding under pressure, it can be formed by casting if desired.

The reinforcing members may be oriented in one or more directions and in one or more planes. Preferably, the reinforcing members will be arranged in two or three intersecting groups of substantially parallel members. Especially good results are obtained by arranging three groups of parallel reinforcing members at right angles to each other. Although the application of any substantial tension to the reinforcing members will improve the mechanical strength and permselectivity of the membrane to some degree, it is preferred that the tension applied equal or exceed the osmotic pressure, that is, the internal swelling pressure of the ion exchange particles incorporated in the membrane.

The osmotic pressure of the ion exchange particles in a heterogeneous ion exchange membrane will normally be in the vicinity of 150 pounds per square inch. Good results therefore will be obtained when the total tension applied to the reinforcing members in each unit of cross-sectional area of the membrane intersected by such members is at least 150 pounds per square inch of cross-sectional membrane area.

In other words, it is preferred that the tension applied to the reinforcing members is such that the product of the number of members per square inch of cross-sectional area of the membrane intersected by such members times the elongating tension per member exceeds 150 pounds per square inch of membrane cross-sectional area, and preferably reaches 400 to 500 pounds per square inch of cross-sectional membrane area.

The reinforcing members can be formed of any material possessing the necessary tensile strength, and they can be monofilamentary or multifilamentary and twisted as in the case of threads, cords, or the like, or they can be in the shape of rods or coiled springs. Woven or felted reinforcing members can also be used. Preferably, the reinforcing members will possess an irregular surface to increase the friction between the bonding plastic and the reinforcing members. Twisted cords provide such an irregular surface and are therefore preferred.

Reinforcing members can be further improved for the purpose of this process by the provision of artificial irregularities such as spaced knots. In the case of rigid reinforcing members integral irregularities in the form of projections can be provided. By way of example, excellent results can be obtained when the reinforcing members are made from cords of spun glass or quartz fibers. Good results are also obtainable with filaments or cords of nylon, rayon, cotton, linen, ramie, jute, synthetic polyester fibers, for example, glycol terephthalate polymer fibers and other materials.

While the reinforcing members will usually be electrically nonconductive, conductive materials such as steel wire or screen can be used, provided that they are electrically and electrolytically insulated. Any electrolytically inert organic plastic can be used to form the bonding matrix. For example, excellent results are obtainable with methyl methacrylate polymers and with styrene-divinyl benzene copolymers.

Polyvinyl Pyrrolidone

M. Mindick and R. Oda; U.S. Patent 3,069,340; December 18, 1962; assigned to Nalco Chemical Company describe a semipermeable membrane which comprises a solid layer of a homogeneous molecular dispersion of a water-insoluble, acid and alkali-resistant thermoplastic film-forming polymer and a polymer of N-vinyl pyrrolidone, the layer having high porosity relative to the porosities of layers of plastic material prepared by conventional methods. The layers providing a semipermeable membrane preferably have a water-holding capacity of at least 45% by weight on a wet basis. In a preferred form of the process the layer contains a substantial quantity of pores over 30 angstrom units in size, and the pore size may be within about 200 angstrom units.

It is necessary for utility as a semipermeable membrane that the layer have a low resistance to the diffusion of materials. Thus, a maximum resistance is provided in the layer of 10 ohms per square centimeter when measured in 0.15 N KCl. Specific applications may require lower resistances. For example, it is preferred to provide a membrane having a maximum of about 6 ohms resistance for dialysis of caustic solutions. It is further preferred in the case of battery separators that the resistance be a maximum of about 2 ohms as measured in the above manner.

High molecular weight vinyl pyrrolidone polymers are most desirably employed, having a minimum molecular weight of about 5,000. It is preferred that the molecular weight be greater than 20,000 and further preferably, greater than 40,000.

Elastomeric Material

P.E. Hoch and P. Robitschek; U.S. Patent 2,978,393; April 4, 1961; assigned to The Hooker Chemical Corporation have found that homogeneous permselective membranes containing these desired properties for use in industrial electrolytic applications may be produced by vulcanizing, in the presence of a vulcanizing agent, a dry composition selected from the group consisting of:

(A) a blended mixture comprised of
  (1) an addition polymer containing groups which may be hydrolyzed to form carboxyl groups, and
  (2) a polymer which is free of potentially active ion exchange groups;

(B) a blended mixture comprised of
  (1) an addition polymer containing carboxyl groups, and
  (2) an olefinically unsaturated monomer free of potentially active ion exchange groups;

(C) a polymerized composition produced by polymerizing a mixture of monomers comprised of
  (1) a mixture of an olefinic monomer containing a group which may be subsequently hydrolyzed to form a carboxyl group and an olefinic carboxylic compound, and
  (2) an olefinic monomer free from potentially active ion exchange groups; and

(D) a polymerized composition produced by polymerizing a mixture of monomers comprised of
  (1) an olefinic carboxylic compound and
  (2) an olefinic monomer free of potentially active ion exchange groups.

The dry elastomeric membrane thus produced is free from water or any liquid solvent and is then treated in a hydrolyzing medium such as aqueous sodium hydroxide until the potentially active ion exchange groups have been converted into carboxyl groups or their salts. It has been found that the membranes have inherently good physical properties, high efficiency in selectively transferring ions and good conductivity. The membranes thus formed comprise an elastomeric addition polymer containing active ion exchange groups homogeneously distributed throughout the membranes so that the strength of the membrane is not adversely affected.

The following examples illustrate the process. Throughout the examples, for the sake of expediency, a number of trade names and proprietary named materials have been referred to. Following is a listing of the formulations of the various materials so used:

PA-21-commercial polyethyl acrylate-alpha-chlorovinyl ether rubber
Rotax - mercaptobenzothiazole
Methyl tuads - tetramethylthiuram disulfide
Duponal ME - fatty alcohol sulfates
Daxad 11 - polymerized mixture of alkylaryl sulfonic acids
Dow 529K - 80% styrene, 20% butadiene copolymer
GRS 2000 - butadiene-styrene rubber containing 44 to 48% sytrene
Dow 512K - 60% styrene, 40% butadiene copolymer
Kralac AP - 85% styrene, 15% butadiene copolymer
Hycar 2007 - 80% styrene, 20% butadiene copolymer

Example 1: 40 g of a commercially available polyethylacrylate rubber designated PA-21 was placed on a rubber mill maintained at 150°F. This material was banded slowly. After banding for a short period, 60 g of Dow 529K was slowly added to the banded polyacrylate rubber. After thorough banding and mixing on the mill, the following ingredients were added: 2 g of stearic acid, 2 g of zinc oxide, ½ g of a Rotax, 1 g of methyl tuad, 30 g of carbon black, 1 g of triethylenetetramine and 2 g of sulfur. These materials were blended thoroughly. The amine and sulfur were added at a lowered mill temperature of 120°F.

After thorough mixing the material was stripped from the mill and was ready for vulcanization. A piece of this material was placed in a flash mold at 167°C for 1 ½ to 2 hr. The cured rubber at this point was then submitted to hydrolysis in hot 13 to 17% caustic for 6 days. At this point the electrical conductivity of the membrane had attained a sufficiently high value.

Example 2: To a preheated (150°F) roller mill was added 75.0 g PA-21 rubber gum. After breakdown and bonding had taken place, 25.0 g GRS-2000 rubber was added. These materials were thoroughly mixed. To this bonded material 50 g of carbon black was added slowly. After mixing 1.0 g of stearic acid was added. To the mix was then added 0.5 g Rotax, 1.0 g methyl tuad and 2.0 g sulfur. Finally 1.0 g triethylenetetramine was slowly added. The batch was mixed, stripped from the mill several times and repassed through the mill. A tacky sheet was obtained.

A sheet of the crude rubber from the mill, as prepared above, was cut and placed in a square flask mold lined with cellophane. This cold mold was placed in a press preheated to 160°C. Pressure was raised slowly to close the mold (during 5 min) and the pressure adjusted to above 20,000 lb platen pressure. The temperature of 160° to 165°C and the pressure was maintained for 40 min. The mold was then cooled and the rubber removed.

A sheet of the cured rubber described above was placed at 90° to 95°C in 17% potassium hydroxide for 5 days. The rubber was considerably swollen. A piece of this material exhibited a current efficiency of 66% in the test cell, and permselectivities of greater than 90%. This material was suitable for use as a permselective membrane.

Example 3: The membrane produced above in Example 2 was placed in a two compartment test cell in order to test its usefulness in producing chlorine and caustic soda. The test cell contained two compartments which were separated from each other by the membrane produced above. Sodium chloride brine was fed continuously at approximately 25.5% sodium chloride and was depleted to approximately 22 to 23% sodium chloride. Water was added continuously to the cathode compartment to maintain a concentration of approximately 25 to 40 g/l sodium hydroxide. The anode consisted of a graphite block and the cathode of a steel screen.

The membrane was spaced from the anode by a ¼" thick soft rubber gasket. The cathode screen was placed in contact with the diaphragm and spaced ½" from the steel back plate of the cell. The active area of the diaphragm was 7.9 in^2. The voltage was adjusted to pass a 5 ampere current through the cell, or a current density of 91 amperes per square foot of membrane. Chlorine was continuously produced at the anode, hydrogen at the cathode, and substantially pure sodium hydroxide in the cathode compartment. The membrane exhibited high current efficiency and long useful life.

The table below shows examples of elastomeric permselective membranes made according to the process described in Examples 1 and 2 and tested as described in Example 3. Various blends were utilized in varying proportions as set out in the table. In every example, the following traditional rubber additives were mixed as described in Examples 1 and 2; 2 g sulfur, 2 g zinc oxide, 0.5 g Rotax, 1 g methyl tuad, 2 g stearic acid and 2 g triethylenetetramine.

Ex. No.	Functional Component, grams		Diluting Component, grams		Filler, grams		Permanent Selectivity, percent	Current Efficiency, percent	Life
4	PA-21	100	–	–	Carbon black	20	83	55	9 hr
5	PA-21	75	GRS-2000	25	Carbon black	20	68.6	60	2 days
6	PA-21	40	GRS-2000	60	Carbon black	30	75	81	11 days
7	PA-21	40	Dow 529-K	60	Carbon black	30	81.6	86	43 days
8	PA-21	40	Dow 512-K	60	Carbon black	30	95.1	81.8	12 days
9	PA-21	40	Kralac-AP	60	Carbon black	30	72	55	41 days
10	PA-21	40	Hycar 2007	60	None	–	72.2	63	21 days
11	PA-21	40	50% butadiene 50% DCDFE*	60	Carbon black	30	61	60	11 days

*Dichlorodifluoroethylene

Filled Silicone Rubber

E. Pungor, J. Havas, K. Toth and G. Madarász; U.S. Patent 3,446,726; May 27, 1969; assigned to Metrimpex Magyar Muszeripari Kulkereskedelmi Vallalat, Hungary describe heterogeneous selective membranes which are suitable for the direct potentiometric determination of the quantity of ions present in solution, for ion exchange and for the separation of ions. The membranes contain an active material of suitable particle size dispersed in a silicone-rubber carrier phase in a proportion that the original mechanical properties of silicone-rubber are unchanged. Particle size of the active material is for pearl-type conduction and for the attainment of chemical equilibria at the two boundary layers are realized.

To the silicone-rubber carrier phase active materials such as inorganic or organic precipitates, metal chelate complexes, ion exchange resins or molecular sieves are preferably added whereas the active material dispersed in the carrier phase acts at the same time as a filler for the silicone rubber as well. The suitable particle size of the inorganic or organic precipitate or metal chelate complex as regards pearl-type conduction is governed by the parameters of precipitation.

However, if active materials composed of an ion exchange resin or a molecular sieve are used, the control of the particle size is achieved by the use of a suitable milling technique. The particle size of 1 to 50 microns of the active material dispersed in the carrier phase has been found preferably in practice.

As a silicone-rubber polymer, preferably a polysiloxane of the molecular weight 15,000 to 30,000 and containing methyl-, phenyl-, vinyl- or other organic substituents and SiOH end groups is applied. It is possible to add solvents to the polymeric mixture containing the active material, and in this case the method of processing in order to obtain membrane layers can be controlled by the quantity of the solvents used.

The membrane layers prepared of the polymeric mixture can be formed by pouring, calendering or rolling. A kneader or roller mill can be used to disperse the precipitate in the polysiloxane. A mixer or rubber roller mill can be used to homogenize the mixture. The usual curing agents and catalysts are added to the homogenized mixture and the film-formation or curing is carried out at ambient or elevated temperatures. Advantages of membranes prepared according to the process are the following:

(1) In water they show but a moderate swelling (1 to 10%) and consequently they can more preferably be employed for the determination of electrode potentials as compared to other membrane electrodes previously known.

(2) Their mechanical strength and elasticity is equal to those of the original silicone-rubber film, and consequently they can advantageously be used; their elasticity is 300 to 500%.

(3) They show a fair temperature resistance up to 270°C. The temperature of application depends only on the properties of the active material dispersed in the carrier phase.

(4) A property of paramount importance from an electro-chemical point of view is their high capacity.

(5) Their preparation is extremely simple and easily reproducible. They can be prepared in any desired thickness or form.

Example 1: The preparation of a heterogeneous membrane for selective determination of iodide activity by means of calendering follows. At first, the silver iodide (AgI) precipitate suitable for the determination of iodide activity is prepared. To 100 ml N/100 silver nitrate 10 drops of a 0.2% p-ethoxychrysoidine solution and a 50% excess of potassium iodide, in the form of an aqueous N/10 solution, are added.

Precipitation is carried out at room temperature with stirring. The precipitate formed is centrifuged and washed three times with distilled water in a room protected from direct sunshine. The washed precipitate is collected upon a filter paper and is dried in a drying cabinet at 80°C to constant weight. The dried AgI precipitate is applied as active material in the preparation of heterogeneous membranes.

In order to form a carrier phase for the heterogeneous membrane, dimethylpolysiloxane is applied. 30 parts by weight of a precipitate prepared according to the above are dispersed by mixing in 70 parts by weight of dimethylpolysiloxane (Silopren K 100, Farbenfabriken Bayer) until a statistically homogeneous suspension is obtained. The curing agent and catalyst mixture is prepared by mixing 1 part by weight dibutyl stannic dilaurate and 2 parts by weight hexaethoxysiloxane.

The curing agent and catalyst is added to the dimethylpolysiloxane containing the precipitate in such a proportion as to ensure perfect curing and the curable mixture thus obtained is calendered between two plastic sheets to the desired thickness. Curing of the mixture, i.e., completion of the reaction, is effected within about 10 min. A piece of a size corresponding to the desired application of the membrane is cut from the elastic film prepared according to the above and cemented to a glass tube by means of silicone cement.

The silver iodide membrane prepared according to the above can be employed for the determination of iodide-activity by pouring a 10^{-3} M potassium iodide solution into the supporting glass tube and in order to form a galvanic contact a calomel half-cell is placed in the solution. The membrane electrode thus obtained indicates the activity of iodide ions present in the solution when measured against a reference electrode. An equilibrium state is reached within a few minutes, the measuring accuracy is ±5 mv. Membrane electrodes prepared according to the above respond to any changes in iodide or silver ion activity with a change in membrane potential. Measuring range: 10^{-1} to 10^{-7} M.

The selectivity of the membrane electrode can be characterized by the fact that it can furnish a value characteristic of the activity of the ions to be measured even in the presence of some other electrolyte of high concentration. As is apparent from the data summarized in the tables below, the iodide activity is changed in the presence of other electrolytes but the measured values are in good agreement with theory even in such cases. Similarly, it can be seen from the data that the accuracy of silicone rubber carrier phase heterogeneous membrane electrodes is equal, or even higher than that of membrane containing paraffin carrier phase.

Iodide-Activity Values Measured in the Presence of 10^{-1} M Potassium Chloride

p_I Calculated	p_I Measured	Δp_I
1.12	1.0	-0.12
2.12	2.05	-0.07
3.12	3.07	-0.05
4.12	4.05	-0.07
5.12	5.20	+0.08
6.12	6.18	+0.06

Iodide-Activity Values Measured in the Presence of 10^{-1} M Sodium Sulfate

p_I Calculated	p_I Measured	Δp_I
1.35	1.17	-0.18
2.35	2.08	-0.27
3.35	3.18	-0.17
4.35	4.40	+0.05
5.35	5.45	+0.10

Example 2: The preparation of a heterogeneous membrane containing barium sulfate active material follows. To a N/100 barium chloride solution ammonium sulfate in the form of 2N solution is added in a 50% excess at a pH value of 2, at ambient temperature, with stirring. 30 parts by weight of the barium sulfate precipitate, processed and dried in the usual manner, are dispersed in 70 parts by weight of silicone rubber polymer and the mixture is processed as described in Example 1. The heterogeneous membrane thus prepared has been used for the determination of sulfate ion activity and the results obtained are summarized below.

Determination of Sulfate Ion Activity

Measured p_{SO_4}	Calculated p_{SO_4}	Δp_{SO_4}
1.24	1.34	+0.10
2.25	2.14	+0.11
3.05	3.05	0
4.10	4.01	+0.09
5.10	5.00	+0.10
5.60	6.00	-0.40

Graft Copolymers by Means of High Energy Radiation

W.K.-W. Chen; U.S. Patent 3,247,133; April 19, 1966; assigned to American Machine & Foundry Company describes ion exchange membranes made from copolymers which have been made with "high energy ionizing" radiation. The ion exchange materials are in the solid phase at room temperature and pressure. They may be used as sheets or tubes or other molded or cast shapes, or may be used as beads or pellets.

When in the form of membranes, they find use as the elements of multiple chamber dialysis cells and in the more finely divided form they are easily used in standard ion exchange columns. They are suitable for all known ion exchange purposes such as demineralization of water, collection or concentration of radioactive material or heavy metals and the purification of protein and sugar solutions. Ion exchange membranes made according to this process can be stored or transported in a dry state, and this is an important commercial advantage.

The inert trunk polymer, which may be internally crosslinked, is preferably a vinyl polymer or copolymer. Polyethylene of densities between 0.91 and 0.95 is a preferred trunk material. Polyvinyl halides such as polyvinyl chloride, polyvinyl bromide, polyvinyl fluoride and polyvinyl iodide are likewise useful.

The active polymer branches which furnish sites for the location of ion exchange groups are formed by grafting monomeric material onto the inert trunk polymer by treatment with ionizing radiation such as styrene, which is a preferred branch material, and styrene derivatives such as alpha alkyl styrenes (e.g., alpha methyl styrene), alpha halogenated styrenes (e.g., alpha chlorostyrene or alpha bromostyrene), haloalkyl styrenes (e.g., chloromethyl styrene or bromoethyl styrene) or esters of styrene sulfonic acid.

The active polymer branches are treated by chemical methods to produce ion exchange groups. For example, cation exchange material can be produced by sulfonation of the phenyl groups in the branch chains. This procedure is easily performed by treatment with chlorosulfonic acid, oleum, concentrated sulfuric acid or sulfame. Amination by a tertiary amine after chloromethylation of the phenyl groups produces very satisfactory quaternary amine anion exchange material.

Some of the active polymer branch material is inherently ionizable and requires no further treatment. Among the branch materials which require no further treatment to form ion exchange groups are vinyl pyridine, vinyl piperidine, vinyl lutidine, vinyl aniline, allyl amine and N,N-dialkylaminoethyl methacrylates. The process is illustrated by the following examples.

Example 1: In a preferred example of a method according to this process, a polyethylene film 4 mils thick and 3 in square was immersed in styrene monomer for about 30 min. Absorption or swelling reached an equilibrium maximum within this time and the whole system was next irradiated by exposure for 5 hr to a cobalt 60 source. The total radiation dose was about 1.5×10^6 roentgens. The film was removed from monomer and washed for several hours with benzene to remove unpolymerized and homopolymerized styrene. The film was dried to constant weight and a 26% weight increase was observed which corresponds to a 21% styrene content in the copolymer. The flexible copolymer film had a tensile strength of about 2,000 psi.

A 0.2 g piece of the copolymer film was sulfonated at room temperature for 40 min by immersion in chlorosulfonic acid. It was then immersed in carbon tetrachloride and washed with water to remove excess acid. Finally, it was heated in about 20% caustic soda for about 15 min at 60° to 70°C. The ion exchange membrane so produced showed a wet tensile strength of about 1,150 lb/sq in, and it was flexible even when dry. The H^+ form of the membrane swelled in distilled water to about 162% of the weight of the dry membrane. Titration with 0.10 normal caustic soda showed a dry capacity of 1.72 meq/g and a wet capacity of 1.06 meq/g. The permselectivity of this membrane was 0.94 in 0.10 N sodium chloride and 0.80 in 0.60 N sodium chloride. The Na^+ form of the membrane showed a specific conductivity of 11×10^{-3} reciprocal ohm per centimeter when equilibrated in distilled water. In 1.0 N NaCl the specific conductivity increased to 13×10^{-3} reciprocal ohm per centimeter. Measurements were made parallel to the membrane surface.

Example 2: A polyethylene film 10 mils thick and 3 in square was immersed in styrene monomer for about 30 min. Absorption or swelling reached an equilibrium maximum within this time and the whole system was next irradiated by exposure for 6 hours to a cobalt 60 source. The total radiation dose was about 1.8×10^6 roentgens. The film was removed from monomer and washed for several hours with toluene to remove unpolymerized and homopolymerized styrene. The film was dried to constant weight and a 47% weight increase was observed which corresponds to a 32% styrene content in the copolymer. The flexible copolymer film had a tensile strength of about 1,500 psi.

A 0.5 g piece of the copolymer film was sulfonated at room temperature for 140 min by immersion in chlorosulfonic acid. It was then immersed in carbon tetrachloride and washed with water to remove excess acid. Finally it was heated in about 20% caustic soda for about 15 min at 60° to 70°C. The ion exchange membrane so produced showed a wet tensile strength of about 600 psi, and it was flexible even when dry. The H^+ form of the membrane swelled in distilled water to about 190% of the weight of the dry membrane. Titration with 0.10 N caustic soda showed a dry capacity of 1.08 meq/g. The permselectivity of this membrane was 0.97 in 0.10 N sodium chloride and 0.90 in 0.60 N sodium chloride.

The Na^+ form of the membrane showed a specific conductivity of 14×10^{-3} reciprocal ohm per centimeter when equilibrated in distilled water. In 1.0 N sodium chloride the specific conductivity increased to 17×10^{-3} reciprocal ohm per centimeter. The conductivity measurements were made parallel to the membrane surface.

Example 3: A synthetic rubber sheet (butadiene and acrylonitrile copolymer sheet) was immersed in styrene monomer containing inhibitors. The swollen gel was placed in a flat aluminum mold and placed in a cobalt 60 source for 100 hours so that the total dosage received was about 30 million roentgens. The sheet when removed from the mold was a slightly elastic plastic. Sulfonation of this sheet yielded a cation exchange membrane.

Solid Ion Exchange Structures Using Phenol Sulfonic Acid-Formaldehyde Polymers

W. Juda and W.A. McRae; U.S. Patent Reissue 24,865; September 6, 1960; assigned to Ionics Incorporated have found that solid, physically stable structures, including membranes, diaphragms, sheets, rods, tubes, vessels and objects of many different shapes (having at least 1 and more particularly at least 2 dimensions greater than ¼ inch) can be prepared presenting water-insoluble, coherent, ion exchange materials of high specific conductivities.

According to this process, such ion exchange materials are of high specific conductivities, exceeding 0.5×10^{-2} $ohm^{-1} \times cm^{-1}$ and of high capacities, and contain substantial amounts of dispersed water, not less than 15% of the weight of the air dry exchanger material and up to such proportions as might mechanically interfere with its uniform, coherent or permselective characteristics.

Such coherent structures may generally be made by casting, molding (including compression molding, if without substantial loss or segregation of the water component) and other conventional means of providing large continuous coherent structures of thermosetting or thermoplastic resins, except that the structures of this process are made and maintained in aqueous media and/or an atmosphere of substantially saturated humidity, so as to prevent evaporation. In this manner, the active groups attached to the polymeric organic matrix and oriented with respect to the interfaces are partially or completely dissociable in the internal gel water into fixed ions of one sign linked to the polymer and into mobile ions of opposite sign. The latter are exchangeable ions and the main or substantially sole carriers of electric current.

In order to obtain the above high conductivities with active groups of relatively low dissociation constants between 10^{-5} and 10^{-3}, when measured in the form of one resin-forming ingredient in aqueous solution prior to condensation of polymerization, it is necessary to provide a high density of such groups, that is, unusually high exchange capacities, exceeding 3 meq/g of dry exchanger. On the other hand, in the case of ion exchangers having the preferred strongly dissociated active groups (i.e., having a dissociation constant above 10^{-3}) smaller exchange capacities, of not less than 0.3 meq/g of dry exchanger have now been found to be entirely adequate to give the above conductivities.

Suitable active acidic functional groups linked to a polymeric matrix include SO_3H, COOH and the like, SO_3H being preferred because of its high dissociation constant exeeding 10^{-3} in suitable resin-forming compounds. The exchangeable hydrogen ion may be partially or completely substituted by other substantially dissociated cations such as the alkali metal ions, the alkaline earth metal ions, namely, calcium, strontium, barium and radium, and also silver, copper, magnesium and ammonium ions.

Typical polymeric matrices to which the functional groups are linked include phenolaldehyde resins, polystyrene-divinylbenzene copolymers and the like. Similarly suitable active basic groups linked to polymeric matrices include quaternary ammonium hydroxides, amino groups, the guanidyl group the dicyandiamidine group, and other nitrogen containing basic groups. Quaternary ammonium hydroxide groups, the guanidine and the dicyandiamidine residue are among the preferred basic groups because of their high dissociation constant exceeding 10^{-3}. Typical polymers to which active basic groups are linked include the urea-formaldehyde type resins, the melamine-formaldehyde type resins and the poly-alkylene-polyamine-formaldehyde resins. The following example illustrates the process.

Example: The preparation of membranes of phenol sulfonic acid-formaldehyde, the impregnating, low molecular weight polymer, was prepared as follows:

	Parts by Weight
Aqueous phenol sulfonic acid (65%)	50
Aqueous formaldehyde (35.4%)	24.7

The acid and the formaldehyde are shaken together and partially polymerized at 50°C in a closed container (to retain the moisture and formaldehyde). This precuring requires 1½ to 2 hr after which the viscous mixture was used to impregnate reinforcing webs such as saran, vinyon, glass cloth, and similar materials resistant to strong acids. It was poured into a mold to form the cast disks. The curing (until the polymer turned dark brown or black) was carried out at 100°C in a closed system and in the presence of moisture. This process required from 2 hr to 2 days depending upon the quantity and geometry of the polymerizing mass.

The special characteristics of the cured material both cast and impregnated have been entered into the following table. Precured material was preserved in a refrigerator at 5°C for weeks at a time with no deleterious effect. Before using, the diaphragm was conditioned by soaking in water to bring the water content to the saturation value and maintained in wet condition throughout, for purposes of testing and use.

In the course of measuring the conductance of the diaphragm, it was found that the specific conductivity (1.4×10^{-1} ohm^{-1} cm^{-1}) of the diaphragm in equilibrium with 1 N hydrochloric acid was greater than that of the 1 N hydrochloric acid (0.36×10^{-1} ohm^{-1} cm^{-1}) itself. It may also be observed that in preparing the ion exchange materials of this example the time and/or temperature and other conditions of curing may be considerably extended without substantially altering the electrical and/or surface characteristics of the resin so long as the water component is maintained.

	Cast	Impregnated Saran
Characteristic concentration potential, mv	+55	+51
Conductivity of structure ($ohms^{-1} \times cm^{-1}$)	$0.9\text{-}1.4 \times 10^{-1}$	$4.5\text{-}5.5 \times 10^{-2}$
Moisture % of bone dry material	114	114
Capacity meq/g bone dry material	1.5 - 1.7	1.5 - 1.7
K avg. (order of magnitude)	7×10^{-2}	7×10^{-2}

S.S. Alexander and V. Cioffi; U.S. Patent 3,412,006; November 19, 1968; assigned to Ionics Incorporated describe ion exchange membranes for use in electrodialysis apparatus whereby certain areas of the membranes, especially the area around the manifold holes, are chemically treated by impregnating the area with a liquid monomer which is polymerized therein to form a crosslinked, substantially nonionic polymer for the purpose of obtaining an area having increased stiffness and increased electrical resistance as compared to the untreated remaining area of the membrane.

Vinyl Alcohol Copolymers Containing Hydroxyl Sulfonyl Groups

A process described by *G. Bourat and R. Margraff; U.S. Patent 3,741,945; June 26, 1973; assigned to Rhone Poulenc SA, France* provides vinyl alcohol copolymers partially etherified by substituted or unsubstituted hydroxysulfonyl organic radicals optionally partially crosslinked by

```
 |     |
HC—O—CH
 |     |
```

linkages. Such copolymers can be used as cation exchange resins possessing excellent mechanical and electrical properties. These copolymers are obtained by reacting the vinyl alcohol copolymer in solution with one or more sultones.

The starting copolymers are vinyl alcohol copolymers with one or more ethylenically unsaturated monomers. Ethylene, vinyl chloride and vinyl acetate are advantageously used as the unsaturated monomers. The percentage by weight of vinyl alcohol (or hydroxyethylene) units $—CH_2—CHOH—$ in the copolymer is usually 5 to 60% and preferably 15 to 30% by weight of the unetherified copolymer. The preparation of such copolymers has been described in U.S. Patent 2,386,347 and in French Patent 1,401,550. It is possible partially to crosslink these copolymers by thermal dehydration or by acetalization.

The sultones which can be used are monosultones or polysultones, generally mono- or di-γ-sultones which introduce hydroxysulfonylalkyl or bis(hydroxysulfonyl) alkylene radicals. Examples of suitable mono-γ-sultones are those substituted by an alkyl radical at the carbon atom linked to the oxygen atom, e.g., propenesultone, which is preferably used, especially because of its high reactivity and introduces the radical: $HOSO_2CH_2CH_2CH_2—$.

Suitable di-γ-sultones are 1,2-bis(p-nitro-o-sulfophenyl)-ethanediol-1,2, methylene di-naphthosultone, di-naphthosultone, and especially the di-γ-sultone of 2,2-bis(hydroxymethyl)-propane-1,3-disulfonic acid referred to hereinafter as spirobi-2,2'-propanesultone, which introduces the radical of formula: $(HOSO_2CH_2)_2C(CH_2—)_2$.

The membranes obtained are generally supple and can be stored dry without losing their mechanical properties. They simultaneously possess excellent permeation selectivity and a low resistance to the passage of an electric current. The combination of these properties makes it particularly advantageous to use them in electrodialysis and in fuel piles of the type usually called "low temperature hydrogen-oxygen piles." The examples which follow illustrate the process.

Example 1: 2 g of an ethylene-vinyl alcohol copolymer, obtained by total saponification of an ethylene-vinyl acetate copolymer containing 40% by weight of acetoxyethylene units and having a reduced viscosity of 87 cm^3/g in a 5 g/l solution in toluene at 25°C, are dissolved with stirring in 30 cm^3 of hexamethylphosphotriamide heated to 115°C. 3 g of spirobi-2,2'-propanesultone are added and the stirring and temperature are maintained for 2 hr. 1 g of propanesultone is then added, the mixture is kept at 115°C for 5 min, and the solution is then spread on a 20 cm x 2 cm glass plate arranged horizontally in a ventilated oven heated to 115°C.

After remaining for 10 hr in this oven, the plate is cooled to 20°C and then dipped in water heated to 80°C. The membrane detaches itself from the plate and is steeped for 8 hr in a normal aqueous potassium hydroxide solution heated to 95°C by a water bath. The membrane is then washed with water until the wash waters are no longer alkaline. Its theoretical exchange capacity as deduced from its sulfur content is 0.9 milliequivalent per gram of dry resin. The dry membrane is translucent and supple even in the dry state.

Its permeation selectivity measured between 2 KCl solutions respectively 0.8 and 0.4 M, is 0.91. Its electrical substitution resistance in an 0.6 M KCl medium is 2 Ω cm^2 (the electrical substitution resistance of a given membrane surface is the variation in electrical resistance of a liquid column if the membrane is substituted for a slice of liquid of the same thickness and the same surface area as the membrane, in a position at right angles to the axis of the column).

Examples 2 through 6: The conditions of Example 1 are used, with the respective contents of vinyl alcohol-ethylene copolymers, spirobi-2,2'-propanesultone and propanesultone being varied. The table which follows indicates these various contents and the results obtained. These membranes are translucent and supple even in the dry state.

Example	Weight of— Copolymer in grams	Disultone in grams	Monosultone in grams	Percent S in dry product	Substitution [1]
2	2.5	3.75	1.25	2.2	2.5
3	2	1	0.5	2.2	6
4	2	0.5	1	2.3	2.5
5	2	2	0.5	4.5	6
6	2	1.5	1	3.8	2.5

[1] Resistance in an 0.6 M KCl_2 medium (Ω $cm.^2$).

Acrylic Polymers Containing Complexed Silver and Copper

A. Stoy; U.S. Patent 3,734,897; May 22, 1973; assigned to Ceskoslovenska Akademie ved, Czechoslovakia describes polymers and copolymers containing silver and/or copper which are useful in manufacturing dialysis and filtration membranes, ion exchangers, light sensitive layers, catalysts and hydrophilic coatings. These are prepared from copolymers or polymer mixtures containing both nitrile groups and strong acidic groups, by bringing them in contact with monovalent silver or copper ions or with their mixtures.

Silver and/or copper are attached to nitrile groups by side valences, forming thus complex compounds. The probable formulae of the complex sidegroups are: $—CN·Ag^+$ or perhaps $—CN·Ag^+·NC—$ and $—CN·Cu^+$ or perhaps $—CN·Cu^+·Cu^+·NC—$, the complex cations forming polymeric salts with strong acidic groups of the same or of an adjoining macromolecule, to which all the nitrile and strongly acidic groups are attached as their side substituents.

Membranes from the polymers, when used for filtering aqueous solutions, reject large molecules and particles without being blocked thereby. Small molecules such as water or small ions can penetrate through such membranes. If there are no large molecules or particles present which could choke up pores, it is possible to use membranes or other forms of the polymers with visible porosity or macroporosity.

Besides known methods for obtaining porous structure, it is possible to obtain porous membranes also by casting polymer solutions with solvents consisting at least partly of low volatile liquids such as dimethylformamide, and evaporate the solvent but incompletely. Such membrane is apparently almost dry, but when washed in water it becomes porous as the low volatile solvent is rapidly dissolved and the very viscous solution coagulated.

Instead of water the still partly wet membrance can be washed in any other solvent capable of dissolving the remaining true solvent but incapable of dissolving the polymer. By reacting such porous membranes with silver or cuprous ions one obtains large surface structures, very effective, e.g., for catalysts or for dialysis of clear solutions.

Example 1: Concentrated aqueous zinc chloride solution (d = 1.95) was gradually added to 2 mols of acrylonitrile and 1 mol of sodium ethylenesulfonate (the latter in the form of a 50% aqueous solution), until the two monomers were uniformly dissolved. A redox polymerization catalyst was then added in the form of 5% aqueous solutions of potassium pyrosulfite and ammonium persulfate (0.1% of each, on the total weight of the monomers).

Finally, 0.0001% of copper sulfate-pentahydrate (in the form of an 0.1% aqueous solution) was added to polymerization activator. The somewhat turbid solution was left to polymerize under carbon dioxide at room temperature. Next day the viscous copolymer solution thus obtained was poured in a thin stream in excessive, vigorously stirred water containing 0.2% of hydrochloric acid.

The fibrous, highly swollen coagulate was then repeatedly washed in 0.5% nitric acid, and water. Excess of water was removed by filtering under suction and the highly swollen copolymer was mixed, without being dried, with threefold volume of dimethylformamide. After three hours of stirring at 60°C the almost clear solution was filtered. It contained about 5% of the copolymer. The still hot solution was degasified by applying a vacuum and poured on an exactly horizontal polished plate glass in a 2 mm thick layer.

By slowly evaporating the solvents a hard, clear membrane was obtained which could be easily removed from the glass when moistened with water. The membrane was repeatedly washed in 0.5% nitric acid, in water, in a 1% aqueous solution of disodium salt of ethylene-diamine tetra-acetic acid and again in water. Finally, the pure membrane was immersed into a 0.2 N silver nitrate solution. The membrane shrank thereby and became opaque and greyish white. After washing out the excess of silver nitrate the membrane, having increased tenacity, could be used, e.g., for filtering or dialysis. It could be easily cold-stretched, whereby its tenacity was further increased.

Silver could be removed from the membrane in an aqueous solution of sodium thiosulfate. The membrane was then more swellable than prior to the removal of silver, but considerably less swellable than the raw unoriented membrane. The difference of length caused by shrinking in 0.1 N nitric acid and swelling in pure water, which was 150 to 215% in the case of the unoriented membrane, decreased to about 10% after the reaction with silver nitrate and to about 20 to 40% after the orientation and removal of silver.

Example 2: Between two pieces of plate glass a sling of steel wire 0.8 mm was firmly held and three sides were immersed for a second into a melt of beeswax and stearine (1:1). After a little while the wax mixture solidified and a flat mold, opened on one side, was obtained, having 0.8 mm thickness. The mold was filled with a 10°C cool monomer mixture through a flat funnel made from polyethylene foil.

Monomer mixture consisted of 10 ml of 50% aqueous solution of sodium ethylene sulfonate, 4 ml of acrylonitrile, one drop of ethylene glycol bis-methacrylate and 8 ml concentrated aqueous zinc chloride solution (d = 1.98). Immediately prior to the casting the mixture was initiated with 5 drops of a 5% potassium pyrosulfite aqueous solution and with same amount of ammonium persulfate.

Gas bubbles were removed by patting on the mold. The mold was then left standing at 15°C temperature. Gel point was reached in about 20 min, after 3 hr the copolymerization was finished. The wax layer on the border was cut and the mold carefully opened under water. The copolymer foil thus obtained, first highly plastified with zinc chloride solution, was then washed alternately in 0.5% nitric acid, water, 5% citric acid and again in water.

Last it was immersed into a 5% silver nitrate solution and then washed again. The membrane thus obtained had similar properties like that in Example 1, except that it was less extensible and more elastic, due to the low crosslinking caused by the small amount of glycol bis-methacrylate added.

Example 3: Water-soluble copolymer of acrylonitrile, acrylamide and ethylene sulfonic acid was prepared in following way: 57.5 g of a 47% aqueous solution of sodium ethylene sulfonate, stabilized against polymerization by 0.2% sodium nitrite, was destabilized by adding 3 g urea and acidified with concentrated hydrochloric acid to pH 2. Liberated gases were repeatedly sucked off. 7.1 g of acrylamide and 13.6 ml of acrylonitrile were added and the whole was dissolved in 76 ml of oxygen-free water. 1 ml of a 5% sodium pyrosulfite solution and the same amount of ammonium persulfate was stirred in and the somewhat turbid solution was left standing under carbon dioxide at room temperature for 24 hr. The viscous solution thus obtained, was poured in 2 liters of anhydrous ethanol. The white coagulate was reprecipitated by dissolving in water again and precipitating in anhydrous ethanol, broken to small pieces and dried at 70°C.

A 20% solution of the copolymer in water was cast on horizontal plate glass and the solvent slowly evaporated. The dry layer was treated with 37% aqueous formaldehyde solution, acidified with 0.2% hydrochloric acid, for 2 hr at room temperature. The formaldehyde solution was then removed, the membrane washed with water, in which it became insoluble as a result of crosslinking with formaldehyde, and cut to 2 pieces. One was left in water as control, the other treated 5 min in a 5% silver nitrate solution and washed. The silver nitrate treated membrane shrank considerably, was opaque and its strength increased.

Amide-Acid Units on Surface of Sulfonic Acid Containing Polymers

Y. Mizutani, R. Yamane, T. Sata and R. Izuo; U.S. Patent 3,647,086; March 7, 1972; assigned to Tokuyama Soda Kabushiki Kaisha, Japan describe a cation exchange membrane which is characterized in being a membranous insoluble and infusible organic high-molecular-weight polymer having chemically bonded cation exchange groups and having the dimension of at least one centimeter in at least two directions, the substantial surface of the membrane being chemically bonded with acid amide units in an amount indicated by the expression [A/(A + B)] 100 where A is the number of acids amide bonds in one gram of the dry membrane and B is the number of cation exchange groups in one gram of the dry membrane, the acid amide units being composed of an amine having at least one amino group containing at least one hydrogen atom bonded to a nitrogen atom and a cation exchange group.

The above expression will hereinafter be referred to as the percent acid amide bonds. There is no particular restriction as to the cation exchange group that is present in the cation exchange membrane as long as it is one possessing a cation exchange capacity. Those preferred, however, are the sulfonic acid, carboxylic acid and phosphoric acid groups, and from the standpoint of economy the sulfonic acid group is most preferred.

Further, in general, the number of sulfonic acid, carboxylic acid and phosphonic acid groups, and other cation exchange groups, that is introduced into the membrane as cation exchange groups, i.e., the exchange capacity, varies depending upon the class of the exchange group as well as the class of the resin.

Generally speaking, when the number of cation exchange groups in the membrane is too great, the mechanical strength is decreased and, on the other hand, when it is too small, the electric resistance of the cation exchange membrane increases and moreover the permselectivity between ions of different signs become lower. In the process, cation exchange membrane usually the number of cation exchange groups is most suitably within the range of 0.5 to 6.0 milliequivalents per gram of the dry membrane.

In the examples and comparisons the electric resistance of the cation exchange membrane is a per unit area value obtained by using a 1,000 cycle alternating current in 0.5 N NaCl solution of 25°C. The transport number is that obtained when the electrodialysis was carried out with a current density of 20 milliamperes/cm^2 and 0.5 N NaCl solutions on both sides of the cation exchange membrane.

On the other hand, the permselectivity designated $T_{M_1}^{M_2}$ was calculated by means of the following equation from ratio of the transport number in the membrane obtained by carrying out the electrodialysis with a mixed salt solution of $CaCl_2$ = 0.200 N and NaCl = 0.200 N. The electrodialysis was carried out in this case at 25°C and a current density of 20 milliamperes per square centimeter with vigorous stirring of the solutions on both sides of the ion exchange membrane. $T_{M_1}^{M_2} = (t_{M_1}/t_{M_2})/(C_{M_1}/C_{M_2})$ wherein t_{M_1} and t_{M_2} each represent the transport numbers of cations M_1 and M_2 in the ion exchange membrane, and

C_{M_1} and C_{M_2} each represent the concentrations of cations M_1 and M_2 in the solutions. Further, the pure salt ratio at the time of concentrating seawater was obtained as follows: pure salt ratio = [(Na) + (K)]/(Cl) x 100. Further, the intramembranous ion exchange capacity of the cation exchange membrane was calculated on the basis of the following experiment.

After three or four washings of a cation exchange membrane alternately in 1 N HCl and 0.5 N NaCl, it is washed four to five times in 1 N HCl to effect a complete conversion of the cation exchange membrane to the H-type. This cation exchange membrane is then thoroughly washed in water until the methyl orange indicator does not turn red for eliminating the excess hydrogen ions.

Next, this cation exchange membrane is placed in pure water and its ion exchange capacity is determined by obtaining the titration curve with a pH meter using 1/10 N NaOH. The cation exchange membrane whose measurements have been completed in this manner is again immersed in a 0.5 N NaCl and brought to equilibrium after which it is taken out, allowed to stand for two hours in an 80°C air dryer and then returned to room temperature in a desiccator. This dried membrane is weighed, after which the ion exchange capacity obtained above is divided by the membrane weight to obtain the ion exchange capacity per one gram of dry membrane.

The amount of acid amide bonds formed in the cation exchange membrane was determined by the microanalytic method of quantitative analysis unless otherwise indicated. For example, in the case of sulfonic acid type cation exchange membrane imparted the acid amide bonds using polyethyleneimine, the amount of acid amide bonds is determined in the following manner.

To a fine powder scraped from the surface of the foregoing cation exchange membrane is added a great excess of acetic anhydride followed by refluxing for two hours. Next, the foregoing fine powder is separated by filtration and thoroughly dried under reduced pressure to completely remove the acetic acid and acetic anhydride. The so obtained sample is submitted to infrared analysis by means of the KBr tablet method and the intensity ratio of the absorption peaks of sulfonic acid amine and carboxylic acid amide is obtained.

In general, the absorption peak of sulfonic acid amide is observable at about 1,160 cm^{-1}, while the absorption peak of carboxylic acid amide which shifts between 1,740 to 1,630 cm^{-1} depending upon the class of the ion exchange membrane usually appears in most cases at about 1,630 cm^{-1}.

Example 1: A pasty mixture of 95 parts of styrene, 5 parts of divinylbenzene, 100 parts of finely divided polyvinyl chloride, 25 parts of dioctyl phthalate and 1.5 parts of benzoyl peroxide was applied to a polyvinyl chloride cloth, which was then polymerized by heating at 110°C for 4 hr to obtain a membranous high-molecular-weight polymer 0.015 cm in thickness.

The divinylbenzene which was of a purity of 50 to 54%, was a mixture of o-, m- and p-divinylbenzene, the rest being ethyl benzene. (This is likewise also in the subsequent examples). Using the so obtained membranous high-molecular-weight polymer as the starting membrane, this was dipped in a mixed solution consisting of 2 parts of chlorosulfonic acid of about 90% purity and 1 part of carbon tetrachloride, where the chlorosulfonation reaction was carried out for 2 hr at 4° to 7°C.

The resulting chlorosulfonated starting membrane was dipped in an aqueous or methanol solution containing 5% of a compound indicated in the following table, having either a primary or secondary amino group under the conditions specified in the table to form the acid amide bonds in the surface of the membrane. After completion of the foregoing reaction, the starting membrane was immersed in aqueous 1 N NaOH solution for 8 hr at room temperature to hydrolyze the unreacted chlorosulfone groups to cation exchange groups.

	- Compound Solution Used -		Reaction - Conditions -		- - - - - - - - Measurement Results - - - - - - -				
Number	Solvent	Class of Compound	Temp. (°C)	Time (hr)	Acid Amide Bonds (%)	T_{Na}^{Ca}	Transport Number	Electric Resistance (Ω-cm²)	Exchange Capacity (meq/g dry membrane)
1	-	-	-	-	-	2.6	0.98	3.5	1.82
2	-	Diethylamine	25	6	9.9	1.6	0.98	10.3	1.64
3	Water	Diethanolamine	25	10	7.6	1.3	0.92	10.0	1.68
4	Methanol	Diphenylamine	25	24	1.6	2.0	0.98	3.6	1.79
5	Methanol	Phthalimide	25	24	3.8	2.1	0.98	3.5	1.75
6	Water	Piperazine	25	3	12.1	1.8	0.88	8.5	1.60
7	Water	p-Aminosalicyclic acid	40	24	2.2	1.5	0.97	2.6	1.78
8	Water	Dodecylamine	25	2	3.3	1.5	0.88	10.5	1.76
9	Water	Tetraethylene pentamine	25	8	3.4	0.9	0.95	8.0	1.74
10	Water	Triethylene tetramine	25	5	7.8	0.7	0.90	9.5	1.68
11	Methanol	m-Phenylene diamine	25	24	12.6	0.5	0.88	11.0	1.59
12	Water	Bismarck brown	25	8	4.9	0.7	0.95	9.5	1.73
13	Water	Auramine concentration	25	8	1.6	1.8	0.97	8.0	1.79
14	Water	Isonicotinic acid hydrazine	25	24	2.2	1.8	0.98	8.0	1.78
15	Water	Chrysoidine crystals	25	8	1.1	2.0	0.98	3.5	1.80

The so obtained cation exchange membrane was then washed thoroughly in 1 N HCl, 1 N NaCl and water in the order given. The cation exchange membrane was thus obtained. The amount of the acid amide bonds and measurement results of the so obtained cation exchange membrane are shown in the above table.

Item No. 1 in the table has been given as a control. The results given here are those of a cation exchange membrane obtained as follows: After the starting membrane has been chlorosulfonated as described above, it is immediately immersed in 1 N NaOH without imparting the acid amide bonds, i.e., without reacting it with a compound having the primary or secondary amino groups, and this is followed by the hydrolysis of the chlorosulfone groups.

Example 2: A pasty mixture was prepared by adding 100 parts of finely divided polyethylene and 2 parts of benzoyl peroxide to a mixture consisting of 90 parts of styrene, 10 parts of divinylbenzene and 25 parts of dioctyl phthalate. This paste was applied to a net made from polyethylene and polymerized by heating for 4 hr at 110°C to obtain a membranous high-molecular-weight polymer.

Using the so obtained membrane as the starting membrane, the chlorophosphate groups were introduced thereinto by immersing it for 12 hr at room temperature in a solution consisting of 25 mols of phosphorus trichloride and 1.2 mols of anhydrous aluminum chloride.

After washing this membrane thoroughly with water, it was dipped for 5 hr at room temperature in a methanol solution containing 5% polyethylene imine to form the acid amide bonds. The membrane was further dipped for 12 hr at room temperature in 1 N NaOH to convert the chlorophosphate groups to phosphorous acid groups to thus obtain a cation exchange membrane. The percent phosphorous acid amide bonds of the so obtained cation exchange membrane was 0.05%. Further, its transport number was 0.98, its resistance was 7.2 ohm cm² and T_{Na}^{Ca} was 0.6.

Sulfonated Polyarylether Sulfones

J.P. Quentin; U.S. Patent 3,709,841; January 9, 1973; assigned to Rhone-Poulenc SA, France describes ion exchange resins which contain units of the formula:

$$(1)\qquad -O-\overset{(Y)_r}{A}-\left[R-\overset{(Y_1)_s}{A}-\right]_m O-\overset{(Y_2)_t}{B}-SO_2-\overset{(Y_3)_u}{B_1}-$$

in which A, B and B_1, which may be identical or different, represent aromatic radicals of which at least one is substituted by a hydroxysulfonyl radical, Y and Y_1, which may be identical or different, represent substituent radicals which are inert to sulfonation, Y_2 and Y_3, which may be identical or different, represent electron acceptor radicals, r, s, t and u, which may be identical or different, represent integers of 0 to 4 inclusive, at least one of them being less than 4, m is 0 or 1, and R represents a valency bond, —CO—, —O—, —SO_2, or a divalent organic hydrocarbon radical.

By way of illustration, the radicals A, B and B_1 can represent phenylene radicals, such as the p-phenylene radical, with at least one of these radicals being substituted by one or more hydroxysulfonyl radicals (—SO_3H, also called sulfonic groups). Y and Y_1 can represent alkyl or alkoxy radicals of 1 to 4 carbon atoms each, or halogen atoms (F, Cl, Br or I). Y_2 and Y_3 can represent nitro, phenylsulfonyl, alkylsulfonyl, trifluoromethyl, nitroso or pyridyl. If R represents a divalent hydrocarbon radical, it can represent an alkylene or alkylidene radical which, for example, may possess 1 to 7 carbon atoms, or a cycloalkylene radical which, for example, may possess 5 to 7 carbon atoms, or an arylene radical such as phenylene or a methylphenylene radical. These ion exchange resins are obtained, in accordance with the process; by sulfonating polyarylethersulfones containing units of the formula:

$$(2)\qquad -O-\overset{(Y)_r}{A'}-\left[R-\overset{(Y_1)_s}{A'}-\right]_m O-\overset{(Y_2)_t}{B'}-SO_2-\overset{(Y_3)_u}{B'_1}-$$

in which Y, Y_1, Y_2, Y_3, R, m, r, s, t, and u are as defined above and A', B' and B'_1, which may be identical or different, represent aromatic radicals, especially phenylene radicals, such as p-phenylene. At least one of the radicals A', B', and B'_1, must differ from the radicals A, B and B_1 respectively, because at least one of the radicals A, B or B_1 must be substituted by a hydroxysulfonyl radical.

When the sulfonation of the polyarylethersulfone of Formula 2 is incomplete, units of Formula 2 of the starting polyarylether sulfone may persist in the ion exchange resin obtained. The starting polyarylether sulfones of Formula 2 are described in French Patent 1,407,301.

Example 1: A solution A is made by dissolving 12.5 g of polyarylether sulfone in 150 cm^3 of 1,2-dichlorethane (DCE). The polyarylether sulfone used consists of a plurality of units of formula:

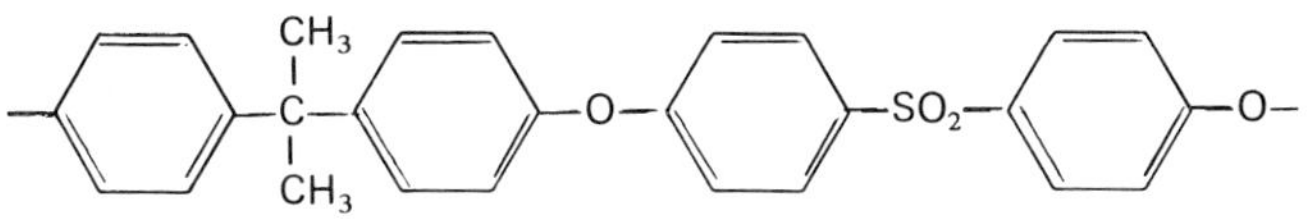

and has a flow speed in the molten state (measured according to standard specification ASTM D 1238) of 0.7 g/min. A solution B is made up by dissolving 3.3 g of chlorosulfonic acid ($ClSO_3H$) in 30 cm^3 of DCE.

The two solutions A and B are simultaneously and gradually (over the course of one hour) introduced, with vigorous stirring (436 rpm), into a 500 cm^3 flask equipped with a stirrer kept at −10°C and containing 50 cm^3 of DCE. The stirring and the temperature are maintained for a further 4 hours. The precipitate is filtered off, washed with 100 cm^3 of DCE and dissolved in 50 cm^3 of dimethylformamide (DMF) at ambient temperatures. This solution is poured into 500 cm^3 of water, the mixture is filtered, and the product is washed three times with 300 cm^3 of water each time, and dried at 60°C under 100 mm of mercury. Six grams of sulfonated polysulfone are thus obtained.

Three grams of this sulfonated polysulfone are dissolved in 20 cm^3 of DMF at ambient temperature. The solution is cast at ambient temperature on a 10 cm^2 glass plate, and dried at 50°C. A membrane M_1 0.2 mm thick is thus obtained, having a theoretical exchange capacity of 0.6 meq/g of dry resin, a selective permeability of 96%, a substitution resistance of 48 Ω cm^2, and a bursting pressure of 6.7 bars (deflection: 9.3 mm). On prolonged immersion in water, the membrane M_1 shows a linear elongation of 1%.

After immersion in 2 N sulfuric acid for 168 hr at 70°C, the membrane M_1 has a selective permeability of 89%, a substitution resistance of 33 Ω cm^2, and a bursting pressure of 5 bars (deflection 12.7 mm). After immersion in normal sodium carbonate solution for 168 hr at 70°C, the membrane M_1 has a selective permeability of 87%, a substitution resistance of 24 Ω cm^2, and a bursting pressure of 4.9 bars (deflection 4.9 mm). After immersion in a 3 N nitric acid bath for 168 hr at 25°C and then for 6 hr at 100°C the membrane M_1 has a selective permeability of 89%, and a substitution resistance of 33 Ω cm^2.

Example 2: Two solutions A and B are prepared as in Example 1. The two solutions A and B are introduced simultaneously and gradually (over the course of one hour), with vigorous stirring, (436 rpm), into a 500 cm^3 flask equipped with a stirrer, kept at 0°C and containing 50 cm^3 of DCE. The stirring and the temperature are maintained for a further 4 hr. The precipitate is filtered off, washed with 100 cm^3 of DCE, and dissolved in 50 cm^3 of DMF. This solution in DMF is poured into 500 cm^3 of water. The mixture is concentrated at 50°C, under partial vacuum (20 mm of mercury) to a weight of 35 g. 100 cm^3 of acetone are then added, and the resulting precipitate is filtered off and washed three times with 200 cm^3 of acetone each time, and dried at 60°C under 100 mm of mercury. 16.2 g of sulfonated polysulfone are thus obtained.

Nine grams of this sulfonated polysulfone are dissolved in 90 cm^3 of DMF at 60°C. The solution is cast onto a 300 cm^2 glass plate and dried at 50°C. The membrane M_2 obtained has a thickness of 0.19 mm, a theoretical exchange capacity of 1 meq/g a selective permeability of 88%, a substitution resistance of 6 Ω cm^2, and a bursting pressure of 4 bars (deflection: 9 mm).

After 168 hr in 2 N sulfuric acid at 70°C, the membrane M_2 has a selective permeability of 84%, a substitution resistance of 3 Ω cm^2, and a bursting pressure of 5.5 bars (deflection: 10.5 mm). After 168 hr at 70°C in normal sodium carbonate solution, the membrane M_2 has a selective permeability of 85%, a substitution resistance of 3 Ω cm^2, and a bursting pressure of 3.5 bars (deflection: 6.5 mm). After immersion in 3 N nitric acid for 168 hr at 25°C followed by 6 hr at 100°C, the membrane M_2 has a selective permeability of 82%, a substitution resistance of 8 Ω cm^2 and a bursting pressure of 4.5 bars (deflection: 5 mm).

According to a process described by *J. Bourganel; U.S. Patent 3,855,122; December 17, 1974; assigned to Rhone-Poulenc SA, France* anisotropic sulfonated polyaryl ether/sulfone membranes having good salt rejection and flow rates making them particularly suitable for use in osmosis and ultrafiltration are provided by casting a solution of the polymer on a support, immersing the supported film in a coagulating bath and recovering the membrane produced. In the process:

(a) the sulfonated polyaryl ether/sulfone used possesses between about 0.1 and 2 meq/g of sulfonic acid groups and a reduced specific viscosity of between about 40 and

200 cm^3/g (measured as a 2 g/l solution in dimethylformamide at 25°C);

(b) the concentration of the casting solution is between about 5 and 60% w/v of solution;

(c) the coagulation bath consists of either water or an aqueous solution containing about 0.1 to 10% by weight of an amine salt or a quaternary ammonium salt which is soluble in water and in the polymer solution;

(d) the temperature of the coagulation bath is between about 0° and 100°C;

(e) the period of immersion in the coagulation bath is between about 30 sec and 60 min;

(f) the immersion in the coagulation bath is preceded by gelling the polymer layer; and

(g) the casting solution optionally contains from about 0.1 to 10% by weight of an amine salt or a quaternary ammonium salt which is soluble in water and in the polymer solution.

Styrene-Butadiene Treated with Phenolsulfonic Acid

A. Nishihara and Y. Oda; U.S. Patent 3,276,989; October 4, 1966 describe cation permselective membranes which are permeable to univalent alkali metal ions and ammonium ions in contrast to polyvalent cations. The process involves cation permselective membranes which are obtained by coating, with a particular resinous compound, the surface of a cation selectively permeable membrane, which is formed by impregnating a copolymer latex of a monovinyl aromatic hydrocarbon compound and a conjugated aliphatic polyene compound in a reinforcing material, drying and then forming a membrane of the copolymer latex, and thereafter, treating the membrane with concentrated sulfuric acid to sulfonate and simultaneously to convert the copolymer constituting the membrane to an insoluble infusible three dimensional network structure and which has a length of at least 1 cm in two dimensions. The resinous compound used for coating is a precondensation product which is produced by cocondensing in an alkaline medium at least one of the compounds of the formula:

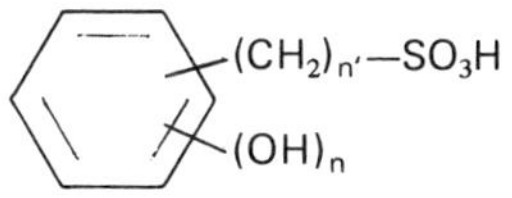

(where n is the integer 1 or 2, and n' 0 or 1) or their salts, phenol and formaldehyde, in a ratio of 1:3.5 to 10:4 to 35.

The effective resistance used in this discussion means the resistance per 1 cm^2 of the membrane, measured in a 0.5 N sodium chloride solution at 25°C. For the purpose of measuring the selective permeability of the membranes, comparison of the permeability to various kinds of ions can be made and for convenience's sake, comparison of the transport number of sodium ions with that of calcium ions (in some cases magnesium ions) was made. The measurements were carried out as follows.

Electrodes made of silver-silver chloride are placed at both ends of an electrolytic cell and three compartments are formed between an anode chamber and a cathode chamber by successively arranging from the anode side a cation permselective membrane, a sample membrane which is the product in the process and is to be tested, an anion permselective membrane and another anion permselective membrane adjacent to the cathode. Starting from the cathode chamber, they are called successively, a sample solution chamber, a concentrating chamber and sodium chloride solution chamber.

A mixed solution of NaCl and $CaCl_2$ (NaCl 0.5 N and $CaCl_2$ 0.02 N) is fed into the sample solution chamber from one side and discharged from the other side. In the concentrating chamber is 0.5 N NaCl solution which is not circulated. In the other three chambers, that is, the anode chamber, the cathode chamber and the sodium chloride solution chamber, a solution of 0.5 N NaCl flows constantly. After an electric current was passed with a current density of 2 amp/dm^2 at a constant temperature of 25°C for 1 hr, the solution in the concentrating chamber is discharged and a fresh solution of 0.5 N NaCl is introduced and this solution is not circulated. After passing the current for another hour, the total solution in this chamber is taken out and the increased amount of Na ion and Ca ion is determined by chemical analysis, The amount of electricity passed for 1 hour is measured by a coulometer, and from this amount of electricity and the increased amount of Ca ions and Na ions the transport number of Na ions and Ca ions (n_{Na} and n_{Ca}) is calculated. The selective permeability to Na ions and Ca ions S_{Na}^{Ca} can be calculated by the following formula:

$$S_{Na}^{Ca} = \frac{n_{Ca}/n_{Na}}{t_{Ca}/t_{Na}}$$

where t_{Ca} and t_{Na} are respectively the transport numbers of Ca ions and Na ions in the sample solution, that is, the mixed solution of 0.5 N NaCl and 0.02 N $CaCl_2$, at 25°C. It is not necessary that the cation and anion permselective membranes used in the above measurement be special. The abovementioned base membrane was used as the cation permselective membrane and used as the anion permselective membrane is a membrane obtained by treating the copolymer film, supported by reinforcing material with $TiCl_4$, chloromethylether and then trimethylamine instead of concentrated sulfuric acid.

S_{Na}^{Ca} of the base membrane of the process which was measured by the abovementioned procedure was 0.8 to 1.2. Furthermore, S_{Na}^{Ca} of two different commercially available products was 0.38 and 0.80. According to the process, it was possible to improve the S_{Na}^{Ca} of the base membrane by 0.1 to 0.7 by the following treatment. Specifically, a precondensation product which is produced by cocondensing in alkaline medium (1) at least one of the compounds represented by the general formula:

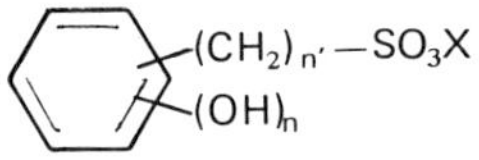

(where X is any one of H, NH_4, Na and K, n is any one of 1 and 2 and n' is any one of 0 and 1) and (2) at least one of phenol and a mixture of phenol and a compound selected from halogenophenol, nitrophenol, cresol, resorcinol and catechol containing less than one-half of phenol in molar ratio, and (3) formaldehyde, the molar ratio of the three components being 1:3.5 to 10:4 to 35, is coated on the surface of the base membrane and then a thin film of the condensation product is formed by heating and curing.

As the compounds represented by the above general formula can be specified a phenol sulfonic acid, resorcinol sulfonic acid, oxybenzyl sulfonic acid and the like, the typical salt of which compounds is the sodium salt. The process is described in the following examples, where parts given are by weight.

Example 1: A styrene-butadiene copolymer latex containing 29% of bound styrene and 60% of the total as solids was diluted with distilled water to a concentration at which the total solids content was 57%. In this latex was dipped a plain fabric of glass-fiber of 0.07 mm thickness and dried. Thus, a membrane of copolymer was obtained in which the weight ratio of the copolymer to the glass fiber in the membrane was 1:1. The membrane thus obtained was first dipped in 96% sulfuric acid at 30°C for 3 hr, after which the sulfuric acid contained in the membrane was successively diluted by dipping the membrane in 80%, 60% and 20% sulfuric acid for 15 min each.

Thereafter, the membrane was rinsed with water, and after the resin was converted to Na-type by dipping the membrane in a solution of 0.5 N NaOH, the membrane was rinsed again with water and stored in 0.5 N NaCl solution. The membrane had excellent properties as a cation permselective membrane, and its effective resistance was 6.2 Ω-cm^2, thickness 0.2 mm, and S_{Na}^{Ca} 0.91.

Separately, 24 parts by weight of the dihydrate of the sodium salt of p-phenol sulfonic acid was dissolved in 210 parts of 35% formalin, and the solution was mixed with 94 parts of phenol and 48 parts of 20% caustic soda solution and heated on a water bath with stirring for 50 min. The viscous liquid thus obtained was coated thinly on the surface of an air dried membrane with a brush and cured by heating at 130°C for 2 hr. The thickness of the film calculated from the increase of weight was 8 μ and effective resistance of the product was 15.4 Ω cm^2 and S_{Na}^{Ca} 0.17.

Example 2: 28 parts of sodium salt of m-phenol sulfonic acid was dissolved in 170 parts of 35% formalin, and the solution was mixed with 94 parts of phenol and 52 parts of 20% caustic soda solution and heated at 85°C for 2 hr. THe viscous liquid thus obtained was diluted with 150 parts of methanol. In this solution was dipped the air dried base membrane as described in Example 1 for 2 min and then withdrawn. After excess resin solution was wiped off, the membrane was dried at 100°C for 3 min and then cured by heating at 160°C for 1.5 hr. The thickness of the film calculated from the increase of weight was 22 μ, and effective resistance of the product was 15.5 Ω-cm^2 and S_{Na}^{Ca} was 1.5.

Example 3: 4[illegible] parts of potassium salt of m-phenol sulfonic acid was dissolved in 240 parts of 35% formalin, and the solution was mixed with 94 parts of phenol and 12 parts of 50% caustic potash solution. The mixed solution thus obtained was heated at 85°C for 3.5 hr and then diluted with 232 parts of dioxane. The liquid was sprayed on an air dried base membrane. This membrane was then dried at 100°C for 3 min and cured by heating at 160°C for 2 hr. The thickness of the film calculated from the increase of weight was 18 μ, and effective resistance of the product was 19.5 Ω-cm^2 and S_{Na}^{Ca} was 0.13.

In related work, *H. Hani, A. Nishihara and H. Hiraga; U.S. Patent 3,276,990; October 4, 1966* describe a process for the production of cation permselective membranes which have a length of more than 1 cm in at least two directions, which have an insoluble infusible synthetic organic solid polymeric structure, a reinforcing material supporting the solid polymeric structure and a dissociable sulfonic group combined chemically with the solid polymeric structure, and which are electrolytically conductive and selectively permeable to cations.

The process comprises impregnating the reinforcing material with a latex of a linear copolymer which consists of monovinyl aromatic hydrocarbon compounds and conjugated aliphatic polyene compounds and contains 20 to 75% of bound monovinyl aromatic hydrocarbon compounds, drying and forming a membrane of the linear copolymer supported by reinforcing material, and thereafter treating the membrane with concentrated sulfuric acid in order to sulfonate and simultaneously to convert the linear copolymer to an insoluble infusible three dimensional network structure.

Styrene-Butadiene with Phenylenediamine Coating

A process described by *H. Hani, A. Nishihara and Y. Oda; U.S. Patent 3,276,991; Oct. 4, 1966* provides membranes which have extremely low permeability to polyvalent anions in comparison with halogen ions during electrodialysis of the solution of electrolytes containing various kinds of ions of different valence.

These anion permselective membranes consisting of strongly basic anion exchange resins generally have the property of being more permeable to monovalent anions such as Cl ions than to divalent anions such as SO_4 ions. If the transport number of SO_4 ion and Cl ion through a membrane respectively is represented by n_{SO_4} and n_{Cl} and that of SO_4 ion and Cl ion through a solution phase is respectively represented by t_{SO_4} and t_{Cl}, the selective permeability of SO_4 ion and Cl ion through an anion permselective membrane ($S_{Cl}^{SO_4}$) can be calculated by the following formula:

$$S_{Cl}^{SO_4} = \frac{n_{SO_4}/n_{Cl}}{t_{SO_4}/t_{Cl}}$$

If $S_{Cl}^{SO_4}$ is less than 1, the membrane is more permeable to Cl ion than to SO_4 ion, while if more than 1, the membrane is more permeable to SO_4 ion than to Cl ion. When selective permeability of SO_4 ion and Cl ion is defined as mentioned above, its value in conventional anion permselective membranes consisting of strongly basic anion exchange resins is usually less than 1 but is not extremely small. The selective permeability of two such membranes that are at present commercially available is 0.40 and 0.30.

To reduce this value further is very advantageous in some cases. For example, in the case where concentrated sodium chloride solution would be obtained by electrodialysis of seawater in an electrodialysis vessel in which cation permselective membrane and anion permselective membrane are placed in parallel and between them a chamber is formed, it is, of course, necessary to use cation permselective membrane through which Ca ion and Mg ion are difficultly permeable, but at the same time it is also very desirable to use anion permselective membrane through which divalent anions such as SO_4 ion are as difficultly permeable as possible so as to improve the purity of the resultant concentrated solution and to prevent an insoluble precipitate such as calcium sulfate from depositing on the membrane. Thus, it is important to decrease permeability of anion permselective membranes consisting of strongly basic anion exchange resins to divalent or higher valent ions as compared to halogen ions.

This process includes anion permselective membranes which are obtained by forming a thin layer of a resin obtained by polycondensation of at least one member selected from the class consisting of m-phenylene diamine, m-aminophenol, m-toluidine and aniline, and formaldehyde, on the surface of a membrane which has an insoluble infusible synthetic organic polymeric matrix and quaternary amine groups bonded chemically to the matrix, has a length of at least one centimeter in two dimensions, is electrolytically conductive, and is anion selectively permeable.

It is well-known to produce anion permselective membranes by casting a mixed solution consisting of aniline or m-phenylene diamine and formaldehyde on a plane surface and then resinifying it. However, these known anion permselective membranes are subject to cracking, their mechanical strength is small, and their selective permeability to anions is low. In addition, another important drawback is that their electric resistance is extremely high in a solution with high pH value due to their weak basicity, and so they are not suitable for use in a solution with high pH value.

Anion permselective membranes have been found which can be used in high pH solutions and which have great mechanical strength as well as high selective permeability to anions and particularly are more permeable to halogen ions among anions but which have extremely low permeability to ions of di- or higher valence, by forming and securing a thin layer of a condensation resin as described above to the surface of a base membrane of the anion permselective type consisting of a strongly basic exchange resin having quaternary amino groups.

The condensation resin, which is secured to the surface of the base membrane, has high electric resistance but it has been discovered that even if the formed layer is so thin that electric resistance of the resulting membrane is low enough for practical use, the permeability of the formed layer to polyvalent anions is considerably reduced. Since the mechanical strength of base membrane is not substantially affected by the film of the condensation resin, it is possible to obtain a product with sufficiently great mechanical strength by adopting a suitable anion permselective membrane as a base membrane.

In addition, because the selective permeability of the product to anions may be determined by that of the base membrane it is possible to obtain a product with high overall selective permeability to anions. The film of the resins coated on the surface of the base membrane as a thin layer has less tendency to cause cracks and even when the membrane cracks the surface of the base membrane is exposed only locally and accordingly, leakage of solution and cations does not occur and substantially no serious difficulty arises. The following examples illustrate the process.

Example 1: 70 parts of copolymer rubber of styrene and butadiene (weight ratio 1:1) and

30 parts of polyethylene were mixed and kneaded together, and then the mixture was rolled by a calender roll into film having a thickness of 0.15 mm. The membrane thus obtained was dipped in a solution which was prepared by dissolving 133 parts of $AlCl_3$ in a mixture of 74 parts of diethylether and 207 parts of dichloroethane and then saturating the resultant solution with hydrogen chloride gas, at 30°C for 2 hr. After converting the copolymer rubber of the membrane into an insoluble infusible three dimensional network structure, the membrane was rinsed thoroughly with methanol and then air dried.

The membrane thus treated was swelled in tetrachloroethane and thereafter, chloromethylated in a mixed solution consisting of 15 parts of chloromethylether, 3 parts of $SnCl_4$ and 82 parts of tetrachloroethane, at 30°C for 3 hr, after which the chloromethylated membrane was rinsed with methanol, aminated by dipping in 0.5 N methanol solution of trimethylamine, and then rinsed with water and stored in 0.5 N NaCl solution. Effective resistance (R) of the membrane thus obtained was 4.2 Ω-cm^2, n_{Cl} 0.96 and $S_{Cl}^{SO_4}$ 0.23, respectively.

After being withdrawn from the NaCl solution, this membrane was rinsed with water and dried and then was dipped in a solution of m-phenylene diamine hydrochloride at 50°C for 5 min, and thereafter, resinified by dipping in a mixed solution consisting of 50 parts of 36% formalin and 5 parts of hydrochloric acid at 50°C for 10 min. After being withdrawn, the membrane was heated at 80°C for 1 hr, washed with water and then stored in 0.5 N NaCl solution. Effective resistance (R) of the membrane thus obtained was 7.4 Ω-cm^2, n_{Cl} 0.97 and $S_{Cl}^{SO_4}$ 0.09, respectively, Effective resistance as used in this example, and the following examples, means the value measured in a solution of 0.5 N NaCl and 0.05 N Na_2SO_4 at 25°C.

Example 2: A mixture of 100 parts of copolymer rubber consisting of methylvinylpyridine and butadiene (weight ratio 25:75), 100 parts of carbon black, 15 parts of sulfur, 3 parts of zinc oxide, 1.5 parts of dibenzothiazol disulfite, 1.5 parts of stearic acid and 5 parts of chloranil was kneaded together and the mixture was rolled into a sheet having a thickness of 0.3 mm with roller.

Then, this sheet was vulcanized by heating at 150°C for 4 hr in a heating press and then treated with methyl iodide to convert the amine group contained in the structure to quaternary ammonium and then rinsed successively with 1 N Na_2SO_4 and 0.5 N NaCl solution. Effective resistance (R) of the membrane was 4.5 Ω-cm^2, n_{Cl} 0.97 and $S_{Cl}^{SO_4}$ 0.23, respectively.

This membrane was dried in air and after being daubed with an aqueous solution containing 30% of m-toluidine hydrochloride and 10% of carboxymethylcellulose, dried for 2 min in air and dipped in an acidic formaldehyde solution in the same way as in Example 1 for 10 min, rinsed with water and stored in 0.5 N NaCl solution. Effective resistance of the membrane 6.2 Ω-cm^2, n_{Cl} 0.97, and $S_{Cl}^{SO_4}$ 0.14, respectively.

Chloromethylated and Aminated Styrene-Butadiene

H. Hani, A. Nishihara and H. Hiraga; U.S. Patent 3,276,992; October 4, 1966 describe a process for the production of anion permselective membranes, which are electrically conductive, selectively anion permeable and physically strong, which have a length of more than 1 cm in at least two dimensions and which include an insoluble and infusible synthetic solid organic polymeric structure, a reinforcing material supporting the structure and anion exchange groups combined chemically with the solid polymeric structure.

The process comprises impregnating a latex of linear copolymer, which consists of a monovinyl aromatic hydrocarbon compound and a conjugated aliphatic polyene compound and contains 35 to 75% of bound monovinyl aromatic hydrocarbon compound based on the weight of the copolymer, to a reinforcing material, drying the reinforcing material impregnated with latex, forming a membrane of the linear copolymer supported on the reinforcing material and then treating the membrane with a Friedel-Crafts reagent, for example,

$AlCl_3$, BF_3, $AlBr_3$, $TiCl_4$, $TiBr_4$, or $SnCl_4$ to convert the linear copolymer into an insoluble infusible three dimensional network structure and treating the membrane composed of insoluble infusible copolymer and reinforcing material with a chloroalkyl ether followed by amination with amines.

The following examples illustrate the process. The transport number referred to in the following examples are those of chloride ion through the membrane, calculated from the membrane potential at 25°C, between 0.5 N and 1.0 N sodium chloride solutions on opposite sides of the membrane, and the effective resistance is the resistance per square centimeter of the membrane measured in 0.5 N sodium chloride solution at 25°C. The bursting strength is measured by Müllen's bursting strength tester.

Example 1: In 500 cc of styrene-butadiene copolymer latex having 66% of bonded or bound styrene and total solid of 48% was dipped a glass fiber cloth of 10 x 10 cm^2 size; 0.07 mm thickness; warp 60 threads/inch; weft 64 threads/inch and an apparent free space content 60% for 12 min, withdrawn and hung from one end to be air dried for 16 hr. The dried copolymer supported on the glass fiber cloth showed 58 volume percent of copolymer content.

This membrane was cut to 10 x 10 cm^2, and the cut membrane was dipped in 300 cc of a solution composed of 134 parts by weight of anhydrous aluminum chloride, 74 parts by weight of diethylether and 208 parts by weight of dichloroethane, for 1 hr at 30°C and then rinsed twice with 200 cc methanol and air dried. Thus, the styrene-butadiene copolymer of linear structure was converted to a three dimensional structure and a dried membrane of insoluble and infusible copolymer supported by glass fiber cloth was obtained.

Thereafter, this membrane was dipped for 3 hr at 30°C in 300 cc of chloromethylating solution composed of 3 parts by weight of anhydrous tin tetrachloride, 15 parts by weight of chloromethylether and 82 parts by weight of tetrachloroethane, whereby chloromethyl groups were introduced into the benzene nucleus of the copolymer. The treated membrane was then rinsed twice with 300 cc of methanol, then dipped in a 0.2 N methanol solution of trimethylamine for 24 hr at 30°C to effect amination, rinsed in running water and thus an anion permselective membrane was obtained. The membrane thus obtained had a size of 10 x 10 cm^2; a thickness of 0.19 mm; a transport number of 0.93; a bursting strength of 160 psi and an effective resistance of 2.3 Ω-cm^2.

Example 2: In 500 cc of the same latex as in Example 1 was dipped a polyethylene terephthalate fiber cloth of 10 x 20 cm^2. The thickness of the cloth was 0.10 mm; warp 75 threads/inch; weft 65 threads/inch; apparent free space content 50%. The copolymer content was increased to 51% by repeating twice the dipping and drying. The copolymer supported on the cloth was treated in the same manner as in Example 1 and thus an anion permselective membrane was obtained. The transport number was 0.92; the effective resistance 2.8 Ω-cm^2; the bursting strength 120 psi; and the thickness 0.17 mm.

Example 3: In 500 cc of a latex of styrene-butadiene linear copolymer having 46% of bonded styrene, was dipped the same glass fiber cloth as in Example 1 and of the same size. It was withdrawn and dried by irradiation on both its sides with two 250 watt infrared ray lamps, and this dipping and drying was repeated twice. The membrane showed a copolymer content of 58%. Thereafter, it was dipped in the same mixed solution of anhydrous aluminum chloride, diethylether, and dichloroethane as in Example 1 for 1 hr to effect cyclization. It was rinsed, dried, subsequently chloromethylated for 4 hr with the same chloromethylating solution as in Example 1, and thereafter aminated similarly. In the thus obtained membrane the transport number was 0.90 and the effective resistance 4.2 Ω-cm^2.

Styrene Monomer Absorbed on Polyethylene

H. Zenftman; U.S. Patent 3,304,272; February 14, 1967; assigned to Imperial Chemical

Industries Limited, England has found that certain monomers, for example styrene, can be absorbed, by a process of solution, in selected polymers, examples being polyethylene, polypropylene and polyvinyl chloride, which act as carrier, and that the monomers can be converted to ion exchange resins in situ in the carrier polymer. Thus in the preparation of ion exchange materials in accordance with the process a selected monomer or mixture of monomers having polymerizable groups, for example, ethylenic groups, and capable of forming polymers that have molecules susceptible to the introduction of ionic groups, is absorbed in a synthetic polymer selected to act as an absorbent and carrier.

The absorbed monomer or mixture of monomers is polymerized in situ and then reacted to introduce into the molecule of the thus formed polymer a group appropriately chosen to confer ion exchange properties; such groups include, for example, sulfonic, amino and quaternary ammonium groups.

In one method of carrying out the process a film of a carrier polymer is immersed in an active liquid monomer containing a polymerizing agent, the film heated to effect polymerization of the monomer absorbed in the film and the resulting product treated by known means to introduce the group selected to confer ion exchange properties on the polymerized monomer.

In another method of carrying out the process granules of a carrier polymer, for example, polyethylene, are placed in a mixing device, the required amount of active monomer, for example, styrene, is introduced and the components mixed. This step is followed by heating to polymerize the monomer in situ in the carrier polymer granules. The resulting product is treated according to known methods to introduce the group selected to confer ion exchange properties on the polymerized monomer. If it is desired to produce ion exchange membranes the granules so obtained are milled or treated by other known means to produce films. The following examples illustrate the process.

Example 1: In the preparation of an ion exchange membrane of the cationic type, a polyethylene film of thickness 0.005" was immersed in styrene monomer containing 1% benzoyl peroxide for 1 min at 75°C and was found to absorb styrene in the proportion of 48 parts of styrene to each 100 parts of polyethylene by weight. The film was heated with exclusion of air for an hour at 125°C. The originally slightly opaque film became clear on heating then regained its opacity to an intensified degree which was found to indicate that the polymerization reaction had occurred.

The film incorporating the polystyrene was immersed for an hour in chlorosulfonic acid at 50°C, then washed with 92% sulfuric acid, neutralized with 2% sodium hydroxide solution and finally washed with water. The ion exchange material was found to swell to a moderate extent only in water and the substituted polystyrene containing ion exchange groups did not leach out in aqueous solutions.

A film prepared by the above method was found to have a resistance of 1.65 ohms/cm^2 as measured with a 1,000 cps conductivity bridge; its selectivity, as determined by measuring the potential set up across the membrane in contact with solutions of 0.1 and 0.2 N potassium chloride, was found to be 0.96.

Example 2: In the preparation of cationic ion exchange membrane a polyethylene film of thickness 0.005" was immersed in styrene monomer containing 1% azodiisobutylnitrile for 5 min at 60°C, during which time the weight of the film increased by about 30%, and treated as described in Example 1. The film was sulfonated in similar manner to that described in Example 1 above. The membrane obtained was found to have a resistance of 4.6 ohms/cm^2 and a selectivity of 0.98.

Cellulosic Paper Containing Ion Exchange Resin

A process described by *C.H. McBurney and E.F. Meitzner; U.S. Patent 2,955,067; Oct. 4, 1960; assigned to Rohm & Haas Company* involves paper products which contain finely divided and substantially uniformly distributed discrete particles of ion exchange resins and which are capable of adsorbing ions from liquids and gases.

The products have the general physical properties, including appearance, porosity and feel of conventional paper products such as crepe, tissue, wrapping and blotting papers, cardboard, chipboard, boxboard, molded or shaped pulp products such as plates, boxes or cartons, and paper containers in general. They are made by dispersing, distributing or incorporating discrete particles of one or more ion exchange resins in a pulp or papermaking stock and thereafter forming a sheet from the stock or otherwise molding or shaping the mixture of pulp and resin.

The process is distinguished from other methods of impregnating paper with thermoplastic or thermosetting resins, such as dipping a paper in a solution of a resin, in that the resins employed are ion exchange resins in the form of fine discrete particles which are insoluble in aqueous media and in practically all organic liquids. The products are materials which contain discrete particles of insoluble resin distributed throughout, and held in place by, the pulp fibers. Since fiber particles are more easily distributed throughout the papermaking stock or pulp and are more efficiently retained, it is recommended that the particles of the resins be at least fine enough to pass through a U.S. Standard sieve No. 50, No. 100. Best results are obtained by employing micropulverized resins.

In the following examples which illustrate the process, hand sheets, containing varying amounts and kinds of ion exchange resins, were made by the standard TAPPI (Technical Association of the Pulp and Paper Industry) process, T218m-48. Thus, a slurry of a pulp at a concentration of 2.5% was beaten to the desired freeness, as measured, for example, by the standard TAPPI procedure, T460m-49.

This was then diluted to a pulp concentration of 1% and particles of ion exchange resins, all of which were so fine as to pass through a U.S. Standard sieve No. 100, were then added as a 10% slurry in water. In some cases alum or sulfuric acid was added to impart a lower pH. Also in some cases a typical urea-formaldehyde resin was added in order to impart enhanced wet strength. Sheets were then laid down from the furnish at a pulp concentration of 0.04%. The sheets were dried on a drum drier for 2 min at 200°F and finally conditioned for 24 hr at 75°F and 65% relative humidity.

Example 1: A paper was thus prepared from a mixture of unbleached kraft pulp, beaten to a Canadian freeness of 485 ml and a pulverized anion exchange resin which had been made by the aminolysis, with diethylenetriamine, of an insoluble copolymer of about 95% ethyl acrylate and about 5% divinylbenzene. The paper contained approximately 30% resin by weight, based on the weight of dry pulp. This had a dry strength of 34 lb/in and a wet strength of 2.8 lb/in. Its densimeter value, which is a measure of porosity, was 13 sec.

A "densimeter value" is a measure in seconds required to displace 100 ml of air through one square inch of area of a paper. Lower values indicated higher porosity; and the procedure of determining the values is the standard ASTM test, D-726. This paper, which had the general appearance and feel of wrapping paper, also had an anion adsorbing capacity for HCl of 1.78 meq/g of dry paper.

The paper also removed stearic acid from a dilute solution of stearic acid in benzene when it was immersed in and was shaken with such a solution. By way of comparison, a paper was made in an identical manner except that the anion exchange resin was omitted. That paper had a dry strength of 34 lb/in, a wet strength of 0.8 lb/in and a densimeter value of 33. Of course, it had no ion adsorbing properties whatever.

Example 2: A paper was made by the process of Example 1, except that the papermaking stock contained 2% of a commercial wet strength resin of the urea-formaldehyde type and 3% alum, based on the dry weight of the pulp, in addition to the pulp and anion exchange resin. The finished product had an anion exchange capacity of 1.82 meq/g (dry), a dry strength of 29 lb/in and a wet strength of 3.9 lb/in, and a densimeter value of 8 sec.

Another paper was made like that described in the preceding paragraph except that it contained approximately 40% of the anion exchange resin. This product had a dry strength of 31 lb/in, a wet strength of 3.2 lb/in and a densimeter value of 8 sec and a capacity of 2.25 meq/g dry.

Antipolarization Membrane

A process described by *P. Kollsman; U.S. Patent 3,227,662; January 4, 1966* involves electrodialysis membranes and is primarily directed towards the reduction or elimination or polarization. The process employs membranes which differ from the conventional permselective membranes in that the membranes contain both anionic and cationic ion exchange groups, so as to be capable of exchanging and conducting electrically driven ions of both polarities in solvated condition in opposite directions.

The anionic and the cationic ion exchange groups are present in such proportions that the total membrane conductance for ions of one polarity exceed the conductance for ions of the opposite polarity. The ions of one polarity for which the membrane is primarily selective are for convenience termed "pol ions" in this description, and the ions of the opposite polarity which pass the membrane in the opposite direction, but in smaller numbers, are termed "depol ions."

The depolarizing effect of a membrane made according to this process, sometimes referred to as depol membrane, depends not only on the ratio of conductance of the depol ions with respect to the conductance of the pol ions, in other words a membrane characteristic, but also on the character of the ionic solution to be processed, particularly on the pH and the mobility of its different ions, an electrolyte characteristic.

A depol membrane therefore has two functions. It firstly operates as a permselective membrane and is capable of being used for all purposes where previously a conventional permselective membrane was used, for example concentration, deionization, fractionation, etc. It secondly operates to reduce or eliminate polarization and its accompanying phenomena, such as pH changes, and may therefore be operated efficiently where operation with conventional permselective membrane would be inefficient or impossible.

The depolarization action of depol membranes is particularly desirable in deionization systems, for example, multi-stage deionization units, where according to the process an increasing degree of depolarization may be provided for as deionization proceeds, as for example in advanced stages. Considering a depol membrane from the standpoint of permselectivity, a depol cation membrane for example has a predominant conductance for its exchangeable cations over its exchangeable anions. This predominance prevails regardless of differences in the mobilities of the ions, such as the cations which have a relatively low mobility and OH ions which have a high mobility.

For this purpose the membrane is being considered equilibrated in aqueous 0.1 N NaOH solution, representing the high pH of 13. For brevity's sake ions of low and high mobility will be referred to as "slow" and "fast" ions, respectively.

Similarly, in a depol anion membrane the conductance for anions exceeds the conductance for its exchangeable cations regardless of differences in ion mobility, the membrane being considered after equilibration in aqueous 0.1 N HCl solution representing the low pH of 1.1. In distinction from conventional permselective membranes which in highly dilute ionic solutions contain exchangeable ions of one polarity only and practically none of the opposite polarity, depol membranes contain a definite ratio of ions of both polarities.

This makes the depol membranes effective for uses where conventional permselective membranes exhibit strong polarization effects. Taking aqueous KCl solution of 0.001 N as a standard of comparison, in a depol membrane in its KCl form the conductance of ions of one polarity exceeds the conductance of ions of the opposite polarity by a ratio of between 100/1 to 2/1.

Considering depol membranes under conditions of different ion mobilities, an average and two extreme conditions will illustrate the membrane characteristics. K and Cl ions are of about equal mobility, H has a mobility of about 5 times the mobility of Cl and OH has about 4 times the mobility of K. Taking a certain depol membrane, equilibrating it in 1 N aqueous solution of HCl representing one extreme condition and subsequently leaching it in pure water, it shows a predominance in the content of exchangeable ions of one polarity over the ions of the opposite polarity of greater than 2 to 1.

The same membrane equilibrated in aqueous 1 N NaOH solution representing the other extreme condition and thereafter leached in pure water still exhibits a predominance in the content of exchangeable ions of one polarity over the ions of the opposite polarity exceeding 2 to 1. This applies to depol anion membranes and depol cation membranes. It will later be explained how the conductances and their ratio may be determined conveniently.

A depol membrane is made from an ion exchange material containing preferably at least 0.3 meq of exchangeable ions per gram of dry membrane material and its solvent content is such that the ratio of equivalent exchangeable ions to the solvent contained in the membrane is at least 1 N. The depolarizing activity of a depol membrane is inversely proportional to the ratio of the membrane conductance for ions of one polarity to the membrane conductance for ions of the opposite polarity.

The result of the depolarization activity is that a depol membrane may be operated at a relatively low potential, considering the number of ions being transferred, and that it may be operated at current densities at which ordinary permselective membranes polarize. At extremely low potentials and current densities the current efficiency of a depol membrane is less than that of a comparable permselective membrane due to the transfer, for example, of a small number of depol ions into a deionization chamber. At substantial operating potentials and current densities, chosen for example with a view of maintaining heat development in a cell within tolerable bounds, the gain due to reduction in the operating potential far exceeds the loss in current efficiency.

Depol membranes may be prepared from known and commercially available materials by employing membrane making techniques known and presently used for making conventional permselective membranes. The known techniques are modified by the necessity of using both anion and cation material in the same membrane. Four representative methods were employed in preparing the membranes listed in the table which follows giving the membrane compositions.

The table is divided into four groups and each group represents a different method. The first group is identified by order numbers beginning with 1, such as 13, the second group by order numbers beginning with 2, such as 21 and so forth. The letter C identifies a depol cation membrane, and the letter A represents a depol anion membrane. The figures indicate parts (in terms of weight). Dynel is a commercially available copolymer composed of six parts of vinyl chloride and four parts of acrylonitrile.

All these membranes meet the requirement that the conductance for mobile ions of one polarity exceeds the conductance for mobile ions of the opposite polarity over the entire pH range from 1.1 to 13 in an aqueous solution of 0.1 N, or at least within the pH range from 1.1 to 7 or in the range from pH 7 to 13. For example, the conductance ratio of a C-12 membrane is 9.9 in 0.1 N NaCl solution at pH 7 and is 42.2 in 0.1 N HCl solution at pH 1.1. The conductance ratio of an A-11 membrane is 16.7 in 0.1 N NaCl solution at pH 7 and 38.1 in 0.1 N NaOH at pH 13. The total exchange capacity for ions of both polarities exceeds 0.3 meq/g of dry membrane material for all membranes. For example, membrane C-14 has an exchange capacity of 0.81 meq/g of dry resin. The material contained 0.36 g of water. Consequently the ratio of equivalent exchangeable ions to water contained is 81/36 = 2.25.

Group 1: Membranes were molded from a highly viscous solution of 40 parts of ionogenic compounds and 60 parts of Dynel and ½ part of benzophenone as catalyst dissolved in enough dimethyl formamide to form a highly viscous solution at 90°C.

The molded membranes were then dried and irradiated for 1 hr at a temperature of 80°C by a 100-watt mercury vapor lamp emitting ultraviolet rays of germicidal wave lengths to cause crosslinking within the membrane material. The irradiated membrane was then activated by successive immersion in aqueous solutions of 1 N HCl, 1 N NaOH and 1 N NaCl at a temperature of 70°C.

The ionogenic compounds used consisted of a mixture of polystyrene sulfonate in the form of its butyl ester and quaternized polystyrene in the form of its propionate in anhydrous condition. Butyl ester of polystyrene sulfonic acid is made by the reaction of an aqueous solution of equivalent quantities of polystyrene sulfonic acid in the H form and butyl alcohol and subsequent dehydration by drying in hot air.

Propionate of quaternized polystyrene is made by the reaction of an aqueous solution of equivalent quantities of quaternized polystyrene in the OH form and propionic acid and subsequent dehydration by drying in hot air. Different ratios of the sulfonate and the quaternized compounds produced the different membranes listed in the table. In preparing membranes C-12 and A-12 dimethyl sulfoxide was used instead of dimethyl formamide.

Group 2: The membranes of this group were prepared in the same manner as the membranes of Group 1, except that Dynel was replaced by 55 parts of polystyrene and 5 parts of divinyl benzene as a matrix forming compound.

Group 3: The membranes of this group were prepared in the same manner as in Group 1, except that Dynel was replaced by 60 parts of polyethylene of the high density type and 5 parts of divinyl benzene as a matrix forming compound. Furthermore, toluene was used as a solvent instead of dimethyl formamide.

Group 4: Membranes were molded from a random mixture of dry anhydrous thermoplastic cation exchange resin particles in K form and thermoplastic anion exchange resin particles in Cl form by heating the mixture to 130°C and applying a pressure of 2,000 lb/in^2. The molded membranes were activated by immersion in aqueous 1 N NaCl solution of 70°C.

Certain membranes were irradiated before activation by irradiating them for 1 hour by two 100-watt germicidal mercury vapor lamps. In order to render the irradiation more effective, ¼ of 1%, by weight of the mixture, of benzophenone dissolved in benzene was added to the mixture. The benzene was removed before molding by drying. Three mixtures were employed.

Mixture A: Particles of between 0.05 mm to 0.1 mm diameter produced by grinding up Amfion cation membranes and Amfion anion membranes.

Mixture B: Particles of fibrous form about 0.1 mm in width and 0.5 mm in length produced by shredding Amfion cation membranes and Amfion anion membranes. A description of the process of making Amfion membranes is found in French Patent 1,777,380 and Republic of South Africa Patent 1900/57 of American Machine and Foundry Company.

Mixture C: Particles in fibrous form, the fibers being substantially straight filaments of 0.02 mm diameter and 0.25 mm in length. The filaments were produced by graft-polymerizing polyethylene filaments with styrene. The graft-polymerized fibers were sulfonated to produce cation fibers, and quaternized to produce anion fibers. Representative test data for the membranes in brackish water solutions are provided in the patent.

Composition of Membrane Groups 1 through 4 in Parts

Membrane	Ionogenic Compound: Sulfonated	Ionogenic Compound: Quaternized	Matrix Compound Dynel
Cation Membranes Group 1:			
C-11	36	4	60
C-12	32	8	60
C-13	28	12	60
C-14	24	16	60
Anion Membranes Group 1:			
A-14	16	24	60
A-13	10	30	60
A-12	8	32	60
A-11	4	32	60
A-10	2	34	60

Membrane	Sulfonated	Quaternized	Polystyrene	Divinylbenzene
Cation Membranes Group 2:				
C-21	36	4	50	5
C-22	32	8	55	5
C-23	28	12	55	5
C-24	24	16	55	5
Anion Membranes Group 2:				
A-24	16	24	55	5
A-23	12	28	55	5
A-22	8	32	55	5
A-21	4	36	55	5
Cation Membranes Group 3:				
C-31	36	4	60	5
C-32	32	8	60	5
C-33	28	12	60	5
C-34	24	16	60	5
Anion Membranes Group 3:				
A-34	16	24	60	5
A-33	12	28	60	5
A-32	8	32	60	5
A-31	4	36	60	5

Membrane	Sulfonated	Quaternized	Mixture
Cation Membranes Group 4:			
C-41*	75	25	A
C-42	65	35	
C-43	90	20	B
C-44*	75	25	
C-45	65	35	
C-46	80	20	C
C-47*	75	25	
C-48	65	35	
Anion Membranes Group 4:			
A-48	35	65	C
A-47*	25	75	
A-46*	20	80	
A-45	35	65	B
A-44	25	75	
A-43*	20	80	
A-42	35	65	A
A-41	25	75	
Cation Anion Membranes Group 4:			
AC-42*	50	50	A
AC-43	53	47	B
AC-44	50	50	C

*Irradiated

Laminar Membrane

F.B. Leitz; U.S. Patent 3,562,139; February 9, 1971; assigned to Ionics, Incorporated describes a method and apparatus for the deionization of electrolyte solutions where alternatingly oriented anion-cation bilaminate ion exchange membranes define the chambers of a multichamber electrodialysis cell and where the anion exchange laminae of each of the membranes bound the salt diluting chambers and the cation exchange laminae bound the salt concentrating chambers. A direct electric current is passed transversely through all chambers and membranes which current is periodically reversed.

The process is dependent upon the unusual properties of certain unique types of laminar ion exchange membranes. Such membranes are depicted in the figures as cationic (**C**) on one face and anionic (**A**) on the other. The unusual properties of such laminar membranes may be observed for example in a Hittorf transport test cell. If current is passed in the direction shown in Figure 5.2 where **C** and **A** indicate the cationic and anionic faces respectively, a substantial fraction of the current is carried by sodium ions (Na^+) (about 44%) "tunneling" through the anion selective resin and then passed through the cationic layer.

The remainder (about 56%) is carried by hydrogen ions through the cation layer and by hydroxyl ions, moving in the opposite direction through the anion layer. These ions are apparently provided by the splitting of water molecules at the interface between the cation and anion regions. When the current flow is reversed as in Figure 5.2b, essentially the entire current is carried by chloride ions (Cl^-) (about 98%) passing through the anionic resin layer and then "tunneling" through the cation selective resin. This behavior appears to be due to the initial accumulation of both types of ions (Na^+ and Cl^-) at the interface between the resin regions which builds up to a concentration at which the Donnan exclusion of the cation resin is overcome.

FIGURE 5.2: CATIONIC-ANIONIC ION EXCHANGE MEMBRANE

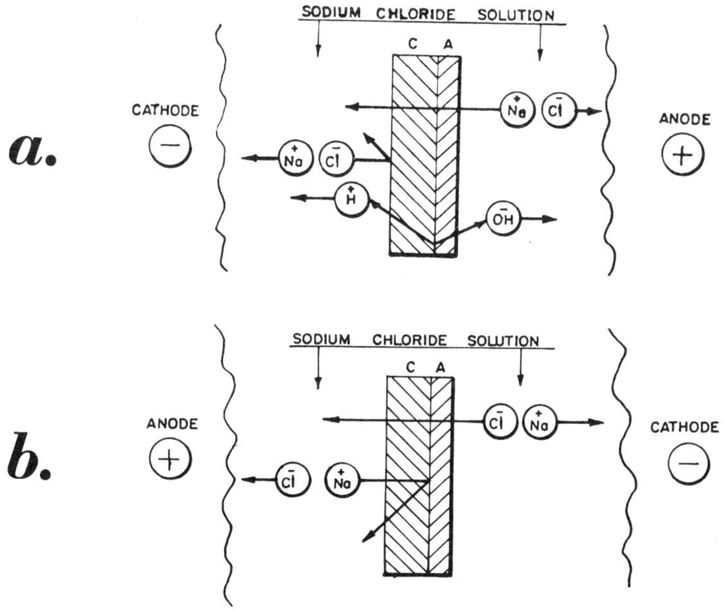

(a) (b) Simple Cell Design

(continued)

FIGURE 5.2: (continued)

c.

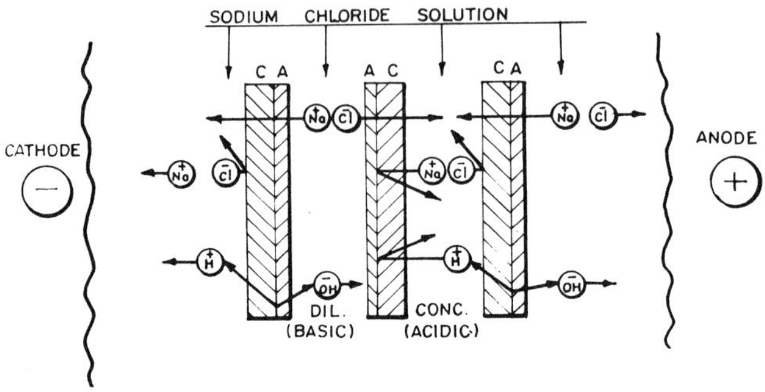

d.

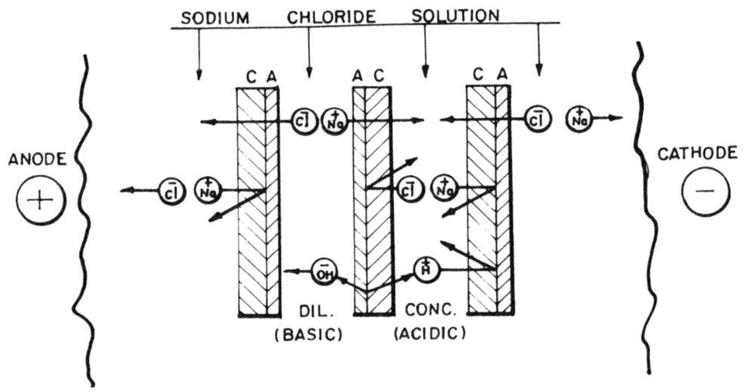

(c) (d) Assembly of Alternately Oriented Composite Ion Exchange Membranes

Source: F.B. Leitz; U.S. Patent 3,562,139; February 9, 1971

The laminar membranes of the process behave asymmetrically, transferring mainly cations when the cationic lamina of the membrane faces the cathode and transferring mainly anions when the anionic lamina faces the cathode. A possible explanation of the mechanism of the operation of such ion exchange membranes under the conditions shown in Figure 5.2a is as follows.

Under the conditions of current flow, anions are initially transferred out of the anion section of the membrane towards the anode, cations out of the cation section of the membrane toward the cathode. This transfer results rapidly in a depletion of mineral ions at the cation-anion resin interface in the membrane. If the electrical potential across the membrane is sufficiently large, current will flow, carried by hydrogen (H^+) and hydroxyl (OH^-) ions from the ionization of water at the resin interface.

The negative hydroxyl ions pass through the anion portion of the membrane towards the positive electrode and the positive hydrogen ions pass through the cation portion of the membrane towards the negative electrode. However, since the mobility of hydrogen ions is considerably greater (about 3 times) than that of hydroxyl ions, a greater potential gradient is set up across the anion membrane. This potential appears ultimately to result in a potential at the anion membrane-solution interface which overwhelms the Donnan exclusion force of the anion resin allowing sodium ions to pass into and through the anion portion of the membrane. In the steady state condition there results a transport of sodium ions from the anode to the cathode side of the membrane. It is to be noted that this particular process is not very efficient, approximately 50% current efficiency being observed with the membranes prepared so far.

Referring now to Figure 5.2b, when the direction of the current is reversed essentially the entire current is carried by the chloride ions. This is due initially to the accumulation of both types of ion (Na^+ and Cl^-) at the resin interface which build up to a concentration where the Donnan exclusion of the cation resin is overcome. The process is more readily understood in that the passage of current in the direction shown in Figure 5.2b causes transport of chloride ions into the anion portion, so that the concentration of sodium chloride at the cation-anion interface increases.

In a very short time (apparently of the order of seconds) a concentration is reached at which the Donnan forces again are overcome, this time by concentration. In the preferred laminar membranes the Donnan forces of the cation resin are overcome so that the net steady state process is the passage of chloride ions through the entire membrane toward the anode. The breakdown of the Donnan exclusion forces by concentration is observed in the case of homogeneous membranes when the current decreases rapidly as the external concentration is increased above a certain value.

The current efficiency of the process of Figure 5.2b, i.e., the transport of chloride ions, is very high (an experimental value of 98% was observed) while the transport of the hydroxyl ions (OH^-) towards the anode is much lower (an experimental value of about 2% was observed). In an electrodialysis cell using such laminar membranes arranged in alternating orientation one would expect an overall current efficiency of about 75%.

Confirmation of the hypothesized cell behavior was obtained as follows. A three compartment cell (cathode, middle and anode) employing silver-silver chloride electrodes was built with the compartments separated by oppositely oriented composite membranes produced as described. Four permutation runs were made (two membrane orientations, two current directions). Using a silver nitrate titration (corrected for hydroxyl in the product) for determination of chloride, the following data were obtained.

Run No.	Chloride Content (meq) of ------Middle Compartment------ Initial	Final	Difference	Percent Current Efficiency	pH
1	1.098	1.241	+0.143	64	2.0
2	1.124	1.240	+0.066	66	3.0
3	1.140	1.074	-0.066	66	11.3
4	1.119	0.999	-0.120	58	11.6

The runs in which the cationic resin faced the middle compartment in (Runs I and II) showed an increase in concentration of chloride in the middle compartment regardless of the direction of the applied direct electric current. The runs in which the anionic resin faced the central compartment (Runs 3 and 4) showed a decrease in chloride in the central compartment regardless of current direction.

Since the quantity of H^+ or OH^- transported is only about one-third of the quantity of chloride, the majority of current carried by positive ions must have been carried by sodium ions. The pH changes are as predicted, the diluting compartment becoming basic and the concentrating compartment becoming acidic. Current efficiencies are roughly constant.

It has been found that the preferred laminar membranes suitable for the purposes of this process have anion resin lamina which are substantially thinner (½ or ¼ or even less) than the cation resin lamina. Preferably, the exchange groups in the anion resin should be predominantly quaternary ammonium or other strongly basic groups and the exchange groups in the cation resin should be predominantly sulfonate or other strongly acidic groups. Finally it was also found that the preferred internal molality of the cation exchange resin should be less (10 to 50% less or even more) than that of the anion resin. For purposes of this process, the internal molality is defined as:

$$\text{Internal molality} = \frac{\text{meq of exchange capacity}}{\text{grains of gel water}}$$

If a group of such membranes is now assembled with alternating orientations as in Figures 5.2c and 5.2d the resulting electrodialytic process can be seen as indicated in the figures. The result of this operation is separation of a feed stream into a dilute (dil) and a concentrate (conc). stream. The only difference between the operation involved in Figures 5.2c and 5.2d is that the position of the anode and cathode have been switched, i.e., the current flow has been reversed.

However, it will be noted that the position of the dilute and the concentrate streams have not been altered and will perform as shown in the drawings. The number of coulombs passed between current reversals should preferably be large in comparison to the membrane capacity. With membranes prepared in accordance with the procedures given below, it was found sufficient to pass 0.25 meq of current per square centimeter of membrane (25 amp seconds or about 1,000 seconds between reversals). This was about 10 times the areal membrane capacity (meq/cm^2). The following example illustrates the process.

Example 1: For the manufacture of cation resin laminae, 20 sheets (each 12 x 13 in) of 3-mil high-density polyethylene are suspended on a rack and immersed in a mixture of 1,000 parts (by weight) of monostyrene, 10 parts of commercial 55% divinylbenzene and 10 parts of benzoyl peroxide. The temperature of the mixture is maintained at 70°C. The sheets are spaced at ¼ in intervals on the rack and the mixture is violently agitated by withdrawing the mixture with a centrifugal pump from the lower half of one side of the tank containing the mixture and returning it through a strainer to the upper half of the opposite side of the tank.

The sheets are removed after 6½ hr, drained and the excess monomer solution removed from the surface with absorbent paper, toweling, or squeegees. The sheets are placed individually between glass plates covered with aluminum foil which has been painted with a dilute solution of stearic acid in ethyl ether. (The sequence is: a glass plate; a piece of aluminum foil; a sheet of styrenized polyethylene; aluminum foil; glass plate; aluminum foil; styrenized polyethylene; aluminum foil; glass plate; etc.). The stack of glass plates, foil and polyethylene is weighted on top with lead bricks and heated in a recirculating hot air oven for about 17 hr.

The resulting cured styrenized sheets are stripped from the foil, and soaked for 15 min in ethylene dichloride at room temperature. The loose polystyrene polymer on the surface of the sheets is brushed off while maintaining the sheets damp with ethylene dichloride. The sheets are again rinsed in ethylene dichloride, mounted on stainless steel racks and immersed in 98% (by weight) sulfuric acid (containing 0.1% by weight of silver sulfate) at 50°C for 3 hr, then removed, allowed to cool briefly, immersed at room temperature for 15 min each in sequence in 75% sulfuric acid, 50% sulfuric acid, 25% sulfuric acid, 10% sulfuric acid, 10% sodium chloride and distilled water.

For the preparation of anion resin laminae, the above styrenization procedure is repeated using 1-mil high-density polyethylene film. Instead of immersing in sulfuric acid, the styrenized sheets are immersed for 7 hr at room temperature in chloromethyl ether containing 20% by weight of anhydrous ferric chloride, then removed and rinsed twice (15 min each time) in ethylene dichloride, twice in methanol and then soaked for 17 hr in 30% by weight of trimethylamine. The sheets are then immersed successively (15 min each time) in 35% hydrochloric acid, 18% hydrochloric acid, 3.5% hydrochloric acid, 10% sodium chloride and distilled water.

The resulting anion and cation laminae are then dried in air at room temperature for 24 hr. A sheet of anion resin is placed on a stainless steel plate coated with Teflon; a sheet of cation resin placed on the anion resin and a second sheet of Teflon-coated stainless steel placed over the cation resin, care being taken to eliminate folds, creases and bubbles. The sandwich of membranes and plates is heated in a press at about 150°C and 400 lb/in^2 for 15 min. The resulting structures are immersed in distilled water for 24 hr and if then substantially free of blisters are suitable for use in the apparatus of this process. It is found that the internal molality of the cation resins prepared as above will be in the range of 50 to 90% of the internal molality of the anion resins.

Example 2: An electrodialysis cell is assembled using the membranes of Example 1 which are cut to a size of 9 x 10 in. Tortuous path spacers of the general type described in U.S. Patent 2,708,658 are used having an active area of about 35.7 in^2, a thickness of 0.040 in and a path width of 0.25 in. Turbulence promoting straps are placed in the path every 0.43 in.

The orientation of the membranes alternates so that one set of alternate chambers is bounded by cation resin and the intervening set by anion resin. A synthetic brackish water is prepared which is approximately 0.054 N in Cl^-, 0.006 N in $SO_4^=$, 0.049 N in Na^+, 0.0085 N in Mg^{++} and 0.0022 N in Ca^{++} and is thus representative of a brackish water formed by intrusion of seawater into fresh water. This solution is pumped at a velocity of about 1.25 ft/sec through the tortuous path spacers; the effluents from anion resin bounded chambers being collected separately from the effluent from the cation bounded chambers.

A direct current density of about 165 milliamperes per square inch of effective membrane area is applied between the electrodes and through the membranes. The electrodes are platinum plated niobium. A separate electrode solution of 0.06 N NaH_2PO_4 is recirculated in parallel through the electrode compartments and adjacent buffer or rinsing compartments. The effluent from the compartments bounded by anion resin has been partially demineralized and that bounded by cation resin has become more concentrated. The latter is distinctly acid. By chemical analysis the current efficiency is found to be between 60 and 70%.

Without stopping the solution flow, the direction of the electric current is reversed after about 30 min and maintained at about 165 amp/in^2. At steady state it is again found that the effluent from the compartments bounded by anion resin has been partially demineralized and that bounded by cation resin has become more concentrated and is still distinctly acid. By chemical analysis the current efficiency is again found to lie between 60 and 70%.

The electric current is reversed every ½ hr and the performance at steady state remains unaltered, that is, the effluent from the anion bounded compartments is partially demineralized and that from the cation bounded compartments becomes more concentrated. Similar results are obtained with 15 min cycles. Upon disassembly the stack is found to be substantially free of scale and precipitate.

Mono-Grain Layer Unit

J.J.C. Oomen and D.R. Wolters; U.S. Patent 3,787,277; January 22, 1974; assigned to U.S. Philips Corporation describe a process which relates to a mono-grain layer membrane comprising a single layer of grains which are embedded in a synthetic resin foil in which part of the surface of the grains on at least one side of the foil is not coated with a synthetic resin. Such a membrane may be obtained as follows.

The grains to be embedded are scattered on a surface provided with a tacky layer. Those grains which are not fixed to the surface by the tacky layer are removed. A solution of a synthetic resin is provided in a thin layer on the single layer of grains left on the surface. The solvent is subsequently removed, for example, by heating the assembly. In this way a synthetic resin foil is formed. It is alternatively possible to use a solution of materials which form a synthetic resin as a result of chemical reaction after removal of the solvent. Heat may be required for removal of the solvent or for the chemical reaction. The synthetic resin film with the embedded grains is subsequently detached from the surface, for example, by dissolving the adhesive layer. The result then is a membrane in which part of the surface of the grains on one side, namely that part of the surface by which the grains have adhered to the adhesive layer, is not coated with synthetic resin.

If membranes are desired in which on both sides part of the surface of the grains is free from synthetic resin the synthetic resin foil is removed to the necessary depth by partly etching or dissolving the resin foil away on that side of the membrane where the surface of the grains is to be freed from synthetic resin. A membrane in which on both sides part of the surface of the grains is not coated with synthetic resin may alternatively be obtained by choosing the circumstances in the manufacture of the synthetic resin foil in such a manner that the synthetic resin retracts between the grains when the solvent is removed.

Dependent on the properties of the grains which are present in the mono-grain layer membranes obtained in accordance with these methods and dependent on the extent of accessibility of grains, a more or less selective transport of molecules, ions and/or electrons through the membrane in a direction at right angles to the plane of the membrane is possible. Membranes of this kind may be used as partitions between two compartments containing media between which an exchange or transport of ions, molecules and/or electrons must take place.

Examples of these uses are partitions in devices for dialysis and electrodialysis, fuel elements, accumulators, batteries, devices for separating gases, for selective permeation as in filtering devices and devices for reverse osmosis. Electron-conducting membranes may be used as electrodes in batteries. Semiconducting membranes may be used in solar cells and other semiconductor devices. Several examples of membranes according to the process are described with reference to the following examples, where dimensions are expressed in microns (μm for micrometers).

Example 1: A piece of wire gauze having square apertures of 60 μm x 60 μm manufactured from nylon wires, diameter of wires: 30 μm, was covered on one side with an adhesive film consisting of a flexible support provided with a layer of polyisoprene rubber. Grains having a diameter of between 37 and 44 μm of a cation-exchange material mainly consisting of a styrenedivinylbenzene copolymer were scattered on the side of the piece of wire gauze remote from the adhesive film.

The grains which did not adhere to the adhesive film were subsequently brushed off. The assembly forming the piece of wire gauze, adhesive film and grains left thereon was subsequently immersed in a solution comprising: 500 g Desmopheen 1200; 600 g Desmodur L and; 400 g ethylacetate. Desmopheen 1200 is a branched polyester obtained by condensation of adipic acid propanetriol and butylene glycol. Desmodur L contains the addition product of 2,2-dioxymethyl butanol-1 including a mixture of 2,4-toluene diisocyanate and 2,6-toluene diisocyanate (65:35).

The assembly was subsequently lifted out of the solution and heated for 16 hr at 100°C. The solvent was then removed and a film of polyurethane resin was formed from the abovementioned components. Part of the surface of the grains was not coated with polyurethane resin. The adhesive film was then removed by immersing the assembly in xylene during which the polyisoprene rubber partially dissolved and lost its adhesive strength. The membrane obtained in this manner had a permselectivity which is better than 99%. The electric resistance was approximately 9 Ω cm^2.

Example 2: A piece of wire gauze consisting of nylon wires (diameter 30 μm) having square apertures of 60 μm was immersed in a polyisoprene solution (10 g of polyisoprene per 100 ml of xylene) and was subsequently retreated from this solution so slowly that a thin, homogeneous layer covered the piece of wire gauze. The solvent was evaporated at 20°C ambient temperature. A piece of wire gauze was obtained whose wires are enveloped by a thin film of polyisoprene. Grains comprising sulfonated styrenedivinylbenzene copolymer having cation-exchange properties were subsequently scattered on the piece of wire gauze. The grains had a size of between 36 μm and 45 μm. The grains were pressed into the apertures of the piece of wire gauze with the aid of a roller. A coherent film was formed after 16 hr of heating at 100°C. Part of the surface of the grains protruded on either side from the film surface.

Softening and Dealkalizing System

A.R. Tejeda; U.S. Patent 3,893,901; July 8, 1975; assigned to J. Vast Associates, Inc. describes a softening or dealkalizing process is accomplished electrodialytically by incorporating in an electrodialytic cell a chamber defined by two spaced ion permselective membranes of the same type between which is disposed an ion exchange material of the same type; or, phrased differently, if the two membranes are cation permselective, then the ion exchange material is a cation exchange material and, conversely, if the two permselective membranes are of the anion permselective type, then the ion exchange material is an anion exchange material. Where all three of the foregoing components are of the cation type, the system will produce softened water, which is to say, water in which the cation of the output is principally sodium, and when all three are of the anion type, the process will produce dealkalized water which is a water in which the anion is a chloride.

The chamber, which as noted may either be a softening chamber or a dealkalizing chamber depending upon the types of materials used, is a chamber which forms part of an electrodialytic cell having a cathode electrode at one end and an anode electrode at the other. Preferably, the permselective membranes are flat and parallel. The ion exchange material, typically, is in the form of a bead-like ion exchange material, or an ion exchange foam, or an ion exchange fabric, or an ion exchange cloth. Such a cell also may include cathode and anode electrode chambers which are so designed as to enable the use of ferrous electrodes, and particularly a ferrous anode.

Thus, the process provides an electrodialytic system which by suitable adjustment of the types of exchange material and permselective membranes employed is capable of either softening or dealkalizing water at a low power consumption and high efficiency.

Electrolytic Conversions

G.W. Bodamer; U.S. Patent 2,921,005; January 12, 1960; assigned to Rohm & Haas Co. describes a process which comprises converting a water-soluble salt of a weak acid into the corresponding weak acid per se by passing a direct electric current through an aqueous solution of the salt while the salt solution is contained between two cationic permselective membranes on the opposite sides of which membranes are aqueous solutions, one a solution of a strong mineral acid and the other a solution of an electrolyte.

The direct current is passed through the three solutions in a direction so that hydrogen ions migrate from the acid solution through one membrane into the salt solution and there combine with the anions of the salt to form a weak acid while the metal ions of the salt migrate through the second membrane into the solution of electrolyte. Because the weak acid is only slightly ionized, having a dissociation constant of less than 0.1, only relatively few hydrogen ions are present to migrate together with the metal ions through the second membrane into the electrolyte solution.

Consequently, the salt is gradually converted into the corresponding weak acid. The process is eminently suited for the preparation of weak acids which are subject to electrolytic oxidation or decomposition. Thus the process is employed to advantage in the production, for example, of the following weak organic acids which are known to decompose to hydrocarbons, carbon dioxide and other products when they are exposed, during conventional electrolysis processes, to the oxidative conditions which surround an anode: acetic, propionic, butyric, itaconic, maleic, fumaric, acrylic, tartaric, citric, sebacic and ascorbic acids. The practical and efficient conversion of salts to acids is accomplished according to the process by electrolyzing an aqueous solution of a water-soluble salt, preferably an alkali metal salt, of a weak acid in a special kind of electrolysis cell. Figure 5.3 is a representation of a simple but typical electrolysis cell which is divided into three compartments by means of a pair of cationic permselective membranes or films.

FIGURE 5.3: ELECTROLYTIC CONVERSION CELL

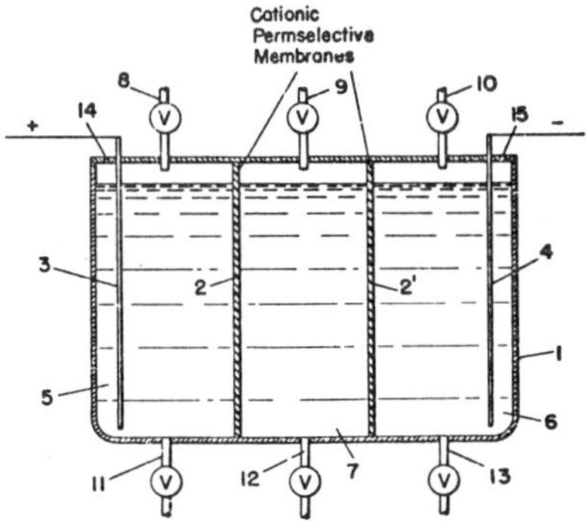

Source: G.W. Bodamer; U.S. Patent 2,921,005; January 12, 1960

In Figure 5.3, the numeral **1** identifies a container which is divided into three compartments **5**, **6** and **7** by two cationic permselective membranes **2** and **2'**, which serve as partitions between compartments **5** and **7** and between compartments **6** and **7**. The membranes are described below in greater detail. Compartment **5** is an anode compartment by virtue of the presence therein of the anode **3**, while compartment **6** is the cathode compartment containing the cathode **4**. The intermediate compartment **7** is that one in which the acid is formed. When the cell is in operation the electrodes **3** and **4** are connected to a source of electric power.

In the simplest form of the process, an electrolysis cell is employed which is like that shown in Figure 5.3. An aqueous solution of a salt of a weak acid is placed in compartment 7, while an aqueous solution of a mineral acid, preferably sulfuric acid, is placed in the anode compartment **5** and an aqueous solution of an electrolyte, preferably of an hydroxide of an alkali metal, is placed in the cathode compartment **6**. A direct current of electricity is passed through the cell and the ions in the three compartments tend to migrate toward the electrodes of opposite charge.

Hydrogen ions resulting from the ionization of the acid in compartment **5** migrate through the cationic permselective membrane **2** into compartment **7** while at the same time the cations of the water-soluble salt of the organic acid pass through the cationic permselective barrier **2'** into cathode compartment **6** where, according to the well established principles of electrolysis, hydrogen is liberated and the hydroxide of the metal is formed. At the same time oxygen is liberated at the anode when sulfuric acid is employed or chlorine is liberated if hydrochloric acid is used.

The anions of the salt in compartment **7** tend to migrate toward the anode but are constrained by the permselective barrier **2**, which, because it is cationically permselective, repels the anions and confines them to the compartment **7**. Hydroxyl ions formed in the cathode compartment likewise migrate toward the electrode but are repelled by the permselective barrier **2'** and remain in compartment **6**. Relatively few hydrogen ions pass from compartment **7** into the cathode compartment **6** because the acids, in contrast to the salts of the organic acids, are either insoluble in the aqueous solution of compartment **7** or are weakly ionized. The net result is that oxygen or chlorine is liberated from compartment **5** through outlet **14** and hydrogen, as ions, leaves the same compartment; a weak acid is formed in compartment **7**; hydrogen gas is liberated and escapes through outlet **15** while the hydroxide of the metal is formed in compartment **6**. It will be noted that no acid, aside from the small amount in compartment **5**, is required when sulfuric acid is employed.

Hydrogen and oxygen are obtained, and the hydroxide of the metal is produced which can, of course, be utilized. The process can be operated batchwise or continuously. When operated continuously water or dilute acid is admitted to compartment **5** through inlet **8** at a rate sufficient to replace that which is lost by electrolysis; a solution of the salt of the weak acid is admitted to compartment **7** through inlet **9** and a solution or slurry of the weak acid is removed from compartment **7** through outlet **12**; and the relatively concentrated solution of metallic hydroxide is removed from compartment **6** through outlet **13** and is replaced by a dilute solution of an electrolyte through inlet **10**.

The preferred cationic permselective membranes are those containing a cation exchange resin, preferably a sulfonated phenol-formaldehyde resin or a sulfonated copolymer of a monovinyl hydrocarbon such as styrene and a polyvinyl hydrocarbon such as divinylbenzene, dispersed as particles in a layer of an alkali resistant matrix. Since the electric current must be carried through the permselective film by cations in association with the cation exchange resin it is ordinarily advisable that the amount of resin constitute a substantial part of the permselective membrane. The following example illustrates the process.

Example: An electrolysis cell of the type shown in Figure 5.3 was employed. It was divided into an anode compartment containing a platinum electrode, a cathode compartment, containing a platinum electrode, and an intermediate compartment between the anode compartment and the cathode compartment and separated by a pair of cationic permselective films or diaphragms. The permselective films were made according to U.S. Patent 2,681,320, by dispersing on a rubber mill a commercially available cation exchange resin in a film of polyethylene. The cation exchange resin, which constituted 70% by weight of the permselective film, was a sulfonated copolymer of styrene and divinylbenzene (Amberlite IR 120), and was itself made by the process of U.S. Patent 2,366,007.

In the anode compartment was placed 75 parts of a 0.1 N aqueous solution of sulfuric acid. In the cathode compartment was placed 75 parts of a 0.1 N aqueous solution of sodium hydroxide. The intermediate compartment was charged with 50 parts of a 10% aqueous solution of sodium acetate.

A direct current was passed through the cell for 3⅔ hr under an impressed potential of 11 to 12 volts, at a current density of approximately 60 amperes per square foot, which current density, however, fell to about 6 amperes per square foot towards the end of the operation. During the electrolysis, oxygen was liberated at the anode and hydrogen at the cathode.

The contents of the three cells were analyzed and it was found that the anode compartment had gained 1.3 milliequivalents of acid as a result of a slight leakage through the permselective film. The amount of sodium hydroxide in the cathode compartment was increased by 63.5 milliequivalents, and 64 milliequivalents of acetic acid was found in the center compartment indicating virtually complete conversion of sodium acetate to acetic acid.

The advantage of the process over an electrolytic process of electrolyzing sodium acetate in a two compartment cell, in which the two compartments were separated by a single permselective film, is evidenced by the fact that 75.8% of the acetic acid was lost by electrolytic oxidation in the anode compartment of the cell. In this process such decomposition is avoided because the organic acid is prevented from coming in contact with the anode.

Cathode Cartridge for Electrolytic Cell

A process described by *G.A. Jensen and H.N. Parsonage; U.S. Patent 3,920,534; Nov. 18, 1975; assigned to The Mead Corporation* is directed to an apparatus useful for performing electrolytic and electrochemical reactions. The apparatus is especially useful in cases where the reaction has a tendency to produce a gas, at least at the cathode.

The preferred form of the apparatus is a closed concentric electrolytic cell. The apparatus includes a modular cartridge for the electrolytic cell comprising a porous tubular support member, a fiber-glass mat overlying the support member, an ion exchange membrane overlying the mat, and a nylon cord wrapped around the membrane for holding the membrane and mat against the support member. The cell has a cathode concentrically disposed within the support member and an anode surrounding the modular cartridge, with the space inside of the tubular support member filled with wetproofed catalyst particles.

The process described as used is a process for oxidizing alkali metal ferrocyanide to alkali metal ferricyanide. In photographic processing a ferricyanide silver bleach is used in most color reversal processing and in some color negative processing. This is a rehalogenating process in which the ferricyanide oxidizes the metallic silver image to silver ion which, in the presence of a bromide salt such as sodium bromide, produces a water-insoluble silver bromide salt.

The basic reactions are $K^+ + Ag + K_3Fe(CN)_6 \rightarrow Ag^+ + K_4Fe(CN)_6$ and $Ag^+ + NaBr \rightarrow AgBr + Na^+$. The bleach bath is followed by a sodium thiosulfate fixing bath which forms a water-soluble silver complex. As indicated, the reactions are not reversible although the build-up of ferrocyanide does not have an applicable adverse affect on the bleaching time, the latter being a function of the decrease in ferricyanide and bromide concentration.

Ferricyanide is one of the more expensive inorganic chemicals used in reversal color photographic processing, and its regeneration from ferrocyanide has economic value. Some of the known regenerating schemes including oxidation by bromine, oxidation by persulfate, oxidation by ozone and peroxide oxidation. While these systems in the main are operative and have been used before, there are drawbacks, e.g., bromine vapors, excessive persulfate creates acid pH with the formation of Prussian blue and free cyanide, potential health and explosion hazards of ozone, and the high cost of peroxide oxidation.

Referring to Figure 5.4a, the apparatus as shown generally with electrolytic cell **10** connected to a liquid-gas entrainment separator **12**, which operates to remove any liquid entrained in the gas exiting cell **10** and allows only air to pass to the atmosphere, and a control panel **14**, which generally will contain the instruments used to monitor the flow of the liquids and gas through the apparatus and may include means to measure and continually monitor the ionic activity of the process stream.

Figure 5.4b more clearly shows the overall concentric electrolytic cell **10**. Cell **10** will generally consist of three concentric, cylindrical elements: the anode **22**, the cathode **24**, and the ion exchange membrane **30**, with its supporting structure. These elements define the reaction surfaces and the flow spaces in the cell. Top cover **16** and bottom cover **18** are generally cylindrical shaped and closed at one end. They are joined together at their open ends by bolts **20**. Together they form a closed system serving to protect cathode **24**, anode **22**, and membrane **30** from dust, chemicals, the outside atmosphere, etc., as well as to define the exterior of the cell. Appropriate means are provided to allow for the necessary flow of fluids, gases and electricity into and out of the cell.

Located concentrically within the walls of bottom cover **18** is anode **22**. Since only the inside surface of anode **22** is usually necessary to carry on the electrolytic reactions in the cell, and since it may be necessary to render the anode impervious to solution migration, the outside surface of anode **22**, which abuts against cover **18**, may be coated with polyvinyl chloride **72**. Any other appropriate plastic or other nonconductive material may be used, since its purpose is to render the unused outside surface of the anode inactive and, when applicable, the anode itself impervious.

FIGURE 5.4: ELECTROLYTIC CELL

a.

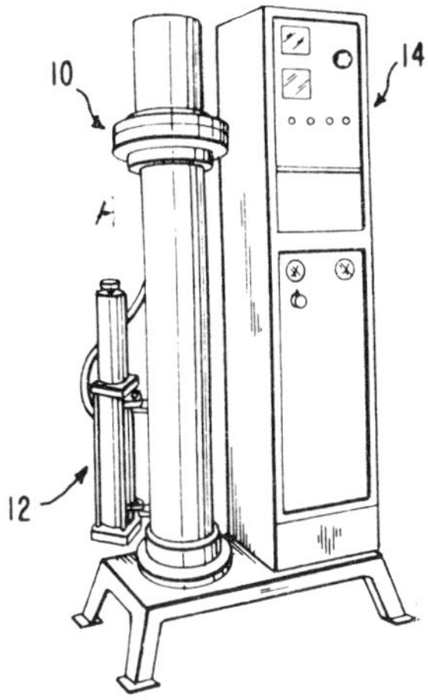

b.

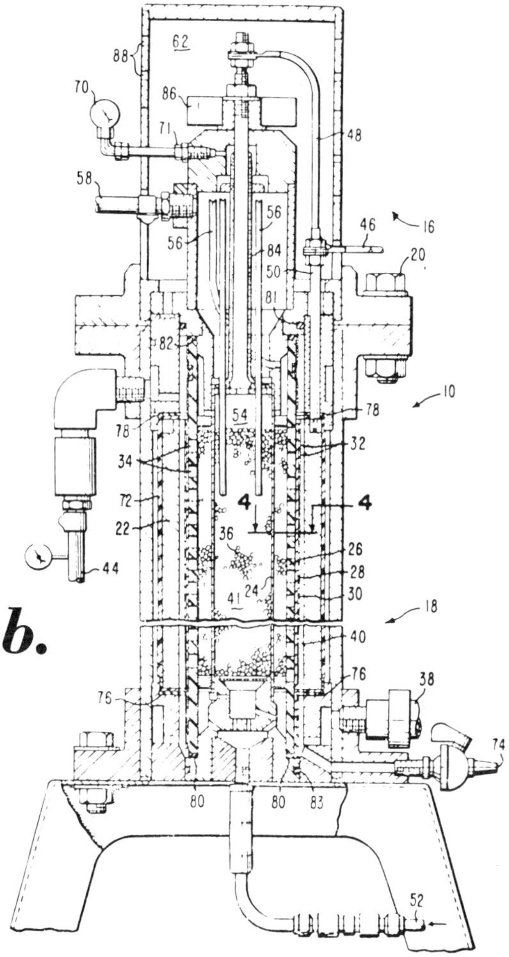

c.

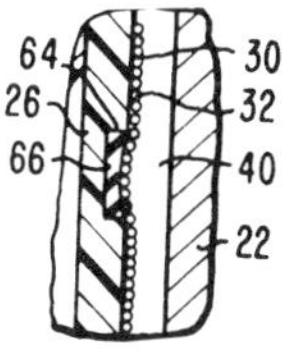

d.

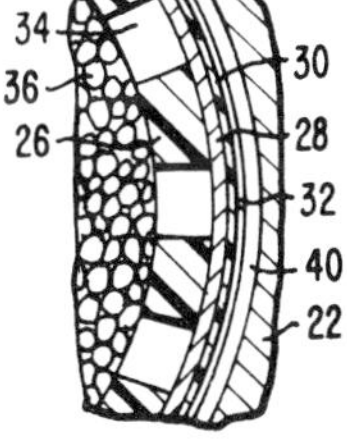

(a) Overall electrolytic system
(b) In section and in elevation with portions broken away, of the electrolytic cell
(c) Enlarged sectional view of the end seals
(d) Enlarged partial cross-sectional view along lines **4—4** of Figure 5.4b showing the ion exchange membrane and its supporting structure

(continued)

FIGURE 5.4: (continued)

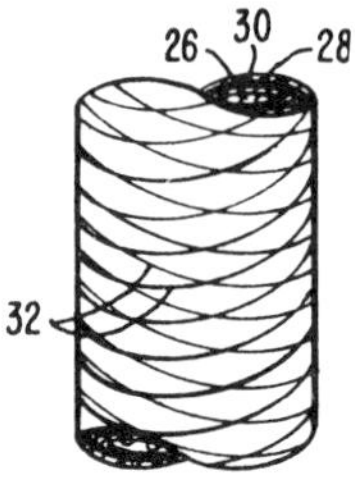

e.

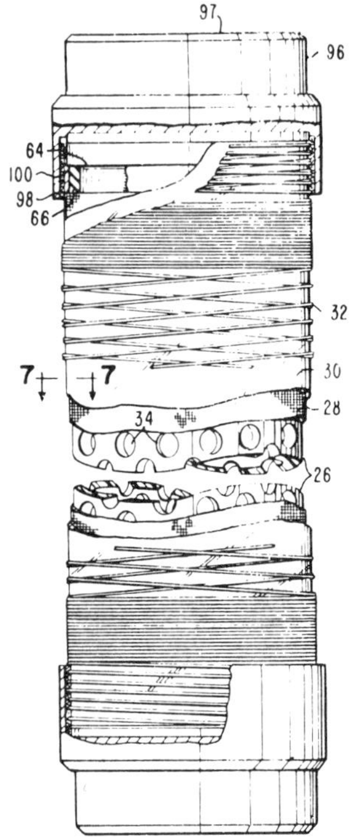

f.

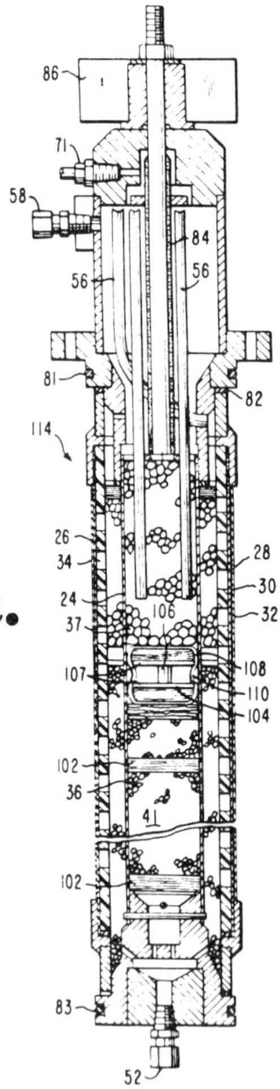

h.

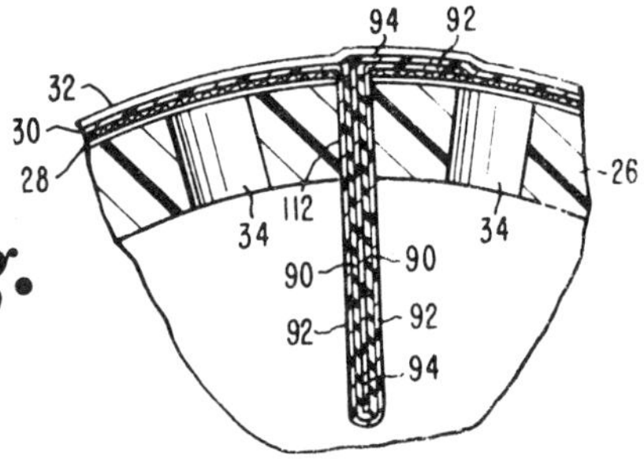

g.

(e) View of cathode cartridge with spiral wrapping of a nylon cord

(f) View partly in elevation and partly in cross section of membrane support member

(g) View of cross section of the membrane support member shown in Figure 5.4f and taken along lines 7—7 and rotated 90°

*(h) In cross section of the modular cathode cartridge

Source: G.A. Jensen and H.N. Parsonage; U.S. Patent 3,920,534; November 18, 1975

This also prevents loss of power and short-circuiting of the anode. In the preferred case the inner, active surface of anode **22** is about three square feet in surface area. This is a cylinder having an inside diameter of about four inches and being three feet high.

Spaced concentrically inward from anode **22** is ion exchange membrane **30**. In the preferred case the membrane is spaced about one-quarter of an inch inward from the active anode surface. Ordinarily, a cation exchange membrane will be used. A preferred and particularly effective cation exchange membrane is Nafion perfluoro sulfonic acid membrane but other, known, cation exchange membranes can be used. Typically, ion exchange membranes are fragile and delicate. Because of this, it is usually necessary to provide support for their protection.

As shown in Figures 5.4b, 5.4d and 5.4f, membrane **30** is supported by fiberglass mat **28** and rigid polyvinyl chloride (PVC) tubular member **26**, PVC member **26** is porous, with pores **34** being formed by drilling holes in the tubular member. Alternative support materials, to the PVC, can be used and these include other plastics, porous ceramic, porous nonconductive metals, and the like. Also, the pores need not be formed by drilling. They could be made by including a pore forming ingredient in the plastic material when it is shaped to form the support member.

PVC support member **26** is generally cylindrically shaped and has a slit **112** (shown in Figure 5.4g) which runs the length of the cylinder. When a sheet-like ion exchange membrane is used the ends of the membrane can be conveniently tucked into slit **112**. Alternatively, tubular or endless ion exchange membranes can be used. When the tubular membrane is slipped over the tubular support there are no loose ends. Thus, the need for the slit is eliminated.

Fiberglass mat **28** need not be employed. But, full use of membrane **30** is achieved when mat **28** is used. Holes **34** in member **26** allow cations to pass through membrane **30**. When membrane **30** is flush against member **26**, only those areas of membrane **30** which overlie holes **34** are available for ion exchange. By using fiberglass mat **28**, membrane **30** is lifted away from the surface of member **26**, whereby the full surface area of membrane **30** is made available for ion exchange. Fiberglass mat **28** also serves to prevent membrane **30** from stretching too far into hole **34** and to prevent the wetproofed catalyst particles located within support member **28** from working their way under membrane **30** where the sharp edges of the wetproofed catalyst particles might poke holes in the membrane. Fiberglass mat **28** is preferably a woven fiberglass fabric, although other forms of fiberglass fabric and woven or nonwoven synthetic materials may be used.

When membrane **30** and mat **28** are in sheet form they are wrapped around support member **26**. Their ends are inserted into slit **112** of member **26**. Their ends are inserted into slit **112** of member **26**. This is best illustrated in Figure 5.4g of the drawings. Figure 5.4g shows a first edge **92** of membrane **30** into and out of slit **112** so as to form a sort of envelope. The envelope formed by edge **92** is then lined with a gasket material **90**. Gasket material **90** is a soft rubber such as Viton rubber. The purpose of the rubber gasket **90** is to cushion membrane **30** and prevent contact between first edge **92** of membrane **30** and second edge **94**, which is tucked into the envelope formed by edge **92**. In this way, the edges **92** and **94** of membrane **30** will not abrade or tear and yet are held firmly in place.

Mat **28** and membrane **30** are also held in place on PVC member **26** by spiral wrapping with a plastic cord **32**. Alternatively, metal bands could be used to hold the membrane in place but the materials must be selected so that they will not corrode. Also, if the middle of the membrane is left unsupported, i.e., is not wrapped, high pressure inside the membrane will cause the membrane to balloon out and into contact with the anode, where it may burn.

By spirally wrapping membrane **30**, ballooning problems should be overcome regardless of how the pressure is controlled. Preferred material for cord **32** is a 100 pound test nylon cord. As shown in Figures 5.4e and 5.4g, cord **32** is wrapped at approximately eight wraps per inch, with two passes, one up the tube at four wraps per inch and the other back down at another four wraps per inch. Other cord materials and spacings may be used as desired. Further, if the cord is wrapped wet, as it dries it will shrink and apply additional pressure to keep membrane **30** in place. As cord **32** tightens, it applies circumferential pressure, which will close slit **112** and provide additional gripping of the ends of membrane **30**. The tightening of the cord, which was wet during wrapping, as it dries, also takes up any slack which may be present in the core as a result of the initial wrapping process.

Cathode **24** is located concentrically inward from membrane **30** and is surrounded by membrane **30**. Both inner and outer surfaces of the cathode **24** are reactive. As stated earlier, the cathode is preferably a two inch outside diameter cylinder which is three feet high. The cathode will have about the same reactive surface area as the anode, i.e., about three square feet.

The dimensions given for the anode, cathode, membrane and their spacings are preferred in that the result is a working, closed electrolytic cell, having maximum electrode surface in a minimum space and able to keep the current density up, while keeping the voltage down. Also, caustic, i.e., hydroxyl ions, produced as a consequence of the depolarization reaction enables the power consumption to be minimized, since a low caustic level results in a high power consumption.

The inside surface of anode **22** and membrane **30** define an annular flow space or anode chamber **40** wherein the anolyte is passed. The space defined by the inside of cation membrane **30**, including the space inside cathode **24** and the annular space between cathode **24** and membrane **30**, is the cathode chamber **41**. This space may be filled with wetproofed catalyst particles **36**, e.g., Contacogen particles, to provide sites for the reaction of the hydrogen ions with oxygen to depolarize the cell, as well as for the production of caustic, when the cell is operational.

In operation, bleach is passed into the cell at inlet **38**, and passes between the anode **22** and cation membrane in annular, anode chamber **40**. As the bleach passes from the bottom of the cell to the top it is oxidized from ferrocyanide to ferricyanide. The alkali ferricyanide exits chamber **40** at exit **42** and flows out bleach outlet **44** to be reused in photographic processing. The bleach flow rate is generally two to four gallons per minute. The cell will handle a flow rate of up to 10 gpm, but efficiency goes down.

As the bleach passes through the cell, air is pumped into cathode chamber **41** via inlet **52**. Cathode chamber **41** contains catholyte, which is essentially water and caustic or hydroxide ions, and catalyst particles. As the bleach is oxidized the hydrogen which is the by-product of the oxidation-reduction reaction, reacts on the wetproofed catalyst particle sites with the air to produce water and additional caustic.

The volume of air necessary for the depolarization reaction is such that, when the cell is started up, the air flows in the form of a slug which travels up the cell and spreads to fill the entire width of cathode chamber **41**. To prevent this initial air from forcing all of the fluid from the chamber, i.e., when the slug flow acts like a piston as previously described, tubes **56** some of which extend into the Contacogen bed in chamber **41** provide a by-pass allowing some of the initial air to flow past the water being pushed in front of the slug. Once the cell is in operation the air continuously introduced into the cell will bubble through the cell with no further problems.

A potential is applied across anode **22** and cathode **24** from a power source via cable **46**. Cable **46**, in turn supplies power to the anode and cathode via cables **48** and **50**. Insulation **84**, such as PVC or other suitable material, is provided on that part of cable **48** which extends above the cell. Insulation **84** prevents that part of the cable from inadvertently becoming a conductor and causing a cathodic reaction, since the vapor flowing past could be reacted.

As bleach is regenerated, hydrogen generated at the cathode **24** is converted to water by the reduction-oxidation reaction of the hydrogen with oxygen in the air which is bubbled through the cell. The formation of water as well as its reaction in part to form caustic occurs on wetproofed catalyst particle sites. The air is pumped into the cathode chamber **41** through inlet **52**. Any unreacted air passes through the cell to space **54**, and of the cell through liquid-gas entrainment separator **12**, Figure 5.4a. Separator **12** removes any liquid carried by the air.

The Contacogen particles generally should not present any problems in terms of air flow through the column. On occasion, it may be desirable to pack part, e.g., the upper one-third, of the column with wetproofed platinized titanium sponge particles while the bottom two-thirds contains another type of Contacogen, i.e., wetproofed activated carbon particles.

Such an arrangement is shown in Figure 5.4h where the bottom part of the cell contains wetproofed carbon particles **36** and the top part contains wetproofed titanium particles **37**. Fibrous packing may be employed to support the wetproofed catalyst particles and prevent them from shifting or settling. The fibrous packing, such as packing **102** shown in Figure 5.4h, can be any conventional nonwoven fabric, such as 3M Scotch Brite or a similar material.

Also, a plug, such as plug **104** shown in Figure 5.4h, may be used to support and to separate the wetproofed titanium catalyst particles **37** from the wetproofed carbon catalyst particles **36**. Plug **104** is a porous plastic, spool-like member comprising a cylindrical core **106** and flanged ends **107** but, having holes in flanged ends **107**. Additionally a fiberglass mat **108** and nylon cord **110**, which holds the fiberglass in place, may be wrapped around the plug. Titanium sponge does not pack as readily as does carbon and therefore assures air flow through the upper part of the column.

Vent tubes may be employed to enable the initial air to by-pass the upper part of the column. It may also be desirable to intersperse porous materials, including the packing mentioned above, throughout the wetproofed carbon particles to prevent the particles from becoming packed too densely and maintain an open system through which the air can pass readily.

When the cell is operated, a great amount of heat is generated by the reaction of hydrogen and oxygen to form water, as well as by the electric resistance in cathode connector **48**. Heat sink **62** is provided for the removal of heat. Fins **86** carry the heat by conduction, from the cathode chamber, and radiate it into heat sink **62**. Holes **88** in cap **87** allow air to pass through and carry the heat away from cell **10** by convection.

One example of the end seals of the process is shown in Figure 5.4c. A groove **64** is made around the circumference of the ends of the polyvinyl chloride tube. The ends are solid, i.e., nonporous. A rubber insert **66** is placed in groove **64**. As nylon cord **32** is wrapped around membrane **30**, the tension of the nylon wrap will force the membrane into the rubber insert and seal off the membrane. The nylon cord used in the seal area may be finer, e.g., 80 pound test line, than that previously described. Additionally an epoxy resin can be coated on to provide additional seal.

An alternative end seal is shown in Figure 5.4f. A plastic cap **96**, made of PVC or a similar material, having a hollow upper end **97** is placed over the ends of membrane **30** and support **26**. The ends of the membrane previously having been wrapped in the manner discussed earlier. A conventional epoxy cement is then placed between the walls **98** of the cap **96** and membrane **30** to seal the ends of the membrane.

Generally, it is not necessary nor economical to regenerate the bleach continuously. Some build-up of alkali ferrocyanide can be tolerated without effecting the photographic processing. This method of operating can be achieved by allowing the bleach to flow continually through the cell at all times, while connecting and disconnecting the power in response to the ionic activity. An alternative, albeit less economical, way is to let the cell run continuously, while only flowing the bleach through the cell when regeneration is indicated as needed by an appropriate process flow controller.

The bromide ion activity in the photographic process is related to the ratio of alkali ferricyanide to alkali ferrocyanide. When the bromide activity grows low, bleach regeneration is required. Monitoring bromide activity will indicate when the bleach regenerator needs to be run. At the same time that the bleach is regenerated, fresh bleach may be added from a bleach replenisher supply. The same control can be achieved by monitoring the ratio of alkali ferrocyanide to alkali ferricyanide itself.

U.S. Patent 3,770,608 describes apparatus to measure bromide ion activity in a photographic processing stream. When the bromide activity is low, as measured by a process controller a switch can be activated to apply voltage to the electrodes of the bleach regeneration cell. When the bromide activity is sufficiently high, the switch is deactivated. As described in the above noted reference, bromide activity is monitored by immersing a pair of ion sensing probes in two solutions, one of which is a reference solution and the other is a sample of the process stream. The probes preferably employ silver bromide membranes with a pressed silver backing.

If aeration is used to depolarize the cathode, a flow sensor may be employed to shut off electrolysis if insufficient air is flowing through the cathode chamber. The system may be interlocked to prevent electrolysis if the bleach recirculation rate drops below a prescribed level. This will eliminate the potential for overoxidation of the bleach solution.

To insure a proper pressure on either side of membrane **30**, pressure control gauges **68** and **70**, Figure 5.4b, can be used in association with conventional valves to control the back pressure throughout the system. The pressure is controlled so that there is a positive, e.g., four psig differential, pressure within the anode chamber (for example, where the anode chamber is at 24 psig and the cathode chamber is at 20 psig). The positive anode pressure will prevent ballooning of the membrane **30**, prevent air from entering the anode chamber and into the bleach solution (the air causes cavitation of the pump which is pumping bleach, with subsequent burning out of the pump) and helps to dewater the anolyte, because the positive pressure causes water to be carried from the anolyte to the catholyte across the ion exchange membrane along the cations. The dewatering is important in that dilution of the bleach solution is a factor in the need to replenish the bleach solution.

Drain **74** is provided to drain the cell of anolyte should it need to be serviced, including the replacement of the Contacogen packing, the ion exchange membrane, or the whole cathode cartridge. Cathode **24** and ion exchange membrane **30**, and its supporting structure, i.e., support member **26**, fiberglass mat **28** and nylon cord wrap **30**, may be combined as a modular unit, filled with Contacogen particles, such as wetproofed carbon **36** and wetproofed titanium **37**.

Such a modular unit is cathode cartridge **114**, which is shown in Figure 5.4h. When the cation exchange membrane or the Contacogen particles need to be replaced, cartridge **114** need only be lifted out of cell **10** and replaced by a new cathode cartridge. Appropriate seals can be provided to render the cartridge fluidtight. O-rings **81** and **83**, as well as pressure packing **80** and **82**, are provided for cathode cartridge **114**. Pressure packings **76** and **78**, Figure 5.4b, are provided for anode **22**.

SOLVENT AND GAS SEPARATION

HYDROCARBON SEPARATION

Butadiene-Acrylonitrile Polymers for Diene Synthesis

J.-P. Brun, G. Bulvestre, M. Juillou, P. Thirion and R. Pautrat; U.S. Patent 3,819,742; June 25, 1974 describe a process for separating mixtures of hydrocarbons, notably conjugated diene hydrocarbons, from mixtures containing the same, by the general technique of permeation or pervaporation through specific membranes. The specific membranes used are formed of a high polymer comprising functional groups chemically combined by covalent bonding within the polymer forming the membrane, the groups possessing a physiochemical affinity, notably a capacity to complex, with respect to one of the hydrocarbons and, substantially, the absence of such a capacity with respect to other hydrocarbons of the mixture. For the separation of dienes, such as butadiene and pentadiene, membranes are used containing active nitrile, pyrrolidone, N-substituted groups, such as N-alkyl groups, notably N-methyl or N-methyl-pyrrolidone. The membrane can be obtained, e.g., from a vulcanized butadiene-acrylonitrile rubber, which can contain about 5 to 60% acrylonitrile, and notably from 20 to 30%.

Styrene-Acrylonitrile Copolymers for Diene Separation

According to a process described by *E. Perry and W.F. Strazik; U.S. Patent 3,773,848; November 20, 1973; assigned to Monsanto Company* dienes such as butadiene and isoprene are separated from organic mixtures by contacting the mixture against one side of a polymeric membrane and withdrawing at the other side a vaporous mixture having increased diene concentration. The polymer membrane is an acrylonitrile-containing polymer which is (1) a copolymer of acrylonitrile or (2) a polymer blend containing acrylonitrile polymer. Exemplary of the polymer membranes are copolymers of acrylonitrile and styrene, vinyl acetate and N-vinylpyrrolidone and blends of polyacrylonitrile and polystyrene, polyvinyl acetate and polyvinylpyrrolidone. Exemplary of the organic mixtures is a mixture of butadiene and butene.

Modified Cellulose Esters

E.C. Martin, R.C. Binning, L.M. Adams and R.J. Lee; U.S. Patent 3,140,256; July 7, 1964; assigned to Standard Oil Company describe a permeation process whereby organic compounds such as hydrocarbons, among others, may be concentrated or separated from mixtures. This process is based upon the concept that permeation membranes comprised of organo-soluble organic derivatives of cellulose, particularly cellulose esters, cellulose ethers, mixtures of cellulose esters and cellulose ethers, can be modified to improve their characteristics as permeation membranes for the separation of organic mixtures such as a mixture of hydrocarbons, by reacting the free hydroxyl groups attached to the anhydroglucose units of the cellulose derivative with an organic reagent which is reactive with the hydroxyl groups. The organic reagent may contain one or more of the same or different functional groups, each of which are reactive with a hydroxyl group. A great variety of organic reagents and modifying techniques may be used.

Aldehydes, mixtures of aldehydes with alcohols, organic diisocyanates, organic monoisocyanates, organic dichlorides such as chlorides of dibasic acids, are examples of some of the organic reagents which may be employed for this purpose. Either a minor portion or else substantially all of the free hydroxyl groups attached to the anhydroglucose units of the cellulose derivatives may be modified by reaction with the organic reagent. A number of examples of the manner in which the cellulose esters and/or ethers may be modified by reaction with an organic reagent reactive with the hydroxyl groups follows.

Aldehyde Modification—The free hydroxyl groups in an organo-soluble cellulose ester and/or ether can be reacted with an aldehyde to improve the permeation rate of the membrane prepared from the cellulose derivative. It is believed that an acetal is formed by the reaction. Any of a wide variety of aldehydes may be employed in this modification, for example, aldehydes having from 1 to 12 carbons may suitably be used. Formaldehyde has been found to be highly effective as a modifying agent, but other aldehydes such as acetaldehyde, propionaldehyde, butyraldehyde, 2-ethylhexaldehyde, ethoxyacetaldehyde, phenoxyacetaldehyde, etc., can be used.

Example 1: 30 parts by weight of ethylcellulose (N-50) which contained an average of approximately 0.52 free hydroxyl group and 2.48 ethoxyl groups per anhydroglucose unit was added to a solvent solution consisting of ⅓ acetone, ⅓ methyl ethyl ketone, and ⅓ methyl isobutyl ketone to obtain a viscous solution. To this solution was added 7.5 parts by weight of formalin (37% HCHO) together with 3 parts of a 5% stearic acid in carbon tetrachloride solution. Thus approximately 1.38 mols of aldehyde per mol of free hydroxyl groups (present in the amount of cellulose derivative used) was employed in carrying out the reaction. The mixture was blended and heated at 50°C until the solution was free of bubbles.

Thereafter films were formed by casting of the solution onto stainless steel panels. The film was dried under heat lamps and then placed in an oven at 80°C for about 0.5 hour to complete the reaction. The aldehyde-modified ethylcellulose film formed was 1.0 mils in thickness. This modified film and an ethyl cellulose (N-50) film of the same thickness which had not been modified by reaction with an aldehyde were employed as permeation membranes and their effectiveness determined.

Figure 6.1 represents in schematic form a small-size permeation apparatus which was used in carrying out the permeation runs. The permeation apparatus consisted of a box-like feed chamber **11** for the feed mixture of liquid hydrocarbons introduced therein, a smaller-size membrane holder **12** of box-like shape having five open faces across which the permeation membranes **13** were sealed, the sixth face having sealed thereto a line **14** for removing the portion which permeates through the membrane into the interior (permeate zone **16**) of the membrane holder. The entire membrane holder was positioned within the feed chamber **11**. Pump **17** passes the charge mixture (which has been heated to the desired permeation operating temperature) through line **18** into feed chamber **11** until the liquid level of the charge mixture is substantially above membrane holder **12**.

The permeated portion is continuously withdrawn from permeate zone **16** through line **14** by means of vacuum pump **19**. Manometer **21** and pressure-regulating valve **22** measure and regulate the pressure in permeate zone **16**. The total surface area provided by the membranes was approximately 22 square inches. In the permeation experiments, the charge to the feed zone consisted of a 50:50 mixture of n-heptane and isooctane. A temperature

FIGURE 6.1: SEPARATION PROCESS EMPLOYING MODIFIED CELLULOSIC MEMBRANES

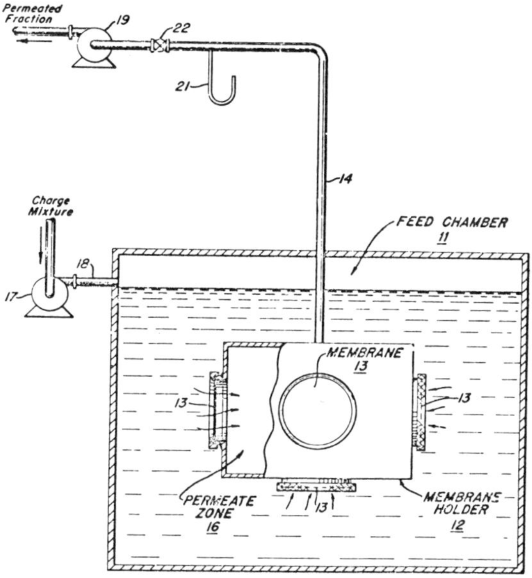

Source: E.C. Martin, R.C. Binning, L.M. Adams and R.J. Lee; U.S. Patent 3,140,256; July 7, 1964

of 70°C and atmospheric pressure were maintained in the feed zone. The permeate zone was maintained at a pressure of 30 mm Hg abs. Hydrocarbons permeated through the membrane were removed from the opposite side of the membrane and into the permeate zone as vapors which were withdrawn continuously and their compositions determined. The results below show that the aldehyde, which did not change the composition of the permeate portion, substantially improved the permeation rate.

Permeation Membrane	Permeation Rate*	Percent n-Heptane in Permeate
Aldehyde-modified ethylcellulose	201	70
Ethylcellulose	144	70

*Gallons/hour/thousand square feet of membrane surface.

Example 2: 25 parts by weight of cellulose acetate-butyrate (AB-504-35) which had 0.96 free hydroxyl group per anhydroglucose unit, 0.44 acetyl group per anhydroglucose unit, and 1.6 butyryl groups per anhydroglucose unit, was dissolved in approximately 200 parts of dioxane. To this solution was added 2 parts of para-formaldehyde and 1 part of toluene sulfonic acid. Thus approximately 0.81 mol of aldehyde per mol of free hydroxyl group (attached to the anhydroglucose units) was present in the reaction zone. The mixture was heated at about 55°C with stirring for about 4.5 hours. Thereafter acetone was added to the solution to decrease its viscosity. A 50:50 mixture of water and acetone was then gradually added with stirring until substantially all of the aldehyde-modified cellulose derivative was precipitated as a gelatinous material.

The liquid was decanted from the gelatinous material, and the gelatinous material was redissolved in acetone. Thereafter the acetone solution was poured into water and firm fibers were precipitated. The fibers were removed and washed several times with water to remove traces of acetone. The fibers were air-dried. The dried fibers were then dissolved in a solvent consisting of 30% benzene, 30% ethanol, 30% ethylene dichloride and 10% dioxane. This solution was then cast upon steel panels to form film which was dried under heat lamps. The film thickness was 1.0 mil. This aldehyde-modified film and a cellulose acetate butyrate film which had not been modified were employed as permeation membranes and their effectiveness determined.

The permeation experiments were carried out in the manner described in Example 1, except that the charge mixture was a 50:50 mixture of benzene and cyclohexane and the permeation temperature was 80°C. The results obtained in the permeation runs were as follows.

Permeation Membrane	Permeation Rate*	Percent Benzene in Permeate
Aldehyde-modified cellulose acetate-butyrate	75	86.8
Cellulose acetate-butyrate	63	84.0

*Gallons/hour/thousand square feet membrane surface.

It will be noted that this particular method of carrying out the aldehyde modification of the cellulose derivative improves the selectivity of the membrane while maintaining essentially the same permeation rate or even improving the permeation rate.

Aldehyde-Alcohol Modification—The free hydroxyl groups in an organo-soluble cellulose ester and/or ether can be reacted with a mixture of an aldehyde and an alcohol. It has been found that permeation membranes prepared from the aldehyde-alcohol-modified cellulose derivatives permit hydrocarbons to permeate the membrane more rapidly than the nonmodified cellulose derivative. This improvement in permeation rate is often obtained without a decrease in selectivity of the membrane.

It is believed that an alkoxyacetal derivative of the cellulose derivative is formed during the modification process, the alkoxy group depending upon the alcohol used and the acetal depending upon the aldehyde which is used. Thus when the alcohol is ethanol and the aldehyde is formaldehyde, the ethoxy formal of the cellulose derivatives is formed. While paraformaldehyde is preferred, the reaction proceeds well with a formaldehyde-yielding substance such as trioxane, trioxymethylenes.

Other aldehydes such as acetaldehyde and higher molecular weight aldehydes may be employed. Any of a variety of aliphatic alcohols containing up to about 10 carbons or so may be used. Thus methanol, ethanol, propanol, butanol, etc., may be employed. The aldehyde and the alcohol may be employed in the modification reaction in amounts as low as 0.1 part of aldehyde and 0.1 mol of alcohol per mol of free hydroxyl group present in the amount of cellulose derivative subjected to the modification. An equal number (or even a great excess) of mols of aldehyde and mols of alcohol per mol of free hydroxyl group (containing in the cellulose derivative present in the reaction zone) can be used.

In related work, *R.J. Lee and J.F. Jennings; U.S. Patent 2,960,462; November 15, 1960; assigned to The American Oil Company* describe the use of laminated permeation membranes for the separation of organic chemical mixtures. The laminated membrane consists of a thin layer or film of a selective membrane material mounted upon a thicker film of a membrane material having a higher permeation rate. The layer of the membrane material having the higher permeation rate ordinarily has a poor selectivity by comparison with the layer of selective membrane material. This laminated membrane is employed so that the film layer having the higher permeation rate is in contact with the mixture of the chemicals to be separated.

It has been found that when the higher permeation rate surface of the laminated membrane is in contact with the mixture to be separated, the selectivity of the laminated membrane is greater than the selectivity of either of the component films from which the laminate is formed. If the laminate membrane were to be reversed so that the more selective membrane material were in contact with the mixture to be separated, the selectivity of the laminate membrane would then be intermediate between the selectivities of the films from which the laminate membrane is formed.

Any of a number of membrane materials can be employed in forming the laminated membrane which is to be employed in the process. The following membrane materials have been found to be suitable for this process, with the first one listed having the highest selectivity and the lowest permeation rate and the last one listed having the highest permeation rate and the lowest selectivity: cellulose esters, cellulose ethers, and polyethylene (both irradiated and nonirradiated). The rate of permeation increases from the first-named membrane material to the last-named membrane material, while on the other hand the selectivity decreases from the first-named membrane material to the last-named membrane material.

R.J. Lee; U.S. Patent 2,947,687; August 2, 1960; assigned to The American Oil Company has found that when a hydrocarbon mixture is permeated through a nonporous membrane in which certain of the hydrocarbons contained in the mixture are more soluble than others, for the purpose of separating the mixture into different components, the rate of permeation can be greatly increased by contacting the membrane during the permeation process with a substituted hydrocarbon which is soluble in and has solvent power for the membrane. The substituted hydrocarbon is an organic compound containing one or more atoms of an element such as halogen, oxygen, sulfur, or nitrogen.

When the substituted hydrocarbon solvent comes in contact with the membrane it enables hydrocarbons to permeate the membrane more readily. The substituted hydrocarbon solvent may be introduced into the permeation step by adding it to the feed zone or the permeate zone in either a liquid or gaseous condition. Ordinarily the physical state of the substituted hydrocarbon solvent should be the same as for the hydrocarbons which are present in the zone to which it is added; it is preferred to introduce the substituted hydrocarbon solvent into the feed zone either by separate introduction or introduction along with the feed hydrocarbon mixture. A desirable method of operating consists of maintaining the feed hydrocarbon mixture and the added substituted hydrocarbon solvent in the liquid phase in the feed zone and recovering permeated hydrocarbons together with permeated substituted hydrocarbon solvent in the vapor phase in the permeate zone.

Cellulose Acetate Butyrate

J.M. Stuckey; U.S. Patent 2,930,754; March 29, 1960; assigned to Pan American Refining Corporation has found that hydrocarbon mixtures may be rapidly and effectively separated according to type, i.e., aromatic, unsaturated, saturated hydrocarbons, and/or molecular configuration and/or boiling point by causing a portion of the mixture to permeate through a nonporous membrane comprised of certain cellulose esters under defined conditions. For obtaining rapid selective permeation, a thin nonporous membrane is employed which is less than 10 mils and preferably 0.2 to 5 mils in thickness and comprised of cellulose esters containing acyl radicals of a type and in a quantity to permit certain of the hydrocarbons to dissolve in the membrane without rendering the membrane so soluble in the hydrocarbons as to rupture the permeation process.

The carboxylic acid esters of cellulose which may be employed in making the membranes for use in this process are those which have at least a portion of introduced acyl groupings which contain more than 2 carbons. Approximately 10% or more of the acyl radicals introduced into cellulose should have more than 2 carbons. The cellulose esters preferably have from 2 to 4 carbons in the introduced acyl grouping, although the number of carbons may be more. Cellulose acetate is not effective for the selective permeation of hydrocarbons since it is practically impermeable to hydrocarbons. However, cellulose acetate-butyrate is highly effective. Other examples of useful cellulose esters are cellulose propionate, cellulose acetate-propionate, cellulose butyrate, cellulose pentanoate, cellulose hexanoate, cellulose acetate-hexanoate, cellulose propionate-butyrate, cellulose benzoate, cellulose acetate-benzoate, and like compounds of single or mixed carboxylic acid esters of cellulose. The permeation process is carried out by contacting one side of the membrane with the hydrocarbon mixture which may be in the liquid or vapor state and removing permeated hydrocarbons from the opposite side of the membrane in either the liquid or vapor state. The hydrocarbon mixture may be continuously introduced into the charging zone so that it contacts one side of the membrane and nonpermeated hydrocarbons may be continuously removed from this zone after having contacted the membrane. Permeated hydrocarbons may be continuously withdrawn from the permeate zone.

It is believed that permeation occurs because the concentration of permeating hydrocarbons at the surface of the membrane on the permeate side is less than the concentration of the permeating hydrocarbons at the surface of the membrane in the charging zone. Thus, the permeated hydrocarbons should be rapidly removed from the surface of the membrane in the permeate zone to insure rapid permeation of the hydrocarbons. This may be accomplished by removing permeated hydrocarbons from the surface of the membrane on the permeate side or by diluting permeated hydrocarbons in this zone with a diluent or sweep gas or liquid, for example, steam and air.

When liquid-liquid systems are employed on opposite sides of the membrane, a diluent or sweep liquid is circulated about the permeate surface of the membrane. When permeate hydrocarbons are obtained in the permeate zone in the vapor state, the permeate hydrocarbons may be removed from the permeate zone by employing a lower pressure than exists on the charging side or by using a sweep gas to dilute and/or remove the permeated hydrocarbons. When the hydrocarbons in both the charging zone and the permeate zone are in the vapor phase, a pressure differential is employed to facilitate the permeation of the hydrocarbons.

Pressure differential of from 10 mm Hg to as high as 100 psig, dependent upon the strength of the membrane and the effectiveness of the manner in which it is supported, may be used. Under such operating conditions, the pressure may be superatmospheric, atmospheric, or subatmospheric, dependent upon the pressure differential existing between the two zones. It is preferred to operate the permeate zone at subatmospheric pressures so that the permeated hydrocarbons when removed in the vapor state are easily evaporated.

The preferred mode of operating the permeation process consists of contacting the feed mixture of hydrocarbons while in the liquid state with the membrane and removing the permeate in the form of a vapor from the permeate zone. This permeate may be removed by employing a reduced pressure in the permeate zone or by sweeping the vapors of the hydrocarbons with a gas such as air. It has been found that when operating the permeation process in this preferred manner it is possible to obtain increased rates of permeation of about 40% higher than those obtainable by having the hydrocarbons in the vapor state on both sides of the membrane. If desired, it is possible to operate with the pressure differential maintained across the membrane to insure rapid permeation of the hydrocarbons and also to use a sweep gas or liquid, whichever is called for by the conditions, to increase the permeation rate.

p-Xylene Separation

R.F. Baddour and A.S. Michaels; U.S. Patent 3,299,157; January 17, 1967; assigned to Amicon Corporation describe the preparation of polymeric membranes specifically designed to display enhanced permeability to a selected single ingredient of any given mixture even when the ingredients are very similar to each other in chemical and physical properties, e.g., are position isomers. The membranes may be employed in a single or multistage diffusion process which will effectively separate the ingredients. The separation may be carried out at any convenient temperature while the ingredients of the mixture are in gaseous, vaporous or liquid form, the diffusion being brought about by maintenance of a pressure or concentration differential between opposite faces of the membrane. The separation may also be carried out while the mixture is dissolved in a liquid solvent, in which case a concentration differential may be used to bring about diffusion by flowing a supply of pure solvent in contact with the downstream face of the membrane to dissolve and carry away the material diffusing through. The polymeric material selected for use as the membrane may be any of the well-known film-forming natural or synthetic polymeric materials which are capable of being sheeted out to form a film or membrane having substantial mechanical strength and having a thickness no greater than about 0.05 inch. In most cases the membrane preferably has a thickness from 0.0005 to 0.010 inch. The polymeric material must also be one in which the ingredients of the mixture will dissolve or be absorbed to the extent of 1 to 30% by weight of the dry membrane at the temperature at which the diffusion separation is to be carried out.

For most conventional polymeric films, this temperature is from -50° to 200°C, preferably from 0° to 50°C. The extent to which a wide variety of fluid materials are dissolved by or absorbed by particular polymeric materials is known. The membrane of polymeric material which has been so chosen must then be subjected to swelling or solvation by one selected ingredient of the mixture of fluid materials which is to be separated. Selection of the ingredient of the mixture which is to be employed for this purpose is determined by consideration of the molecular cross-sectional areas of the components comprising the mixture.

If it is desirable to separate from such a mixture, in high purity and with optimum efficiency, the ingredient or component possessing the smallest molecular cross-sectional area, then that component should be employed for treatment of the membrane. If, on the other hand, it is desired to separate from the mixture some other component, treatment of the membrane with any component having a molecular cross-sectional area smaller than that of the desired component may be carried out, although treatment with a component having a molecular cross-sectional area intermediate in size between that of the smallest component and that of the deisred component may often be preferred.

The swelling or solvation of the polymeric membrane by the selected ingredient or substitute material must be carried out at a temperature at least 5°C higher than the temperature at which the ultimate diffusion separation is to be carried out. However, the swelling or solvation temperature must be less than the temperature at which the polymeric membrane dissolves in the selected ingredient to such an extent that the membrane loses its integrity. In general, this step is carried out at a temperature at which the membrane absorbs from 1 to 50% by weight of the ingredient or substitute material, preferably from 5 to 15% by weight. The polymeric membrane is allowed to remain immersed in the ingredient for a sufficiently long period of time to ensure that equilibration between the two has occurred.

The membranes which have been equilibrated in this fashion are found to exhibit substantially increased permeability for the selected ingredient while permeability of the remaining ingredients of the mixture has been either substantially reduced or has been increased only to a slight extent, much less than the permeability for the selected ingredient has been increased. The following example illustrates the process.

Example: The fluid material which was chosen for separation consisted of a liquid mixture of ortho-, meta-, and para-xylene. It was determined by simple test that a conventional 1-mil film of high density (0.9570 g/cc) polyethylene was capable of absorbing about 10% by weight of the mixture at 50°C. It was also determined from molar refraction measurements that the molecular volume of each of the three xylenes was as shown in the following table.

	Molecular Volume A^3/Molecule at 20°C	Maximum Linear Molecular Dimension, A	Molecular Cross-Sectional Area, A^2
Ortho-xylene	59.7	7.80	7.65
Meta-xylene	59.8	8.33	7.20
Para-xylene	59.9	8.67	6.92

The maximum molecular linear dimension in Angstrom units of each of the xylenes was determined by constructing a Stuart and Briegleb model and measuring it. The quotient of these two values is the molecular cross-sectional area or apparent diffusion cross-sectional area of the molecule, as shown in the above table. It is clear from the values set forth in the table that the para-xylene has the smallest molecular cross-sectional area of the three ingredients in the mixture, and it was therefore selected as the ingredient for treating the polymeric membrane.

A sample of the high-density polyethylene film was immersed in a bath of pure liquid para-xylene and heated at 100°C for 24 hours, cooled, and dried at 25°C. This treated film was then mounted in a conventional diffusion apparatus, a mixture of liquid xylenes containing, by weight, 30% ortho-, 65% meta-, and 5% para-xylene being supplied to one face of the membrane at a temperature of 30°C, while the opposite face of the membrane was maintained under a vacuum less than 0.1 mm of mercury. The mixture which diffused through the membrane was analyzed and found to contain 20.9% ortho-, 72.1% meta-, and 7.0% para-xylene. This represents a 40% increase in the proportion of para-xylene in the mix after only a single pass, while the proportion of ortho-xylene has decreased by 30% and the proportion of meta-xylene has increased by only 11%.

The same treated film was used for diffusion separation, under the same conditions, of a mixture containing, by weight, 90% ortho-, 5% meta-, and 5% para-xylene. After a single pass through the membrane, the mixture contained 83.4% ortho- (a decrease), 7.4% meta- (a 48% increase), and 9.2% para- (an 84% increase).

The results agree very closely with the calculated results based upon measurements of individual permeability coefficients of each of the three isomers separately through the treated film or membrane. Such permeability coefficient measurements show that a marked increase in permeability of para-xylene through the film is achieved by treating the film with para-xylene at temperatures as low as 35°C, when the permeation step is carried out at 30°C (although optimum results are obtained by treating at 50°C), while at the same time the permeability of ortho- and meta-xylene is increased to a markedly lesser extent by the treatment. In addition to the improvements in separation of the xylene isomers achieved by use of these treated membranes, it is also found that polyethylene membranes treated by the method described show markedly enhanced total or overall permeability to the liquid mixtures.

In a continuation of the above work, *A.S. Michaels; U.S. Patent 3,320,328; May 16, 1967; assigned to Amicon Corporation* has found that further improvement in the characteristics of the membrane for diffusion and separation processes may be brought about by stretching the membrane prior to the treatment with the selected compound, the latter treatment being carried out with or without the membrane being mechanically constrained from elastic recovery. Conveniently the improvement is brought about by uniaxially pulling the membrane to elongate it substantially, typically in excess of 100% of its original dimension.

After it has been stretched, and with or without its being held in a stretched condtion, the membrane is immersed in the selected compound, at a temperature more than 10°C below its melting or dissolution temperature, until the desired amount of the compound has been absorbed. Following treatment with the selected compound the membrane is allowed to dry with or without constraint, typically at room temperature, and is then mounted in suitable conventional diffusion separation equipment.

Separation of Azeotropic Mixtures Using Irradiated Polyethylene

A process described by *R.C. Binning; U.S. Patent 2,981,680; April 25, 1961; assigned to Standard Oil Company* involves the separation of mixtures of different molecules and it particularly concerns an improved permeation process for effecting the separation. This process is based on the principle of using two or more dissimilar (having different compositions) permeation membranes, each of which dissimilar membranes are preferentially selective for permeating different components of the charge mixture. In the process, the dissimilar permeation membranes separate the feed zone from separate and noncommunicating permeate zones. The number of dissimilar permeation membrane compositions is at least two and is preferably the same as the number of components contained in the mixture to be separated and the same as the number of permeate zones. The effective (i.e., utilized) surface area of the dissimilar permeation membrane compositions are employed in ratios with respect to each other such that the composition of the mixture of molecules in the feed zone is the same as the composition of the charge introduced to the feed zone. A wide variety of mixtures of different molecules containing two or more components can thus be partially or completely separated, using various permeation membranes. In this manner, mixtures of different molecules existing in the form of molecular solutions, as distinct from suspensions or colloidal solutions, may be resolved.

One form of the process is highly useful in separating azeotropic mixtures in a continuous manner. For example, an azeotropic mixture of components A and B can be resolved to recover the individual components A and B in essentially pure form by introducing the azeotropic mixture into the feed zone of the permation apparatus which contains two non-communicating permeate zones A_1 and B_1 which are separated from the feed zone by two dissimilar permeation membranes A_2 and B_2. Permeation membrane A_2 permits the permeation into permeate zone A_1 of component A in preference to component B, but does not prevent permeation of the latter component. Permeation membrane B_2 permits the permeation into permeate zone B_1 of component B in preference to component A, but does not prevent permeation of the latter component.

The permeated portions are separately withdrawn from permeate zones A_1 and B_1 and sent to separate fractionators A_3 and B_3 respectively. In the fractionators, the permeated portions are fractionated to produce two fractions. One of the fractions consists of a component of the azeotropic mixture in essentially pure form, and the other fraction consists of a mixture of components A and B. The separate fractions of essentially pure components are withdrawn from the fractionators, and the fractions consisting of mixtures of components A and B are recombined and recycled to the feed zone of the permeation apparatus. The recombined portion has a composition which approximates the composition of the azeotropic mixture. Continuous operation of the permeation unit and the fractionators in the manner outlined above maintains the composition of the mixture in the feed zone of the permeation apparatus at approximately the composition of the azeotropic mixture and thereby secures the advantages of the process.

R.J. Lee; U.S. Patent 2,984,623; May 16, 1961; assigned to Standard Oil Company describes the use of irradiated polyethylene membranes. This process is based upon the discovery that by employing irradiated polyethylene membranes the permeation process can be carried out at a higher temperature with an unexpectedly high increase in permeation rates without causing the membrane to rupture; whereas if nonirradiated polyethylene is used at permeation temperatures of above about 75°C, the membrane ruptures.

In the process, an irradiated polyethylene membrane is employed as the plastic membrane in a permeation apparatus. The polyethylene is that which has been subjected to an irradiation dosage level sufficient to cause crosslinking of the polymer and thus improve its temperature stability and resistance to solvents but not to that dosage level which causes degradation of the polyethylene. A dosage level of about 1 to 200, suitably 10 to 100, megaroentgens per square centimeter of membrane surface is satisfactory. In the permeation process a mixture of compounds which are permeable through the membrane can be separated by taking advantage of their differing solubilities within the membrane and their differing permeation rates through the membrane, the compound which is more soluble in the membrane being contained in the mixture which has permeated through the membrane in a higher concentration than in the initial mixture.

The mixture to be separated, which may be a mixture of organic chemicals, a mixture of hydrocarbons, a mixture of water and an organic chemical, etc., is introduced into the feed zone of the permeation apparatus. It is there maintained in the liquid state. A portion of the mixture is permeated through the membrane and into the permeate zone where it exists in the form of vapors. The pressure in the feed zone should be sufficient to maintain the mixture undergoing separation in the liquid phase. The pressure in the permeate zone should be such that the permeated mixture is maintained in the vapor phase, subatmospheric pressure being preferred. Since the rate of permeation of the mixture increases with increasing temperature, the higher operating temperatures, e.g., 75° to 200°C, are preferred. The rate of permeation is increased by 10 to 20 fold by using the irradiated polyethylene under higher permeation operating temperatures in place of conventional polyethylenes under permeation operating temperatures which are the maximum that the latter material can withstand without rupturing.

E.V. Kirkland and J.F. Jennings; U.S. Patent 3,225,107; December 21, 1965; assigned to Standard Oil Company have found that a polymeric film used in a permeation separation process can be improved with respect to either increasing selectivity for the organic chemicals being separated or increasing the rate of permeation, or both, by grafting a polymeric coating upon the one or both surfaces of the polymeric film. Also, it has been discovered that the polymeric film when used alone has improved permeation properties if, prior to use, it has been irradiated. Further, the properties of a permeation membrane can be improved by irradiating a polymeric film having a grafted polymeric coating thereon or irradiating the polymeric film and thereafter grafting upon the surface a polymeric coating.

As an illustration of the process, polyolefin films which have been modified by certain techniques are employed as the permeation membranes in the separation of organic chemicals by permeation. The polyolefin film, which may be a polymer or copolymer of ethylene, propylene, butene-1 and/or pentene-1, etc., is either modified by grafting thereon a polymerizable monomer, irradiated with ultraviolet light having a wavelength in the range of 2000 to 4000 A, or given both treatments. It may then be used as the permeation membrane. In forming the graft-coated membrane the polyolefin film, which is believed to have free radical reactive sites as a result of one of the treatments, may be contacted with a monomer which reacts in the presence of free radicals and grafts itself onto the film surface.

In this process a permeation apparatus which is comprised of a feed zone that is sealed or separated from a permeate zone by the modified polyolefin film is used. The permeation apparatus is maintained under conditions causing permeation to occur when the feed mixture of organic chemicals is introduced into the feed zone. A portion of the mixture of the organic chemicals in the feed zone permeates through the modified polyolefin film and passes into the permeate zone. The permeated portion, which has a higher concentration of one of the components of the feed mixture than the concentration of this same component in the feed mixture, is rapidly removed from the permeate zone. The nonpermeated portion is withdrawn from the feed zone.

Radiation Crosslinked, Swelled Membranes

R.W. Coughlin and R.D. Siegel; U.S. Patent 3,720,321; March 13, 1973; assigned to Research Corporation describe a semipermeable membrane of enhanced permeability or permselectivity which is formed by swelling a crosslinkable polymer film in a nonreactive, nonpolymerizable swelling agent for a time sufficient to uniformly penetrate and swell the film, and thereafter subjecting the swollen film to high-energy ionizing radiation for a period of time and at a dose rate sufficient to cause crosslinking of the polymer.

It is believed that radiation causes some degree of crosslinking of the polymeric chains thereby binding the chains into dense clusters. Large crystallites in the polymeric chain formed by the dissolution and recrystallization during the annealing or solvation procedure are locked in or fixed in the dense clusters so that they are not easily redissolved in the changed environment. It is theorized that the formation of the large crystallites in the polymeric chain during the solvation procedure permits easy transport of the foreign molecules through the surrounding amorphous regions. Where the films have been annealed, but not irradiated, it is suspected that the crystallites redissolve and become smaller during use of the membrane. Whereas much of the enhanced permeability and permselectivity is lost within a few hours of use of the annealed, nonirradiated membrane, where the film has been further treated by irradiation, the enhanced properties remain for periods exceeding 500 hours and are believed to be relatively permanent.

Examples 1 through 24: The separation of organic materials from liquid-liquid mixtures was demonstrated using a polyethylene film membrane having a thickness of 5 mils. The solvent-swollen membranes were subjected to gamma rays from a Co^{60} source at a dose rate of about 10^6 rads per hour. Exposure periods of up to 24 hours were investigated. Toluene and cyclohexane were used as swelling solvents and the permeability and permselectivity were tested with toluene, cyclohexane and benzene. The experimental procedures and apparatus used to test the permeability of the membrane are further described in *AICE Journal* 15 (2), 208 (1969). The films were presoaked at 30°C and during the irradiation procedure the temperature fluctuated between 33° and 38°C. The results are shown in the table below.

Ex. No.	Solvent Used for Measuring Permeability	Radiation Treatment, hours*	Solvent Used for Swelling Film	Permeation Medium	Permeability, $cm^2/(sec \times 10^7)$	Remarks
1	Toluene	0	-	Cyclohexane	6.28	-
2	Toluene	0	-	Cyclohexane	6.17	-
3	Toluene	3	Cyclohexane	Cyclohexane	7.73	Not dried between irradiation and permeation.
4	Toluene	0	-	Toluene	4.14	Preswollen in toluene, no irradiation.
5	Toluene	0	-	Toluene	4.87	Preswollen in cyclohexane and dried before immersion in toluene for permeation studies.
6	Toluene	0	-	Toluene	4.81	Preswollen in cyclohexane and dried before immersion in toluene for permeation studies.
7	Toluene	1	Cyclohexane	Toluene	4.79	-
8	Toluene	1	Cyclohexane	Toluene	5.31	Rerun of film from 7 after soaking in toluene for 550 hr.
9	Toluene	3	Cyclohexane	Toluene	5.80	-
10	Toluene	3	Cyclohexane	Toluene	5.16	Rerun of film from 9 after soaking in toluene for 500 hr.
11	Toluene	8	Cyclohexane	Toluene	7.00	-
12	Toluene	24	Cyclohexane	Toluene	7.91	-
13	Toluene	24	Cyclohexane	Toluene	7.61	Rerun of film from 12 after soaking in toluene for 520 hr.
14	Toluene	0	-	Toluene	4.14	-
15	Toluene	3	Toluene	Toluene	4.96	Not dried between irradiation and permeation.
16	Toluene	3	Air	Toluene	4.37	Soaked in cyclohexane and dried before irradiation in air.
17	Toluene	3	Air	Toluene	4.10	Irradiation in air.
18	Benzene	0	-	Toluene	5.45	-
19	Benzene	0	-	Benzene	4.56	-
20	Toluene	3	Benzene	Toluene	6.12	Dried for 6 hr between irradiation treatment and permeability measurements.
21	Benzene	3	Benzene	Toluene	6.71	Dried for 6 hr between irradiation treatment and permeability measurements.
22	Benzene	3	Benzene	Benzene	5.40	Dried for 6 hr between irradiation treatment and permeability measurements.
23	Benzene	3	Toluene	Benzene	6.71	Dried for 6 hr between irradiation treatment and permeability measurements.
24	Benzene	3	Toluene	Toluene	8.08	Dried for 6 hr between irradiation treatment and permeability measurements.

*Dose rate of 10^6 rads/hr

As the results of the table indicate, permeability is enhanced not only in comparison with a nontreated polyethylene film, but also with respect to the solvent-swollen membrane which has not been irradiated. Moreover, the enhanced permeability remains substantially constant even after periods of up to about 550 hours of use.

It is also clear from the table below that permselectivity is substantially enhanced by the techniques of this process. The brackets in the table are used to show that these ratios are computed from these values, for example, 5.45/4.14 = 1.31, 8.08/4.96 = 1.63.

Example Number	Solvent Used in Measuring Permeability	Radiation Treatment, hr	Solvent Used for Swelling Film	Permeation Medium	Permeability, $cm^2/(sec \times 10^7)$	Ratio*
1	Benzene	0	-	Toluene	5.45	1.31
2	Toluene	0	-	Toluene	4.14	
3	Benzene	3	Toluene	Toluene	8.08	1.63
4	Toluene	3	Toluene	Toluene	4.96	

*Permeability of benzene/permeability of toluene

Polymer films can also be cast from a solvent or swelling media and thereafter irradiated with high-energy ionizing radiation before drying, while in a swollen condition. For example, hydroxyethylcellulose film can be cast from a 6 to 8% solution of 7% NaOH in water onto a smooth plate, partially dried to form a self-sustaining film swollen with water, washed, irradiated for 3 hours at a dose rate of 10^6 rads per hour and thereafter dried in an oven at 180°F.

Cellophane films and membranes can be formed by coagulation of a viscose solution whereby the viscose solution is extruded in the form of a sheet into an acidified aqueous bath. The hydrated film is then irradiated at a dose rate of about 10^6 rads per hour and thereafter dried at a temperature of about 100°C.

Similarly, polystyrene or polyvinyl chloride film can be cast from a solution of methyl ethyl ketone, dioxane, tetrahydrofuran, or ethylene dichloride. Polyvinylidene chloride film can be cast from solutions of tetrahydrofuran; polyvinyl acetate films can be cast from water; cellulose acetate films can be cast from acetone; nylon can be cast from formic acid; or polyolefins can be cast from hot toluene. The cast films are then partially dried, irradiated, and then the solvents are removed by drying or otherwise.

The semipermeable membranes of the process can be used for a wide range of separation applications. For example, these membranes can be used for separating azeotropic or close boiling mixtures. They can also be used for removing water from a chemical reaction mixture in which water is a product of the reaction so as to shift the reaction equilibrium, or they can be used for dehydrating organic liquids. They further can be used for enriching an isobutane-butene alkylate in the gasoline-boiling range, as well as for a variety of other liquid-liquid separation procedures.

WATER REMOVAL

Mixtures of Acrylonitrile-Acrylic Acid Copolymers with Formaldehyde

R. Chiang and E. Perry; U.S. Patent 3,750,735; August 7, 1973; assigned to Monsanto Company describe a process for the separation of water from mixtures of water and organic or inorganic compounds, soluble in water which involves contacting the mixtures against a nitrogen-containing polymeric membrane, and withdrawing at the other side of the membrane a mixture having a higher concentration of water. The feed solutions may also contain additional components, e.g., methanol with formaldehyde-water, or sodium chloride with hydrochloric acid-water, or butanol in ethanol-water systems.

Example 1: This example shows the concentration, i.e., water removal from formaldehyde, using membranes of a copolymer of acrylonitrile (AN) and acrylic acid (AA), made according to the general procedure of Example 6. This example is directed to the concentration of formaldehyde via the permeable membrane. Formaldehyde concentration is determined by the quantity of NaOH liberated by reaction of HCHO with sodium sulfite (in excess) to form the formaldehyde-bisulfite addition product according to the equation:

$$HCHO(aq) + Na_2SO_3 + H_2O \longrightarrow NaOH + CH_2(NaSO_3)OH$$

The percentage of HCHO is calculated by the equation: percent formaldehyde equals (cc HCl x normality HCl x 3.003)/weight sample. The separation factor (SF) is defined as the ratio of the concentrations of the two substances A and B to be separated divided into the ratio of the concentrations of the corresponding substances in the permeate: SF = (C_A/C_B) in permeate/(C_A/C_B) in permeant where C_A and C_B are the concentrations of water and formaldehyde, HNO_3, HCl or ethanol, respectively.

A formalin solution containing 36.95% HCHO, 53% H_2O, 0.05% formic acid and 10% CH_3OH is concentrated into a slurry containing 62% HCHO at room temperature. As shown in the following table, the separation factor increases from 30 for a 41.3% formaldehyde solution at the beginning of the permeation to 48 for a solution containing 62.0% HCHO. At such a high concentration, paraformaldehyde separates out as a solid and the permeation is terminated. The formic acid is also selectively permeated (relative to formaldehyde) during this operation.

Permeation of Formaldehyde Solution Through an 80:20 Weight Ratio AN:AA Membrane (25°C)

Period of Permeation, hours	Grams Permeate During Time Interval	Grams Removed from Feed Cell	---- Wt % HCHO in --- Feed Cell	Permeate	SF
16.5	1.394	0.6	37.0	-	-
24.1	0.861	-	41.8	2.3	30
24.5	0.729	-	42.2	1.2	60
22.3	0.533	-	-	-	-
71.1	0.934	-	46.9	-	-
46.1	0.492	-	49.6	-	-
23.4	0.359	0.383	50.5	2.68	37
46.5	0.478	0.214	53.9	3.22	35
72.0	0.72*	0.18	54.8	-	-
-	0.183	0.17	56.5	-	-
27.0	0.282	0.27	59.0	2.89	48
24.1	0.231	0.132	62.7	3.9	42
42.3	0.301	0.187	62.0	3.4	46

*Estimate

Examples 2 through 5: Poly(vinyl butyral), which can be considered as a condensation product of polyvinyl alcohol and butyraldehyde, contains an acetal linkage. The acetal oxygen is hydrophilic and weakly anionic. The membrane prepared from poly(vinyl butyral) is also selective, giving a separation factor (SF) of 14 to 18 for formaldehyde solutions. The following table compares the separation factors obtained with various membranes for permeation of formaldehyde solutions.

Example	Membrane	SF
2	Cellulose acetate	2 to 8
3	Silicone polycarbonate (M-2/3 GE)	3
4	84.4:15.6 AN:AA (by weight)	30 to 48
5	Poly(vinyl butyral)	14 to 18

Example 6: It has been found that the membranes prepared from copolymers of AN and AA which are already characterized by high selectivity, may be further improved as to permeability. The rate of permeation is remarkably improved by replacing the weaker anion (namely the acrylic acid with a pK = 5) with stronger anionic groups, namely styrene sulfonic acid, (SSA) with a pK = 1. Acrylic acid has a relatively low degree of ionization when compared to styrene sulfonic acid which is virtually completely ionized. Membranes prepared from copolymers of AN and sodium styrene sulfonate (SSS) are prepared as follows. The polymerization is carried out according to the proceudre of *Macromolecular Synthesis,* Volume 2, J. Wiley, 1966, page 78. In this procedure acrylonitrile (36 grams) with 1.985 grams of sodium styrene sulfonate is dissolved in distilled water (0.4 liter) and charged to a reaction vessel having a reflux condenser. The flask is maintained at 50°C, under nitrogen pressure, while the catalyst is added (catalyst solution composed of 25 ml

of water, 4 ml of 0.1 N sulfuric acid, and 0.001 gram of ferrous ammonium sulfate hexahydrate). There is also added a solution of 25 ml of water containing 0.10 gram of potassium persulfate, and then 50 ml of water containing 0.50 gram of sodium metabisulfite. The polymerization is continued for one hour during which time a slurry develops. The reaction is then stopped by the addition of a 1% solution of sodium carbonate to reach pH 7.0. The slurry is filtered, washed with water and acetone, and dried at 100°C.

The membranes are prepared by casting films from a 5% dimethyl sulfoxide (DMSO) solution of the polymer as described below, containing varying amounts of the acid group, on a glass plate heated on a steam bath. After drying one hour on the steam bath, the membrane is conditioned by immersion in water followed by treatment with the permeating solution in the cell at least a day prior to measurement. The thickness of the membrane in water varies from 1.0 to 1.2 mils. The data below summarize the results of concentrating 37% formalin solution using a copolymer of acrylonitrile and sodium styrene sulfonate.

Mol % SSS	Temperature, °C	SF
0.54	25	39
0.54	50	40 to 80
1.27	25	17 to 23
1.27	50	37
3.44	25	20
3.44	50	13

The results obtained on the permeation of formaldehyde solutions with the AN-SSS membranes are given in Example 6. While the SF changes in a relatively minor way (in the range of 15 to 50), the permeation rate (not shown) increases with the concentration of the acid group as well as with the temperature.

Hydrolyzed Polyvinyl Acetate

R.C. Binning, J.F. Jennings and E.C. Martin; U.S. Patent 3,035,060; May 15, 1962; assigned to Standard Oil Company describe a permeation process for removing the water from a water-organic chemical mixture. The process employs as the permeation membrane a hydrolyzed polyvinyl acetate membrane which has at least 98% of its acetyl groups removed by hydrolysis.

The stability toward water which is displayed by the hydrolyzed polyvinyl acetate membrane can be improved by curing the hydrolyzed polyvinyl acetate membrane at a temperature of at least about 100°C, for example, 100° to 150°C or higher.

The curing is most advantageously carried out in the substantial absence of oxygen. The process provides a method for removing water from a wide variety of organic chemicals such as alcohols, for example, ethanol, isopropanol, sec-butanol, etc., dioxane, pyridine, acrylonitrile, etc., which are mixed with water.

Modified Vinyl Polymers-Organic Acids

E.V. Kirkland; U.S. Patent 3,182,043; May 4, 1965; assigned to Standard Oil Company describes a permeation process for separating water from organic acids utilizing permeation membranes. It has been found that a particular series of modifications gives a highly improved vinyl chloride-vinyl acetate copolymer which gives excellent permeation rates and selectivity when formed into a membrane and used in the permeation of aqueous solutions of monocarboxylic acids.

This series of modifications is conducted by first hydrolyzing the acetate groups in the copolymer, next cyanoethylating a portion of the hydroxyl groups followed by crosslinking a portion or all of the remaining hydroxyl groups. The excellent permeation rate and selectivity of the modified copolymers make them useful in the commercial practice of membrane permeation.

GAS SEPARATION

Plasma Discharge of Benzonitrile on Polymeric Siloxane Films

According to a process described by *A.F. Stancell and A.T. Spencer; U.S. Patent 3,657,113; April 18, 1972; assigned to Mobil Oil Corporation* gaseous or liquid components are concentrated or separated from others differing in molecular size or shape and/or solubility characteristics in homogeneous multicomponent mixtures (e.g., concentrating hydrogen in a hydrogen-methane mixture) by permeation through an ultrathin crosslinked plasma-discharge polymerizate layer of high permselectivity and controlled thickness supported on or bonded to a face of a fluid-permeable substrate.

Membranes are prepared by ionizing and polymerizing vaporized monomeric organic compounds (e.g., benzonitrile) at low subatmospheric pressures by a plasma glow discharge which deposits and bonds unbroken crosslinked coatings of about 0.1 to 0.5 micron total thickness onto thicker preformed substrates (e.g., films of polymeric siloxanes or polyphenylene oxide) of adequate strength.

As substrate materials, polyphenylene oxide and the siloxane polymers and copolymers are often preferred as they are noncrystalline materials with flexible backbone polymer chains. Moreover, polyphenylene oxide is a strong resin with a tensile strength of 10,000 psi and about half of that strength is retained at temperatures of 250°F. With due regard for the characteristics of specific fluid substances to be separated, process conditions and adequate substrate permeability for the desired permeant, many other substrate materials may be employed in separating the components of various fluid mixtures. Other substrate resins for various permselective operations are listed in the table below along with permeability data for hydrogen and methane.

	Permeability*	
Substrate	H_2	CH_4
Polyethylene (low-density)	0.53	110.15
Polybutene	0.54	0.06
Polyvinyl chloride	0.75	0.15
Fluorinated ethylene-propylene copolymer (Teflon, FEP)	0.86	–
Cellulose triacetate	0.87	–
Polystyrene	1.23	–
Cellulose acetate butyrate	1.26	–
Polyphenylene oxide	5.9	0.25
Siloxane-carbonate copolymer	9.2	10.5
Polydimethylsiloxane	19.8	26.4

*x 10^{-7} cc (STP) cm/cm^2-sec-atm at 25°C

The listed materials of low permeability are of relatively little interest for the specific application of separating hydrogen and methane on any substrate scale.

Examples 1 through 23: The membrane substrates used in the tables of the following examples are films fabricated in conventional manner from known polymers or copolymers, such as a cast film of a siloxane-carbonate copolymer (resin XD) derived from dimethyldichlorosilane and 4,4'-dihydroxydiphenylpropane, an extruded low density (0.92) polyethylene film, a cast polyphenylene oxide film and an extruded polydimethylsiloxane film.

To illustrate a suitable method for preparing the selective composite membranes employed in the process, a 1.5-mil (0.0015-inch) thick film of the siloxane-carbonate copolymer is suspended between two parallel electrodes spaced one inch apart in a vacuum chamber and the pressure in the vessel is reduced to about 0.1 torr (absolute pressure in mm Hg at 0°C) and maintained at that level for 10 to 15 minutes to expel any gaseous material from the film. After the pressure in the chamber is increased to 1 torr by admitting benzonitrile, a plasma glow discharge is initiated between the electrodes by means of a capacitive device operating at 100 watts power and 350 KHz frequency. The discharge is sustained for a period of 5 seconds; then the material is allowed to cool for 5 seconds to avoid overheating the film, and benzonitrile is supplied as necessary to maintain the stated pressure. These alternating deposition and cooling operations are repeated to provide a total treating or deposition time of one minute. The product of this treatment has ultrathin crosslinked polymer layers of uniform thickness completely covering and firmly bonded to the exposed areas of both sides of the substrate film. The results obtainable in permeation tests of this particular composite membrane are listed in tabulated Example 1.

For another example providing the results tabulated as Example 17, the permselective membrane is prepared by subjecting the same film base to 6 plasma glow discharges (100 watts and 350 KH_z) of 5 seconds' duration with two different monomer vapors alternately present. In this instance, a plasma discharge of 5 seconds' duration in an atmosphere of 1,3,5-trichlorobenzene at 1 torr pressure is followed by a similar treatment in acrylonitrile vapor after removal of the chlorobenzene compound and cooling of the substrate, and this sequence of operations repeated twice more for a total plasma discharge treating time of 30 seconds.

		Plasma treatment		Permeability [1]		Selectivity H_2/CH_4 moles	Coating thickness
Ex.		Min	Torr	H_2	CH_4		
A	Substrate: XD siloxane-carbonate copolymer			9.2	10.5	0.875	None.
	Coating precursors:						
1	Benzonitrile	1.0	1.0	7.26	0.222	32.7	Heavy
2	Cyanogen bromide	1.0	1.0	2.74	0.075	36.6	Do
3	Nicotinonitrile	1.0	1.0	3.96	0.113	35.0	Do.
4	3-butenenitrile	1.0	1.0	8.87	0.39	22.8	Moderate.
5	Methacrylonitrile	1.0	1.0	9.04	0.55	16.3	Do
6	1,3,5-trichlorobenzene	1.0	1.0	8.38	0.62	13.5	Heavy
7	Styrene	1.0	1.0	8.46	0.68	12.4	Moderate
8	Naphthalene	1.0	1.0	8.1	0.44	18.4	Light
9	Divinylbenzene	1.0	1.0	9.85	1.33	7.4	
10	Benzene	1.0	1.0	9.57	1.33	7.2	Light
11	Picolene	1.0	1.0	10.7	1.62	6.6	
12	Mesitylene	1.0	1.0	9.43	1.72	5.5	
13	1,3-di(trifluoromethyl)benzene	1.0	1.0	9.92	5.98	1.7	
14	Acrylic acid	1.0	1.0	8.15	0.33	24.7	Light.
15	Allyl bromide	1.0	1.0	8.37	0.66	12.7	Moderate.
16	Vinyl acetate	1.0	1.0	10.78	2.29	4.7	
17	1,3,5-trichlorobenzene and acrylonitrile	0.5	1.0	6.13	0.13	47.2	

[1] $\times 10^{-7}$ cc. (STP) cm./cm.^{2}-sec.-atm. at 25° C.

		Plasma treatment		Permeability [1]		Selectivity
Ex.		Min.	Torr	H_2	CH_4	H_2/CH_4 moles
B	Substrate: Cast polyphenylene oxide			5.97	0.31	19.2.
	Coating precursors:					
18	Benzonitrile	1.0	1.0	4.47	0.066	68.
19	Styrene and cyanogen bromide	0.5	0.5	4.28	0.032	134.
20	Cyanogen bromide	0.5	0.5	3.41	0.0114	298.
21	do	0.5	0.5	2.59	0.0093	280 at 0° C.
22	do	0.5	0.5	5.06	0.035	161 at 45° C.
23	do	0.5	0.5	[2] 7.32	[2] 0.11	68 [2] at 64° C.

[1] $\times 10^{-7}$ cc. (STP) cm./cm.^{2}-sec.-atm. at 25° C.
[2] In comparison, the permeabilities are 8.53 for hydrogen and 0.55 for methane and the selectivity is 15.5 at 64° C. for the uncoated polyphenylene oxide substrate.

The tabulated examples of permeation through the various single layer and composite membranes are carried out under essentially the same conditions for purposes of comparison. Undiluted hydrogen and methane of commercial purity are employed separately rather than as mixtures in these permeation runs in order to simplify the procedure and eliminate analytical determinations. Thus, the selected gas, hydrogen or methane, is introduced at 6 psig and 25°C into a closed vessel which is divided into an inlet chamber and an outlet chamber by means of a membrane facing the inlet chamber and clamped onto a fibrous support and sealed at the edges so that all of the gas passing from the inlet chamber to the outlet chamber must pass through the permeable membrane.

The exit chamber is maintained at atmospheric pressure, and the permeation rate is determined by measuring the volume of gas in cc displaced from the exit chamber over a suitable measured time interval of typically 60 minutes. In the tables, the results obtainable with controls in the form of uncoated substrate films are designated by the letters A and B and numbered examples illustrating the process are indented by listing the vaporized monomer precursors used in forming the plasma polymerizate coatings on the same films. Unless otherwise stated, all membrane permeabilities were measured at 25°C and are set forth as the corrected volumes of a specific gas (e.g., 10^{-7} cubic centimeters at STP) passing through the various membranes per second per centimeter of membrane thickness per square centimeter of membrane surface exposed to the incoming undiluted gas per atmosphere of pressure drop across the membrane; the selectivity or permselectivity of a particular membrane is defined here as the ratio of hydrogen permeability to methane permeability; coating treatments in preparing the films are specified by monomer or alternating monomer atmospheres and absolute pressure (torr) as well as the total time in minutes of the repeated plasma glow discharges, and temperatures are given in degrees centigrade (°C).

In the foregoing examples, it will be noted that the hydrogen selectivity was greater in every case with the composite membranes than with the uncoated control membranes (substrate films) regardless of the chemical composition of the glow discharge polymerizate coating. This is due to the methane permeability being severely restricted by the crosslinked polymerizate coatings in almost all examples with the corresponding substantial improvement in selectivity favoring the permeation of hydrogen, and this results in extremely high permselectivity for hydrogen in many instances. Moreover, it is apparent from these data that even less permeation of these composite membranes by ethane and other higher hydrocarbon vapors is to be expected by reason of their having molecules of larger size than that of methane.

It is also noteworthy that the reduction in hydrogen permeability with composite membranes is usually either insignificant or relatively modest, even when extraordinary increases in selectivity are obtained. In a few examples there are indications of slightly higher hydrogen permeabilities with coated membranes than with the substrate film alone; however, these improvements are regarded as apparent rather than real and may be attributed to experimental errors or limitations as to reproducibility attendant in such experimentation including variations in the thickness of the particular substrate film employed and possible thermal effects therein resulting from unobserved minor overheating of the material during the plasma discharge treatment.

Examples 20 through 23 provide a study of the effects of permeation temperatures upon the permeabilities of the test gases with the selectivity rising somewhat from 0° to 25°C and then dropping sharply at temperatures of 45°C and higher; also the hydrogen permeability increases over the entire range of 0° to 64°C particularly in the upper part of this range. Generally, similar but much less pronounced effects occur in similar operations with a control or uncoated polyphenylene oxide over the same temperature range, for the permeabilities of the two gases does not even double and the selectivities remain much more constant with a maximum of less than 25 and a minimum above 15.

Upon subjecting numerous different representative composite membranes of the types listed in the tables to infrared spectroscopy employing the attenuated total reflectance (ATR) technique, the substrate spectrum is found to be visible in all cases and there is evidence of a high degree of crosslinking in the polymeric coating layers. The indications from these data are that all of the composite membranes have a total coating thickness (i.e., the sum of the thickness coatings on the two faces of each film) within the range of about 0.1 to 2.0 microns and usually less than 1.0 micron, even when a heavy coating thickness is listed in certain of the foregoing examples on the basis of visual observation of the color of the membrane.

p-Xylylene Polymers

According to a process described by *M.A. Spivack, D.D. Stewart and F.R. Tittmann; U.S. Patent 3,830,733; August 20, 1974; assigned to Union Carbide Corporation* diffusion membranes of linear p-xylylene polymers having controlled permeability properties are prepared by condensing vaporous polymer-forming diradicals with vaporous nonpolymerizable filler materials, and then removing the condensed filler materials from the resulting membranes to provide the desired degree of permeability. Linear p-xylylene polymers are usually prepared by condensing in a condensation zone, vapors of diradicals which can be produced by the pyrolytic cleavage, in a pyrolysis zone, of one or more cyclic dimers represented by the following structure:

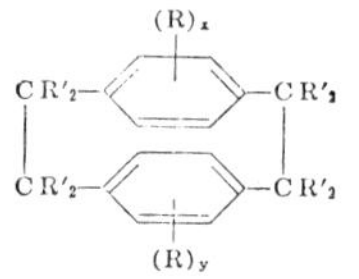

where R is an aromatic nuclear substituent, x and y are each integers from 0 to 3 inclusive, and R' is H, Cl and/or F. The thus-formed vaporous diradicals have the structure:

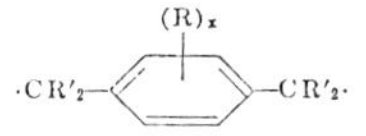

and

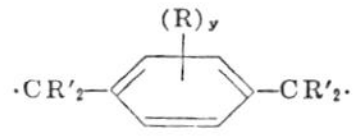

Thus, where x and y are the same, and the aromatic nuclear substituent on each diradical is the same, and all the R's are the same, two mols of the same p-xylylene diradical are formed, and when condensed, yield a substituted or unsubstituted p-xylylene homopolymer. When x and y are different or the aromatic nuclear substitutents on each diradical are different, or the R's are different, condensation of such diradicals will yield copolymers as hereinafter set forth.

Examples of the R substituent groups which may be present in the dimers and diradicals are organic groups such as alkyl, aryl, alkenyl, cyano, alkoxy, hydroxy alkyl, carbalkoxy and like radicals and inorganic radicals such as hydroxyl, halogen and amino groups. COOH, NO_2 and SO_3H groups may be added as R groups to the polymer after it is formed. The unsubstituted positions on the aromatic rings are occupied by hydrogen atoms.

The particularly preferred substituent R groups are the C_1 to C_{10} hydrocarbon groups, such as the lower alkyls, i.e., methyl, ethyl, propyl, butyl and hexyl, and aryl hydrocarbons such as phenyl, alkylated phenyl, naphthyl and like groups, and the halogen groups, chlorine, bromine, iodine and fluorine.

Condensation of the diradicals to form the p-xylylene polymers can be accomplished at any temperature below the decomposition temperature of the polymer, i.e., at <250°C. The condensation of the diradicals will proceed at a faster rate, the colder is the substrate on which the condensation is to take place. Above certain temperatures, which might be defined as a ceiling condensation temperature, the diradicals will only condense at rates which are relatively slow for commercial applications. Each diradical has a different ceiling condensation temperature. For example, at 0.5 mm Hg pressure the following condensation and polymerization ceilings are observed for the following diradicals.

	°C
p-Xylylene	25 to 30
Chloro-p-xylylene	70 to 80
Cyano-p-xylylene	120 to 130
n-Butyl-p-xylylene	130 to 140
Iodo-p-xylylene	180 to 200

Thus, homopolymers may be made by maintaining the substrate surface at a temperature below the ceiling condensation temperature of the particular diradical species involved, or desired in the homopolymer. This is most appropriately termed homopolymerizing conditions. The diffusion membranes may be prepared so as to provide membranes with total thicknesses of about 0.05 to 1,000 microns. The diffusion membranes provide at least a 2 to 1,000-fold improvement in gas diffusion properties, as contrasted to the gas diffusion properties of films made entirely, or substantially, of unfilled p-xylylene polymer. Examples of these improved properties are shown by a comparison of the properties of five various membranes in the table below. The membranes compared in this regard were all essentially composed of porous strata of polymeric matrix. Each film was made by pyrolyzing the dimer in question and condensing it to form a film as described in Examples 1 through 5 below, respectively.

	Membranes made from—	Thickness of film, microns	Percent by weight of extractables in film [a]	Nature of extractable	Permeability constant of film for—		
					N_2 [b]	CO_2 [b]	H_2O [c]
1	Poly-chloro-para-xylylene.	25	1.0	Di-Cl-di-paraxylylene.	0.6	7.2	0.7
2	do	18	13	do	1.4	17.2	
3	do	22	12	do	3.8	29.9	
4	do	17	40	do	1.0	42.5	
5	do	25	43	do			33

[a] Percent extractables in boiling benzene for 2 hours at 75° C.
[b] ASTM D-1434-63T—constant reported at 23° C. as cc.·mil/100 square inches in 24 hours per atmospheric pressure.
[c] ASTM E-96-63T—constant reported at 37° C. as gram·mil/100 square inches in 24 hours at 90% relative humidity.

Film No. 1 was made with less than 1% by weight of extractable dimer as a filler and it had poor gas diffusion properties. Films 2 through 5 were made with 12 to 43% by weight of extractable dimer as a filler and they had good to excellent gas diffusion properties. The filler used in Films 1 through 5 was an excess of the dimer from which the polymeric matrix was formed.

The preferred membranes are those made of polymer formed from the unsubstituted dimer, di-p-xylylene, and the chlorinated substituted dimer, dichlorodi-p-xylylene, and the dimer in which R is H and R' is F, as shown in the structure above, i.e., α-octafluoro-di-p-xylylene. The p-xylylene polymers generally have nitrogen permeability constants of >0.6 cc-mils/100 in^2 in 24 hours per atmosphere of pressure and carbon dioxide permeability constants of >6 cc-mils/100 in^2 in 24 hours per atmosphere of pressure.

The preferred membranes of poly(p-xylylene) i.e., made from unsubstituted di-p-xylylene, have a nitrogen permeability constant (at 23°C) of at least 8, and preferably of about 16 to 8,000 cc-mils/100 in^2 in 24 hours per atmospheric pressure, and a carbon dioxide permeability constant (at 23°C) of at least 215, and preferably of about 430 to 215,000 cc-mils/100 in^2 in 24 hours per atmospheric pressure. These membranes preferably also have moisture vapor transmission constants or values (at 37°C and 90% RH) of at least 1.5 and preferably of 3 to 1,500 gram-mils/100 in^2 in 24 hours.

The membranes of poly(chloro-p-xylylene) have a nitrogen permeability constant (at 23°C) of at least 1, and preferably of about 2 to 1,000 cc-mils/100 in^2 in 24 hours per atmospheric pressure, and a carbon dioxide permeability constant (at 23°C) of at least 7.5, and preferably of about 15 to 7,500 cc-mils/100 in^2 in 24 hours per atmospheric pressure. These membranes preferably also have moisture vapor transmission constants or values (at 37°C and 90% RH) of at least 0.6 and preferably of 1.2 to 600 gram-mils/100 in^2 in 24 hours.

Examples 1 through 5: The five membranes having the properties described in the table above were prepared in these examples. In each example a film was first prepared which had unpolymerized dimer therein as the filler. Each of the films was prepared by vaporizing a charge of about 100 grams of dichlorodi-p-xylylene at a temperature of about 180° to 300°C. The system was evacuated to a pressure of less than 100 millitorr. The dimer vapors were then passed through a pyrolysis zone which was heated at about 630° to 660°C. The dimer vapors were thus caused to pass through the pyrolysis zone at the rate of about 1.5 to 7 g/min.

In each case a portion of the dimer passed through the pyrolysis zone without being pyrolyzed into the corresponding diradical. The diradical vapors and dimer vapors, which were admixed were then passed into a deposition zone (12 inches in diameter by 30 inches long) and codeposited on 8 x 8-inch glass plates. The plates were at a temperature of about 25°C. The diradical vapors polymerized into a film of poly(chloro-p-xylylene) which was about 25 μm thick, and in which the unreacted dimer was embedded. The pressure in the system was then raised to atmospheric pressure and the films were mechanically removed from the glass substrates.

The films were weighed and them immersed in benzene at 75°C for 2 hours. The films were then washed with fresh benzene and dried in a vacuum oven, set at 80°C for 24 hours. The dried films were then weighed to ascertain the amount of dimer removed therefrom by the benzene. The table below lists, with respect to each of the five films thus prepared, the vaporization temperature and pyrolysis temperature used, and the rate of feed of dimer into the pyrolysis zone.

	Vaporization temperature, ° C.	Pyrolysis temperature, ° C.	Dimer feed, grams/minute
Film number:			
1	180	660	1.5
2	250	660	4.5
3	230	660	4
4	300	640	7
5	300	630	7

Aromatic Polyimides, Polyesters and Polyamides

H.H. Hoehn and J.W. Richter; U.S. Patent 3,899,309; August 12, 1975; assigned to E.I. du Pont de Nemours & Company describe a gas separation membrane of which at least 50% by weight consists essentially of a polymer whose main chain has a repeating unit containing at least one group selected from the group consisting of aromatic imide, aromatic ester and aromatic amide in which the repeating unit

- (a) contains at least one rigid divalent subunit, the two main chain single bonds extending from which are not colinear,
- (b) is sterically unable to rotate 360° around one or more of the main chain single bonds, and
- (c) more than 50% of the atoms in the main chain are members of aromatic rings.

These criteria define predominantly aromatic polymers whose molecules are unable to pack densely because of having within the repeating unit of the polymer chain at least one main chain single bond which makes an angle with at least one next adjacent main chain single bond and around which the polymer molecule is sterically unable to rotate freely. It is considered that configurations as defined above render polymer molecules containing them incapable of packing as densely togehter as polymer molecules without such configurations. Specifically, the bend in the polymer chain caused by the noted angle cannot be accommodated in packing by free rotation around the bond. The structure of the solid polymer is thus kept permanently more open to the passage of small gas molecules, resulting in higher flux rates for the passage of such gases.

The polyimides from which membrane materials of this process are selected may be represented generally as polymers in which the repeating unit is as shown in the formula:

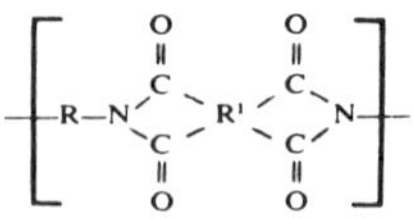

where R and R^1 are, respectively, divalent and tetravalent organic radicals, i.e., with their bonds stemming from carbon atoms. The polyesters from which the membrane materials are selected may be represented generally as polymers in which the repeating unit is as shown in the formula on the following page.

$$\left[R^2-O-\overset{\overset{\displaystyle O}{\|}}{C}-R^3-\overset{\overset{\displaystyle O}{\|}}{C}-O \right]$$

In the formula, R^2 and R^3, alike or different, are divalent organic radicals, i.e., with their bonds stemming from carbon atoms. The polyamides from which the membrane materials are selected may be represented generally as polymers in which the repeating unit is as shown in the formula:

$$\left[R-\overset{\overset{\displaystyle R^7}{|}}{N}-\overset{\overset{\displaystyle O}{\|}}{C}-R^3-\overset{\overset{\displaystyle O}{\|}}{C}-\overset{\overset{\displaystyle R^7}{|}}{N} \right]$$

where R and R^3 are defined as above and R^7 is hydrogen, lower alkyl, or phenyl. The particular polyimides, polyesters and polyamides useful as membranes in this process are selected on the basis of the three criteria noted above.

Requirement (a) specifies that the repeating unit of the polymer contain at least one rigid divalent subunit, the two main chain bonds from which are not colinear. The rigid subunits in a polymer chain are those atoms, groups of atoms, or cyclic structures which are joined to other units in the main chain by single bonds between two atoms. The single bond jucntion points in a polymer main chain are readily recognized from the structural formula of the polymer repeating unit and these points are the demarkation points between rigid subunits.

Such a subunit is rigid because the angle between the two single bonds extending from the subunit is fixed. The two bonds from a rigid subunit are colinear (L) if they form an angle of about 180°, (particularly 170° to 180°) or if they are parallel and offset not more than 2A. Otherwise, they are noncolinear (N). Preferred polymers have 2 to 10 main-chain rigid subunits in the polymer repeating unit.

Requirement (b) specifies that the polymer chain contain at least one bond between rigid subunits around which bond the polymer chain is sterically prevented from rotating 360°. This determination is based on the well-known textbook rules of stereochemistry.

The determination of requirement (c) that over 50% of the atoms forming the backbone chain of the repeating unit of the polymer be aromatic groups can readily be made from a two-dimensional representation of the polymer repeating unit. The main chain atoms which are counted are those in the single atom and cyclic rigid subunits. In cyclic subunits in which the two single bonds stem from different atoms, all member atoms in the basic ring of the subunit are counted, e.g., p-phenylene counts 6 atoms.

Side chain atoms such as hydrogen, carbonyl oxygen, alkyl groups, haloalkyl groups, carboxyl groups, ester groups, halogen substituents and other pendant groups are not counted. If both single bonds from a cyclic rigid subunit stem from the same atom, only that one atom is counted, e.g., 1,1-cyclohexylene counts 1 atom, the remaining pentamethylene being a pendant group. Aromatic rings include not only the hydrocarbon aromatic rings such as benzene, naphthalene, anthracene, phenanthrene, pyrene, chrysene, naphthacene, indene, and the like, but also those heterocyclic rings commonly acknowedged to have aromatic character such as furan, benzofuran, dibenzofuran, thiophene, pyrazole, indole, benzimidazole, pyrazine, carbazole, pyridine, quinoline, acridine, imidazole, isoimidazole, and the like.

The selectivity of a membrane in separating a two-component fluid mixture is defined as the ratio of the rate of passage of the more readily passed component to the rate of passage of the less readily passed component. Selectivity may be obtained directly by contacting a membrane with a known mixture of gases and analyzing the permeate or approximately first by setting up the ratio of the rates of passage of the two components determined separately on the same membrane. Rates may be expressed as GTR or as cB units.

Gas Transmission Rate (GTR): One characterization of membrane permeability used in this disclosure is the gas transmission rate; GTR data represent the steady state rate of gas transmission through a membrane. GTR values are not normalized for membrane thickness. For homogeneous membrnaes the GTR is inversely proportional to the sample thickness. When the thickness of the active part of the membrane is not known, e.g., in asymmetric membranes, the GTR is still a valid permeability characterization. GTR values determine the value-in-use of the membrane in a permeation device. Derivation of the GTR equation follows. The volume of gas transmitted through a membrane is directly proportional to the area, time, and pressure of the permeation test as shown below.

$$\text{GTR} = \frac{\text{Volume}}{\text{Area x Time x Pressure}}$$

The units selected for volume, area, time, and pressure are cm^3 (STP), 100 in^2, 24 hours, and atmosphere, respectively. Substituting these units in the permeation test above gives the results below.

$$\text{GTR} = \frac{\text{cm}^3\text{ (STP)}}{100\text{ in}^2\text{ x 24 hr x atm}}$$

Except in special cases, all measurements of transmitted gas as cm^3 (STP) were made at 30°C. GTR values were usually measured at gas pressures of 39.7, 114.7, 314.7, 614.7 and 1,014.7 psia.

Centibarrer Permeation Coefficient (cB): The standard unit for the permeability coefficient in observing the permeability of polymer films to gases is defined as the barrer which is equal to

$$\frac{10^{-10}\text{ cm}^3\text{ (STP) x cm}}{\text{cm}^2\text{ x sec x cm Hg}}$$

in which cm^3 (STP) is the volume of permeated gas at standard temperature and pressure, cm is the thickness of the film, cm^2 is the area of film, sec is the time, and cm Hg is the pressure. (ASTM Test D 1434-66, 1970 Edition, Part 27, pages 447 and 453.) In this process permeabilities are reported in centibarrers (cB), a unit which is 1/100 of the barrer as defined above. Centibarrer values can be calculated from the relationship:

$$\text{cB} = \text{GTR x film thickness in mils x 0.6}$$

In the illustrative examples which follow, parts and percentages are by weight unless otherwise specified.

Example 1: Part A – A polyester from isophthaloyl chloride and 2,2-bis(3,5-dichloro-4-hydroxyphenyl)propane is prepared as follows. A mixture of 183 grams of 2,2-bis(3,5-dichloro-4-hydroxyphenyl)propane, 1 liter of s-tetrachloroethane, 0.72 gram of $AlCl_3$ and 101.6 grams of isophthaloyl chloride was heated and stirred under nitrogen at reflux for one hour. The resulting viscous polyester solution was allowed to cool and 1 ml of water was added with vigorous stirring. The resulting solution contained about 17% polyester. The polyester was precipitated in hexane with vigorous agitation, filtered and dried.

The repeating unit of the polyester prepared as above was checked against requirements (a), (b), and (c). The repeating unit has 2 (L) and 6 (N) rigid subunits, 2 bonds with restricted rotation, 2 of the (N) subunits have at least one bond with restricted rotation, and 18/23 (78%) of the atoms in the chain are aromatic. This polymer thus satisfies the requirements set out above.

Part B — The polyester of Part A was dissolved in tetrachloroethane in an amount to give a 16% solution. The solution was filtered through a 5.0 μ silver membrane. The filtrate was degassed to remove bubbles. The filtrate was cast on a glass plate which had been coated with a low molecular weight polytetrafluoroethylene wax dispersion (Vydax) and heated to 110°C. A 15-mil doctor knife was used to spread the solution. The film was then covered to protect the solution or the partially dried tacky film from dust. The cover was equipped with vents. After allowing the film to dry for 5 minutes at 110°C vents in the cover were opened and the film was allowed to dry another 10 minutes. The 1.5-mil film was then stripped from the glass plate and tested as a permeation membrane for a mixture of oxygen and nitorgen using conventional test apparatus. The film permeated O_2 at 140 GTR and 126 cB. The film permeated nitrogen at 25 GTR and 22 cB. The selectivity, S_{O_2/N_2}, was 5.6.

Example 2: The procedure of Example 1, Part B was repeated using a 15-mil doctor knife. The film was covered and dried at 100°C for 5 minutes after which vents were opened and drying continued for 10 minutes. The clear, bluish, smooth, crisp, 2.4-mil film was stripped from the plate and tested for oxygen and nitrogen permeability. The film permeated O_2 at 115 GTR and 166 cB and N_2 at 21 GTR and 30 cB. The S_{O_2/N_2} was 5.5.

Example 3: Part A — The polyamide from m-phenylenediamine and a 70/30 mixture of isophthalic and terephthalic acid chlorides were prepared using the procedure shown in U.S. Patent 3,567,632. This polyamide is referred to as MPD-I/T (100-70/30).

Part B — This example concerns a semipermeable membrane made from a blend of a polyester which satisfies requirements (a), (b), and (c) with a polyamide. A casting solution was prepared using 10 parts of a solution containing 15 weight percent MPD-I/T (100-70/30) in tetrahydrofuran, 10 parts of a solution containing 15 weight percent of the polyester of Example 1, Part A, in dimethylacetamide, and 1.2 parts Santicizer 8. This solution was cast on a Vydax-coated glass plate at room temperature with a 15-mil knife. The film was dried for 15 minutes at 110°C. The 1.55-mil film permeated H_2 at 572 GTR and 532 cB, and CH_4 at 6 GTR and 6 cB. The S_{H_2/CH_4} was 95.

Example 4: Part A — A polyamide from metaphenylenediamine and isophthaloyl chloride is prepared as follows. Under nitrogen a solution of 2.36 mols of metaphenylenediamine in 32 mols of N,N-dimethylacetamide was stirred and cooled at -10° to 0°C and 2.36 mols of molten isophthaloyl chloride was added in small portions. During the addition the temperature was allowed to rise to 20°C. The reaction was completed by heating the resulting viscous solution to about 50°C. The solution was diluted to 9% polyamide by adding dimethylacetamide. The polymer was isolated from this solution by drowning in crushed ice and ice water under vigorous agitation. The polymer was recovered by filtration, washed with water and dried under vacuum at 80°C.

Part B — A solution of 3 grams of the polyester of Example 1, Part A, in 17 grams of tetrahydrofuran was mixed with a solution of 3 grams polymetaphenylene isophthalamide (prepared as in Part A) in 17 grams of dimethylacetamide. The combined solution was filtered through a 0.8 μ silver membrane. The filtrate was cast on a Vydax-coated glass plate at room temperature in a dust-free cabinet with a 15-mil doctor knife. The film was allowed to dry for 15 minutes at room temperature in the dust-free cabinet and was then transferred to a hot plate maintained at 110°C and allowed to dry for 5 minutes. The film was stripped, air-dried, and then dried in a vacuum at room temperature. The 1.15-mil film permeated H_2 at 689 GTR and 475 cB and CH_4 at 111 GTR and 77 cB. The S_{H_2/CH_4} was 6.2.

In related work *H.H. Hoehn; U.S. Patent 3,822,202; July 2, 1974; assigned to E.I. du Pont de Nemours & Company* has shown that the heat treatment of the above-described polyester, polyimide and polyamide membranes results in improved gas separation efficiencies.

Arylene Oxide Polymers and Polycarbonates

According to a process described by *S.G. Kimura; U.S. Patents 3,852,388; December 3, 1974; and 3,709,774; January 9, 1973; both assigned to General Electric Company* an asymmetric microporous polymer membrane structure is produced having a graded porosity, i.e., graded pore size progressing from one major surface of the membrane to the other major surface, by the use of a proper solvent system. The structure may (depending upon the conduct of the process) present the porosity thereof as graded interconnected pores either (a) ranging from an effective diameter of about 100 A at one membrane face to a smaller undetermined size adjacent a dense nonporous layer forming the second face, or (b) ranging from an effective diameter of about 100 A at one membrane face to some very small finite value at the opposite face of the membrane through which gas can freely pass.

The method for the preparation of such membranes involves purely physical phenomena beginning with the selection of a polymer, which can be cast from solution and is soluble to different degrees in two mutally miscible solvents. The better of the two solvents for the polymer must be more volatile than the poorer of the two solvents. Next, a precipitating agent is selected having the properties of being miscible with both the mutually miscible solvents and being a nonsolvent for the polymer.

The method steps involve preparing a solution of the polymer in the better solvent and then adding to this solution a quantity ranging from about 10 to 100% by volume (of the first solution) of the poorer solvent to prepare the casting solution. A membrane is cast from this polymer solution and is allowed to desolvate for a short time before immersion into the precipitating agent. Immersion time should be long enough for the membrane structure to develop, that is, of the order of one minute or longer. After the immersion the membrane is dried.

With many polymers that exhibit crystallinity it is possible, depending upon the desolvation time employed, to produce either an asymmetric microporous membrane having one thin dense major surface (or skin) or an asymmetric microporous membrane without such a dense skin. Among the thermoplastic polymer materials from which asymmetric microporous membranes may be prepared by the process are the arylene oxide polymers described in U.S. Patent 3,350,844, polycarbonate resins such as are described in U.S. Patent 3,256,675, polyvinyl acetate resins, polyalkyl methacrylate resins, polysulfones and polymers of monovinyl aromatic hydrocarbons. The following examples illustrate the process.

Example 1: A 10% solution of polyxylylene oxide in chloroform was prepared, chloroform being the good, volatile solvent. Next, a solution was prepared consisting of 2 parts of the polyxylylene oxide solution to 1 part dichlorobenzene, the poor, nonvolatile solvent. A membrane was cast in air at ambient temperature and pressure on a glass plate using a doctor blade with a blade setting of 10 mils. The cast membrane was allowed to desolvate for 30 seconds and was then immersed in methanol, the precipitating agent. The membrane was dried.

The resultant product was an opaque microporous membrane of about 1.3 mils in thickness. On visual inspection it appears that the side away from the glass plate during casting was slightly shinier than the side toward the glass plate. On subsequent testing for permeating properties the O_2 and N_2 permeabilities for the dried membrane were, respectively 18.5×10^{-9} and 4.22×10^{-9} (cc gas RTP, cm thick)/(sec, cm^2, cm Hg ΔP).

The permeability values for normal nonporous polyxylylene oxide membranes for O_2 and N_2, are respectively, 1.7×10^{-9} and 0.35×10^{-9} (cc gas RTP, cm thick)/(sec, cm^2, cm Hg ΔP). A comparison of the above permeability values establishes that the effective thickness of the whole asymmetric polyxylylene oxide membrane was about 2.7 microns, which indicates a skin thickness of 2.7 microns or less. The oxygen-to-nitrogen permeability ratios for both the membrane prepared in Example 1 and normal nonporous polyxylylene oxide membranes is substantially constant thereby verifying the presence of a nonporous skin.

Example 2: Polycarbonate resin was dissolved in methylene chloride to form a 20 weight percent solution. The solution is then mixed with toluene in a volume ratio of 2 to 1. The methylene chloride is the good, volatile solvent and the toluene is the poor, nonvolatile solvent. A membrane was doctor-blade-cast in air (ambient conditions) on a glass plate at a liquid thickness of 20 mils and was allowed to desolvate in air for one minute. The glass plate and membrane were then immersed in methanol and the membrane allowed to gel. The resultant product was a white, opaque membrane 4 mils thick with a nitrogen permeability of 0.62×10^{-9} (cc gas RTP, cm thick)/(sec, cm^2, cm Hg ΔP) and $Pr_{O_2}/Pr_{N_2} = 4.5$. The accepted dense membrane values for polycarbonate are $Pr_{N_2} = 0.037 \times 10^{-9}$ and $Pr_{O_2}/Pr_{N_2} = 4.6$.

Thus the effective thickness of the 4-mil asymmetric film is calculated to be 0.24 mil. As is indicated in Example 1, this is an indication of the skin thickness. The balance of the membrane was made up of microporous (graded porosity) crystalline structure.

Polytetrafluoroethylene

According to a process described by *N. Murayama; U.S. Patent 3,843,570; October 22, 1974; assigned to Kureha Kagaku Kogyo Kabushiki Kaisha, Japan* a porous material comprising polytetrafluoroethylene and having micropores is obtained by polymerizing a polymerizable monomer capable of forming a thermoplastic resin, which has been adsorbed or attached to fine particles of polytetrafluoroethylene, milling the resulting polymeric mixture of the polytetrafluoroethylene and the thermoplastic resin at a temperature higher than the softening point of the thermoplastic resin to form a shaped article and then removing the thermoplastic resin.

This resin is utilizable as gas filter for manufacturing semiconductors and as water separator. A porous material of this resin is used as partition membrane for an oxygen-generating electrolysis and edible salt electrolysis or as partition membrane for fuel cells, sulfuric acid cells and alkali cells. Further, such porous material is used not only as protecting material required for permeation of gas but also for separation of isotopes by way of gas diffusion.

It is a characteristic feature of the process that as the minute structure of fine polytetrafluoroethylene particles is fixed by the coated resin to secure perfect dispersion, mutual adhesion of the polytetrafluoroethylene particles is prevented and the minute structure is not lost during the milling treatment, thus making it possible to provide a porous body having very fine pores. Any monomer which forms a thermoplastic resin by polymerization can be used as the polymerizable monomer. Especially preferable are methyl methacrylate, styrene, vinyl chloride, vinylidene chloride, and copolymers.

Example: To 500 grams of fine powders of unbaked polytetrafluoroethylene for use in shaping (apparent specific gravity: Ca 0.45) obtained by dispersion polymerization were added 500 grams of methyl methacrylate containing 1 gram of azo-bis-isobutyronitrile as polymerization initiator and 1.5 grams of dodecylmercaptan as chain transfer agent, the methyl methacrylate having been treated with an aqueous caustic soda solution to remove any polymerization inhibitor and washed well with water. Almost all of the methyl methacrylate was adsorbed to the fine powders of polytetrafluoroethylene.

The mixture was placed in a 2-liter autoclave and, after replacing air therein by nitrogen, maintained at 80°C for 24 hours. A polymeric mixture of polytetrafluoroethylene and poly(methyl methacrylate) was obtained as a solid mass. The mixture, when milled with a milling roll at 170°C, was hardly coiled around the roll at the initial stage. As the milling proceeded, however, the fine powdery polytetrafluoroethylene was gradually stretched and obtained in the form of a sheet strong in the direction of the roll if the mixture was taken out on the way in the form of a sheet.

When the milling treatment was continued without taking out the mixture on the way, the mixture became strong also in the transverse direction and simultaneously was hardly coiled around the roll. The milling treatment was stopped at this stage and the mixture was taken out in the form of a sheet having a thickness of about 0.4 mm. This sheet was shaped by means of a press operated at 230°C to form a sheet having a thickness of about 0.2 mm, which was milky white and more or less tended to become translucent. 10 grams of this sheet was dipped all day in tetrahydrofuran, washed for 8 hours with tetrahydrofuran at its boiling point to extract poly(methyl methacrylate) and dried to yield 5 grams of a white opaque sheet which was not contracted in thickness, width and length in comparison with the original sheet and was tough and flexible like leather.

The permeation rates of this sheet measured by air (P_{air}) (mol mm/cm^2 min cm Hg) were as follows: $P_{air} = 4.7 \times 10^{-6}$ in the case of an average pressure being 10 cm Hg, $P_{air} = 4.8 \times 10^{-6}$ in the case of 20 cm Hg and $P_{air} = 4.9 \times 10^{-6}$ in the case of 30 cm Hg. Deviation of the permeation rate by pressure was extremely small. This sheet had a strength of 2.0 kg/mm^2. The gas permeability of this sheet measured by helium and air was: $P_{He} = 1.24 \times 10^{-5}$ and $P_{air} = 4.8 \times 10^{-6}$. This result led to $P_{He}/P_{air} = 2.6$. In addition, the following results were obtained for hydrogen and deuterium: $P_{H_2} = 1.92 \times 10^{-5}$, $P_{D_2} = 1.37 \times 10^{-5}$ and $P_{H_2}/P_{D_2} = 1.40$. The sheet had highly uniform micropores and was useful for diffusive separation of isotopes.

An IR-absorption analysis of this sheet showed that it contained a trace amount of remaining poly(methyl methacrylate). If it is necessary to remove entirely the remaining adsorbed polymer, the following treatment may be carried out: This sheet is placed in an electric furnace heated at 280°C and held for 12 hours while recycling air to obtain a white sheet having a permeation rate variable only within the range of experimental errors. When this sheet was pressed at room temperature with a 100 kg/cm^2 press and then subjected to IR-adsorption analysis, it was found that no poly(methyl methacrylate) remained in the sheet.

Metal Oxides for Gaseous Diffusion

C. Eyraud, M. Prettre, Y. Trambouze, J. Lenoir, G. Thomas and P. Plurien; U.S. Patent 3,022,187; February 20, 1962; assigned to Commissariat a l'Energie Atomique, France describe processes for producing membranes of very fine porosity. These methods are characterized in that, in the interior of a rigid support having relatively large pores, a membrane is constructed of very fine porosity by filling the relatively large pores with submicronic grains having very fine channels between them. Deposit of the grains may be obtained by suction of a solid-gas or solid-liquid suspension through the support. If desired, the assembly may be covered with a thin metallic film having a thickness of from 500 to 5000 A which protects the fine grains without impairing porosity. This film may be obtained by any known metallization process.

The required external mechanical strength (rigidity) of the membrane is provided by a very permeable strong support, having pores of a large radius, for example, a sintered metal such as steel, nickel or bronze. Within the pores of the sintered metal, which may have diameters of from ten to several tens of microns, a membrane of very fine porosity is constructed by passing through the relatively large pores of the support a solid-gas or solid-liquid suspension containing submicronic grains of a metal oxide which are monodispersed as much as possible.

The order of magnitude of these grains is from one micron to several hundredths of a micron. Better surface cohesion is provided to obtain more efficient protection against the external agents which might tear away the fine grains. This may be done by metallizing the active surface of the membrane. Experience has shown that when the metal is suitably fixed, the pores are not obtained and the material is deposited substantially normal to the fine grains of the surface.

After this treatment, the active portion of the membrane is formed by the deposit of the fine grains and by the addition of the metal film. The thickness of this active portion may be on the order of one to several tens of microns. The main advantage of the membranes of the process is that they have excellent mechanical strength, are very thin and have great permeability. The radius of the pores may be determined as desired by choice of the radius of the grains in the suspension. The membranes are particularly suited for ultrafiltration and for separation by gaseous diffusion of the constituents of a mixture. Suitable choice of the different components of the membranes provides resistance to the various types of corrosion.

Diffusion of Multivalent Cations into Polyelectrolyte Sols

O. Meyer-Berge and U. Heyse; U.S. Patent 3,565,259; February 23, 1971; assigned to Kalle AG, Germany describe a process for the manufacture of porous membranes having an oriented capillary structure, by diffusion of multivalent cations into polyelectrolyte sols. It has been found that the time-consuming and technically expensive dehydration of the gel membranes by means of a series of solvent treatments can be eliminated by a simple procedure. In the process, layers of the capillary gel obtained are dehydrated on porous supports, if desired with suction, and then dried. After the excess of electrolyte solution has been removed, the cut, capillary membranes in the form of the gel are placed on supports having a strongly dehydrating action, such as plates of clay, porcelain, plaster of Paris, bentonite, kieselguhr and chalk.

Porous supports with a smooth surface, such as sintered glass, plastic material or metal, also may be used, applying a vacuum to one surface. In both cases, the gel structure condenses quickly by dehydration and the membrane shrinks to 20% of its original thickness. Practically no shrinkage occurs in the two other directions, so that the uniform capillary structure, the diameter of the individual capillaries, and the ratio of pore area to the total area of the membrane remain practically unchanged.

The shrinkage of the gel structure is irreversible, so that no swelling or increase in thickness takes place upon contact with water. Due to the compression of the gel, considerably thinner porous membranes can be produced than were previously possible owing to the technically limited cutting thickness of the gel membranes. By the dehydration procedure and the ensuing drying process, the structure of the membranes is condensed and solidified to such a degree that, in the case of many applications where no substantial mechanical stresses occur, it is no longer necessary to stabilize the gel membranes by the additional introduction of polymerizable components.

The membranes produced in accordance with the process which have a thickness of 50 to 400 μ, have the advantage over previously known membranes of 10 times this thickness, in that, due to their smaller thickness, the size of their pores is much more accurately defined and deviations in the direction of the capillaries are no longer noticeable. The thin membranes preferably are used as special sieves and filters with a high throughput per cross section, for the filtration of gases and liquids, further as catalyst screens and supports, and also as film screens for reproduction purposes. Due to their small thickness, the application of a metal layer by currentless or galvanic metallization or by vapor deposition is much easier and offers further possibilities of application for the material. The following examples illustrate the process.

Example 1: By diffusion of copper ions for 1 N copper nitrate solution into a 1.8% by weight sodium alginate sol, a capillary gel having a thickness of 15 mm is produced. By means of a horizontally swinging knife, the uppermost layer, which is free from pores, is removed, and then gel layers of a thickness of 2 mm are cut off in the same manner. The gel layers are placed on highly absorptive clay plates, and after 3 minutes have lost 80% of their thickness by shrinkage. When the gel layers are placed on porous plates of sintered glass and subjected to reduce pressure from one side, a shrinkage of 80% takes place within 5 minutes. The porous membranes shrunk and substantially dehydrated in this manner are dried, either between perforated metal plates at temperatures between 60° and 80°C, or, after being clamped onto a glass cylinder, at room temperature. Porous membranes of a thickness of 200 μ are obtained.

Example 2: The membrane is dehydrated and caused to shrink as described in Example 1 and then placed in acetone. When it is dried at room temperature while clamped onto a glass cylinder, the drying time is less than 10% of the time required when drying from an aqueous phase.

Example 3: A membrane which has been produced as described in Example 1 is placed in a solution of glycerol in water and cautiously squeezed off. After drying under the conditions stated in Example 1 or 2, an elastic membrane of a high bending strength is obtained. Instead of a solution of glycerol in water, it is also possible to use a solution of glycerol in a water-miscible solvent, such as ethanol or acetone.

CATALYST REMOVAL AND HEAVY METAL SEPARATION

OLEFIN SYNTHESIS

Transition Metal Complexes

According to a process described by *M.T. Westaway and G. Walker; U.S. Patent 3,617,553; November 2, 1971; assigned to The British Petroleum Company Limited, England* transition metal complexes are separated from mixtures containing organic components of lower molecular weight by contacting the mixture with one side of a cellulosic membrane. The applied pressure is greater than the pressure on the opposite side of the membrane and the pressure differential is greater than the osmotic pressure of the system. The permeate has a reduced concentration of complex. The membrane is preferably a cellulose acetate membrane. Applied pressures are from 500 to 2,500 psig and complexes separated may be those of rhodium, nickel, vanadium and other metals of groups VIIa and VIII of the Periodic Table. Organic components may be hydrocarbons or mixtures from the dimerization of olefins or from the hydroformylation of propylene.

The cellulosic membrane referred to is preferably a cellulose acetate membrane. The molecular weight of the cellulose acetate may be within the range 10,000 to 100,000. Cellulose acetate having a molecular weight of from 10,000 to 50,000 is preferred. The membrane may be prepared from a solution of the appropriate cellulose derivative in enough solvent to ensure that the solution is not too viscous, together with a pore-forming agent and/or water. The pore-forming agent may, for example, be a perchlorate salt of formamide.

Cellulose acetate membranes obtained from Sartorius GmbH Germany and reported by Sartorius to contain pores in the range 50 to 100 A have also been found to be suitable.

The cellulosic solution is cast on a plate and solvent allowed to evaporate until a thin surface skin forms on a film of the solution. The film is immersed in water at low temperature so as to produce a macroporous structure beneath the surface skin, and the membrane thus produced is conditioned for use in the presence of hydrocarbons by equilibrating it with a series of solutions containing a hydrocarbon component and a component miscible with water, the solutions having successively greater proportions of the former and smaller proportions of the latter component. Finally the film is immersed in a mixture of the feedstock to be used and the hydrocarbon component. Alternatively, the membrane may be conditioned by permeating isopropyl alcohol.

In the case of cellulose acetate membranes suitable thicknesses for the membranes have been found to be in the range 1 to 5 x 10^{13} inches.

After preparation, but before conditioning, the film may be heated by immersion in hot water preferably at not less than 80°C. This can improve selectivity and complex retention.

Preferably the film is laterally restrained during conditioning and the above immersion in hot water. The nature of the film will also be affected by the conditions of casting, including the nature of the pore-forming agent, and the rate of evaporation of solvent. The actual method of preparation of the membrane film and its subsequent treatment may be established by experiment taking into account the system in which the membrane is to be used.

In Figure 7.1a feedstock was supplied from a reservoir **1** at atmospheric pressure, to which dry nitrogen could be supplied if necessary via line **8**. The feedstock was pressurized by means of a hydraulic ram pump **2** and supplied to the cell **3**. Permeated material left the cell to collection via line **4**. Nonpermeated material returned to the reservoir **1** via a felt filter **5** and a pressure control valve **6**. A cell bypass line **7** was provided containing a pressure relief valve **9** operating at 3,000 psig. Pressures were read from gauge **10**.

In Figure 7.1b a membrane **11**, diameter 6.5 cm, was supported on a sintered stainless steel disc **12**, of pore size 60 μ, which in turn rested on a mild steel baseplate **13**. By screwing the baseplate into stainless steel head **14** the membrane was sealed against nitrile rubber O-ring **15**. Feedstock entered the cell through a 1 mm orifice **16** so as to create turbulence, and left the cell via outlet **17**. The permeated material was collected in vessel **18**.

The second form of the cell differed only in the baseplate, and the alternative baseplate **19** is shown in the lower part of Figure 7.1b. This consisted of a stainless steel plate **20** in which inclined channels **21** were cut mounted on a mild steel base **22**. The permeated material drained to the collector through a stainless steel pipe **23**.

The capacity of the cell above the membrane was 24 ml. Other methods of mixing than that referred to above may be used in a cell such as is described, for example a magnetic stirrer may be employed. Pressurization may also be carried out by other means than are described, for example by using nitrogen.

In Figure 7.1c a 150 ml, stainless steel, nitrogen pressurized cell comprises three parts, top plate **14** and plate **13** and barrel **35** held together by four rods **34**. The membrane **11** having a permeable area of 11.4 cm^2 is supported on a sintered stainless steel disc **12** of pore size 6 μ seated in the baseplate **13**. The seal between the membrane **11** and the barrel **5** is made with a 1 mm Teflon gasket **38** and between plates **14** and **13** and barrel **35** by nitrile O-rings **15**. The top plate **14** contains a relief valve **31** and a connector opening **39** to the nitrogen supply. Turbulence in the cell is maintained by magnetic stirrer **36**.

In Figure 7.1d a stainless steel high pressure cell is shown. A membrane **11** 7.8 cm in diameter (permeable area 23 cm^2) is cushioned by filter paper **28** and is supported on a sintered stainless steel disc **12** (of 6 μ pore size) resting in a circular recess **40** in the cell base **13**. Gramophone grooves **41** cut into the bottom of the recess facilitate permeable drainage through the drain pipe **42**. Six Allen capped screws **43** secure the cell top **14** and base **13** together, thereby sealing the membrane **11** against a nitrile or Teflon O-ring **15** carried in the cell top **14**. An outer ring O-ring prevents permeate leaking to the outer edge of the sintered disc **12** and returning into the feed. To create turbulence in the cell, feedstock enters via a feed inlet opening **16** and then through a 1 mm orifice **16**. Feed leaves the cell via outlet **17**. The dead volume of the cell is 3.5 ml.

The flow system and apparatus described were used to investigate the separation of three different catalyst complexes from their respective reaction mixtures, and conditions and

FIGURE 7.1: CATALYST ULTRAFILTRATION PROCESS

a.

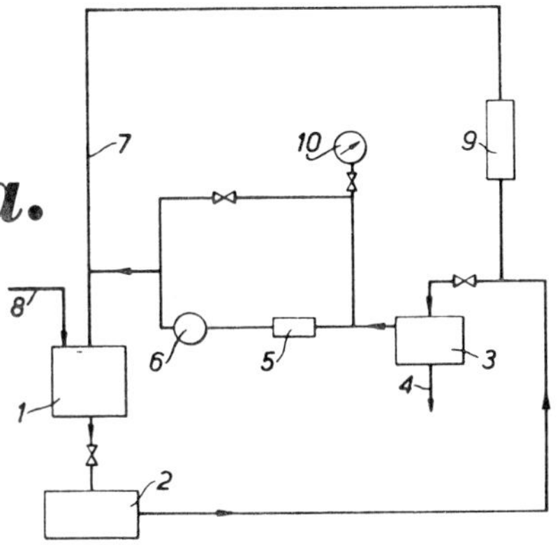

Schematic of the High-Pressure Flow System

b.

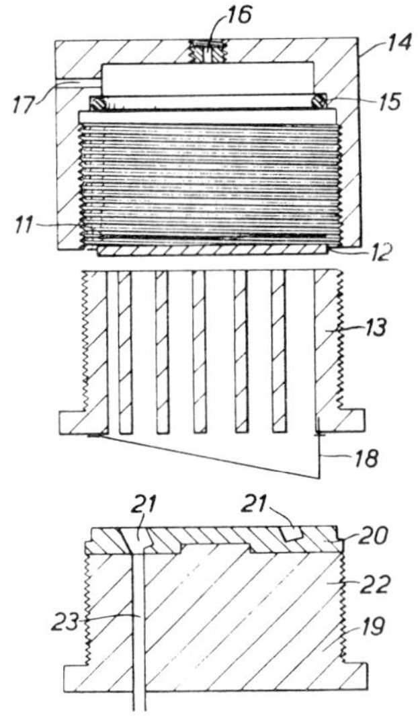

Side Elevation, and in Expanded Sectional Form Two Forms of a Typical High-Pressure Cell

c.

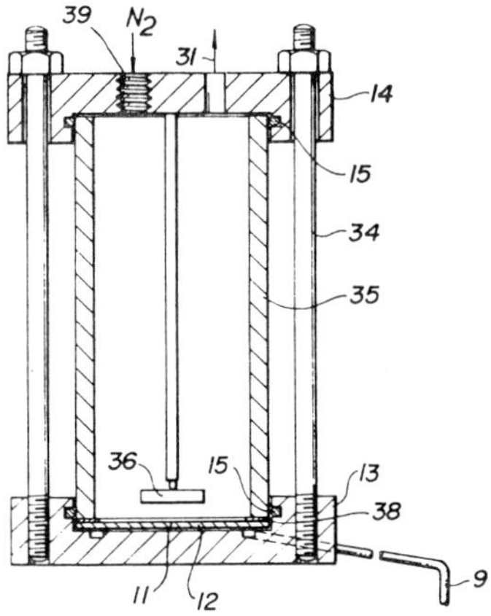

d.

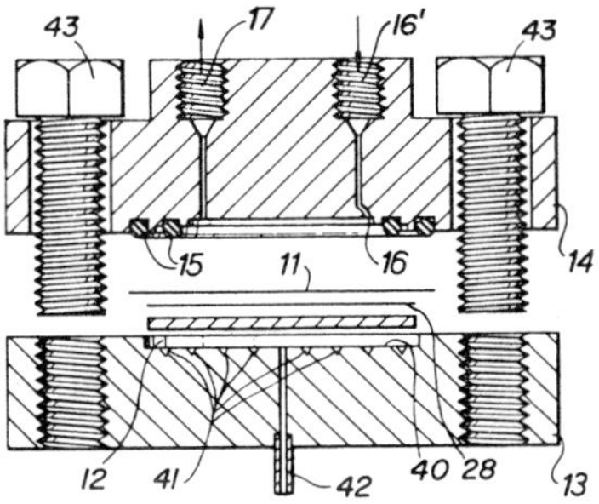

Other Forms of the High-Pressure Cell

Source: M.T. Westaway and G. Walker; U.S. Patent 3,617,553; November 2, 1971

the results obtained are given in the following examples. In these investigations rates of permeation (PR) per unit area of membrane were determined by measuring the volume of permeate collected over known periods of time. The metal contents of the permeates and feedstocks were measured by x-ray fluorescence so that the resistance of the membrane to catalyst permeation could be expressed in terms of catalyst retention (CR), where CR was the ratio, expressed as a percentage, of the difference between the metal content of the feedstock and the permeate to the metal content of the feedstock.

Metal weight balances were made in selected cases so as to determine the quantitative nature of the separation. In these cases the metal content of the membrane itself was determined after dissolution in dioxane. The nature of the catalyst complex remaining in the nonpermeated material was further investigated by means appropriate to the nature of the complex.

Example 1: This example shows the separation of two complexes of rhodium from their reaction mixtures. The rhodium complexes were $Rh(Bu_3P)_2CO(prop)$ MW 609, (catalyst 1) and $Rh(Bu_3P)CO(acac)$(catalyst 2), where acac = acetylacetonate, Bu = n-butyl and prop = propionate. The MW of the latter complex is 433. These catalysts are used in the hydroformylation of olefins with carbon monoxide and hydrogen to form aldehydes and alcohols. The catalysts are both stable at room temperature and atmospheric pressure.

The reaction mixtures investigated were those resulting from the hydroformylation reaction and contained 0.002 mol percent of the rhodium complex, 5 to 10 mol percent hexenes and 90 to 95 mol percent linear and branched aldehydes. Reaction conditions were a temperature of 80°C and pressure of 500 psig.

The membranes used were made as previously described using formamide as the pore-forming agent. The YBP membranes were prepared from a casting dope comprising 25% cellulose acetate, 30% formamide and 45% acetone. The cellulose acetate used had an acetyl content of 39.8 wt % and an ASTM viscosity of 3.

The dope was cast onto a glass plate using a doctor knife for spreading and the film allowed to stand in the atmosphere for ½ to 1½ minutes so that the acetone partially evaporated to form a thin film. The film was then doused in cold water (0° to 30°C) for 1 hour. The film was then floated off the glass plate. Some of the membranes were heat-treated, as shown in Table 1, and all were conditioned using the solutions and immersion times outlined in Table 2, the conditioning being carried out with the membrane clamped in the cell.

TABLE 1

Reference	Thickness x 10^3 Inches	Heat Treatment
1	5	None
2	5	None
3	5	In water*
4	5	In water*
5	5	In water*
6	1.5	None

*At 80° to 90°C

TABLE 2

Solution, volume percent	Immersion Time (min)
90 ethanol, 10 n-heptane	2
50 ethanol, 50 n-heptane	1
10 ethanol, 90 n-heptane	1
100 n-heptane	1
50 n-heptane, 50 feed	1

Results obtained are given in Table 3 below. These were obtained at a temperature of 22°C using the cell shown in Figure 7.1b.

TABLE 3

Membrane	Catalyst	Pressure (psig)	Typical % CR	Permeation Rate x 10^2 gfd
1a*	1	200	96.0	116
1a	1	2,250	98.0	678
1b*	1	750	99.0	390
2a	1	10-15	97.0	9.8-19.6
2a	1	15-25	95.0	19.6-39.2
2b	1	250	98.0	63
2b	1	750-1,250	98.0	-
3	1	30	94.0	29.4
4	1	40	94.0	24.5
5	1	50	94.0	$<$0.5
6	1	40	89.0	12.3
6	2	40	89.0	21.6

*a and b refer to two different membranes cut from the same original sheet.

In all cases the catalyst was unchanged after the separation process. Rhodium weight balances showed the separation to be quantitative while the phosphorus content of the complex remained in constant ratio to the rhodium content. The activity of the catalyst in the concentrated feedstock was unimpaired. Membrane surfaces were not contaminated with precipitated catalyst.

In Examples 2 through 6 the membranes were conditioned prior to use by permeating 20 to 30 ml of isopropanol under 60 to 500 psig applied nitrogen pressure. The operating temperature was room temperature unless otherwise stated.

Example 2: $Rh(Bu_3P)CO(acac)$ was separated from mixtures obtained from the hydroformylation of propylene in which the reaction conditions were 120°C and 800 psig. The first mixture, designated light ends (LE) was the depropanized reaction product. The second mixture, designated heavy ends (HE) was the light ends from which the low boiling components (boiling $<$120°C) had been removed by vacuum distillation and rotary film evaporation. The heavy ends comprised 1 to 2% of the light ends.

Cellulose acetate membranes of two types both of 5 X 10^{13} inches in thickness were used: (a) as previously described and designated YBP; (b) as supplied by Sartorius Membranfilter GmbH Germany. These latter membranes described as very dense are reported by Sartorius to contain pores 50 to 100 A in diameter. Analyses show the cellulose acetate to contain 38 wt % acetyl and to have a number average MW of 46,000 as determined by osmometry.

TABLE 4: LIGHT ENDS*

Membrane	Pressure (psig)	% CR	PR (gfd)
Sartorius No. 1	2,000	94	2.62
	1,500	93	2.00
	1,000	93	1.32
	500	93	0.73
Sartorius No. 2	500	86	4.3
	1,000	87.5	4.8
	2,000	86	3.9

*Circulating system at 40 to 50 ml/min; catalyst concentration 380 to 480 ppm; system as in Figure 7.1a; cell as in Figure 7.1d.

TABLE 5: HEAVY ENDS

Membrane	Pressure (psig)	% CR	PR (gfd)	Catalyst Conc. (ppm)	Cell as in Figures 7.1c or 7.1d
Sartorius No. 3	400	73	0.72	924	3
Sartorius No. 4	500	90	0.43	11,000	3
Sartorius No. 5	1,500	93.5	0.38	11,000	4

(continued)

TABLE 5: (continued)

Membrane	Pressure (psig)	% CR	PR (gfd)	Catalyst Conc. (ppm)	Cell as in Figures 7.1c or 7.1d
Sartorius No. 6	1,500	73	0.42	656	4
YBP 1	400	87	0.28	924	3
YBP 2	1,500	93	0.14	656	4

Example 3: This example demonstrates the effect of temperature on the separation of $Rh(Bu_3P)CO(acac)$ from heavy ends. The nitrogen pressurized cell shown in Figure 7.1c was used. Catalyst concentration was 540 ppm.

TABLE 6

Membrane	Temp. (°C)	Pressure (psig)	% CR	PR (gfd)
Sartorius No. 7	24	510	90	0.17
	41	510	96	0.55
	56.5	510	82	1.01

Example 4: The effect of toluene dilution on permeation of heavy ends at 1,500 psig was measured. The circulation system used was 40 ml/min; the cell was as in Figure 7.1d.

TABLE 7

Membrane	Dilution Toluene/Feed (% v/v)	Kinematic Viscosity (cs)	Overall PR (gfd)	PR of HE (gfd)	% CR	Catalyst Conc. (ppm)
Sartorius No. 8	11.4	4.9	1.0	0.84	93	10,080
	19.4	3.35	1.11	0.82	95	9,120
	34.1	1.83	1.31	0.68	97	8,240
	0	8.3	0.35	0.35	97	11,000

This example shows the effect of dilution by light end product, pressure at 400 psig, cell used as shown in Figure 7.1c.

TABLE 8

Membrane	Dilution LE/Feed (% v/v)	Kinematic Viscosity (cs)	Overall PR (gfd)	PR of HE (gfd)	% CR	Catalyst Conc. (ppm)
Sartorius No. 9	0	12.4	0.72	0.72	73	908
	13	6.4	1.63	1.42	75	812
	20	4.9	1.90	1.52	71	768
	30	3.4	2.22	1.56	76	716

Example 5: This example shows the separation of $Rh(Bu_3P)CO(acac)$ from 2-ethylhexanal and also from a n-heptane/heptaldehyde mixture.

TABLE 9

Membrane	Solvent	Rh (acac) Conc. (ppm)	Pressure (psig)	% CR	PR gfd
Sartorius No. 10 (cell as Fig. 7.1c)	50/50 v/v n-heptane/ heptaldehyde	305	95	88	1.0
Sartorius No. 11 (cell as Fig. 7.1d)	2-ethylhexanal	3,240	-	95.0	12.50
		3,580	-	94.0	12.30
		5,060	1,000	97.5	10.45
		10,000	-	97.5	8.40
		17,120	-	96.0	8.75
		18,420	-	97.0	7.50

Example 6: Example 6 demonstrates the separation of different rhodium complexes from toluene. The rhodium complexes were made into solution with toluene up to 200 ppm. Separation was effected using a nitrogen pressurized cell (Figure 7.1c) at 400 psig.

TABLE 10

	PR, gfd	% CR
$Rh(Et_3P)_2COCl$	3.4	27
$Rh(i\text{-}Pr_3P)_2COCl$	3.4	31
$Rh(Bu_3P)_2COCl$	3.4	36

In the formulas Et represents ethyl and i-Pr represents isopropyl.

Silicone Rubber Membrane

According to a process described by *A. Goldup, M.T. Westaway and G. Walker; U.S. Patent 3,645,891; February 29, 1972; assigned to The British Petroleum Company Limited, England* an organometallic compound of a metal of Group VIII or VIIa or Va of the Periodic Table according to Mendeleef is separated from a solution of the compound in an organic component. The process involves bringing the solution into contact with one side of a silicone rubber membrane at an applied pressure greater than the pressure on the opposite side of the membrane, the pressure differential being greater than the osmotic pressure of the system and the molecular cross section of the organometallic compound being greater than that of the organic component and collecting as a liquid the organic component which permeates the membrane, the permeate having a reduced content of the organometallic compound.

Typical organic components from which the organometallic compounds may be separated are heptaldehyde, toluene, benzene, xylene, n-hexane, n-heptane, butanol, isopropyl alcohol, butylamine, thiophene, etc. In general the silicone rubber membranes have been found to be more permeable to nonpolar solvents, e.g., hydrocarbons, than to polar solvents, e.g., alcohols, of comparable molecular weight and size.

The process is particularly suitable for the separation of transition metal coordination complexes from their solutions in organic solvents, for example the separation of rhodium complexes from the reaction mixture resulting from the hydroformylation of olefins.

Thus the processes can be employed in systems in which the organic components are alkenes having up to 20 carbon atoms per molecule and preferably 5 or less carbon atoms per molecule, in which hydrogen and/or carbon monoxide or carbon dioxide may be present and in which the transition metal complex contains one or more of metals of Groups VIII, Va and VIIa.

The complex may also contain at least one biphyllic ligand and a bidentate ligand coordinating through at least one oxygen atom. Thus, for example, a suitable mixture comprises the reaction product of the hydroformylation of a lower olefin in the presence of carbon monoxide and hydrogen, a typical catalyst is a rhodium complex in which the biphyllic moiety is tri-n-butyl phosphine and the bidentate moiety is acetylacetonate, that is, $Rh(Bu_3P)CO(acac)$. The products from this reaction are alcohols and/or aldehydes, depending on the catalyst concentration. Another suitable system is that in which a lower olefin is dimerized in the presence of a nickel acetylacetonate complex promoted by aluminum diethylethoxide.

Other typical rhodium complexes which can be separated from solution in organic components are $Rh(Bu_3P)_2CO$ propionate, $Rh(Et_3P)_2COCl$, $Rh(i\text{-}Pr_3P)_2COCl$ and $Rh(Bu_3P)_2COCl$ where i-Pr is an isopropyl and Bu a n-butyl radical.

The membranes may also be used to separate nickel and vanadium containing complexes, for example, the etioporphyrins, from hydrocarbon solvents such as toluene.

URANIUM-PLUTONIUM

Removal of Radioactive Ions

In a process described by *H.S. Perl, W.F. Rittner and O.M. Glemser; U.S. Patent 3,563,890; February 16, 1971; assigned to Sartorius-Membranfilter GmbH, Germany,* the known effect of a heterogeneous isotopic exchange is used for decontaminating radioactive solutions in that microcrystalline, difficultly soluble salts are produced on a support, which salts are isotopic with the dissolved ions to be exchanged, and the thus impregnated support is charged into a column in which the liquids to be decontaminated flow so that a rapid decontamination is effected.

In a heterogeneous liquid-solid system, the ions of the crystal lattice of a precipitate are in a state of thermodynamic equilibrium with the isotopic ions which are present in a solution (liquid phase) in contact with the precipitate (solid phase). There is a continuous exchange of ions. Ions which were originally in solution occupy sites in the lattice of the solid phase and ions of the lattice are dissolved. The process enables a rapid and complete exchange of isotopes and an ultimate reduction of the radioactive matter to a very small volume. The following significant aspects are considered.

(a) Difficultly soluble salts of various systems can be prepared in microcrystalline form and fixed in a fine three dimensional distribution (impregnation).

(b) Whereas the isotopic exchange has previously been carried out in a batch process, it can be conducted as a column process whereby the dynamic effects tending to establish an equilibrium are so changed that a complete isotopic exchange can be effected within very short time.

(c) A fixation of difficultly soluble crystals on cellulose wadding as an inert support (impregnation) enables a high flow rate in the loose structure. Besides, the cellulose wadding laden with the exchanged radioactive ions can be converted into ash except for a small residue.

(d) The wadding which is impregnated with the difficultly soluble salts can be stored and does not lose its exchange capacity and may be used without need for preceding chemical operations such as precipitation or the like.

(e) The impregnated wadding has a high exchange capacity so that it can be used in a column process for decontaminating large amounts of liquid.

(f) When wadding consisting of cellulose esters is used, such as collodion wool or cellulose acetate, the support which has been laden with radioactive salts can be dissolved in organic solvents, such as methyl acetate. The radioactive salts can then be separated from the carrier by filtration.

The preparation and effectiveness of cellulosic supports impregnated with difficultly soluble salts is described in the following examples.

Example 1: 5 grams cellulcotton are suspended in 300 ml H_2O and continuously stirred. A solution of 0.001 mol $Sr(NO_3)_2$ in 10 ml H_2O is first added with agitation. Then a solution of 0.001 mol Na_2SO_4 in 50 ml H_2O is added in drops. A precipitate is formed at the moment when the drop reaches the surface of the cellulose water suspension, but is immediately dispersed by the intense agitation. The precipitation with equimolar amounts results in a formation of approximately equivalent crystals, which are free of a surplus of cations or anions at their surface. This prevents the formation of a double electrical layer at the crystal solution phase boundary; such double electrical layer could result in undesired secondary effects.

Example 2: The wadding is impregnated as in Example 1 but $YCl_3{\cdot}6H_2O$ and Na_2HPO_4 are used instead of $Sr(NO_3)_2$.

Example 3: The wadding is impregnated as in Example 1 but $Sr(NO_3)_2$ and $YCl_3{\cdot}6H_2O$

are added at the same time and are precipitated simultaneously with a solution of Na_2SO_4 and Na_2HPO_4.

Example 4: The wadding is impregnated as in Example 1 but $AgNO_3$ and an NaI solution are used.

Example 5: Decontamination — 0.5 gram of a cellulose wadding impregnated according to Example 3 was suspended in H_2O and introduced in a column 30 cm long and 2.5 cm in diameter. The cellulose column had a height of 5 cm. A cock and a superimposed frit slab were provided at the lower end of the column. In other experiments, the frit plate in the column was replaced by glass wool so that a high flow rate was obtained.

A ^{90}Sr stock solution was provided which contained 23 mg $Sr(NO_3)_2$/100 ml as a carrier, 0.5 ml was pipetted from said stock solution and filled up to 10 ml. The activity was measured in a liquid counting tube and amounted to 14,678 pulses per 30 seconds as an average of 10 measurements. These 10 ml were diluted to 250 ml and the diluted solution was passed through the prepared column at a flow rate of about 200 ml/min. Various samples of 10 ml each were taken from the effluent and measured. It was found that the activity of the total volume was about 40 pulses per 30 seconds, which is less than 0.3% of the original activity. The column of cellulose was then removed from the tube and divided into 10 discs 0.5 cm thick each. When the activity of each disc was ascertained, it was found that the activity had been retained entirely in the first disc.

Uranyl Nitrate Separation

R. Bloch, D. Vofsi, O. Kedem and A. Katchalsky; U.S. Patent 3,450,631; June 17, 1969; assigned to Yeda Research and Development Company, Israel describe a method for the separation of ions of a preselected species from ions of one or more additional species having the same charge and a similar ionic diameter. The method involves permeating the preselected species through an ion specific membrane comprising a supported polyvinyl chloride polymer film plasticized by an organo phosphorus compound. It is known that specific solvents are being used for metal ion separations in hydrometallurgical extraction processes. Particularly well known is the case of extraction of uranyl nitrate, while in aqueous solution in a mixture with ions, such as iron, aluminum and others, by means of tributyl phosphate.

Example 1: A solubility membrane was prepared by applying by knife-coating a uniform layer of about 30 μ of a mixture of 3.5 parts of tributyl phosphate and 1 part polyvinyl chloride to parchmentized kraft paper of high wet strength of 60 g/m^2. The curing of the applied coating was effected at a temperature of 140°C for 30 seconds, resulting in the solidification of the plastisol. The membrane was clamped between two half cells made of methyl methacrylate, defining two compartments of 5 cm diameter and 0.8 cm width. One of the compartments was filled with an aqueous solution, the concentration of which was 0.02 molar in uranyl nitrate, 1.0 molar in iron nitrate and 1.0 molar in aluminum nitrate. The second compartment was filled with distilled water.

Dialysis for 1 hour resulted in the selective removal of 52% of the uranium, while after 2½ hours the entire quantity of uranium had passed into the second compartment. No iron or aluminum passed through the membrane. The lifetime of the membrane was tested in a cell which allowed continuous replacement of the two solutions contacting the membrane. The membrane was clamped between two rubber devices each covered by a Lucite plate. The membrane was 1 x 100 cm in area. This geometry was chosen in order to prevent a flow pattern with dead areas. A grid of netlon prevented the membrane from attaching at the walls of the cell. With a peristaltic pump equal flows of the uranium mixture and water were pumped through both compartments.

An identical membrane as described above was tested for several days in continuous operation. With prolonged use the permeation rate for uranium decreased, however, without decline in selectivity. After 48 hours of continuous operation at a flow rate of 150 ml/hr, the permeation rate dropped to 20% of its initial value. In the course of this operation 11 grams of uranium permeated the membrane per membrane area containing 1 gram of tributyl phosphate.

Example 2: A solubility membrane was prepared by applying a coating of about 30 μ thickness of a mixture of 3.5 parts triethylhexyl phosphate and 1 part of polyvinyl chloride and curing under identical conditions as in Example 1. This membrane was used in the same cell and with the same solution as in Example 1. After 1 hour 56% of the uranium passed into the second half cell, while after 4 hours the first half cell was free of uranium. No iron or aluminum passed into the second half cell.

Example 3: A plastisol coating, having a thickness of 40 μ and consisting of 3.5 parts dibutyl butyl phosphonate and 1 part polyvinyl chloride, was cured for 60 seconds at 140°C on top of a kraft paper sheet. A membrane was thus obtained which, when tested under the conditions of Example 1 in a one hour permeation experiment, 49% of the uranium, initially present in the uranium mixture, was transported across the membrane, while no iron and no aluminum permeated into the second half cell.

Recirculation Technique

According to a process described by *J.S.H. Wu and B.L. Baker; U.S. Patent 3,718,583; February 27, 1973; assigned to the U.S. Atomic Energy Commission* the concentration of heavy metal cations, such as uranyl or plutonium cations, from ionic solutions separated by an ion selective membrane, where the difference in chemical potentials of the various ion species provides the driving force to effect a transfer across the membrane, is enhanced by separating a portion of the extract solution being concentrated in heavy metal cation product and recirculating that portion along with the initial strip solution into contact with the membrane. The preferred ratio of the recirculating portion to the initial strip solution is between about 1 to 1 and 5 to 1.

A method is provided for suppressing the inherent osmotic diffusion of water across cation selective membranes for the improved concentration and separation of heavy metal cations from aqueous solutions. This is accomplished by removing from the ion selective membrane separation system a portion of the solution being concentrated in heavy metal cations (extract solution) and recirculating or recycling that portion. It has been found that recirculating a portion of the solution that is being concentrated in heavy metal cations (extract) (i.e., returning it to the initial strip solution) tends to suppress the osmotic diffusion of water across the membrane thereby increasing the concentration of the heavy metal cations in the strip solution without increasing the loss of heavy metal cations to the raffinate.

Thus, in a method for the concentration and recovery of heavy metal cations using an ion selective membrane which includes continuously contacting an aqueous feed solution containing the heavy metal cations with one side of the cation selective membrane and countercurrently contacting an aqueous strip solution with the opposite side of the membrane to effect a transfer and concentration of heavy metal cations in the strip solution, separating a portion of the solution that is being concentrated in heavy metal cations and recirculating that portion into contact with the membrane tends to suppress the osmotic diffusion of water across the membrane.

The reduction in osmotic flow across the membrane may be caused by the substantial conversion of the membrane to the heavy metal cation form over its entire length. It is known that the membrane in the hydrogen form swells more than in the uranyl form, thereby providing larger pores that can permit more transfer of water across the membrane with less resistance. It is believed that by providing more heavy metal cations in the strip solution the membrane is maintained in the heavy metal cation state and the effect is to suppress the transfer of water across the membrane.

Suppression of the inherent osmosis of the membrane using this method results in a significantly enhanced concentration of heavy metal cations without an increased loss of these same desired cations to the raffinate solution. The method has been found to be particularly useful for the separation of heavy metal cations, such as actinide and lanthanide cations.

At least three significant advantages result from this recirculation: (1) Better mixing of the solutions in the stripper solution, (2) avoidance of high acid concentration differences between the raffinate and the stripper stream, and (3) maintenance of the entire membrane length in the most suitable cation form for efficient concentration.

Oxalic Acid Cleaning Technique

W.S. Higley; U.S. Patent 3,700,591; October 24, 1972; assigned to the U.S. Secretary of the Interior describes a method for cleaning and rejuvenating a used reverse osmosis membrane by treating the membrane in situ with a solution of oxalic acid. Not only are deleterious deposits removed from the membrane, but the original osmotic properties of the membrane may be partially restored. The membrane properties may be further improved by a subsequent treatment with a swelling agent or a hot water solution.

Example 1: As an indication of the relative efficacy of oxalic acid solutions, tests were conducted with used cellulose acetate membranes which were in operation with brackish water feed for 13 months resulting in the accumulation of a heavy brownish deposit in the membranes. After immersion of this membrane in 1% oxalic acid solution for 1 hour, almost all of the deposit had been removed.

Similarly, with a 1% solution of a solution of approximately 60% sodium bisulfite and 40% sodium hydrosulfite (RoVer, Hach Chemical) the removal action was somewhat slower, but was complete within 24 hours. A solution of 1% citric acid was also tried, but a 24 hour cleaning period and a water wash were required in order to remove all visible deposits. Solutions of 1% ethylenediaminetetracetic acid, tetrasodium salt (EDTA) and 1% hydrogen peroxide were also tried but demonstrated no visible removal of deposit from the membranes after a 24 hour treatment period.

The oxalic acid solution, therefore, was faster in removing these deposits than solution of the same concentration of the other reagents.

Example 2: To determine the ability of oxalic acid to remove iron deposits from a reverse osmosis membrane, the following experiment was conducted. A standard cellulose acetate reverse osmosis membrane was utilized at 800 psi on a 5,000 ppm sodium chloride feed solution. The initial properties of the membrane comprised a salt rejection of 88.5% and a water flux of 24.2 gfd (gallons/foot^2/day). Ferrous sulfate was added to the feed solution at a concentration of 50 ppm and desalination was continued with air bubbling through the feed solution to oxidize the ferrous sulfate to the hydrous oxide. After about 1 hour the flux had declined to 21.4 gfd, although the salt rejection remained constant. After 21 more hours of operation the flux had further decreased to 20.9 gfd, but the salt rejection continued to remain the same. The membrane, which visually showed a moderate deposit, was analyzed and found to contain 0.072 weight percent iron.

Three-inch circles were cut from the membranes and treated by immersing four of them in each of two different solutions for 24 hours. The cleaning solutions used were 1% citric acid, and 1% oxalic acid.

The membranes treated with oxalic acid solution showed a dramatic visual removal of deposit after only 1 hour of treatment. The other membranes treated in citric acid showed visible deposits even at the end of the treatment, although some of this remaining deposit could be removed with a water rinse.

After all the membrane samples were water rinsed and dried, an analysis was made and the efficacy of the cleaning solutions determined. In 1% citric acid 82% of a 1% iron deposit was removed, and in 1% oxalic acid 93% was removed.

Example 3: This experiment was performed to demonstrate the capability of the oxalic acid cleaning technique for improving used membranes under practical in situ regeneration conditions. The used membranes were evaluated in a test cell at 800 psi and 5,000 ppm sodium chloride feed solution to determine their individual osmotic properties.

After this initial evaluation the membranes were treated with an aqueous 0.5% oxalic acid solution at near atmospheric pressure by allowing the solution to flow serially over the membranes in the unit from a three foot high reservoir. The effluent from this treatment was reddish brown in color. The treatment, which lasted about 3 hours, was controlled by means of a valve in the exit line from the third cell. After this cleaning treatment the membranes were washed with water using the same reservoir. The membranes were again tested at 800 psi and 5,000 ppm sodium chloride feed solution. The results given in the following table show an increase in both flux and salt retention in each case, which can be attributed primarily to the removal of the deposit and the consequent improved boundary layer conditions.

To reswell the membrane and increase the salt rejection, the membranes were subject to a mild hot water treatment. This heat treatment was performed by recycling hot water through the cells in series from a thermostatically controlled hot water bath. The water was circulated at a sufficiently fast rate to avoid significant heat losses from noninsulated portions of the apparatus. The temperature of the hot water was maintained at about 48° to 52°C. The membranes were then tested at the conditions previously used and the results are reported in the table below.

Used Membrane		Cleaned Membrane		Heat-Treated Membrane	
Flux, gfd	Salt Rejection, %	Flux, gfd	Salt Rejection, %	Flux, gfd	Salt Rejection, %
21.5	58.2	25.1	70.5	23.5	76.9
14.3	74.1	16.5	76.4	16.4	79.3
19.2	65.5	24.6	69.5	23.3	77.0

As can be seen from the above data, the final heat treatment did cause a slight loss in the flux of the membranes. However, the salt rejections were improved.

The cumulative effect of the cleaning and heat-treatment did cause a slight loss in the flux of the membranes and salt rejection improvement ranging from 7 to 32%.

OTHER PROCESSES

Separation of Homogeneous Catalysts from Nitrile Solutions

L.W. Gosser; U.S. Patent 3,853,754; December 10, 1974; assigned to E.I. du Pont de Nemours and Company describes a process for the permselective separation of an organic solution containing (a) a metal-organo complex containing a metal selected from the group consisting of Ni, Pd, Pt, Co, Rh, Ru and B, and (b) at least one organic hydrocarbon derivative having at least one nitrile group and up to 12 carbon atoms. The process consists essentially in contacting one surface of an asymmetric permselective membrane with the organic solution under pressure and recovering from the other side of the membrane fluid which has passed through the membrane, the membrane consisting essentially of an aromatic polymer having recurring —C(O)N= groups in the polymer chain.

Exemplary nitriles which may be mentioned are acetonitrile, propionitrile, β-chloropropionitrile, adiponitrile, dodecanedinitrile, acrylonitrile, 3-pentenenitrile, 4-pentenenitrile, 2-methylbutenenitrile, the isomeric dicyanobutenes, mucononitrile and benzonitrile.

Figure 7.2a is a representation of a permeator apparatus **10** which may be used in the process. Base section **11** and upper section **12** are machined from blocks of corrosion resistant metal. Film **13**, the reverse osmosis membrane, is a disc mounted on a layer of filter paper **14** against a stainless steel wire screen or mesh **15**. When upper section **12** of the cell is bolted to lower section **11**, synthetic elastomer O-rings **16** seat firmly around the periphery of the membrane and against the metal. Inlet **17** for feeding fluid into the cell is near the membrane. The fluid is agitated by a magnetically driven stirrer blade **18**, positioned by support **19** and controlled by external and internal magnets **20** and **21** to ensure contact of fresh fluid with the membrane surface at all times. Bypass of a portion of the feed fluid is provided through exit **22**. Fluid passing through membrane **13** passes through metal frit **23** and into exit pipe **24** and collected. The rejected solution is collected separately or recycled.

Figure 7.2b shows a pumping system for providing circulation of feed fluid and maintenance of pressure inside the permeator. Fluid is circulated from reservoir **30** by pump **31** through permeator **32**, which can be the permeator of Figure 7.2a, through pressure regulator **33**, flow meter **34**, and back to reservoir **30**. Regulator **35** and flow meter **36** permit excess fluid from pump **31** to bypass permeator **32** and return to the reservoir. Pressure is monitored by gauge **37**. Conventional piping is, of course, supplied to connect the units of the control system as indicated.

FIGURE 7.2: SEPARATION OF HOMOGENEOUS CATALYSTS FROM NITRILE SOLUTIONS

a.

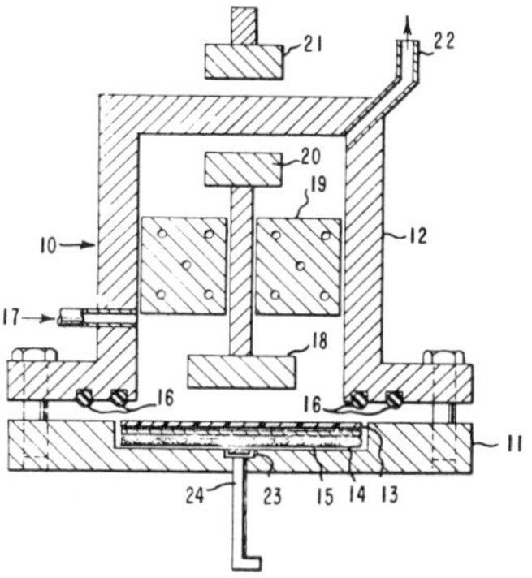

Permeator Apparatus

b.

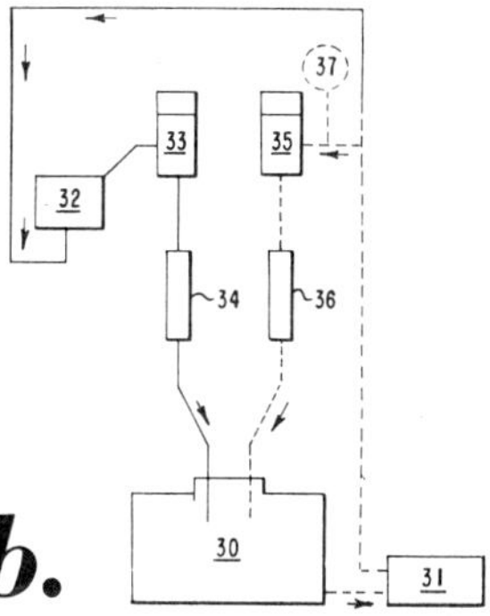

Pumping System

c.

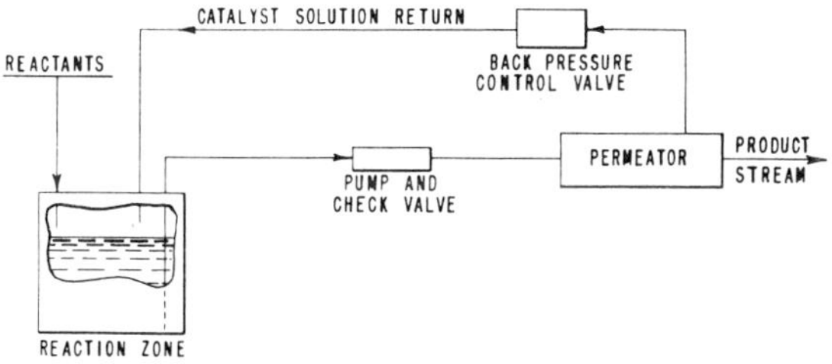

Flow Sheet for Continuous Processing

Source: L.W. Gosser; U.S. Patent 3,853,754; December 10, 1974

The process can also be adapted to a continuous liquid catalytic process as shown in the flow sheet of Figure 7.2c. Here a portion of the reaction mixture is continuously pumped to the permeator where product nitriles, some of the solvent and other solutes are removed and the remaining concentrated catalyst solution is returned to the reaction zone along with fresh quantities of the reactants.

When the amounts of feed liquid available are limited, the process may be carried out by mounting the permeator of Figure 7.2a with the membrane at the bottom, placing the feed liquid on the membrane and pressuring the space above it with an inert gas such as nitrogen. This can be done by closing off passageway **17** after the feed fluid has been admitted, and then supplying the gas from a source through an appropriate valve (not shown) at passageway **22**. Further, the feed liquid may be pressurized by a reactant gas such as hydrogen or ethylene, for example. Where suspended matter is present, it is desirable to first clarify the feed liquid by filtering to avoid plugging of the membrane.

In one aspect the process is an improvement in the method of preparing organic nitriles having up to 12 carbon atoms comprising the step of permeating a nitrile solution through a membrane to reduce the concentration of soluble metal organic catalysts. Among the species of nitrile preparations are: (a) the hydrocyanation of olefins and unsaturated nitriles, (b) the dimerization and hydrodimerization of acrylonitrile, (c) the isomerization of pentenenitriles, (d) the hydrogenation of unsaturated nitriles such as dicyanobutene to form adiponitrile, and (e) the coupling of ethylene and acrylonitrile.

The conditions of temperature and pressure may be selected in keeping with the particular species of the nitrile process. In general the rate of permeation (gfd — gallons per sq. ft. of membrane surface per 24 hours) increases with an increase in temperature or pressure. Both the aromatic character and the —C(O)N= linkages in the polymer chain of the applicable membranes contribute to their utility over a broad range of temperature and pressure. Thus, temperature of 110°C and higher, and pressures as high as 2,500 psi may be employed.

Example 1: Part A — An asymmetric polyimide membrane was prepared starting from a 20% solids solution of the polyamic acid from pyromellitic dianhydride and 4,4'-diaminodiphenyl ether in dimethylacetamide, prepared by the method of W.R. Sorenson and T.W. Campbell, *Preparative Methods of Polymer Chemistry,* 2nd Ed., Interscience Publishers, 1968, pp. 170-1. This solution was diluted to 10% polyamic acid by addition of anhydrous dimethylacetamide, and 10% (by weight of polyamic acid) of cetylpyridinium bromide was dissolved therein. A 15-mil film of this solution was cast on a glass plate under a doctor knife. The assembly was immersed in a bath containing 1 molar acetic anhydride and 1 molar triethylamine dissolved in benzene. The bath was then heated at about 60° to 75°C for 15 minutes.

The assembly was then washed successively at room temperature in benzene, ethanol containing 2% benzene, and water. The membrane was separated from the glass plate in the water bath and was air-dried. It was yellow, opaque, shiny on the side away from the glass, dull on the side against the glass. In cross section the membrane was asymmetric, having a thin, dense "skin" on the shiny side.

Part B — In the apparatus of Figure 7.2a a membrane as described above was mounted in the permeator with the skin side toward the feed solution. The feed solution contained 0.5 gram of $[(C_6H_5)_3P]_3RhCl$ in 230 ml of adiponitrile. Reverse osmosis was conducted at 450 psi pressure for 70 hours to obtain a 2.7-ml sample of permeate. Colorimetric analysis showed an 85% rejection of the complex rhodium catalyst by the membrane, i.e., the catalyst concentration in the permeate was only 15% that in the feed solution.

Example 2: In the permeator of Figure 7.2a an asymmetric polyimide membrane, prepared as in Example 1, was mounted in the permeator with the skin side toward the feed solution. The feed solution contained 5 ml of a 49.5% solution of L_4Ni in L, where L is the mixed cresyl phosphite $(CH_3C_6H_4{-}O)_3P$, 250 ml of adiponitrile and 50 ml of toluene.

The feed solution was filtered under nitrogen prior to injection into the nitrogen-swept reservoir. Reverse osmosis was carried out at 55 psi to produce a flow rate of 0.99 gfd through the membrane. X-ray fluorescence analysis showed an 84% rejection of the nickel catalyst.

Example 3: An asymmetric polyimide membrane was prepared using the method of Example 1, Part A, except that the cetylpyridinium bromide was replaced by an equal weight of silver trifluoroacetate, $AgOOCCF_3$ and the final wash was methylene chloride instead of water. This membrane was mounted in the permeator of Figure 7.2a with the skin side toward the feed solution. The feed solution contained 5 ml of the L_4Ni in L solution of Example 2, 250 ml of 3-pentenenitrile and 50 ml of toluene. Reverse osmosis was carried out at 20 psi to produce a flow rate of 4.5 gfd through the membrane. X-ray fluorescence analysis showed an 81% rejection of the nickel catalyst by the membrane.

Example 4: The procedure of Example 3 was repeated except that a pressure of 480 psi was used, giving a flow rate of 87 gfd through the membrane. X-ray fluorescence analysis showed a 93% rejection of the nickel catalyst by the membrane.

Gold, Silver and Copper Cyanide Solutions

L.W. Gosser; U.S. Patent 3,816,587; June 11, 1974; assigned to E.I. du Pont de Nemours and Company describes a reverse osmosis process for concentrating gold, silver and copper in aqueous cyanide solutions employing membranes of nitrogen-containing aromatic condensation polymers.

The process is applicable to any operation involving gold, silver and copper cyanide complexes in aqueous solutions wherein the complexes are sought to be concentrated. The process is particularly applicable to a leaching operation comprising obtaining aqueous solutions of the soluble complex cyanides of gold, silver and copper (from, say, gold-bearing ores) by employing aqueous alkali and/or aqueous alkaline earth metal cyanides as leachants. The leach liquid in such an operation contains the gold, silver and copper cyanides as well as one or more alkali or alkaline earth cyanides.

A semi-permeable membrane is employed in concentrating the cyanides. Particular nitrogen-linked aromatic condensation polymers employed as the semi-permeable membrane(s) allow for the reverse osmosis concentration by a process comprising contacting one surface of said membrane with the leach liquid at an applied pressure greater than the osmotic pressure, and concentrating the gold, silver and copper complex cyanides on the side of the membrane first contacted by the leach liquid some of which flows through said membrane.

This reverse osmosis process means that the metal ions gold, silver and copper are retained by the membrane while the solvent passes through. This process is basically different from dialysis processes employing dialysis-type membranes and passing the metal ions through the membranes.

The particular cyanide complexes that are concentrated by this process are one or more of the following: dicyanoaurate(I), $Au(CN)_2^-$; dicyanoargentate(I), $Ag(CN)_2^-$; dicyanocuprate(I), $Cu(CN)_2^-$; tricyanocuprate(I), $Cu(CN)_3^{=}$; tetracyanocuprate(I), $Cu(CN)_4^{\equiv}$. In leach solutions obtained by leaching with aqueous sodium cyanide, potassium cyanide or calcium cyanide, the gold, silver and copper complex ions are balanced in solution with corresponding sodium, potassium, or calcium ions.

Particular aromatic polymers found to be useful herein are (1) aromatic polyamides, (2) aromatic polyacyl hydrazides, (3) aromatic polysemicarbazides and (4) aromatic polyureas. The following examples illustrate the process. In the examples, parts are by weight unless otherwise specified. The abbreviation gfd is used to designate the rate of permeate flow and represents gallons per square foot per day.

Example 1: Part A — A polymer was prepared by condensation of a mixture of 80 parts of 3-aminobenzhydrazide and 20 parts of 4-aminobenzhydrazide and a stoichiometrically equivalent amount of a mixture of 70 parts of isophthaloyl chloride and 30 parts of terephthaloyl chloride in dimethylacetamide solution. This polymer was isolated by pouring the condensation mass into ice water, washing free of acid, and drying.

Part B — The polyamide hydrazide of Part A was formed into an asymmetric membrane by the following procedure. A casting solution was prepared with the following composition by weight:

	Percent
Polymer as in Example 1A	15.5
$LiNO_3$	4.65
2,11-Diacetyl-2,11-diaza-5,8-dioxadodecane	6.2
Triethanolamine	0.08
Dimethylacetamide	73.57

The solution was filtered through a 0.45 μ silver filter. The resulting solution had a viscosity of 20,000 cs at 25°C. A seamless tubular membrane was formed by depositing the polymer solution at ambient temperature inside a 1.003" i.d. vertical glass column with the aid of a conical spreader. The latter was moved through the glass column with air pressure at a velocity which avoided entrapment of air bubbles inside the solution layer. Wet film thickness was 0.015 inch.

The glass tube with the polymer film on the inside was placed in a vertical heater. Air heated to 80°C was passed through the tube at 800 fpm for 25 minutes. Electric heaters were set to give a steady state tube wall temperature of 100°C. The glass tube was then removed from the heater and allowed to cool for 5 minutes while a stream of ambient air flowed through the tube at 800 fpm. The salt, the plasticizer and residual dimethylacetamide were removed by extraction with methanol. The extract stream was circulated through the glass tube at 6.9 fpm for 60 minutes. The tubular membrane was then slipped out of the glass tube into water and stored wet.

Part C — A section of the wet tubular membrane from Part B was slit open and mounted flat in an apparatus such as has been described with the skin side of the film on the feed side of the cell (permeator). A feed solution was prepared by dissolving 0.50 gram of KCN, 0.30 gram of $K_3Cu(CN)_4$, 0.30 gram of AgCN and 0.10 gram of KAu(CN), in 250 ml of distilled water.

This was placed in the permeator under a pressure of 400 psi. Under these conditions, permeate passed through the membrane at a rate of 7 gfd. When the volume of the feed solution had decreased about 10%, samples of the feed and the permeate solutions were analyzed by atomic absorption. Concentrations of the metals in parts per million (ppm), along with the percent passage calculated therefrom are shown in the following table.

Example 2: The procedure of Example 1 was repeated except that in the permeation step a different pressure and therefore a different flow rate was employed as set out in the table along with the corresponding analyses and percent passage calculations.

Example 3: Part A — A dimethylacetamide solution was prepared containing 15% by weight of the polyamide hydrazide of Example 1, Part A, 4.5% of $LiNO_3$ and 6% of 2,11-diacetyl-2,11-diaza-5,8-dioxadodecane. A 0.015-inch film of this solution was spread on a glass plate heated at 100°C. After standing in air for 5 minutes at 100°C, the coated plate was immersed in water at 5°C. After thorough extraction with water, the film was asymmetric, i.e., it had a thin skin of dense polyamide hydrazide supported by a less dense form of the same polymer. The dimethylacetamide, $LiNO_3$ and 2,11-diacetyl-2,11-diaza-5,8-dioxadodecane were substantially all leached out of the film.

Part B — The film from Part A was used in a permeation procedure like that of Example 1,

Part C except that a different pressure was used, resulting in a different flow rate and with the analyses and percent passage calculations as shown in the table below.

Examples 4 and 5: The procedure of Example 3 was repeated except that in the permeation step different pressures and therefore different flow rates were employed as set out in the table along with the corresponding analyses and percent passage calculations.

Example 6: Part A — An aromatic polyamide copolymer was prepared as follows. A resin kettle was swept with dry nitrogen and was charged with 32 mols of N,N-dimethylacetamide and 2.36 mols of meta-phenylenediamine. The solution was cooled to a temperature of 0° to -10°C at the start and the solution was maintained below 20°C while the bulk of a molecularly equivalent amount of molten 70% isophthaloyl chloride-30% terephthaloyl chloride mixture was added in about 0.5-mol increments at 5-minute intervals with agitation. The size of the increments was decreased as the reaction progressed. Finally cooling was stopped and the temperature of the solution was allowed to rise to 40° to 50°C. Completeness of reaction was checked by spot testing with p-dimethylaminobenzaldehyde (an aromatic amine end group indicator) in dimethylacetamide until the intense yellow color, which indicates the presence of unreacted amine groups, no longer appeared.

The polymer was isolated by diluting it with dimethylacetamide to a polymer content of about 9% and placing the solution in a high speed Waring Blendor. Crushed ice was added slowly until precipitation began, after which the mixture was stirred rapidly and additional ice was added. The precipitated polymer was filtered and washed with water. Drying was accomplished in a vacuum oven at 80°C until the water content was below 4%.

Part B — A dimethylacetamide solution containing 15% by weight of the polyamide of Part A and 5.25% of LiCl was cast as a 0.015-inch film on a glass plate heated at 100°C. After standing in air for 5 minutes at 100°C, the coated plate was immersed in water at 75°C. After thorough extraction with water, the resulting asymmetric film was separated from the plate and stored in water until used.

Part C — A section of film from Part B was mounted in the apparatus of Example 1 with the skin side of the film on the feed side of the permeator. A feed solution as shown in Example 1, Part C, was placed in the permeator under a pressure of 750 psi. Under these conditions, permeate passed through the membrane at a rate of 2.7 gfd. When the volume of feed solution had decreased about 10%, samples of the feed and permeate solutions were analyzed by atomic absorption. Concentrations of the metals in parts per million along with the percent passage calculated therefrom are shown in the table below.

Example		K	Cu	Ag	Au	Flow Rate (gfd)	Pressure (psi)
1	Feed, ppm	1,750	430	365	250	7	400
	Permeate, ppm	133	11	6	29		
	Percent passage	7.6	2.6	1.6	11		
2	Feed, ppm	1,500	440	310	265	20	1,050
	Permeate, ppm	218	1.5	1	13.6		
	Percent passage	15	0.3	0.3	5		
3	Feed, ppm	1,550	560	334	252	10	750
	Permeate, ppm	153	3	4	7		
	Percent passage	10	0.5	1.2	3		
4	Feed, ppm	1,675	378	1,500	415	23	1,230
	Permeate, ppm	75	2	4.6	12.5		
	Percent passage	5	0.5	0.3	3		
5	Feed, ppm	1,920	322	1,560	400	24	1,250
	Permeate, ppm	61	2	4.6	7		
	Percent passage	3	0.6	0.3	2		
6	Feed, ppm	1,690	520	480	240	2.7	750
	Permeate, ppm	226	28	2	43		
	Percent passage	13	5	0.4	18		

*The summary data in this table reflect the actual metal content of the feed solutions at the time of testing.

Metal Ion Separation Using Vinyl Membranes

R. Bloch, A. Katchalsky, O. Kedem and D. Vofsi; U.S. Patent 3,450,630; June 17, 1969; assigned to Yeda Research and Development Co., Ltd., Israel describe specific membranes which contain uncharged functional groups which are adapted to selectively bind one or more predetermined anions or cations out of a number of such anions or cations, from which these are to be separated.

The separation of ionic species by means of solvent extraction is due to a specific interaction between the ions and functional groups of the selective solvent, e.g., ketone groups, ester groups, ether groups, hydroxyl groups, etc. It is essential that such groups be part of the process membrane. These groups may be chemically bound to the structure of a polymeric membrane or a selective barrier may be constructed, consisting of an inert matrix where such solvents are held by forces such as capillary action, adsorption or gel formation.

As an example of functional groups chemically bound to a membrane structure there may be mentioned such membranes as those formed from polymers like polyvinyl chloroethyl ether, polyepichlororhydrin, polymethyl vinyl ketone, etc., containing as active groups etheric oxygen groups in the polymer side chains, etheric oxygen groups in the polymer main chain and carbonyl groups in the polymeric side chains, respectively.

As examples of the selective barriers containing the active functional groups as parts of solvents held within inactive matrices there may be mentioned such combinations as for instance selective solvents soaked into porous polyvinyl chloride sheet plasticized with tributyl phosphate; yet another example of such a selective barrier would be a cross-linked rubber swollen with a suitable solvent, forming a gel. The following Examples 1 through 3 illustrate the process of separation of aluminum from iron.

Example 1: A sheet of porous polyvinyl chloride, 1 mm thick, was soaked in β,β'-dichlorodiethyl ether and then clamped between two conventional half cells of polymethylmethacrylate of 5 cm diameter and 0.6 cm width. Platinum wires were used as electrodes. Initially the cathodic compartment contained 10 ml of a 1.5 N HCl solution in which there was dissolved 0.2 gram Fe as $FeCl_3$ and 0.4 gram Al as $AlCl_3$. The anodic compartment contained 10 ml 1.5 N HCl. A voltage of 20 v was applied across the cell. The initial current was 8 ma/cm^2, which fell during the separation to 0.4 ma/cm^2. After the passage of 763 coulombs through the cell, the cathodic compartment contained the total initial quantity of aluminum and only 0.2 mg of iron.

Example 2: A separation was effected as in Example 1, with identical solutions. The membrane was porous polyvinyl chloride (1 mm thick) soaked in isobutyl methyl ketone. The current density was 7.5 ma/cm^2. The passage of 157.5 coulombs resulted in the transport of 6.75×10^{-4} mol of iron while no aluminum was passed through the membrane.

Example 3: A separation was carried out as in Example 1, but with a 1 mm porous polyvinyl chloride sheet soaked in tributyl phosphate. The current density was 10 ma/cm^2 and a passage of 167 coulombs resulted in a transport of 8.4×10^{-4} mols of iron, while no aluminum was passed through the membrane.

MEDICAL AND OTHER APPLICATIONS

HEMODIALYSIS MEMBRANES

Polyvinylpyrrolidone-Cellulose Acetate

H.F. Shuey; U.S. Patent 3,847,822; November 12, 1974; assigned to the U.S. Secretary of Health, Education and Welfare describes hemodialysis membranes formed of polyvinylpyrrolidone-cellulose acetate blends. The membranes were found to have inulin permeation values while almost completely reflecting albumin.

The casting solution is made by dissolving the blend of polyvinylpyrrolidone and cellulose acetate in sufficient solvent system to reduce the viscosity of the casting solution so that it can be easily handled. Generally, this will fall into a range of about 200 to about 500 parts of solvent per 100 parts of polymer blend. The membranes are prepared by casting a film of the solution on glass at ambient temperature. The film is cast to a thickness of about 8 mils and gelled in ice water. A short drying time, up to about 60 seconds, could be used, but in the preferred method the film is immediately gelled without any drying time.

The film cast on the glass is allowed to set in the ice water for a time sufficient to allow the same to soak loose from the glass plate. This time could range from about 10 to 20 minutes. The membrane is then tested by comparison with Cuprophan. An asymmetric membrane was prepared from a polyvinylpyrrolidone-cellulose acetate blend according to the foregoing procedure using a casting solution formulation containing polyvinylpyrrolidone 360 (Mann Research Laboratory), cellulose acetate E-400-25 (Eastman Kodak), acetone and glacial acetic acid in 2, 8, 40 and 5 parts by weight, respectively.

Films of 8 mils thickness were cast on a flat glass plate at room temperature (25°C) and immediately gelled in ice water. A 2.76 mil thick membrane prepared as described was evaluated under conditions of ultrafiltration at 25°C and 14.7 psig using a feed solution consisting of 150 ppm inulin (molecular weight of about 5,000) and 1,000 ppm human albumin (molecular weight of about 67,000) in buffered saline solution. The buffered saline solution contained 8,500 ppm sodium chloride in 0.01 M phosphate buffer of pH 7.2. The hydraulic permeability and permeation were determined and compared with typical values for Cuprophan. As can be seen by the following table, compared to Cuprophan, the membranes according to the process exhibited an inulin permeation value of from about 5.5 to 7.5 times higher and an ultrafiltration rate of about 4 to 7.5 times that of Cuprophan while being equally effective in reflecting albumin.

	Cuprophan	Polyvinyl-pyrrolidone-cellulose acetate
Thickness, mils	0.9-1.0	2.76
Hydraulic permeability, ml./min.-m.²-mm. Hg (×10²)	2-4	14.9
Inulin permeation, percent	11-15	83
Albumin permeation, percent	0-0.6	0.2
Burst pressure, p.s.i.g	7-9	9.6

Polyethylene Imine Grafted onto Nylon and Cellulose Acetate

R.P. Crowley; U.S. Patent 3,857,782; December 31, 1974; assigned to Abcor, Inc. describes a permeation process particularly useful for separating low molecular weight solids, such as urea and salts as in osmotic and dialytic separation processes, which employs a semipermeable membrane material characterized by having polyaminoethylene grafts. Aminoethylated polymeric materials, such as cellulose and polyamides or polyethylene imine are employed in capillary fiber form in permeation processes and devices which materials exhibit improved permeation rates, particularly for low molecular weight solids in aqueous solutions.

Example 1: A polyethylene imine polymeric material useful as a semipermeable membrane in having an improved permeation rate for separating low molecular weight solids such as urea and salts from a water solution, for example, as in an artificial kidney is prepared by: reacting a 20-gram sample of a commercial nylon 6–6 ground to pass a 20-mesh sieve with an excess of about 100 ml of ethylene imine in a Parr bomb, the reaction mixture then heated to about 60° to 80°C for 40 hours. After cooling the excess ethylene imine is allowed to evaporate and the remaining material is washed several times with water in a Waring Blendor and then dried.

The polymeric product produced is granular and somewhat rubbery and contains from about 40 to 60% by weight of the combined ethylene imine. The polyethylene imine graft nylon so prepared may then be formed into film form of approximately 2 to 4 mils in thickness for testing its permeation properties. The permeability of the thin film formed from such polyethylene imine graft nylon when tested against an unreacted 6–6 in similar film form exhibits a higher water permeation rate when the films are exposed to a sodium chloride and urea solution.

Example 2: A polyethylene imine graft polymer of cellulose acetate which contains approximately 10 to 20% by weight of the combined ethylene imine is prepared by reacting 10 grams of cellulose acetate having an acetyl content from about 5 to 30% with about 50 ml of ethylene imine for 7 to 20 hours. The resultant polymeric material exhibits improved water permeation rates as compared to unreacted material, thereby permitting the separation of salts and urea from a water solution containing the salt and urea. Cellulose and cellulose acetate containing the polyethylene imine grafts, which polymers are placed in capillary fiber form and in membrane devices similar to that set forth in U.S. Patent 3,228,876 are particularly effective in use for artificial kidneys.

Cuprammonium Regenerated Cellulose

M. Isuge, S. Uematsu, Y. Ono, S. Eiga and M. Makita; U.S. Patent 3,888,771; June 10, 1975; assigned to Asahi Kasei Kogyo KK, Japan describe an improved dialyzing hollow fiber of cuprammonium regenerated cellulose, showing a substantially similar network structure of fine gap passage routes having a maximum gap size of 200 A when seen at any cross-section and longitudinal section, having skinless structure at its inside and outside surface of the hollow fiber and representing a superior water permeability.

When the hollow fiber is manufactured from cuprammonium regenerated cellulose, it has been found that the overall structure of the fiber represents a several times increased dialyzing performance in comparison with that of the prior art representative hollow fibers, thanks to its even and well-formed network of gap passage routes.

The hollow fiber represents no minor deposits on the inside wall surface which is highly smooth and shows no fine undulations, thereby providing no problem of safety, thus providing an ideal hollow fiber highly adapted for dialytic purpose.

Example: A conventionally prepared cuprammonium cellulose spinning solution, cellulose content 10.0 weight percent, ammonium content 7.0 weight percent and copper content 3.6 weight percent, viscosity of 2,000 poises, was extruded through an outer ring-shaped orifice, 5 mm in diameter at a delivery rate of 20 ml/min. At the same time, perchloroethylene was extruded through a central core orifice, 1 mm in diameter, of the same nozzle at a delivery rate of 5 ml/min, so as to fill the core space of the ring-shaped liquid string.

This sheath and core composite string was let to fall under gravity by a vertical length of 300 mm in open air atmosphere, the outside diameter of the composite liquid string being thereby reduced to 600 μ under the influence of the gravity stretch. The liquid string was then introduced vertically into a coagulation bath comprising an aqueous NaOH solution of 11 weight percent concentration kept at 25°C and extending horizontally for a length of 8 m. The traveling speed of the filament through the coagulation bath was set to 100 m/min. The degree of the normannrization amounted to about 30%.

The thus coagulated hollow fiber of filament was taken out on a revolving hank frame, not shown, and kept thereon for about 2 hours. The hank was suspended from a bar and sufficiently rinsed with fresh water by means of a shower. Then, the hank was treated in a diluted sulfuric acid bath, of 3 weight percent concentration, for the regeneration purpose, and rinsed with fresh water. The thus treated hank was suspended from a moving frame and passed through a tunnel drier for drying in a hot air atmosphere, 130°C. The hank was finally and mechanically cut into proper lengths for recovery of the contained core liquid solvent and kept in a calm room at normal temperature for several hours.

The thus obtained hollow fibers represented an outer diameter of 290 μ with a wall thickness of 25 μ which was found substantially precisely constant at any cross-section along the axis of each of the hollow fibers. These hollow fibers were swollen with water and observed under an electron microscope, 20,000X, showing a network of even gap passage routes structure, being substantially similar at the cross-section as well as the longitudinal section of the fiber, representing gaps of maximum 200 A. At each cross-section no skin formation was observed. Observation of the sections of the hollow fiber at 8,000X showed almost no undulations on the inside and outside wall surface and no deposits of hemicellulose particles on the inside wall surface. The fine gap network was of highly uniform distribution throughout the whole mass of the hollow fiber, when seen in the cross-section, as well as longitudinal section. Comparative test data of the hollow fiber and a representative dialyzing conventional one, made of cellulose triacetate, are given in the following table, concerning their structure, tensile strength and tensile elongation.

Item	Unit	Conv.	Process
O. D.	μ	285	290
Wall Thickness	μ	25	25
Tensile Str. (wet)	g	23.0	103.2
Elongation (wet)	%	27.3	32.7
Tensile Str. (dry)	g	92.0	277.4
Elongation (dry)	%	14.4	24.3

As is clear from the foregoing data, the hollow fiber according to this process represents about three times higher tensile strength than that of the conventional one. In the following table, comparative test data in model liquid dialysis are shown, especially of the removal rate of albumin and urea and those of ultrafiltration velocity. As the model liquid use was made of an aqueous solution containing 0.1 weight percent of albumin and 0.167 N of urea in these tests. From the data shown in the following table, the process hollow fiber has superior performance for removal of protein (albumin) and for allowing the passage of urea, comparative to tests of conventional comparative hollow fibers.

It should be stressed, however, that the hollow fiber has a 2.5 times or still higher ultrafiltration velocity than those of conventional comparative fibrous products. In addition, it may be seen that the fiber represents a substantially superior water permeability over conventional fibers. According to test results, it was found that there are no pin holes on 50,000 hollow fibers according to this process, each being of 135 mm.

	Item	Unit	Conventional	Process
Main Sizes of Sample	O. D.	μ	285	290
	Wall Thickness	μ	25	25
Dialyzing Conditions	Effective Filtering Area	m^2	1.0	1.0
	Length of Hollow Fiber	mm	135	135
	Blood Passage Rate	ml/min	200	200
	Buffer Passage Rate	ml/min	500	500
	Press. Differential	mm Hg	−200	−200
Dialyzing Performance	Removal of Urea	%	63.5	64.2
	Removal of Albumin	%	0	0
Ultra-Filtration	Treating Condition (Pressure)	mm Hg	+400	+400
	Filtration Rate	ml/m²·Hr	430	1020

Acrylonitrile Dimethylaminoethyl Methacrylate Copolymers

According to a process described by *W.M. Muir; U.S. Patent 3,616,930; November 2, 1971; assigned to National Research Development Corporation, England,* membranes for extracorporeal hemodialysis are prepared from copolymers of acrylonitrile (80 to 90%) and dimethylaminoethyl methacrylate (20 to 10%) by portionwise addition of reactants to a reactor. These have superior transfer properties for urea, creatinine, uric acid, and other blood components as compared with cellulosic membranes, and are prepared as blood envelopes for use in Ross-Muir and other dialyzers.

Acrylonitrile-Vinyl Acetate and Acrylonitrile-Methylvinylpyridine Copolymers

According to a process described by *I.O. Salyer, J.S. Tapp, W.E. Weesner and R.L. Leonard; U.S. Patent 3,799,356; March 26, 1974; assigned to the U.S. Secretary of Health, Education and Welfare,* nonthrombogenic hemodialysis hollow fiber membranes useful in artificial kidneys are prepared from quaternized hollow fiber membranes of polymers made from monomeric substances of which at least a sufficient amount of a vinylpyridine is included to provide membranes which on heparinization are nonthrombogenic, particularly polymers made from monomeric substances of which from 2 to 10% by weight of the total monomer is a vinylpyridine and at least 80% by weight is acrylonitrile. Both copolymers and blends of polymers can be used.

A particularly suitable copolymer blend for use in making the hollow fiber membranes is a blend of a 95% acrylonitrile and 5% vinyl acetate copolymer with a 50% acrylonitrile and 50% methylvinylpyridine copolymer in a ratio of 88:12, respectively, of the copolymers in the blend.

Example: The hollow fiber membranes of this process are made according to the following parameters.

Dope: 25% polymer blend in dimethyl acetamide + 0.0633 g benzyl chloride/g of polymer. Temperature 120°C. Rate 1.2 cc/min.

Spinerette: Hollofil 22-12-8 (OD_0-OD_t-ID_t) mils. Temperature 70°C. Inject air at 2.0 cm pressure of Carbowax-400.

Air Gap: Spin through 4 inch unheated chimney attached to spinnerette and touching coagulation bath (to control solvent evaporation).
Coagulation Bath: Water at 23°C.
Wash Rolls: Water at 22°C; 25.5 fpm
Doff Time: 90 minutes (time of collecting fiber on bobbin).

Bobbin	- Fiber Wet, microns - OD	ID	Fiber Air Dried, microns OD	ID
1*	440	250	300	180
2 through 4	450	280	310	210

*Injection air increased after the first bobbin.

The spinnerette consisted of two hollow concentric cylinders and the air was injected to the central channel of the inner concentric cylinder with the spinning dope being injected to the annular space between the concentric cylinders. Alternatively, quaternization can be carried out subsequent to the formation of the fibers; however, it is preferred to quaternize the polymer before the fibers are formed since under the conditions employed fiber spinning is facilitated by the presence of the benzyl chloride.

Hollow fiber membranes are made having urea permeability in the range of 0.005 to 0.075 cm/min and ultrafiltration rates in the range of 3 to 300 ml/min/m^2 at 25 mm Hg transmembrane pressure. Adjusting process conditions such as drying time, quenching conditions, etc. changes urea permeability and ultrafiltration rates and can give permeability and ultrafiltration outside the ranges specified.

Porosity of the hollow fibers is lost if the fibers are allowed to dry out, yet the fibers cannot be successfully potted with epoxy compound into the artificial kidney structure when wet with water, so normally the water on the fiber is exchanged with glycerine, ethylene glycol, ethanol or other suitable solvent prior to potting of the fibers, then the potted fibers are washed free of glycerine with water so the fibers can be heparinized.

Heparinization of the hollow fiber membranes can be carried out as described in U.S. Patent 3,441,142. Heparinization can be carried out prior to the formation of the hollow fiber membrane by heparinizing the polymer in solution, but since heparin type materials are so expensive, it is preferred to carry out the heparinization of the hollow fiber membrane itself treating only the membrane surface which will contact blood. The heparinization is carried out on the potted hollow fibers with only the inner surfaces of the hollow fibers being heparinized since these are the only surfaces which will contact the blood.

According to a related process described by *I.O. Salyer, D.D. Bump and W.E. Weesner; U.S. Patent 3,799,355; March 26, 1974; assigned to the U.S. Secretary of Health, Education and Welfare,* nonthrombogenic hemodialysis flat membranes useful in artificial kidneys are prepared from quaternized membranes of polymers made from monomeric substances of which at least a sufficient amount of a vinylpyridine is included to provide membranes which on heparinization are nonthrombogenic particularly polymers made from monomeric substances of which from 2 to 10% by weight of the total monomer is a vinylpyridine and at least 80% by weight is acrylonitrile. Both copolymers and blends of polymers can be used.

n-Butyl Methacrylate-Acrylic Acid Copolymers

According to a process described by *W.M. Muir; U.S. Patent 3,616,927; November 2, 1971; assigned to National Research Development Corporation, England,* membranes for extracorporeal hemodialysis are prepared from copolymers of n-butyl methacrylate (55 to 75%) and acrylic acid (45 to 25%) by portionwise addition of reactants to a reactor. These have superior transfer properties for urea, creatinine, uric acid and other blood components as compared with cellulosic membranes, and are prepared as blood enveloped for use in Ross-Muir and other dialyzers.

Polysulfone

A process described by *R.A. Cross; U.S. Patent 3,691,068; September 12, 1972; assigned to Amicon Corporation* involves subjecting a dialyzable liquid to dialysis by providing an anisotropic microporous polysulfone polymer membrane having a barrier layer at a surface thereof and a more porous support layer integral with the barrier layer, the barrier and support layers being a continuous uninterrupted polymer phase, which membrane has a dialytic resistance to sodium chloride less than about 50 min/cm and a total ultrafiltration flux between about 3×10^{-5} and 1×10^{-6} ml/cm^2 x min x mm Hg), contacting the barrier layer surface of this membrane with the dialyzable liquid, and contacting the other side of the membrane with a solvent for the dialyzable components of such liquid.

The membrane may be formed by forming a solution comprising the polysulfone polymer dissolved in an organic solvent, forming a barrier layer on one side of a thin film formed of this solution by preferentially contacting that side of the film with a diluent characterized by a high degree of compatibility with the organic solvent and a sufficiently low degree of compatibility with the polymer to effect precipitation of the polymer immediately on contact of polymer and diluent, maintaining the diluent in contact with the film until substantially all of the organic solvent has been replaced with diluent, drying the film until at least about 90% (preferably at least about 95%) of the diluent is removed, and rewetting the dried film.

The membrane may then be sterilized in an autoclave. The membrane is preferably less than 6 mils thick, but of sufficient thickness to be used unsupported. The membrane may take the form of a thin sheet, or of a hollow tubular member (with the barrier layer on the inside). Preferred hollow tubular members have an inner diameter of about 3 to 30 mils, and a wall thickness not greater than about 2.5 microns.

The dialysis operation is otherwise conventional. The liquid to be dialyzed passes along the barrier layer surface of the membrane, and dialyzable components of the liquid pass through the membrane into the solvent located on the other side of the membrane. In hemodialysis, this solvent is typically an aqueous solution of salt and glucose, of concentration selected to achieve the desired ultimate salt concentration in the blood, as well as to dissolve certain microsolutes such as urea. Applying a pressure drop (typically, 200 to 300 mm Hg) across the path of the liquid being dialyzed will also result in ultrafiltration of some water from the blood across the membrane into the saline solution. The dialyzed liquid is collected separately from the dialysate, i.e., from the dialyzed components and solvent.

In a particular hemodialysis apparatus, a plurality of tubular membranes as described are arranged in parallel and blood is fed axially through the membranes. The dialyzable components of blood pass through the membrane walls into a saline solution in contact with the outer membrane walls. It is believed that the surprisingly low dialytic resistance to salt of such anisotropic membranes is due in part at least to the flow of solvent through, as well as along the outside of, the porous support layer of the membrane, thereby providing a salt diffusion path length which may be substantially smaller than the total membrane thickness.

DIALYZER UNITS AND EQUIPMENT

Folded Membrane Dialyzer

A process described by *F.W. Markley; U.S. Patent 3,788,482; January 29, 1974; assigned to the U.S. Atomic Energy Commission* relates to a dialyzer which employs a semipermeable membrane in sheet form but does not require an adhesive material to effect an adhesive seal between the membrane and the housing. Instead, the edges of the membranes are sealingly embedded in a plastic material which constitutes at least a portion of the housing. The semipermeable membrane is folded into a large number of very closely spaced pleats, support members being inserted into all those pleats on one side of the membrane while no supports are placed in the pleats on the opposite side of the membrane. The membrane assemblage is disposed within the housing so that the membrane undulates back and forth across the width of the housing. One type employs a preconstructed casing with the two side walls of the casing fitting snugly against the pleats in the assemblage and each of the two side walls being provided at each end with access ports to the interior of the housing. A plastic material within the preconstructed casing constitutes a portion of the housing and surrounds the edges of the membrane sealingly embedding these edges so as to form a separate fluid chamber on each side of the membrane.

Alternatively, no preconstructed casing is used but instead a housing is formed in place about the membrane assemblage. The housing is formed by surrounding the membrane assemblage with a plastic material which constitutes the entire housing and which also surrounds the edges of the membrane and sealingly embeds these edges so as to form separate fluid chambers on each side of the membrane which undulates back and forth across the width of the housing. Access ports are sealingly mounted on the housing over openings to the interior of the housing which were created by the removal of studs which were appropriately positioned against the membrane assemblage as it was surrounded by the plastic material. In each case, the access ports provide means for passing one fluid through one of the separate fluid chambers in a direction generally parallel to the creases of the pleats and a second fluid through the other fluid chamber on the opposite side of the membrane countercurrent to the first fluid.

A process described by *R.J. Leonard; U.S. Patent 3,862,031; January 21, 1975; assigned to Baxter Laboratories, Inc.* relates to a mass transfer device of the type having a plurality of layers formed from a flexible folded semipermeable membrane. Devices of this type have been successfully used as extracorporeal oxygenators of blood in which both oxygen and CO_2 are transferred in opposite directions across a membranous barrier separating the blood and the oxygen.

In this process a flexible semipermeable membrane for a mass transfer device, such as a diffusion cell of the type characterized by a plurality of membrane layers having two separated sets of oppositely directed interconnected openings forming plural paths on opposed membrane surfaces for separated flow of a first fluid and a second fluid, respectively, is secured to one surface of a stiff backing in overlying relationship. To obtain this structure, a membrane and the backing are folded into a plurality of pleats to form two sets of oppositely opening pockets. The opposite ends of the pockets are then sealed; and one set is connected to inlet and outlet means for a first fluid and the other set of pockets is connected to inlet and outlet means for a second fluid.

Complete details are provided for the production of these membrane units for use with an oxygenator which is adapted for connection with a human circulatory system to enable extracorporeal oxygenation of blood.

Folded Membrane Cast on Grooved Plate

In a process described by *R.J.W. Streeton; U.S. Patent 3,578,173; May 11, 1971; assigned to United Kingdom Atomic Energy Authority, England,* a semipermeable membrane is made by casting upon a grooved or ridged casting plate, thereby providing grooves or ridges on the rear surface of the membrane.

Such a membrane is adequately supported by a nonporous backing plate, the grooves in the membrane defining channels for the liquid flow to or from the rear surface of the membrane and the membrane being held on the backing plate by a positive differential liquid pressure applied to its front surface. Preferably the membrane is folded upon itself at an angle to the grooves and two edges of the folded membrane sealed together forming a pocket requiring no separate mechanical support, the liquid flowing to or from the membrane by the unsealed edge. Most simply, the grooves are at an angle of 45° with the fold line and are all straight parallel grooves. Any form of groove can be adopted which provides a free passage for the liquid. Alternatively, a long strip of membrane can be formed and pleated so as to provide a large area of surface within a small volume. If the membrane itself has unidirectional properties, or if it is desirable to minimize the volume on one side of the membrane, or if there is a very large pressure differential, these factors are taken into account in deciding which is to be the rear surface of the membrane or what liquid is to be on the rear side of the membrane. Thus in the case of a blood dialysis machine, it is desirable to keep the blood volume at a minimum while the perfusion liquid into which the waste products are extracted from the blood may easily occupy a large volume. In this arrangement, the blood conveniently contacts the rear of the membrane.

Referring to Figure 8.1a, a casting plate **10** was prepared having grooves **11** which were straight and parallel and at an angle of 45° with the longer edges of the plate. The grooves were 0.004 inch deep and had an interior angle of 60° being on a pitch of 0.012 inch. Using this casting plate, a membrane **12** was cast from a mix comprising 25% of cellulose acetate, 30% of formamide and 45% of acetone (all percentages were by weight). After casting at room temperature, the membrane was allowed to stand for half a minute to allow surface evaporation of the solvents and was then immersed in ice water and left for 16 hours. The membrane was then stripped from the casting plate and was cured for five minutes in water at 75°C.

As is known, this treatment (which is a variant of the technique proposed by Loeb) produces a membrane suitable for reverse osmosis and having a thin active layer on its front surface, i.e., the surface marked **13** in Figure 8.1a. The membrane was then folded about its transverse centerline **14** (Figure 8.1b) with the rear surface inwards so that the grooves on the contacting rear surface were at right angles to each other and the two longitudinal edges **15a, 15b** and **16a, 16b** sealed together to define a pocket. The edge of the pocket remote from the fold line was left open.

The membrane was tested for two hours at a pressure of 500 psi and proved to be an effective desalination membrane, the salt solution being applied to the front surface **13.** No mechanical support was necessary and after the test the grooves, though deformed, still provided adequate passage for the fresh water which escaped from the open edge remote from the fold.

FIGURE 8.1: FOLDED SEMIPERMEABLE MEMBRANE

a.

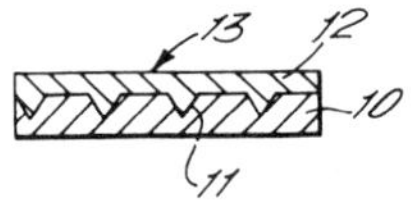

Membrane on Casting Plate

b.

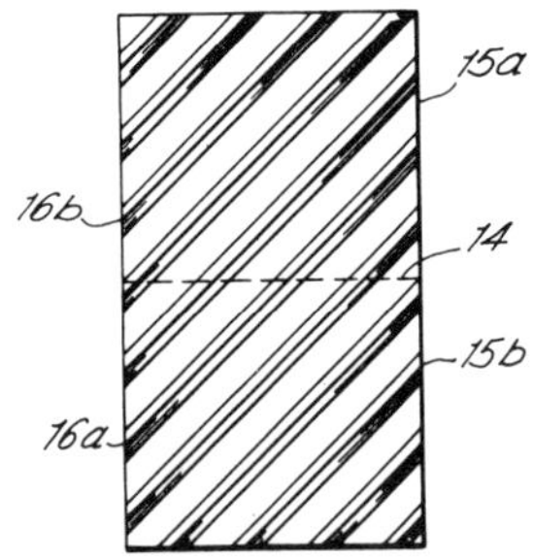

Rear Surface of Membrane

Source: R.J.W. Streeton; U.S. Patent 3,578,173; May 11, 1971

Edge Sealed Stack of Accordian Pleated Membranes

According to a process described by *F.W. Markley; U.S. Patent 3,864,265; February 4, 1975*, a semipermeable membrane is folded to form a stack of accordian pleats of which the edges are sealed to and within a housing that provides ports for flow of blood through channels formed on one side of the membrane and provides ports for flow of a dialyzate through channels formed on the other side of the membrane. End edges of the folded membrane are sealed by a unique end fold sealer plate that positions the end edges of the membrane for encapsulation within a solidifiable liquid and also blocks passage of the sealing material into the liquid flow ports formed in the sides of the case.

The case is formed with enlarged end portions that facilitate the flow of the solidifiable liquid sealer material for improved encapsulation of side edges of the stack of membrane pleats. The case is made in two mating sections in a configuration readily adapted to vacuum forming. The parts are arranged so that they may be assembled in a substantially dry form and substantially all of the solidifiable liquid sealant may be injected into the case after all parts are assembled. Thus, the edge sealing encapsulation of side and end edges is performed upon the positioned and assembled parts. Complete preparative and operational details are provided for hemodialysis type units.

Textured Membrane

According to a process described by *D.M. Ryon; U.S. Patent 3,724,673; April 3, 1973; assigned to General Electric Company*, a myraid of small undulations or projections are formed permenently in a thin membrane to give it a textured surface. The membrane may be permselective for use in fluid exchange devices such as blood oxygenators and artificial kidneys or dialyzers. When such membranes are superposed on each other, a thin film of blood or other fluid flowing between them is gently agitated by the textured surface to enhance the absorption of a gas such as oxygen which is transported through the membrane. The textured surface prevents adjacent membrane layers from adhering.

The method of forming the cones or fine undulations constituting a texture includes the step of placing a smooth membrane over a die on which a field of cones stands out in relief and then evacuating the space between the die and membrane so that air pressure may deform the latter in correspondence with the cone pattern. Heat is applied shortly after evacuation is initiated and is discontinued before evacuation is terminated so that upon cooling the plastically deformed cones will remain in the membrane.

Filtration Rate Control System

R.A. Crane, B.J. Lipps and J.A. Sargent; U.S. Patent 3,795,318; March 5, 1974; assigned to The Dow Chemical Company describe a system for carrying out hemodialysis with high flux membranes without incurring the risk of excessive ultrafiltration rates or reverse ultrafiltration. The transmembrane pressure differential is held constant, despite changes in pressure on the blood side of the membrane. The actual differential is sensed, compared to a desired positive magnitude and the pumping rate of a positive displacement, discharge pump adjusted to alter the proportion of the dialyzate which passes through a restricted shunt path around the pump, thereby altering the back pressure on the dialyzate side of the membrane until the differential has the desired magnitude. Complete details of the system as used to limit ultrafiltration in a clinical investigation of the effect of higher molecular weight solute removal during blood dialysis with highly permeable hollow fibers are provided.

Gas Purge Device

P. Riely and R. Skyles; U.S. Patent 3,631,654; January 4, 1972; assigned to Pall Corporation and Baxter Laboratories, Inc. describe a gas purge device which includes a filter element that contains both liquid-wetted and liquid-repellent parts, interposed across and screening separate outlets for liquid and gas. The liquid-wetted parts will pass the liquid, and the liquid-repellent parts will not be wetted by liquid, and will therefore remain open for passage of gas therethrough. The liquid-wetted and liquid-repellent parts open into separate outlets, the outlet downstream of the liquid-repellent part being the delivery passage for gas-free liquid from the device. In this way, the device is capable of separating gases and liquids, and of either venting the gas or delivering it to a gas collection device, while at the same time providing a gas-free supply of liquid. Blockage of the system by the buildup of a gas lock is avoided, while at the same time the entrained gas is entirely eliminated from the liquid. Thus, this process is particularly adapted for medical applications, where air must be vented from the line, and must also be absolutely prevented from reaching a patient receiving an injection of the liquid.

In a preferred embodiment, the filter employed has pores less than about 0.5 micron. If harmful microorganisms are to be filtered out from the fluid, the pores preferably should be less than about 0.3 micron, and then both the liquid and the gas passing through the device are sterilized at the same time. The device has the further advantage that the liquid-wetted or hydrophilic and liquid-repellent or hydrophobic parts can both be provided on the same filter element, thus facilitating servicing, and simplifying the construction.

The gas purge device of Figures 8.2a and 8.2b comprises a disc of filter medium **13** installed in the housing **12**. The disc contains a liquid-repellent or hydrophobic segment **14** and a liquid-wetted or hydrophilic segment **15**. Each such segment occupies one-half of the disc. The filter disc **13** is so installed that the hydrophobic segment is located contiguous to a semicircular annulus **16** and the hydrophilic segment is located contiguous to another semicircular annulus **17**. The filter disc is bonded to the housing along its entire circumference at **19** and along the line that divides its two segments at **20**. This prevents external leakage of fluid through the air vent port **18** and internal bypassing of either segment.

Both the inlet **11** and the housing **12** can be formed or molded of transparent polymethyl methacrylate resin, and are bonded together along the entire circumference at **23** by a solvent produced seal or an epoxy resin. The end fittings **21** and **22** are formed such that they may be inserted into standard tubing used in medical applications and provide a leak-proof connection therewith.

FIGURE 8.2: GAS PURGE DEVICE

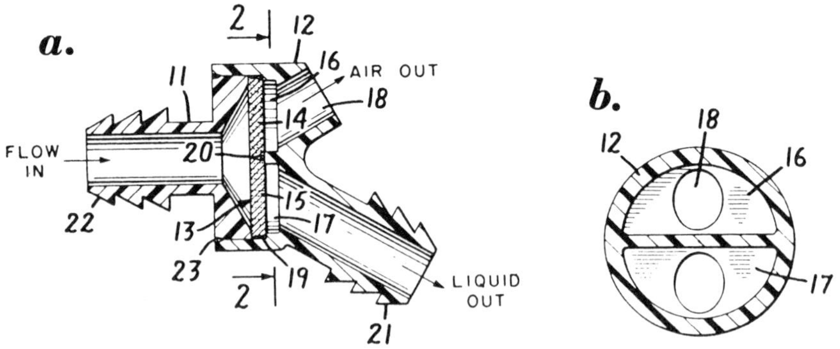

Cross-Sectional View of a Gas Purge Device with a Housing Having Three Passages Arranged in a Y-Configuration

Cross Section Taken Along Line 2–2 of Figure 8.2a

Source: P. Riely and R. Skyles; U.S. Patent 3,631,654; January 4, 1972

The filter medium is prepared as follows: A microporous filter material in sheet form is prepared following the procedure of Example 1 of U.S. Patent 3,353,682. The average pore size is 0.1 micron and the maximum pore less than 0.35 micron as determined by 100% removal of the bacteria, *Serratia marcescens.*

This material is then cut into a disc and one-half is treated with General Electric's RTV-112 silicone resin, to render it water-repellent or hydrophobic. The treatment is carried out by impregnation using a 5% solution of RTV-112 silicone resin solution in perchloroethylene, followed by evaporation of the solvent, and curing the resin at 40% relative humidity and at 25°C for 18 hours. The deposition rate is approximately 0.1 cc of solution per square centimeter of filter material, extending to the opposite side of the material. The treated portion of the disc becomes the hydrophobic segment **14** and the untreated portion becomes the hydrophilic segment **15**.

In use, fluid containing both gas and liquid enters via the inlet fitting **22**. Fluid wets the hydrophilic liquid-wetted segment **15**, and as soon as the pores of this material are filled, gas can no longer pass through. On the other hand, liquid does not wet the liquid-repellent segment **14** and gas is consequently free to pass through this segment, reaching the annulus **16** on the other side thereof, and being vented to the atmosphere through the air vent port **18**. Liquid passing through the liquid-wetted segment **15** enters the annulus **17**, whence it is delivered from the device through the outlet fitting **21**.

The gas purge devices are useful to separate gases from liquids in any type of medical and chemical application. They can, for instance, be used both to clear a line of air and to prevent the introduction of air into a patient receiving an injection of any type of fluid medicament, such as a parenteral fluid, blood transfusions, blood plasma, intravenous feeding solutions, and the like. Such fluids can be delivered to a patient under gravity pressure, or under higher pressures, such as encountered when the fluid delivery is effected by means of a syringe pump, and will prevent the introduction of air into the patient, at all pressures below the bubble point of the liquid-wetted filter material that is used, both at the beginning of the introduction of the liquid, even when the line before the device contains air, and after delivery of fluid has exhausted the supply. The separated gas need not be vented but can, if desired, be connected into a gas reclamation line leading to a concentrator or storage tank.

Fluid Transfer Unit

R.G. Shaw and J.S. Byck; U.S. Patent 3,764,018; October 9, 1973; assigned to Union Carbide Corporation describe a membrane which comprises a thin-walled permeable polymeric film having a plurality of first orifices circumscribed by the inner surfaces of oblong shaped portions and by the outer surfaces of connecting portions of the film joining the oblong shaped portions and having a plurality of second orifices circumscribed by the outer surface of the oblong shaped portions and by the inner surfaces of connecting portions of the film joining the oblong shaped portions at both ends.

Referring to Figure 8.3a, there is shown the membrane of polymeric film **1** where the heavily shaded arrows represent a typical flow pattern for two fluids A and B, the fluids being separated by the film. Again referring to Figure 8.3a, there is shown a polymeric film having a plurality of first orifices **2** (typical) the first orifices circumscribed by the inner surface **7** of oblong shaped portions (viz, frustum of a cone) **8** of the film and by the outer surfaces of connecting portions **9** (typical) joining the cylindrically shaped portions **8** and having a plurality of second orifices **3** (typical) circumscribed by the outer surface **4** of the cylindrically shaped portions **8** and by the inner surfaces **5** and **6** of connecting portions **9** of the film joining the oblong shaped portions **8** at both ends.

Referring to Figure 8.3b there is shown the membrane held between retaining members **10** to demonstrate a typical single membrane unit. Figure 8.3b depicts the mode in which the retaining members **10** ensure separation of the fluid streams A and B. This typical membrane unit may be stacked in multiple units to provide any desired degree of fluid transfer. The membrane units may be arranged for series or parallel flow or combinations thereof as desired. For any given application, the number of oblong shaped flow orifices per square inch of membrane is generally from about 2,500 to 90,000 in number, but may be more or less if desired.

FIGURE 8.3: FLUID TRANSFER MEMBRANE

a.

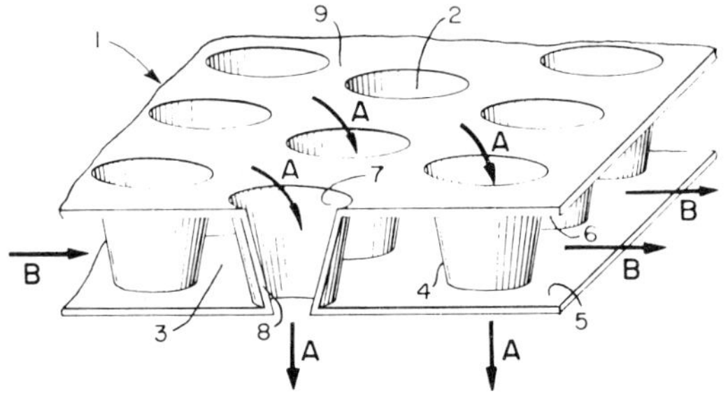

Enlarged Fragmentary Sectional Oblique View of the Membrane

b.

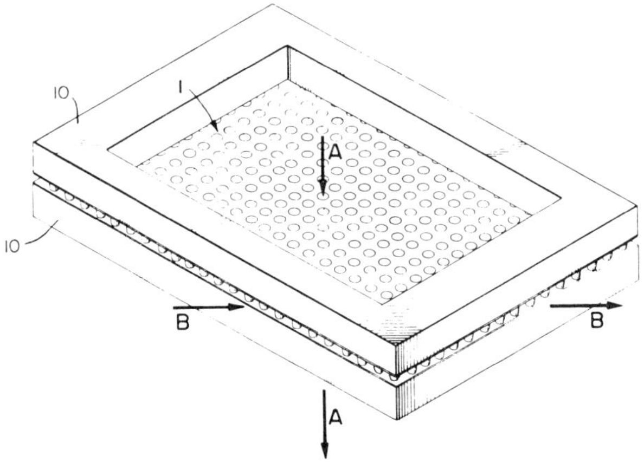

Isometric View of an Assembled Membrane Unit (Orifices Shown Disproportionately Enlarged)

Source: R.G. Shaw and J.S. Byck; U.S. Patent 3,764,018; October 9, 1973

The membrane is broadly useful in chemical, biomedical and pharmaceutical fluid transfer operations. A specific example of such application is in the removal of toxic metabolites from blood and oxygenation of the blood, such as in the case in an artificial kidney or artificial lung operation. The membrane may be used for the exchange of oxygen and carbon dioxide in a blood stream so as to remove the waste gaseous carbon dioxide and subsequently introduce oxygen into the bloodstream. The membrane is peculiarly suited for gaseous transfer in a blood stream, specifically the red corpuscles therein. In a gaseous transfer in blood during an operation, it is believed necessary that the residence time of a red blood cell in a fluid transfer membrane be minimized. By minimization it is meant that minimal period required to effect a gaseous transfer. To achieve this residence time minimization, it is necessary to have a membrane wherein the blood flow orifices are small in diameter to ensure gaseous transfer in a one-pass operation while being large enough to permit flow without an excessive pressure drop across the membrane.

It was known that the diameter of a red blood cell was about 8 to 9 microns. And it was determined that oblong shaped orifices of the membrane of this process having diameters of from 10 microns to about 100 microns with an oblong depth of from about 20 to about 150 microns achieved the desired minimum residence time for red blood cells, in a one-pass operation.

In oxygenation, blood flows through the oblong shaped orifices while oxygen flows through the other orifices of the membrane of this process. To remove carbon dioxide from the blood, the blood flows through the oblong shaped orifices while a carbon dioxide solution medium, such as an inert gas, flows through the other orifices of the membrane of this process. The preferred permeable polymeric film materials are para-xylene polymers and polyethylene. The following is an example of the method of manufacture for the fluid transfer membrane of this process.

Example: A 2 mil brass shim stock was coated with Kodack KPR photoresist. The coated shim stock was covered with a Type A By-Chrome 150 dots per inch screen tint mask, which mask served as a negative. The masked and coated shim stock was placed in a vacuum frame and exposed to a carbon arc. The mask was then removed. The photoresistive coated brass shim stock was placed in a developing medium for the Kodak KPR photoresist rendering the light struck areas insoluble. The shim stock was then removed by etching in an ammonium persulfate bath thereby removing that material coincident with the dots so as to form a template having a plurality of cylindrically shaped holes.

The template was then dip coated by vapor depositions with polypara-xylene. After coating the shim stock was cut at the edges to expose the brass metal. The brass metal shim was then removed by etching in an ammonium persulfate solution which solution did not attack the polypara-xylene. The resultant film defined a fluid transfer membrane. The above example was repeated with the polypara-xylene being vapor deposited in lieu of dip coating. The resultant membrane was again achieved.

ADDITIONAL MEDICAL APPLICATIONS

Urine Treatment

R.A. Zeineh; U.S. Patent 3,591,493; July 6, 1971 describes a method and apparatus for the treatment of biological fluids including primarily the separation and microconcentration of macromolecular structure substances from fluids in which they occur by the utilization of a semipermeable membrane which only permits the flow of smaller molecular structure substances through while retaining the larger molecules for eventual recovery.

Preferably, a biological fluid such as urine is introduced into a serpentine type fluid pathway formed by a porous membrane such as cellophane and because of a pressure differential on either side of the pathway, the components of the biological fluid of smaller molecular structure pass through the membrane to a collector. The remaining material such as protein of macromolecular structure is carried through the serpentine path to its terminus where it may be readily recovered.

Referring to Figure 8.4a the overall apparatus and method of operation may be seen in basic form. The cell structure **2** is shown as a housing in which is supported a tubular or other configured fluid pathway **4** formed by semipermeable or porous membrane **6** having pore openings **8** of such size as to only permit the passage of specific selected components therethrough. Cell **2** has connections **10** and **12** communicating to chambers **14** and **16** on either side of membrane **6**.

FIGURE 8.4: TREATMENT OF BIOLOGICAL FLUIDS

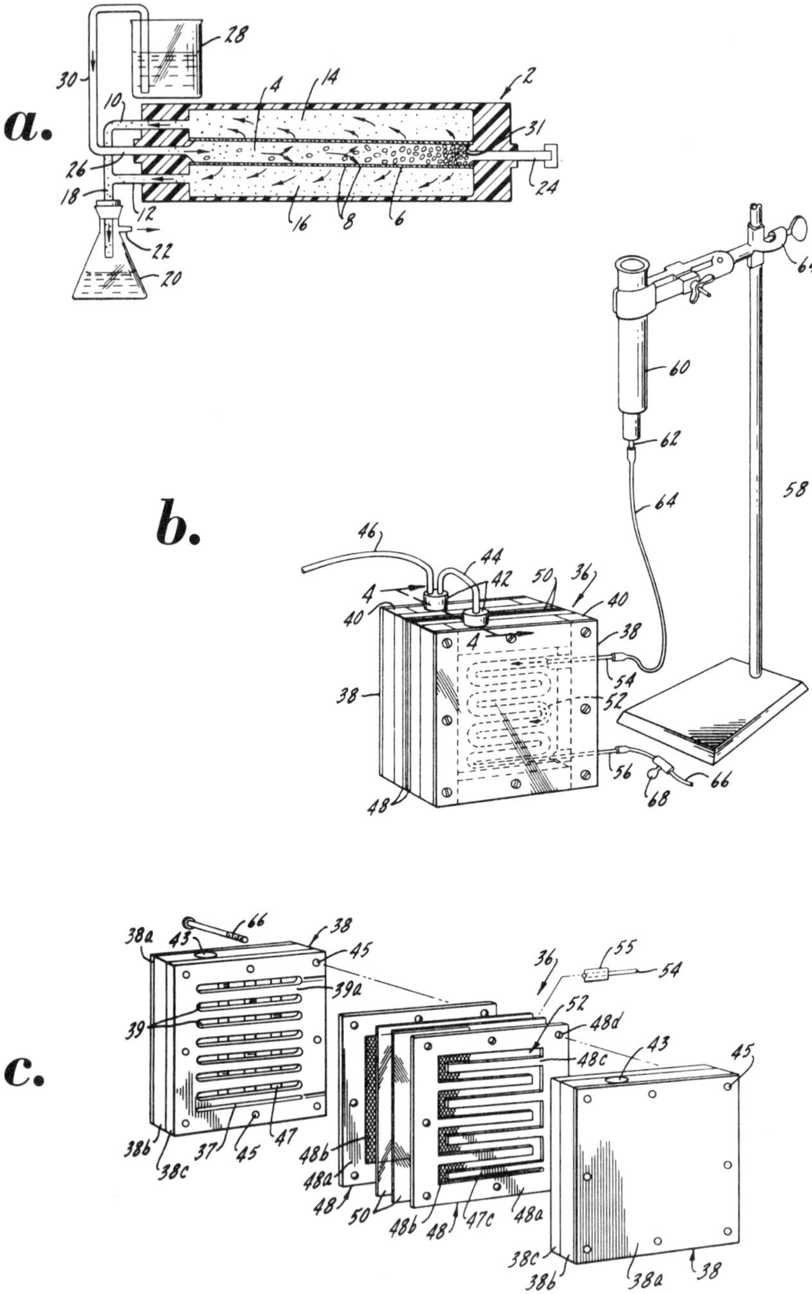

(a) Schematic of Basic Principles
(b) Assemblage of Equipment
(c) Detail of Cell Construction

(continued)

FIGURE 8.4: (continued)

d.

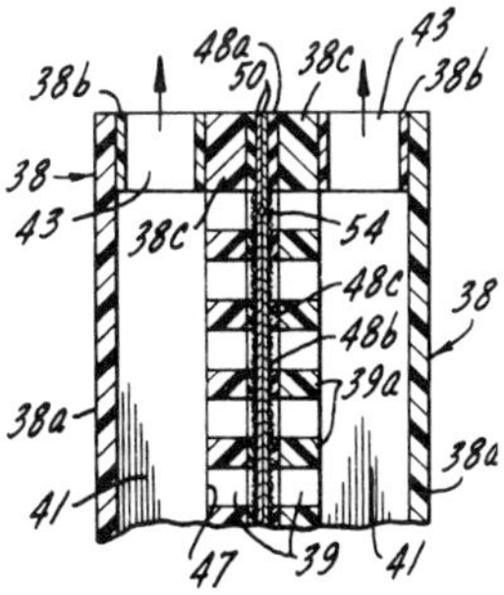

Enlarged Section Along Line **4–4** of Figure 8.4b

Source: R.A. Zeineh; U.S. Patent 3,591,493; July 6, 1971

Connections **10** and **12** merge into a common conduit **18** terminating in flask or receptacle **20** having means **22** for connection to a vacuum source. Fluid pathway **4** terminates in outlet **24** by which the concentrated material may be removed from fluid pathway **4**. An inlet **26** communicating with the fluid pathway connects to a dilute biological fluid specimen supply contained in beaker or flask **28** by means of conduit **30**. The dilute biological fluid contains, for example, macromolecular material such as protein and other components of smaller molecular structure. Because flask **28** is at a higher level than cell **2**, the dilute protein solution contained within beaker **28** will flow through conduit **30** through the fluid pathway inlet and thence along the fluid pathway.

During the flow of the dilute protein solution into cell **2** the suction source is activated to thereby create a pressure differential in the chambers **14** and **16** on either side of fluid pathway **4**. Because of the vacuum and because the porous membrane allows for the passage of only relatively smaller molecular substances as opposed to the larger protein molecules, the fluid comprising the small molecular structure substance passes through the pores **8** into the chambers **14** and **16** and thence by connections **10** and **12** to collecting flask **20**. Left behind within the fluid pathway are the macromolecular substances such as protein which are collected at the terminal end **31** of the fluid pathway.

As can be seen, the fluid pathway **4** has a slight slope so as to facilitate the collection of the protein material at the terminal or outlet end of the fluid pathway **24**. The effects of gravitation, of course, provide for the migration of the protein material to the outlet end. The pressure differential or vacuum maintained in chambers **14** and **16** also aids the flow of the dilute protein solution from the beaker **28** and also provides an assurance that the fluid pathway is maintained in the expanded position as shown. Without the presence of fluid or the vacuum in chambers **14** and **16** the fluid pathway would be in the collapsed position.

Referring to Figure 8.4b the overall assemblage of working components of the separator and microconcentrator is indicated. The specific details of the separator and microconcentrator **36** will be thoroughly described with reference to the remaining figures. It is seen that cell **36** comprises two end plate sections **38** having hollow grooves or cavities formed on the inside faces. The plates have on their upper surfaces **40** a passageway or aperture communicating to the interior of the vacuum chambers referred to above. In passageways **40** are stoppers **42** such as the rubber ones normally found in labs, having a communicating conduit **44** and conduit **46** which leads to a vacuum source not shown. Disposed between plates **38**, shown here as being made of a transparent plastic material, are two grid-like plate structures **48**, each interior surface or face supporting a semiporous membrane such as a cellophane sheet **50**.

The grid support plates **48** have extending portions which for the purpose of this process suffice to say form a serpentine path such as **52** between the cellophane sheets **50**. An inlet into the serpentine path is formed by means of a thin needle or conduit-like connection such as **54** disposed between the porous membranes in fluid-tight relation. It will be noted that the inlet to the serpentine path is of larger volume than the outlet at **56**, the outlet having a similar connection as the inlet, i.e., a thin member disposed between the porous or semipermeable membranes.

A stand such as **58** supports container **60** suitable for containing biological specimen fluid such as cerebrospinal fluid or the like and having an outlet **62**. A hose connection **64** between the outlet and the inlet **54** provides means of introducing the biological specimen into the cell **36** and more specifically into the serpentine fluid pathway **52**. It will be noted that it is possible by the leveling components of the support stand to maintain the specimen at any desired height so as to obtain gravity flow into the cell. An outlet conduit **66** having thumb valve **68** connected to outlet conduit **56** of the cell completes the basic components of the overall apparatus for separating and microconcentrating specific macromolecular structure substances such as protein.

The mode of operation for the apparatus depicted in Figure 8.4b is in principle identical for that described for Figure 8.4a with the single major difference being that the biological specimen is made to flow along a serpentine path. Generally, speaking, in laboratory use, the cell **36** will be filled with a fluid such as water in order to maintain the membranes in the most desirable state and also for purposes of sterility. To empty the apparatus of this water one simply connects the outlet tube **66** to an air supply source, opens the valve **68** and drives the water back through the serpentine path **52**, through the inlet **54**, through the hose connection **64** and thence into the flask **60** where it is accumulated and then discarded. So as to insure that there are no air bubbles within the cell it is preferable to take a portion of the biological fluid which is to be separated and concentrated and inject it through the outlet hose to likewise be driven through the unit to drive out any air bubbles, etc. that might have been trapped within the cell.

Once the cell and other components of the apparatus have been readied for use, a biological fluid specimen is simply placed within flask **60** and the leveling mechanism **64** positioned so that the flask or reservoir is about two to five inches above the cell **36**. Preferably the cell is placed flat on a supporting base such as a laboratory bench tilted in order to insure gravitational flow to the outlet **56**. Each of these measures will insure proper gravitational flow through the serpentine path. It goes without saying, however, that where desired one could apply a slight vacuum to the line **66** in order to obtain continued, positive flow.

Once the biological fluid sample has been introduced into the flask **60** a vacuum is applied via hose **46** which is connected to a vacuum source. A vacuum in the area of about -10 to -25 psig should be continued over the entire period of separating and microconcentration. Preferably, negative pressures in the upper ranges are desirable for faster flow-through of materials. This elapsed time will normally be about four minutes for each milliliter of sample or an overall time period of 15 to 20 minutes after the last portion of the sample leaves the reservoir or flask. As the fluid flows through the hose **46** into the inlet **54** and is introduced into the initial portion of the serpentine path **52**, the fluid components of smaller molecular structure will be passed through the semipermeable or porous membrane of the fluid path and will eventually be drawn out through hose connections **44** and **46** to a collector.

During travel through the serpentine path, the macromolecular structure substances, such as protein, become more and more concentrated until ultimately at the outlet **56** they may be collected by merely opening valve **68** and allowing the sample to flow through the connection **66** to whatever collecting means might be desired, e.g., capillary tubes. From there on, of course, the capillary tubes with the sample of the concentrated protein therein may be frozen until time arises for their use or subjection to other diagnostic studies.

The apparatus just described may be utilized in column chromatography. For instance, should it be desired to conduct a stepwise column chromatography procedure, one merely fills the serpentine concentration path and part of the reservoir or biological fluid supply flask with distilled water while allowing vacuum in the system to build up gradually. The effluent from the column then flows in dropwise fashion into the flask reservoir. For a microconcentrator of the type illustrated in Figure 8.4b, the rate of concentration is about 12 to 18 ml/hr for -15 and -25 psig respectively and, of course, will vary depending upon the magnitude of the vacuum.

If the effluent from the column is less than the rate of concentration the cell will adjust itself to that rate by trapping air bubbles within. Otherwise the excess of column effluent will accumulate in the reservoir which will be later concentrated by continuation of the separation and microconcentration after the column chromatography procedure has been terminated. Overconcentration at the outlet side of the serpentine concentration path may be avoided by partial bleeding. Obviously, reversing the inlet and outlet of the cell will lower the degree of overconcentration.

Referring to Figures 8.4c and 8.4d, more details of construction of the cell **36** may be observed. End plates **38** are illustrated as being formed of three plate-like sections. The first involves a substantially flat nonporous plate **38a**, the second plate member **38b** having a substantially cutout center section thereby forming a rectangular band-like member and an end interior plate **38c** having spaced slots **39** with plate portions **39a** between the slots which are intended to abut in side relationship rib-like members of support grids. Where the materials of construction are such as to be deformable under pressure, it is desirable to provide backup or support members **47** to keep plate portions **39a** from flexing. Here the members are shown as being spaced and rectangular and being the entire width of the chamber **41**.

Obviously, other configurations of plate components and supports may be utilized. It will be noted that all of the slots **39**, with the exception of the lowermost one **37**, are of approximately equal size. The reason for slot **37** being smaller will become apparent. It is, of course, obvious that the three sections when placed and secured together form unitary plate structure **38** having an interior cavity or chamber **41** through which fluids are collected and withdrawn therefrom. For fluid withdrawal, apertures **43** located in each of the upper surfaces of the end plates are provided and communicate to each of the chambers **41** of each end plate.

While the plate structures **38** have been shown as being fabricated of the individual components **38a, 38b** and **38c**, this is not critical to the operation of the device. In some instances it may be desired to machine or otherwise fabricate the plate sections as a unitary member rather than as a fabricated one. Spaced-through holes **45** provide means for ultimate assembly of the components of the cell **36** as by means of through-bolt **66** for maintaining all of the plate-like structures in substantially rigid relationship.

The grid-like, membrane support structures **48** comprise a deformable edge portion **48a** with extending portions of a material such as rubber or plastic, which is suitable for fluid-tight engagement with the interior abutting face of plate **38**. Disposed in the center of support **48** is screen or grid-like interior member **48b**. The edge portion **48a** has extending rib-like spaced portions such as **48c** equally spaced from one another except for **47c** which is spaced to form with the upper inside edge of **48a**, a smaller area than that formed by the upper extending portion **48c** with the lowermost, inside edge of the opposite parallel edge of **48a**. The size and spacing of the extending portions **48c** are intended to coincide and be congruent with plate portions **39a** between slots **39** of plate section **38c**.

To obtain this relationship, upraised ridges may be provided in each extending portion to be received by a comparable groove in the material portions **39a** of the plate between slots. Each of the two membrane supports or grids **48** are of identical construction. The side of the grids having the screen face each other and the extending portions **48c** mate together to form the serpentine fluid pathway when assembled with two sheets of semipermeable or porous membranes **50**. The serpentine path allows for communication throughout the chamber

41 even where intermediate, spaced supports **47** are provided. It is preferred when assembling the apparatus that the plate portions **39a** between slots **39** of the end plates **38** be of the same or slightly larger size as the extending rib-like portions **48c** so as to provide for a tight abutting fit to prevent, for instance, substantial lateral movement of the support grids **48** with respect to the membranes **50**. Grid supports **48** have apertures **48d** so as to be capable of receiving retaining means such as bolt **66**. The inlet **54** is shown as having a rectangular thin portion **55** of a deformable material such as plastic for ease of insertion between the cellophane membranes and likewise the outlet **56** may have an identical construction. In each instance, the rectangular portion is of a material which is slightly compressible so as to be able to provide fluid-tight engagement between the cellophane membranes.

After the components are assembled, an adhesive or other fluid sealant may be placed on the peripheral outside edges of the cell **36** so as to insure fluid retention solely within the serpentine path **52** formed by the pair of cellophane sheets **50** and grid support structures **48**. It is obvious, of course, that once a vacuum is created in the chambers **41**, the cellophane sheets deform somewhat until they are up against the screen **48b** and it is through the fluid pathway thereby formed by the extending portions **48c** that the fluids of small molecular structure are passed, the larger molecules, being unable to pass through the membrane, being swept towards the end of terminus of the serpentine path for ultimate collection.

As earlier noted, the extending portion **47c** is in close proximity to the bottom edge of the peripheral portion **48a** so as to form a smaller pathway near the collecting point and this is merely to create a small volume in which the separated macromolecular substance such as protein may be easily retrieved.

The cell structure as illustrated particularly in Figures 8.4b, 8.4c and 8.4d will generally be about 10 to 12 cm square with vacuum chambers formed in the end plate approximately 2½ to 4 cm deep, the peripheral gasket material of the grid support being from 0.16 to 0.5 cm thick with the total square area of the screen surface being within the area from 100 to 144 square cm, with the extending portions of the support grids being about 0.63 cm thick and from about 0.16 to 0.5 cm deep or high and being approximately 6.25 to 8.13 cm long. Generally, the total volume of a serpentine fluid pathway, for example, will be about 1.4 to 1.8 ml while its overall length may be from about 32 to 64 cm.

A cell as depicted in Figure 8.4b ideally has a concentration path or fluid pathway with a 1.8 ml holdup volume and removes fluids at a rate of 18 ml/hr under 25 lb/sq inch vacuum (12 ml/hr at 10 lb/sq inch). The unit is about 12 x 12 x 3 cm in dimension. Other units may have a concentration path of 1,280 cm long, 0.6 cm wide and 0.05 cm deep with a 40 ml holdup volume and having dimensions of 17 x 17 x 5 cm will have a fluid withdrawal rate of 360 ml/hr under 25 lb/sq inch vacuum.

For reasons of ease of fabrication and for other reasons such as cleanliness, transparency, etc. the end sections have been shown as being fabricated of a plastic material such as Lucite but it is possible to utilize Plexiglas, Teflon, vinyl or any other type material. It need not be plastic and may be of another material such as metal, opaque in nature. The only criteria is that it be of sufficient strength to withstand the pressures involved and of a material inert to the biological and treating fluids. The plates may be formed in any suitable manner as by cutting, molding or otherwise but as indicated earlier, they have been, for experimental purposes, of the various plate and other components configuration. The following examples illustrate the process.

Example 1: Utilizing a cell similar to that depicted in Figure 8.4b where the fluid pathway had a total volume of 1.8 ml and the cell employed a pair of cellulose acetate membranes, (available under the trademark Cuprophan) each membrane having a thickness of 12.5 microns, a pore size of 15 A and being about 20.43 cm x 20.43 cm in size, under a vacuum of 15 to 25 psig, serum protein was separated and microconcentrated. To 200 ml saline solution was added 1 cc of blood to provide a dilute protein solution. 30 ml of this dilute solution having 0.037 grams protein/cc was introduced into the cell. At first single drops were individually collected and stored in separate test tubes from the cell's outlet. Subsequently, additional drops were recovered in separate capillary tubes. To the first collected single drop was added 0.5 ml of distilled water for purposes of analysis. The ultraviolet wave band absorption was determined with a DU spectrophotometer and protein concentration calculated as 8.6 gram percent. Thusly, a dilute protein solution having 0.037 gram percent protein was concentrated to one having a high degree of protein concentration.

Example 2: Using the same type of apparatus and procedure as set forth in Example 1, 30 ml of blood serum was concentrated and desalted by the passage of distilled water through the cell after the serum solution had passed through the cell. Using usual laboratory procedures, selected components of the concentrated serum and the filtrate were determined and compared to that of the original serum and the data obtained is set forth in the following table.

Serum	Total Protein g/100 cc	Alkaline Phosphatase, units	Acid Phosphatase, units	Creatinine, mg/100 cc	Uric Acid, mg/100 cc	Cl^-, milliequivalents per liter	Na^+, milliequivalents per liter
Original serum	6.8	2.3	0.4	1.4	6.2	106	142
Concentrated and desalted	23.1	7.6	1.4	-----	-----	1	2
Filtrate (deproteinized)	-------	-----	-----	1.3	6.0	110	138

Prior art methods of separation and concentration lack the speed of the process. The time element is important so as not to permit denaturation of the protein content of a biological fluid. A series of tests were conducted and content of various fluids determined by electrophoresis to ascertain whether or not denaturation had occurred using the method and apparatus of this process. The averaged results of these tests are set forth in the following table.

Serum	Runs*	Albumin, %	Globulin, percent			
			Alpha 1	Alpha 2	Beta	Gamma
Undiluted	3	52.8	5.0	12.6	11.0	18.6
Diluted with physiological saline NaCl (0.85 g %) solution and then conc.	3	53.1	4.7	14.1	9.8	18.3
Diluted with native urine and then conc.	5	54.2	4.9	13.2	9.9	17.8

*Data is average of runs indicated

From the data it is obvious that protein denaturation did not occur using the process, differences in percent results being attributable solely to experimental error. Because of the small amount of cerebrospinal fluid normally available, as opposed to other biological fluids, a series of separating and microconcentration tests were conducted using such fluid samples and the end products analyzed by electrophoresis to determine the components. The data is shown in the following table.

Number of Runs	Total Protein, mg/ml	Prealbumin, percent	Albumin, percent	Globulin, percent			
				Alpha 1	Alpha 2	Beta	Gamma
5	36.1	3.9	62.0	5.2	7.3	13.2	8.6

Electrophoretic Serum Analysis

J.L. Dwyer, R.A. Paine and C.W. Souza; U.S. Patent 3,553,067; January 5, 1971; assigned to Millipore Corporation describe a laminated sheet which includes a plastic base sheet and a membrane sheet bonded to one side of the base sheet. In addition to these sheets, a thin flexible sheet is bonded to the other side of the membrane sheet. Thus, the porous plastic membrane is sandwiched between the base sheet and the thin flexible sheet. The thin flexible sheet has one or more holes or slits for the application to the porous plastic sheet of the chemicals to be evaluated. The bonding of the sheets together is effected with a thermoplastic adhesive without impairing the capillary transport of the liquids in the pores.

The base sheet is preferably a clear stiff but flexible plastic sheet having the requisite chemical stability to be inert to the chemicals being tested and which permits good bonding to the porous plastic membrane sheet. The porous membrane must have such critically proportioned pores that bonding of the base sheet and the thin plastic sheet can be effected without impairing the pores, and which have such surface characteristics as to permit diffusive fluid travel and also electrophoretic mobility.

The thin flexible plastic sheet is slotted or suitably provided with holes prior to bonding and performs a very important function. It restricts the diffusion of the liquids applied to the porous plastic membrane through the openings within the pores of the plastic membrane; most importantly it prevents drying of the liquids and can be removed to wash out unreacted chemicals so that the insoluble reacted chemicals can be evaluated. The plastic sheet is sufficiently thin that it can be peeled off the membrane sheet at such a sharp angle that the arc radius at the edge of release is so small that the severance is virtually along a line whereby the thin sheet can be removed without removing any appreciable amount of adhering porous plastic membrane.

The laminated sheet shown in Figure 8.5a through 8.5c comprises a clear plastic sheet **B**, a microporous plastic membrane sheet **2** bonded to the plastic sheet **1** with a thermoplastic adhesive **3** and a thin plastic sheet **4** which is also bonded to the membrane sheet with a thermoplastic adhesive **5**. The section is greatly enlarged the pores being disproportionally enlarged, for the purpose of illustration. The laminated sheet is preferably produced of the materials and by the technique as follows. The base sheet is a plastic material having the requisite physical and chemical properties. It is preferably clear and is stiff and resilient so that when used for electrophoresis in this device it will hold the bowed position used in the device and may vary from 6 to 12 mils, preferably 10 mils, thick.

It must be a nonconductor of electricity under the conditions of use. As for the chemical properties, the base sheet must be unaffected by the buffer solution used in the receptacle. Various plastic materials fulfill the above requirements, a polyester such as Mylar, the Du Pont 1000D being especially effective. A polyamide or polycarbonates may also be used. The microporous plastic membrane sheet **2** may be formed of a mixture of cellulose nitrate and cellulose acetate or entirely of cellulose acetate according to U.S. Patent 1,421,341 or U.S. Patent 2,944,017. The sheet is preferably formed of cellulose acetate. An effective sheet is about 5 mils thick and has average pore sizes varying from 0.5 to 1.5 microns. In electrophoresis testing the sheet transports the buffer solution by capillary action and enables the electrophoretic transport to take place.

The bonding or adhesive material **3** must also meet rather critical specifications. The adhesive is advantageously a thermoplastic material, preferably one that does not require a catalyst or a curing agent and has a sealing temperature below the heat distortion point of the porous plastic membrane sheet **2**. A Du Pont material known as a modified synthetic rubber adhesive (Du Pont 56016) has been found to be very effective. The thickness of the finished dry coat of adhesive deposited on the base sheet must be within close limits. Optimum thickness range is 0.25 to 0.35 mil. If less than 0.25 mil is applied to the base sheet the amount is insufficient to bond to the porous membrane sheet. If more than 0.50 mil is applied, some of the adhesive when heated to seal the two sheets is absorbed into the pores and interferes with the diffusive properties.

The sheet **4** is a clear nonporous impermeable plastic such as a polyester preferably about 0.002 inch thick and is bonded to the membrane sheet **2** with a thermoplastic adhesive **5** as described above. Prior to bonding, patterns of holes **6** or slots **7**, or other geometric perforations may be cut into the sheet so that the materials used in the evaluations can be applied directly to the exposed portions of sheet **2**. In forming the laminated sheet in a preferred operation, a base sheet **1** of indefinite length is coated with adhesive **3** as with a doctor device to, say, 0.35 mil thickness and the coated surface is led into contact with a moving porous membrane sheet **2**, also of indefinte length. The thin plastic sheet **4** of indefinite length, also coated with an adhesive about 0.35 mil thick is led into contact with the opposite side of sheet **2**. The superposed sheet as a sandwich is fed over an endless metal belt and between the belt and a heated press roller which heats the laminated sheet to about 250°F for about 1.5 minutes at a pressure of about 60 psig. This operation effectively seals the three sheets together without plugging or impairing the pores.

The lamination of impermeable sheet **1** and film sheet **4** to both surfaces of the microporous membrane **2** serves to block passage of liquid to or from these faces. Liquid is, however, free to diffuse freely through the porous structure of membrane **2** between the laminated interfaces. Through the openings cut in the top thin laminating sheet, liquids can be introduced for controlled migration processes such as diffusion, chromatography, or electrophoresis. The top covering lamination prevents evaporation of any components from the microporous membrane sheet **2** at any point save at the openings in the top sheet **4**.

The use of this laminated sheet in suitable strip dimensions may be illustrated by the technique of immuno-electrophoresis. The top flexible sheet **4** is a little shorter than the overall length of the laminated strip as shown in Figure 8.5a such that either end of the porous membrane sheet **2** is exposed for contact with electrodes that are required in the electrophoretic migration. When such a strip is used in a device, the ends are immersed in a buffer solution and the ends are placed in contact with the electrodes. The strip may also be first stood on edge in a shallow tray or buffer solution. Since the microporous membrane sheet is exposed at the edges of the strip, it tends to chromatograph or wick up between the two impermeable sheets. When the strip is thoroughly permeated with the buffer solution, one microliter of serum sample is applied to the cellulose acetate through the center hole **6** in the sheet **4**.

FIGURE 8.5: LAMINATED SHEET–ELECTROPHORETIC ANALYTICAL SEPARATION

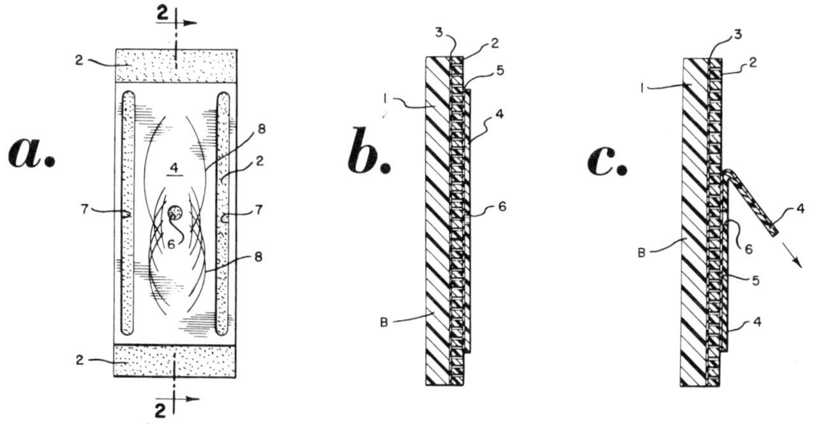

(a) Laminated Sheet
(b) Sectional View Along Line **2–2** of Figure 8.5a
(c) View of Figure 8.5a Showing Top Sheet Partially Removed

Source: J.L. Dwyer, R.A. Paine and C.W. Souza; U.S. Patent 3,553,067; January 5, 1971

The strip is placed in an electrophoresis cell and a voltage is applied at the ends of the strip. The serum runs between the two impermeable sheets during migration and in a direction parallel to the long dimension of the strip. The serum sample is allowed to electrophorese until a suitable separation has been obtained. After electrophoresis has been carried on for a suitable length of time, the strip is removed from the cell. An appropriate amount of antiserum is applied uniformly in each of the long troughs **7**. For most commercial antisera it has been found that the optimum amount applied to each trough is approximately 40 microliters. This quantity will, of course, vary with the strength of the antisera, the amount of serum sample applied in the center hole **6**, and other particulars of the separation.

It has been found that in applying the antiserum to the troughs with a hypodermic syringe a sufficiently uniform concentration of antiserum along the lengths of the troughs can be obtained. Having applied the antiserum in this fashion, the strip is immediately placed in a saturated humidity chamber, to prevent evaporation from those exposed faces of the membrane sheet **2**. Diffusion of the serum components and antiserum components toward one another begins and is allowed to continue for a suitable amount of time, generally 24 to 48 hours. During this time, precipitin arcs **8** begin to form where components of the serum and antiserum react with one another. Since serum and antiserum are relatively colorless, these precipitin arcs are not visible through the transparent top sheet **4**.

At the end of the diffusion period, the strip is removed from the incubation chamber. With a sharp object, one corner of sheet **4** is lifted so that one can grasp the tab between his fingers and peel it off the membrane sheet. The flexible sheet **4** is pulled backward at a sharp angle so that the stresses of peeling are concentrated at the adhering interface between the sheet **4** and the membrane sheet **2**. As a result, the relatively delicate cellulose acetate of which sheet **2** is composed remains uniformly in place rather than being peeled away with the sheet **4**. If a thicker sheet were used for the thin flexible sheet **4**, the desired peeling might not occur or might become irregular with chunks of cellulose acetate being torn away with the stiffer sheet. After stripping, the whole face of the cellulose acetate membrane is exposed.

Precipitin arcs such as those shown in Figure 8.5a reside within the porous matrix of the cellulose acetate membrane as well as unreacted serum and antiserum components. These unreacted components are first rinsed out of the cellulose acetate membrane by immersing it in warm saline. The porosity of the microporous membrane facilitates a much faster wash-out of these unreacted components than is possible with gel structures used for immuno-electrophoresis.

One or two hours suffice for a complete wash-out of unreacted components. The washed strip is then placed into a staining solution that will stain the precipitated protein. The strip is stained for a time sufficient for binding of dye molecules to the protein complexes. This may be anywhere from a few minutes to several hours, depending on the particular stain used.

Following staining, the strip is rinsed in an appropriate rinsing medium. Dye molecules which did not bind to the protein rinse free of the microporous membrane sheet, leave the distinctive pattern of the precipitin arcs against a relatively clear white background of the cellulose acetate membrane. Although the immediate application for the laminated sheet is in the area of immuno-electrophoresis and immuno-diffusion it has other analytical uses such as for other diffusional processes or chromatographic processes.

For example, certain areas of the microporous plastic membrane could be impregnated with specific chemical reagents before the top plastic sheet is laminated on. Substances chromatographed through the plastic membrane could be made to pass through these impregnated layers, thus developing reactions, colors, or appropriate indications. By blocking certain areas within the porous plastic membrane one could define the diffusional migration paths.

Nonthrombogenic Membranes

In a process described by *P.D. Oja; U.S. Patent 3,441,142; April 29, 1969; assigned to The Dow Chemical Company* nonthrombogenic, permeable membranes are prepared from film-forming materials having hydroxyl groups, such as cellulose, by converting at least a portion of the hydroxyl groups to oxy alkali metal groups, then reacting with a haloalkyl dialkyl amine, then quaternizing by reacting with an alkyl halide and subsequently reacting with an alkali metal salt of an antithrombogenic compound such as sodium heparin. The following examples will illustrate the process.

Example 1: Ten portions of approximately 1 mil thick cellophane (regenerated cellulose) film weighing a total of 2.4 grams (0.015 mol equivalent based on anhydroglucopyranose ring) was mercerized for 30 minutes at 0°C with 0.6 gram (0.015 mol) NaOH dissolved in 6 ml water. Following mercerization, a solution of 2.58 grams (0.015 mol) of 2-chloroethyldiethylamine hydrochloride in 24 ml water was added, followed by a further 0.6 gram of NaOH dissolved in 6 ml water. The reaction mixture was heated on a steam bath to 75°C for 40 minutes. The films were then washed in deionized water and let stand in fresh deionized water overnight.

The following day, the films were immersed at 25°C for three hours in a mixture of 25 ml each of dimethyl formamide and water in which approximately 3.0 grams of CH_3Br had been dissolved. The films were then washed in deionized water, followed by three washes in fresh 0.1% aqueous sodium heparin solution, followed by one wash in 0.9% aqueous NaCl. The films were left standing in a 0.06% aqueous sodium heparin solution containing excess heparin. A sample of the film taken before the heparin wash and partially dehydrated showed the following elemental analysis.

	. . . . $\approx C_{44}H_{77}BrNO_{35}$ (percent)	
	Found (GW 12629)	Atomic Ratio
C	41.4	43.8
H	6.1	77.0
Br	7.4	1.18
N	1.1	1.0
O	44.0 (by diff.)	35.0

This analysis corresponds to the introduction of 1 alkonium group per 6 anhydroglucopyranose groups. Samples of these heparinized membranes were thoroughly rinsed with physiologic saline and placed over the concave surface of a watch glass 16 mm in diameter with a central depression of 7 mm. One ml of venous blood from a human being was placed on each membrane watch glass preparation. The preparation was then tilted at various intervals and observed for evidence of coagulation. In addition, a 20 gauge steel needle was stroked across the surface of the blood and the time recorded when a fibrous strand was noted to be adherent to the needle as it left the blood. For a control a plain watch glass as described above was employed without any of the heparinized regenerated cellulose film. The observations were as follows.

Watch Glass Control

2 minutes lapsed–Coagulation evident at edge glass interface.
10 minutes lapsed–Definite layer of fibrin on watch glass base.
13 minutes lapsed–Fibrin layer continues to thicken; no fibrin with steel needle streaking.
16 minutes lapsed–Fibrin strands adherent to steel needle.
48 minutes lapsed–Clot very solid; watch glass can be inverted.

Heparinized Regenerated Cellulose

0 minutes lapsed–The blood has spread out over the film which appears to be slightly hydrophilic.
23 minutes lapsed–No evidence of coagulation.
40 minutes lapsed–First sign of fibrin (approximately 40 minutes delay of clotting).
43 minutes lapsed–Fine fibrin strands adherent to steel needle.
63 minutes lapsed–Jelly-like clot has formed.

Example 2: A nylon membrane is reacted with ethylene oxide to produce a membrane containing 55% combined ethylene oxide. This membrane is heparinized and placed on a watch glass to which is added human blood and observed according to the general procedure described in the foregoing Example 1. The inhibition of blood clotting is significantly improved over that of the corresponding membrane without the heparinization treatment.

Example 3: A bundle of 100 10-inch long hollow fibers of cellulose triacetate are melt spun from cellulose triacetate plasticized with tetramethylene sulfone. These fibers have an outside diameter of 310 microns and an inside diameter of 275 microns. The bundle of fibers are then simultaneously swelled and acetylated by a 4-hour treatment at 50°C with a 25% excess over the stoichiometric deacetylation requirement of 1% NaOH solution in 30/70 by volume tetramethylene sulfone/water. The resulting swollen cellulose fibers are then washed with deionized water and converted to the sodium form, to the amino ether, quaternized and heparinized in turn following the general procedure described in Example 1.

The heparinized fibers are then tested for their ability to inhibit blood-clotting in a manner similar to that described in the foregoing Example 1. Blood clotting is not observed when in contact with these fibers at the end of one hour. Additionally, these fibers are found to have excellent permeability properties including high water permeability and the ability to freely pass low molecular weight salts and urea.

Sterilized Packaged Articles

D.J. Beecher; U.S. Patent 3,598,517; August 10, 1971; assigned to American Sterilizer Company describes a process and apparatus for sterilizing articles sealed in packages made of semipermeable membranes, and with or without porous overwrap, in a sealed chamber in an atmosphere of permeable gas. A control is provided for regulating the pressure in the chamber to limit the differential pressure between the pressure inside the package and the pressure in the chamber to a predetermined value so that the pressure inside the package will never be such as to rupture the package and the chamber pressure will not exert a pressure on the outside of the package which might damage its contents. The sterilizing gas is an epoxide, preferably ethylene oxide.

Polydiorganosiloxane-Polycarbonate Block Copolymers for Internal Organs

S.W. Kantor and P.C. Juliano; U.S. Patent 3,781,378; December 25, 1973; assigned to General Electric Company describe block copolymers of polydiorganosiloxane and polycarbonate having improved heat distortion which are useful as high temperature membrane material. The polydiorganosiloxane-polycarbonate block copolymer membranes can be steam sterilized at elevated temperatures without altering membrane structure.

As shown by U.S. Patent 3,189,662, organopolysiloxane-polycarbonate block copolymers can be made by phosgenating a mixture of a dihydric phenol and the reaction product of a halogen terminated polydiorganosiloxane and a dihydric phenol. Depending upon the block size of the polydiorganosiloxane and the relative amount by weight of dihydric phenol to polydiorganosiloxane employed, a wide variety of useful products can be obtained. Among the uses to which the polydiorganosiloxane-polycarbonate block copolymers can be employed are high strength elastomers, films and parts which can be made by conventional extrusion methods.

In addition, these materials can be employed as insulating coatings for wires and rods, binding material for laminates, in adhesive formulations, etc. As shown by U.S. Patent 3,419,634 improved hydrolytic stability can be achieved in block copolymers of polydiorganosiloxane and polycarbonate by employing polydiorganosiloxane having certain chain stopping siloxy units, such as hydroxybutyldimethylsiloxy units in place of chlorodimethylsiloxy units providing block copolymers having Si—C linkages joining polydiorganosiloxane blocks and polycarbonate blocks.

It has been found that when utilized in the form of a film, the block copolymers of the above mentioned patents exhibit unusual selective permeability for providing for the separation of particular gases from gaseous mixtures and as a permselective membrane for liquids and for ion transfer. In applications requiring steam sterilization of such membranes in an autoclave, it has been found that under superheated conditions, these membranes can readily distort, rendering these materials unsuitable in many medical areas, such as in artificial lung applications, for example the Peirce Lung.

This process is based on the discovery that polydiorganosiloxane-polycarbonate block copolymers having superior heat distortion resistance, and all of the valuable utilities possessed by the block copolymers in the above patents are obtained by substituting certain dihydric phenols derived from cyclic ketones in the phosgenation mixture, in place of the dihydric phenols, such as 2,2-bis(4-hydroxyphenyl)propane, (bisphenol-A) employed in the above described methods.

The polydiorganosiloxane-polycarbonate block copolymers can be described as consisting essentially of about 1% to 75% by weight of polycarbonate blocks having an average of at least about 2 and up to about 100 chemically combined carbonate units of the formula,

$$(1)\qquad -\mathrm{OZO}\overset{\overset{\displaystyle \mathrm{O}}{\|}}{\mathrm{C}}-$$

and from 99% to 25% of polydiorganosiloxane blocks consisting essentially of diorganosiloxy units of the formula,

$$(2)\qquad -\underset{\underset{\displaystyle \mathrm{R}}{|}}{\overset{\overset{\displaystyle \mathrm{R}}{|}}{\mathrm{Si}}}\mathrm{O}-$$

where Z is a divalent organo radical derived from a cyclic organic ketone, and R is a monovalent organo radical selected from the class consisting of hydrocarbon radicals, halogenated hydrocarbon radicals and cyanoalkyl radicals. The block copolymers can be made by phosgenating a mixture of polydiorganosiloxane of the formula,

$$(3)\qquad \mathrm{Y}\left(\underset{\underset{\displaystyle \mathrm{R}}{|}}{\overset{\overset{\displaystyle \mathrm{R}}{|}}{\mathrm{Si}}}\mathrm{O}\right)_n\underset{\underset{\displaystyle \mathrm{R}}{|}}{\overset{\overset{\displaystyle \mathrm{R}}{|}}{\mathrm{Si}}}\mathrm{Y}$$

and a dihydric phenol of the formula

$$(4)\qquad \mathrm{HOZOH}$$

where Z is as previously defined, n is an integer equal to 1 to 500 inclusive, Y is a member selected from

$$-\mathrm{OZOH}\qquad -\mathrm{R'OH}\qquad -\mathrm{R'N}\overset{\overset{\displaystyle \mathrm{O}}{\|}}{\mathrm{C}}\mathrm{OZOH}\qquad -\mathrm{R'}\overset{\overset{\displaystyle \mathrm{O}}{\|}}{\mathrm{C}}\mathrm{OH}$$

$$-\mathrm{R'}\overset{\overset{\displaystyle \mathrm{O}}{\|}}{\mathrm{C}}\mathrm{OZOH}\qquad -\mathrm{R'O}\overset{\overset{\displaystyle \mathrm{O}}{\|}}{\mathrm{C}}\underset{\underset{\displaystyle \mathrm{H}}{|}}{\mathrm{N}}\mathrm{R'}\underset{\underset{\displaystyle \mathrm{H}}{|}}{\mathrm{N}}\overset{\overset{\displaystyle \mathrm{O}}{\|}}{\mathrm{C}}\mathrm{OZOH}\qquad -\mathrm{R'}\underset{\underset{\displaystyle \mathrm{H}}{|}}{\mathrm{N}}\overset{\overset{\displaystyle \mathrm{O}}{\|}}{\mathrm{C}}\underset{\underset{\displaystyle \mathrm{H}}{|}}{\mathrm{N}}\mathrm{R'}\underset{\underset{\displaystyle \mathrm{H}}{|}}{\mathrm{N}}\overset{\overset{\displaystyle \mathrm{O}}{\|}}{\mathrm{C}}\mathrm{OZOH}$$

etc., and R′ is selected from divalent hydrocarbon radicals and halogenated divalent hydrocarbon radicals. Radicals included by Z are for example,

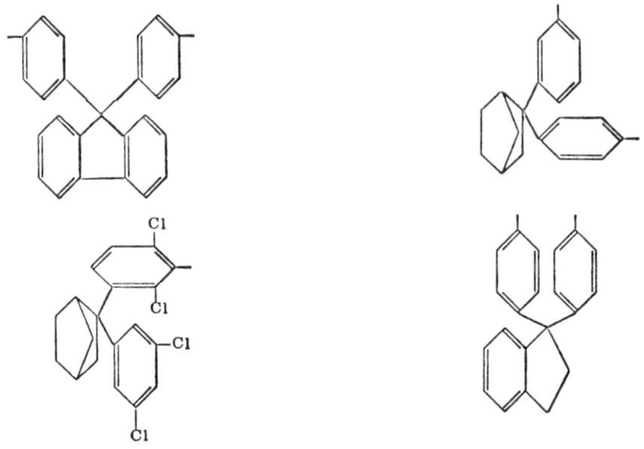

Included by the dihydric phenols of Formula 4, are for example,

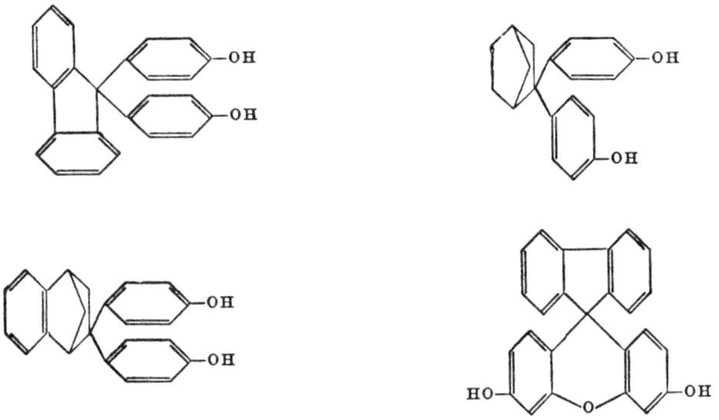

Example: There was added under substantially anhydrous conditions 14.31 parts of a chlorine terminated polydimethyl-siloxane having an average of about 19 chemically combined dimethylsiloxy units to a mixture of 10 parts of 9,9-bis(4-hydroxyphenyl)fluorene, 37.5 parts of dry methylene chloride and 10.9 parts of dry pyridine. The addition took approximately 30 minutes during which time the mixture was vigorously stirred. There was then added phosgene at a rate of 0.74 part per minute until a precipitate of pyridine hydrochloride appeared. The phosgene flow rate was then reduced to 0.06 part per minute, and the mixture was stirred until a maximum viscosity was achieved.

A solution of 5 parts of methanol and 25 parts of methylene chloride was then added and the solution was stirred for 5 minutes. Another 38 parts of methylene chloride was added to reduce the viscosity of the mixture. There was then added 50 parts of a 2.4 N hydrochloric acid solution and the mixture was stirred to remove excess pyridine. Product was recovered from the mixture which was washed with water until it was free of chloride ion. Polymer was then precipitated by pouring the resulting product into a tenfold excess of methanol, filtering and drying in a vacuum oven overnight at 80°C. There was obtained an 85.3% yield of product. Based on method of preparation the product was a block copolymer having the average formula,

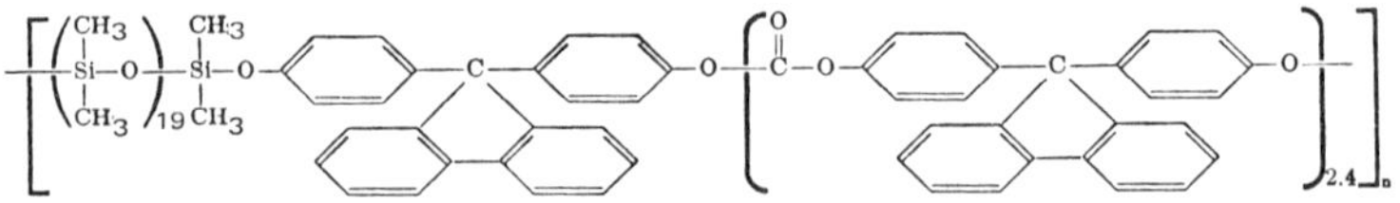

The identity of the above block copolymer was further confirmed by elemental analysis. Its intrinsic viscosity in chloroform was 0.93 at 25°C. A polydimethylsiloxane-polycarbonate block copolymer was prepared in accordance with U.S. Patent 3,189,662 having approximately the same percent by weight and average block size of the polydimethylsiloxane and about the same average block size of polycarbonate as the above described polydimethylsiloxane-polycarbonate block copolymer. Films were cast from the respective block copolymers. Test slabs were then prepared from each of the respective films to compare the physical properties of the respective block copolymers at 25°C and 100°C.

In the table below there is shown the results obtained from the respective block copolymers. Each of the block copolymers had about 55% by weight of polydimethylsiloxane with an average block size of about 19 dimethylsiloxy units. The block copolymer prepared in accordance with the process (bisfluorene) is compared to the block copolymer prepared in accordance with the above described prior art copolymer bisphenol A where T is tensile strength (psi) at break, E is elongation percent at break, expressed at both 25° and 100°C.

	25° C.		100° C.	
	T	E	T	E
Bis-fluorene	3,620	390	1,120	>610
Bisphenol-A	2,900	410	92	>600

1.2 ml films were cast from methylene chloride solutions of the above respective block copolymers to determine the permeability coefficient of these materials for a variety of gases. The following table shows the results obtained where the permeability coefficient was determined at 25°C, the values shown can be converted to permeability parameters expressed as cc(RTP)-cm/cm^2-sec-cm Hg, by multiplying these values by 10^9.

	25° C.	
Gas	Bis fluorene	Bisphenol-A
N_2	6.4	7
O_2	15	16
CO_2	87	97

A peel test was conducted on test slabs obtained from 15 mil sheets cast from each of the above block copolymers. The test slabs were 4" x ¼", which were used to determine whether the slabs would fuse together in areas of contact while being autoclaved at 250°F for 30 minutes. Composites of each of the respective slabs were made which were separated on one end with a ¾" by 0.001" Teflon tape for the peel test allowing for sufficient slab contact on the other end. Microscope cover glasses were also placed on top of the respective composites to prevent the strips from sliding in the autoclave. The composites were placed on Teflon tape strips to prevent any interaction with the surface of the autoclave.

After autoclaving, the samples were tested at a cross-head speed of 0.02" per minute in an Instron tensile tester with a full scale deflection of 2 lb. The test was designed to measure whether any force was required to peel the two strips apart, particularly where the slabs contacted each other. It was found that the block copolymer made from bisphenol A required a force of 6 lb per inch of contact to separate the two strips in the composite, indicating fusion had taken place. The composite of the bisfluorene block copolymer strips did not register any deflection on the scale.

If a Peirce Lung, described in *Trans. Am. Soc. of Artificial and Internal Organs,* No. 14, 1968, p 220, relating to high permeability copolymer membranes, were made with membrane derived from bisphenol A, block copolymer, it could not be autoclaved or sterilized satisfactorily. If autoclaving were attempted, the bisphenol A block copolymer membrane would be expected to fuse in area of contact between block copolymer, resulting in an increase in pressure drop across the lung due to a sealing of the lung membrane.

A decrease in blood flow rate, also would result, necessitating an increase in pump pressure which would likely cause blood damage. However, as shown by the above results, a membrane made from bisfluorene block copolymer in accordance with the process, would not exhibit any change in membrane characteristics and could be readily sterilized under the above described autoclaving conditions.

In Vitro Resorption Model of the Gastrointestinal Tract

The effectiveness of a drug in the system of the human body is, inter alia, a function of its concentration at the place of effective action and thus also a function of its ability to reach that place. Most of the perorally administered drugs must first overcome the barriers of the gastrointestinal tract, i.e., they must be resorbed, before they can be transported by the blood to the place where they exert their therapeutic action. Various in vivo methods for following the passage of chemo-therapeutics from the gastrointestinal tract into the blood circulation system are known.

For instance, the decrease in the concentration of the chemo-therapeutic substance with passage of time in various parts of the gastrointestinal tract after peroral administration can be determined by analytical methods in the tied-off stomach, colon or small intestine of rats. In humans, the change in concentration of various dissolved organic acids and bases in the stomach over a period of time can also be followed by means of an esophageal sound lowered into the stomach.

Likewise, in humans as well as animals, the blood level values at various times after peroral administration of drugs can be determined, from which rate of resorption can then be calculated. Such in vivo methods, however, require a substantial expenditure of work and money, and the results obtained therefrom are often subject to wide variations with respect to accuracy and reproducibility.

In vitro methods employing artificial membranes, on the other hand, represent a valuable supplementation of the in vivo determinations. Thus, by means of greater series tests and more accurate concentration determinations, the kinetics of diffusion can be very accurately ascertained, so that the effects of the interchange action between the active ingredients among themselves or with the pharmaceutical excipients can already be evaluated and taken into consideration in determining the most favorable make-up of the pharmaceutical dosage unit compositions. Moreover, by employing in vitro methods, the number of in vivo tests can be significantly reduced.

H. Stricker; U.S. Patent 3,753,501; August 21, 1973; assigned to C.H. Bolhringer Sohn, Germany describes a process which provides a membrane for use in an in vitro resorption test apparatus, which is capable of maintaining the pH gradient between two aqueous solutions substantially constant over a period of time of about 50 hours and exhibits a specific permeability for acid, basic and amphoteric drug molecules.

The process involves impregnating a solid porous carrier material with a liquid lipoid phase. The porous carrier material may be a commercial membrane filter material made of a cellulose ester. The diameter of the pores must be sufficiently small so that the pores, due to their capillary activity, remain closed by means of the liquid impregnation component; otherwise, the liquid impregnation phase could be dissolved out and the membrane would become useless. Therefore, the diameter of the pores should be from 50 to 1,000 μ, preferably 100 μ. Examples of suitable membrane filter material are the commercially available products Millipore VC and Sartorious Sm 11309.

The liquid lipoid phase must resemble, with respect to its physical properties, such as its relative solution capacity for acid, basic and amphoteric drugs, and its distribution, as closely as possible its natural counterpart, i.e., the stomach and intestinal walls, so that the relative resorption rates of various drugs can be accurately simulated in the in vitro apparatus. It has been found that these requirements are met by a mixture consisting of (1) a higher fatty acid, preferably caprylic acid, or dioctyl sodium sulfosuccinate, and

(2) a neutral lipoid component, such as a higher fatty alcohol of 12 or more carbon atoms, preferably lauryl alcohol. The mixture of these two components must be liquid at the test temperature of 37°C. The membrane is prepared by immersing the porous filter material into the liquid mixture, removing the impregnated porous material from the liquid mixture and allowing excess liquid mixture to drip off, and blotting the impregnated porous membrane between filter paper until its weight is constant. A preferred form of the membrane is obtained by using as the liquid impregnating component a mixture of 80 parts by volume caprylic acid and 20 parts by volume lauryl alcohol, or a mixture of 97 parts by volume lauryl alcohol and three parts by volume dioctyl sodium succinate.

In contrast to common membranes of this type, membranes in accordance with the process have the advantage that they are capable of maintaining the pH gradient between two aqueous solutions substantially constant (± about 0.3 pH units) over a period of about 50 hours and exhibit a specific permeability for acid, basic and amphoteric drug molecules. The diffusion rates under physiological conditions correspond substantially to the values known from in vivo tests.

The combination of a higher fatty acid or dioctyl sodium sulfosuccinate and a neutral lipoid component with a cellulose ester filter membrane as a carrier material proved to be a system which meets the requirements very well. The effect of the fatty acid or sulfosuccinate component is that, corresponding to its pK_a value and the pH of the adjacent solution, it is partly present as an anion and acts as a proton-acceptor or -donor with respect to the dissociated organic acids or bases. The drugs are thereby transformed at the membrane surface into the undissociated form and are thus capable of migrating through the liquid phase of the membrane, whereas the hydrogen and hydroxyl ions can pass through only very slowly because of the solubility conditions.

Detoxification of Saponins

According to a process described by *G.L. Dorn; U.S. Patent 3,883,425; May 13, 1975; assigned to J.K. & Susie L. Wadley Research Institute and Blood Bank*, saponin material extracted from plants is purified and made nontoxic to microbial organisms by removing constituents from an aqueous solution of the saponin extract which have an apparent molecular weight of less than about 600 in the aqueous solution. In a preferred example, an aqueous solution of saponin is prepared and then passed through a filter membrane having average pore sizes no smaller than about 11 A in diameter and no larger than about 24 A in diameter while adding amounts of pure aqueous solvent to the solution approximately equal in volume to the liquid passing through the membrane. The toxic material will be in the liquid passed through the membrane.

Self-Cleaning Ultrafilter for Blood Protein

P.N. Rigopulos; U.S. Patent 3,488,768; January 6, 1970; assigned to Amicon Corporation describes an assembly useful with conventional centrifugal separation apparatus and comprising an anisotropic membrane coupled with a fenestrated or macroporous support member, the membrane being so arranged with respect to the assembly that centrifugal force which makes the filtration possible also tends to keep the filtration surface of the membrane clean, thereby avoiding the expected problem whereby the extremely small membrane pores would be plugged by solids or sludge in the sample being subject to filtration.

To achieve this effect, upwardly facing primary membrane filter areas should be arranged at a relatively steep angle with respect to the vector of the centrifugal force, advantageously at an angle of less than 22½°, and most advantageously less than 15° with the vector of the centrifugal forces acting on the assembly. This steep angle provides means for assuring a major vector of the centrifugal forces are directed downwardly along the membrane surface. These forces cause sludge and macromolecular material, which would otherwise tend to plug or stick on the membrane surface, being retained to slide downwardly immediately on contact with the microporous membrane surface on the wall of the apparatus, thereby achieving a continuous cleaning action during use of the membrane assembly.

For convenience fenestrated and macroporous support members will be referred to by the generic term foraminous. Membranes which are useful in the apparatus are highly anisotropic, submicroscopically porous, membranes formed of polymers having good mechanical integrity, most advantageously those crystalline and/or glassy thermoplastic polymers known to the art. By crystalline and glassy polymers are meant those materials which possess from 5 to 50% by weight crystallinity as measured by x-ray diffraction techniques known to the art and/or a glass transition temperature (Tg) of at least 20°C.

Particularly advantageous are polymers of inherently low water sorptivity which unlike the cellulose acetate materials known to the membrane art may be allowed to dry during storage without losing their beneficial mechanical and processing characteristics. These polymers are those having water-absorptivities of less than 10% by weight of moisture at 25°C and 100% relative humidity. Such membranes are of particular advantage because, unlike any of the anisotropic membranes commonly known, they may be stored under relatively dry conditions; they need not be shipped, nor stored between uses, under water.

Among such polymers are the modacrylic polymer known as Dynel, polycarbonates such as that called Lexan and polyvinyl chloride. The submicroscopically-porous anisotropic membranes useful in the process consist of a macroscopically thick film of porous polymer, usually more than 0.002" and less than 0.050" in thickness. One surface of this film is an exceedingly thin, but relatively dense barrier layer of skin of from 0.1 to 5.0 micron thickness of microporous polymer in which the average pore diameter is in the millimicron range, for example, from 1.0 to 500 millimicrons, i.e., $\frac{1}{10}$ to $\frac{1}{100}$ thickness of the skin.

The balance of the film structure is a support layer comprised of a much more coarsely porous polymer structure through which fluid can pass with little hydraulic resistance. When such a membrane is employed as a molecular filter with the skin side in contact with fluid under pressure, virtually all resistance to fluid flow through the membrane is encountered in the skin, and molecules or particles of dimensions larger than the pores in the skin are selectively retained. Because the skin layer is of such extraordinary thickness, the overall hydraulic resistance to fluid flow through the membrane is very low; that is, the membrane displays surprisingly high permeability to fluids. Furthermore, tendency of such membranes to become plugged or fouled by molecules or particles have been found to be low if sufficiently effective steps are taken to stop accumulation of matter on the surface.

Photolithographic Process for Tissue Culture

S.L. Simons; U.S. Patent 3,303,254; February 7, 1967 describes a filter membrane having discrete openings of minute size disposed in regulated or patterned positions each to each in which the size of the openings is likewise regulated and/or uniform. The resulting membrane is preferably made of a resinous or plastic material which should be nonreactive or nonirritating with respect to the materials to be filtered. A significant development within the scope of this process is the provision of a filter membrane especially adapted for use in the realm of viable tissue culture.

The method for making the membrane having discrete openings of defined size and shape and selective placement comprises selectively exposing a photolithographic plate to light in patterned dispostion to create hardened spots on the plate in positions corresponding to points of light impingement, chemically etching the plate to remove plate materials from nonhardened surfaces whereby hardened posts will be disposed in bas relief above the plane of surrounding depressions.

The process further involves depositing a film of resin plastic moldable material on the etched face of the plate to fill the depressions to a depth less than the height of the posts, curing the resin material, and separating the plate and resin material whereby a filter membrane is obtained having spaced openings in positions corresponding to the position of the posts on the face of the photolithographic plate mold.

OTHER APPLICATIONS

Wool-Scouring Liquor and Centrifuge Effluent

J. Del Pico; U.S. Patent 3,847,804; November 12, 1974; assigned to Abcor, Inc. describes a process of treating wool-scouring liquor and wool-scouring liquor centrifuge effluent, both containing wool-grease emulsion, by the utilization of membranes to concentrate the liquor before centrifuging, or to eliminate the centrifuging step by concentrating the wool grease in the scouring liquor to a concentrate level through the use of ultrafiltration membranes. The process permits the efficient and economic recovery of the wool grease from the membrane concentrate fraction, and provides for reuse or disposal of the aqueous-permeate fraction.

In the process, one or more membrane units, particularly ultrafiltration membrane units, are employed, which membrane units operate at a pressure of from about 10 psig to 200 psig; for example, 40 to 100 psig. The membrane unit includes a semipermeable membrane adapted to separate the emulsion-scouring liquor as desired, and typically, may comprise any polymeric material which permits the passage of water and dissolved low-molecular weight materials (e.g., less than 100 molecular weight), if an ultrafiltration membrane unit is used, while preventing the passage of oil-in-water emulsion particles, as well as suspended particles in the scouring liquor.

The employment of an ultrafiltration membrane device results in the concentration of wool grease, dirt, etc., and in the generation of a nonturbid aqueous-permeate effluent. The permeate effluent is often of such quality that it may be directly disposed of by discharge into the waterways without contributing to pollution, or if desired, further membrane reverse osmosis (i.e., high pressure) processing of the permeate effluent may be employed to reduce further the suint (potassium carbonate), and low-molecular weight organics, and thereby substantially reduce the polluting aspects of the associated permeate effluent. More particularly, all or a part of the permeate may be further employed in the wool-scouring process, such as, for example, be redirected into the cold or hot wool-scouring tanks.

Example 1: In a series of tests to determine the efficacy of the concentration of scouring liquor centrifuge effluent, both containing a wool-grease emulsion by ultrafiltration, the liquor in the liquor effluent was treated in an ultrafiltration unit containing approximately 4.4 square feet of membrane area. An emulsion-scouring liquor was obtained from scrubbing raw wool with a nonionic surfactant solution, with the scouring liquor containing approximately 2.5% wool grease, the wool grease composed of alcohol and fatty acids ranging from about C_5 to C_{30}, and ranging from liquids to solids at approximately 25°C. The wool grease had a content of approximately 30% of a lanolin mixture composed of lanosterol and cholesterol, the mixture melting between about 36° and 42°C. The raw wool so washed resulted in a 2,000-gallon per hour wash water stream of the emulsion-scouring liquor.

The emulsion-scouring liquor was then fed directly to a continuous centrifuge yielding approximately 50 gallons per hour of wool grease, 600 gallons per hour of dirt, and 1,300 gallons per hour of scouring liquor effluent containing most of the nonionic surfactant, and about 1 to 2% wool grease. The centrifuge effluent liquor was then treated with approximately 3% of sulfuric acid, which reduced the pH level to about 3 to 3.5, resulting in breaking of the emulsion. Grease was then recovered, and the wool-grease solvent extracted. The emulsion-scouring liquor having a pH of approximately 7.2 was then concentrated in an ultrafiltration unit employing Abcor ultrafiltration membrane tubes to a concentration of approximately 7% wool grease.

The emulsion-scouring liquor was then treated in pH by the addition of an acid to a pH of 2.10 and concentrated in an ultrafiltration unit to an emulsion concentrate of about 7% wool grease. Additionally, the scouring liquor centrifuge effluent was also concentrated in an ultrafiltration unit to a wool grease concentrate of 1½ to 9%.

It was found that in the concentration of scouring liquor of Example 1, the flux was extremely temperature-sensitive between 30° and 40°C, and it is believed that this temperature sensitivity is due to the transition from the solid-to-liquid particles in the emulsion; that is, from a solid-liquid emulsion to a liquid-liquid emulsion within this temperature range. Example 1 was carried out at a temperature of approximately 40°C; that is, a liquid-liquid emulsion, so that the initial flux is approximately 100 gfd. During the membrane concentration and after concentration of the emulsion to about 1.6 full concentration, the flux dropped to 50 gfd.

It is believed that the drop in flux was associated with the reaching of the solubility limit of some of the components in the water phase, or that the emulsion breaking occurred, resulting in increased membrane concentration polarization which would reduce the flux rate. The permeate effluent recovered in Example 1 was crystal-clear, having a dark-straw color. Acid was added to the permeate fraction, and turbidity in the permeate fraction then occurred at a pH of approximately 2.5, which was believed to be due to the acidification of low-molecular weight acids which had passed through the membrane and into the permeate fraction.

Example 2: Example 2 was carried out employing a pH-adjusted scouring liquor, and the permeate fraction recovered was again crystal-clear, yet much lighter in color than that of the fraction of Example 1, indicating that the acidification of the scouring liquor is desirable in acidifying of the low-molecular weight dissolved components in the prior permeate fraction. The concentrate emulsion sample from Example 2 indicated a distinct liquid-oil layer, indicating that the ultrafiltration membrane unit had concentrated the emulsion-scouring liquor to a concentration level sufficient to destabilize the emulsion, resulting in an emulsion breaking.

Aqueous Protein Solutions

K. Ishii, S. Konomi and Y. Ishiguro; U.S. Patent 3,804,754; April 16, 1974; assigned to Daicel Ltd., Japan have found that, in the concentration treatment of an aqueous solution of a protein of natural origin, such as cheese whey, fish-meat-treating water or the thawed white of an egg, by the reverse osmosis method of the ultrafiltration, the separation characteristics of the membrane can be improved by adding a surfactant or surfactants to the aqueous protein solution, being treated, thereby to prevent major reduction in the water permeation rate.

All cationic, anionic, amphoteric and nonionic surfactants are effective in the process and, therefore, they can be selected freely depending upon the purpose. For example, when an edible concentrate is to be produced, a nonionic edible surfactant is preferred. If the main purpose is to treat wastewater to obtain purified water, any of the above surfactants may be used. Monoglycerides, polyoxyethylene alkyl ethers, alkylbenzene sulfonates, sugar esters and sorbitan esters are among suitable surfactants.

Example 1: Operation pressure: 42 kg/cm^2
Semipermeable membrane: Cellulose acetate membrane prepared by the formamide process
Surfactant: Nonionic surfactant [polyoxyethylene cetyl ether, hydrophile-lipophile balance (HLB) 10.7]
Amount: 0.05 g/l of test liquor
Test liquor: Water used for treating fish meat.

Reverse osmosis procedures were carried out on samples containing and not containing surfactant. The following results were obtained.

	Values after—	
	1 hour	8 hours
Without surfactant:		
Water permeability	0.68 m.^{3}/m.^{2}/day	0.20 m.^{3}/m.^{2}/day.
Rate of removal of protein	97.0%	95.3.
With surfactant:		
Water permeability	0.74 m.^{3}/m.^{2}/day	0.62 m.^{3}/m.^{2}/day.
Rate of removal of protein	97.0%	96.2.

Example 2: Reverse osmosis tests were carried out under the same conditions as in Example 1 except that the same amount of glycerol monostearate (a nonionic surfactant, HLB 3.6) was used in place of polyoxyethylene cetyl ether. The following results were obtained:

	Values after—	
	1 hour	8 hours
With surfactant:		
Water permeability	0.88 $m.^3/m.^2$/day	0.76 $m.^3/m.^2$/day.
Rate of removal of protein	96.8%	96.4.

Detergent Removal

A process described by *F.E. Martin; U.S. Patent 3,291,730; December 13, 1966; assigned to Aerojet-General Corporation* involves the removal of detergents from water by the reverse osmosis technique. The following example illustrates the process.

Example: A casting solution was prepared by mixing 100 grams of cellulose acetate (E-398-3), 300 grams of acetone, 60 grams of water, and 5 grams of magnesium perchlorate. A film was cast 10 mils thick on glass at -12°C, allowed to dry 3 minutes, then set in ice water. The membrane was allowed to soak loose from the glass plate, then heated to 84°C for 1 hour in water, cooled, and stored in deionized water. The membrane was mounted in a pressure cell with the air-dried surface of the membrane in contact with a solution which contained 115 ppm dodecylbenzene sodium sulfonate.

This solution was circulated over the membrane at a linear velocity of 1 foot per second and at a pressure of 1,500 psi. A product was collected at a product rate of 30.2 gallons per square foot of membrane per day and contained 3 ppm dodecylbenzene sodium sulfonate. Highly effective removal of synthetic detergent was obtained using the technique of reverse osmosis. In the example shown above, the solution contained 115 parts per million of dodecylbenzene sodium sulfonate. In practice, however, only about 5 to 10 parts per million synthetic detergent are present in common wastewater.

Removal of Microbiological Flora

C. Wallis and J.L. Melnick; U.S. Patent 3,836,458; September 17, 1974; assigned to The Carborundum Company describe a filter system for the purification of water, whereby particulates and viruses may be economically and reliably removed. The system comprises a filter means for removal of silts and colloidal materials, a treated carbon virus filter, and a specially treated honeycomb cartridge for removal of microbiological flora, e.g., bacteria, fungi, molds, and algae one micron or larger. The fabric filter is treated with an inorganic hydrolyzing composition of matter containing sodium. The hydrolyzing composition can be contacted with the medium through any system that produces sodium hydroxide. Preferably the hydrolyzing composition is an aqueous solution of sodium hydroxide.

Example 1: A cellulose acetate filter cartridge of the type having the fiber wound on to the core in a honeycomb fashion is soaked in about a 1.0 N aqueous solution of sodium hydroxide. After about 10 minutes of soaking in the aqueous sodium hydroxide, the filter cartridge is removed and washed with a dilute aqueous solution of hydrochloric acid followed by a water wash. The cartridge is air-dried. Water is passed through the treated cartridge at 50 psi for 20 minutes. The result is compared with the flow rate through the same filter cartridge at 50 psi for 20 minutes prior to treatment.

The flow rate through the treated filter cartridge is 20% less than the flow rate through the untreated filter cartridge. In a similar manner, a cellulose acetate filter cartridge, which includes a glass fiber core cover with a honeycomb winding is treated with sodium hydroxide and tested for a change in flow rate. The decrease in flow rate as compared to the untreated cartridge was 56%. Similar results are obtained by treating a cellulose acetate filter cartridge with a 1.0 N aqueous solution of acetic acid for one hour.

Example 2: Raw animal serum which contained about 7,000 bacteria per ml is passed through the filter cartridge treated with aqueous sodium hydroxde in a manner similar to Example 1. The bacteria in the effluent after passing through the cartridge is undetectable. The entire filtration process is completed within a few minutes. The bacterial counts in the serum before and after filtration are made by inoculating blood agar plates with 0.1 ml of serially diluted serum, streaking the inoculum with a bacteriological loop, and counting the colonies.

Example 3: Five gallons of tap water is passed through a Millipore membrane. A red-brown deposit forms on the surface of the membrane. The membrane is connected in series with a sodium hydroxide treated glass fiber filter cartridge, and tap water is passed through the filter cartridge prior to passing through the membrane. About five gallons of water passes through the filter cartridge and membrane as connected above, and a white-gray deposit forms on the membrane, indicating that the treated filter cartridge has adsorbed the red-brown deposit contained in the water.

Example 4: A water purifying system for processing about 12 gallons of water per hour is constructed. Tap water at a pressure of about 60 psi enters the system and the water pressure is reduced by means of a valve to about 45 psi. The water passes a pressure gauge and enters a conductivity cell. The tap water of the city of Houston, Texas, contains about 420 parts per million dissolved solids. The water passes through a polyethylene tubing to an activated carbon cartridge treated with sodium hydroxide. The filter cartridge has an inner portion of activated carbon and an outer cover of honeycomb wound fiber. The filter cartridges in the system are contained in Fulflo holders. The water is then passed through two glass fiber cartridges which have been treated with sodium hydroxide in a manner similar to Example 3.

These latter two cartridges include a center core filled with Amberlite resin which is packed into the center core of the filter cartridge. The resin adsorbs nondegradable phosphates and carbonates. The water then passes through a cellulose acetate filter cartridge treated in a manner similar to Example 1, so that the filter cartridge has an effective porosity of about 0.25 microns. The water then passes through a reverse osmosis unit (rated 250 gal/day) and then through polyethylene tubing into a conductivity cell. At this time the concentration of solids in the water has been reduced to about 19 parts per million dissolved solids.

Next, the water passes through a mixed resin bed which reduces the solids content to about 0.01 part per million dissolved solids. Analysis of the water obtained from the system has shown dissolved solids in the area of 0.01 to 0.03 part per million, no detectable chlorides, no detectable organics, and no detectable solids. The water from the system has been used to make media for growth of cultured cells for support of virus growth, and media for growth of bacteriological and other microbiological flora, and has given better results than similar cultures prepared with triple distilled water.

Bacteria Filter

A. Goetz; U.S. Patent 2,926,104; February 23, 1960 describes membranes or films formed of cellulose esters and having extremely small pores, these films being usable for any of various known purposes, as for instance for filtering out bacteria or very minute particles from a fluid stream. According to the process, the film forming solution is first applied to a carrier, and is then contacted by a moist atmosphere which flows past the film in a gelation zone within which the temperature, humidity, and rate of flow are very accurately controlled. These conditions are controlled in a manner such that all portions of the films have substantially identical temperature, humidity and air flow histories during the gelation process, i.e., at any particular stage during gelation of a certain portion of the film, that portion is subjected to air of exactly the same temperature, humidity and rate of flow as that to which the other portions of the film are subjected at the same stage in their formation. This is true even though the conditions may actually vary along the length of the gelation zone, due to changes in the quantities of solvent and moisture in the air stream resulting from absorption of solvent into the air and moisture into the film.

This uniform gelation history may be maintained by relatively moving the film and the air directing means in a manner such that the film relatively advances through the gelation zone, from a film inlet end of the zone to a film outlet end. Preferably, this relative advancement is effected by moving the film while the means defining the air path and gelation zone remain stationary. Also, the relative advancement of the film is desirably in countercurrent relation to the flow of moist air which is passed over the film. As the film advances, the solvent from the film vaporizes into the circulating air, while moisture from the air is at the same time absorbed into the film. By the time the film leaves the gelation zone, the film has been converted to a coherent polymerized structure having uniformly distributed minute filtering pores.

The specified countercurrent movement of the film and air has proven extremely desirable for assuring the maintenance of optimum conditions during gelation of the film. By virtue of this countercurrent arrangement, the moist air stream, as it is directed into contact with the film, first passes over that portion of the film which is about to leave the gelation zone. Subsequently, the same air advances into contact with portions of the film which have been in the zone a shorter period of time, ultimately to contact the film which is first entering the zone. The solvent progressively evaporates into the air, as the film advances, and moisture from the air progressively enters the film.

The relative humidity, temperature, and other conditions of the air are of course most accurately controllable at the point of introduction of the air into the gelation zone. The described counterflow arrangement allows this accurately controllable portion of the air stream to contact the film just before the film leaves the zone, so that the final part of the treating process in the zone is very accurately controlled, with resultant accurate control over the product structure. Also, this incoming air, which contacts the film just before it leaves the gelation zone, is free of solvent and rich in moisture, so that the final contact with the film in the gelation zone attains a maximum removal of solvent from the film, and adds a maximum of moisture to the film. As the air moves toward the film inlet location, the air progressively acquires more and more solvent and progressively loses its moisture.

However, as the air thus gains solvent and loses moisture, it contacts portions of the film which are also rich in solvent and poor in moisture, so that the air is effective along its entire course of flow to cause a substantial transfer of solvent and moisture in the desired directions between the air and film. Following such contact with a moist atmosphere, the film may be heated to the lower portion of its thermoplastic range, to stabilize the film in its coherent but porous state, and to dispel any residual volatile components and moisture from the film. This heating may be effected by passing over the film a second flow of air, heated to a proper temperature, this air preferably being passed over the film in countercurrent relation in the same manner as the first air stream.

Ionization Pump

A process described by *T.W. Snouse; U.S. Patent 3,601,503; August 24, 1971* involves an ionization process vacuum pump. Ions produced in the pump are accelerated to energies sufficiently high to enable them to pass through a thin membrane and thus out of the system being evacuated. The membrane prevents return of low energy gas molecules from the high pressure side of the pump.

Referring to Figure 8.6 there is shown a vacuum pump incorporating an evacuable envelope **11** adapted to contain gas molecules; a transparent anode **12** as a means to transfer energy to the ions and electrons participating in the ionization process, and a means **13** for allowing energetic ions to pass from the discharge region **14** of the pump. Envelope **11** is rectangular both from the front and side and may be of any dimension consistent with the considerations of good vacuum conductance and of economy in the production of magnetic field in the discharge region. Envelope **11** is made of gas impervious material, for example, stainless steel. At the top it is apertured for receiving, in vacuumtight manner, a high voltage feedthrough **15** which is in turn connected to the anode **12**. The envelope is adapted to be connected in vacuumtight manner to a structure to be evacuated through the intermediary of a high conductance passage **16** and a vacuum flange **17**, and is also adapted to be connected in the same manner by means of conductance **18** and flange **19** to a backing vacuum pump (not shown).

FIGURE 8.6: THIN MEMBRANE IONIZATION PUMP

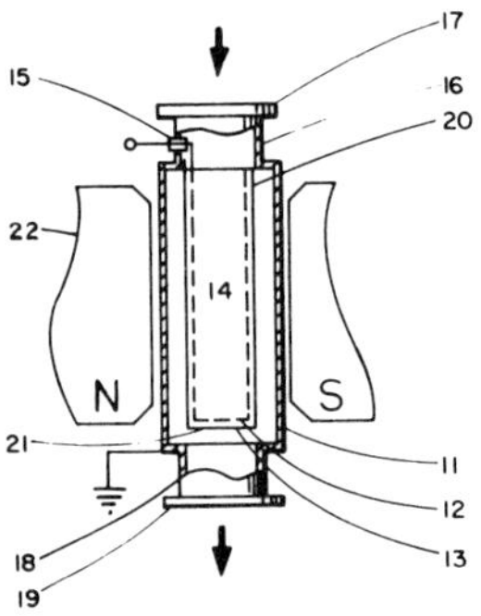

Source: T.W. Snouse; U.S. Patent 3,601,503; August 24, 1971

The working pressure of the backing pump is typically 10 to 10^5 decades higher than that of the membrane pump; commercial ion pumps or diffusion pumps would be suitable. The cathode **13** is composed of any of several materials known to have a low vapor pressure at elevated temperature in vacuum and capable of being formed into self-supporting hole-free membranes of thicknesses 400 A (1.6 microinches) or less. Materials suitable for this are carbon or aluminum oxide, with aluminum oxide as the preferred choice. It is suggested that the aluminum oxide be formed on an integral grid of aluminum for strength and rigidity.

This is done as follows. An aluminum member is shaped into a rectangular can with an open top and then etched to form a multitude of small (about 3⁄16 inch square), thin (about 0.001 inch thick) areas on the sides **20** and bottom **21**. These areas are closely and regularly placed to leave a gridwork of thicker aluminum as a supporting structure. The thinned areas are then treated by any of several methods, e.g., according to that of L. Harris, *J. Opt. Soc. Am.*, 45, 27 (1955) which forms an aluminum oxide membrane. The cathode **13** is joined electrically and physically in a vacuumtight manner to the envelope **11** near the top. This seal may be effected in any of several ways, as for example, by means of an expansible inside collar. The envelope (and by extention, the cathode) is connected to ground. A magnet or magnets **22** providing a field strength of 1,000 gauss, more or less, are placed flush with the outside walls of the envelope.

The anode **12** is a single cell and is constructed of very fine wires strung at large spacing over a supporting framework. It is made as nearly transparent as possible to afford the longest possible electron paths. It is hollow for the same purpose and for the purpose of placing the maximum feasible volume between the cathode membranes **13** at a high potential such that the majority of the ions are formed in this region.

The pump is operated in the following manner. It and the system to be evacuated are pumped to an intermediate pressure, typically 10^{17} to 10^{110} torr by means of other vacuum pumps such as suitable trapped mechanical and diffusion pumps, or by sorption and sputter ion pumps. The pump is bypassed during pump-down to avoid mechanical stress on the membrane by the significant force differentials found at higher pressures. When the desired forevacuum is reached the bypass is closed and a high voltage typically on the order of a positive potential of 15 to 20 kv DC and produced by an external power supply (not shown), is applied to the anode **12** by means of the feed-through **15**.

This voltage initiates a Penning-type discharge one result of which is the acceleration of ions toward the cathode. Most of these ions possess energies sufficient to enable them to pass completely through the membrane [A. Van Wijngaarden and H.E. Duckworth, *Can. J. Phys.,* 40, 1749 (1962), T.W. Snouse, *J. Appl. Phys. Lttrs.* 5, 122, (1964)]. Thus the ions are completely removed from the pumping volume and are later disposed of by the backing vacuum pump. Gas molecules existing in the backing pump, on the other hand, are prevented from entering the membrane pump because their energy is insufficient to allow their direct passage through the membrane.

Thus a net pumping action results, extending to very low pressures. The ultimate pressure will be limited by extinction of the discharge, or, if the backing vacuum is not good, by the slow diffusion of gases back through the membrane.

This pumping mechanism is particularly suited to the lighter gases such as hydrogen and methane which are principal residual gases found in extreme high vacuum systems. It is not well suited to gases heavier than carbon dioxide and sources of such gases such as elastomeric seals and vacuum oils and greases should not be employed. Additionally, both sides of the cathode membrane are subject to erosion by the bombarding ions.

Since the number of the ions is directly proportional to pressure, it is prudent, in terms of long membrane life, to operate the pump at the lowest pressures possible where membrane lifetimes on the order of a year in continuous operation are to be expected. Aside from this consideration, there is no reason why the pump could not be operated at much higher pressures.

COMPANY INDEX

The company names listed below are given exactly as they appear in the patents, despite name changes, mergers and acquisitions which have, at times, resulted in the revision of a company name.

INVENTOR INDEX

U.S. PATENT NUMBER INDEX

NOTICE

PRACTICE OF DESALINATION 1973

R. Bakish — Editor

This down-to-earth practical book is the first publication developed under the auspices of the St. Croix Corrosion Installation West Indies Laboratory, College of Science and Engineering, Fairleigh Dickinson University.

This particular volume is based upon a series of practical instructional papers presented as a special topics course by various authorities at St. Croix, U.S. Virgin Islands, during December 1971. This course was sponsored jointly by Fairleigh Dickinson University and the Association of Carribean Desalination Plants Owners and Operators.

The book therefore reflects the present state of the art of the American desalination industry in both its underlying principles and their applications which should remain valid for some time to come. A complete table of contents follows:

Profusely illustrated, this book is a valuable guide for professional engineers, technicians, and operators; also for business and municipal managements.

ISBN 0-8155-0478-0 **273 pages**

WATER-SOLUBLE RESINS AND POLYMERS 1976 Technology and Applications

by Yale L. Meltzer

Chemical Technology Review No. 57

An intense competitive struggle for markets has developed among water-soluble synthetic polymers, the semisynthetic cellulose ethers and starch derivatives, and natural gums.

Modified starches hold an advantage over synthetics and natural gums, because they are less expensive. Starch products, however, can cause stream pollution in various industrial applications, while the true synthetics, with their lower BOD, produce considerably less polluting effluents.

Strongly favoring the synthetics, moreover, is the great versatility which they have displayed in their applications. In addition, the industrialized countries have developed elaborate marketing and product information systems for these substances.

Production of the natural gums is concentrated in the developing countries of Africa and Asia. Government officials there have expressed grave concern over the loss of markets to the synthetics.

Starch is obtained in quantity in many parts of the world. Corn starch holds the biggest share and the USA is its largest producer.

Over 200 patent-based manufacturing and end-use processes are described in this book. What makes it most valuable are the market survey data that precede the technological information. A much shortened, partial table of contents follows here. Chapter headings and some important subtitles are given. Numbers in parentheses indicate the number of processes per topic.

ISBN 0-8155-0601-5 **371 pages**

INDUSTRIAL WATER PURIFICATION 1974

by Louis F. Martin

Pollution Technology Review No. 16

The Federal Water Pollution Control Act (now made into law) will have far-reaching effects on all industry. The pretreatment standards, about to be published as a result of this act, will exert increasing constraints on industrial process design and expansion plans in the years ahead.

The *Guidelines for the Pretreatment of Discharges of Publicly-Owned Works* prohibit the introduction of industrial pollutants which would pass through a municipal facility "inadequately treated," therefore, all existing industrial facilities and new plant designs must include specific downstream water purification processes, before the effluent is sewered or discharged into public systems.

This book describes over 160 recently devised processes for the treatment of contaminated industrial waters.

A partial and condensed table of contents follows here. Numbers in parentheses indicate the number of processes per topic, chapter headings are given, followed by examples of important subtitles.

While conventional separation processes for solid-liquid and water-oil effluents are applicable to many industrial streams, the process literature clearly highlights the areas of greatest immediate concern as shown by this table of contents:

ISBN 0-8155-0554-X **300 pages**

COFFEE SOLUBILIZATION 1975
Commercial Processes and Techniques

by Nicholas D. Pintauro

Food Technology Review No. 28

Soluble or instant coffee is brewed coffee that has been dehydrated. The water is removed under the most carefully controlled conditions. The color and particle structure of the dry residue are made to resemble roasted and ground coffee. The finished product dissolves with ease in hot water, ideally without foam or scum. The resulting hot solution is intended to have the color, aroma and flavor of freshly brewed coffee. To achieve this goal, even the original roasting of the coffee beans has to be modified.

Technically it is a most difficult task to dehydrate an aqueous extract of any aromatic natural product without degradation of the essential oils and other chemical changes that affect color, taste and consumer acceptance.

All these difficulties become even more complex in the manufacture of decaffeinated soluble coffee. The caffeine is usually removed from the green coffee beans by solvent extraction which also removes important oils and flavor constituents. Yet decaffeinated coffee powder seems to be the fastest growing product of this industry.

This book is based on U.S. patents and thus condenses the enormous amounts of basic research, technological developments and engineering achievements associated with the manufacture of soluble coffee powders. A much shortened table of contents follows. Numbers in parentheses indicate the number of processes per topic. Chapter headings and some important subtitles are given here.

ISBN 0-8155-0598-1 **256 pages**

FRUIT AND VEGETABLE JUICE PROCESSING 1975

by J. K. Paul

Food Technology Review No. 21

The large market for fruit and vegetable juices, because of their relatively low cost and high nutritional value, coupled with modern diet appeal, accounts for the sustained interest in these products and the continued development of improved methods of technology.

It is a primary object of this technology to prepare fruit juices, vegetable juices and related beverages that are able to maintain their natural flavor and aroma characteristics under ordinary, unspecified storage conditions over prolonged periods of time.

Considerable progress has been made in this direction: practically every process in this book is oriented toward the ultimate goal of every juice manufacturer—to retain natural flavors and aromas and to retard indefinitely the formation of unnatural flavors, aromas or colors.

Gone are all those preservatives of the past which, while stabilizing a beverage, also introduced undesirable off-flavors. Approved modern additives not only maximize stability, they also enhance the natural flavors. How to use them in connection with advanced extraction and concentration processes, which were specially designed for juice processing, is the know-how of this book: 164 patent-based processes are described.

A partial and condensed table of contents follows here. Chapter headings are given, followed by examples of important subtitles. Numbers in parentheses indicate the numbers of processes per topic.

ISBN 0-8155-0565-5 **277 pages**

POLLUTION DETECTION AND MONITORING HANDBOOK 1974

by Marshall Sittig

Environmental Technology Handbook No. 1

This handbook contains methods for the detection and monitoring of pollutants in industrial effluents, notably air and water.

Such methods are prerequisites for any kind of management of environmental quality. Of equal importance are the data handling systems that must be used to collect and interpret the measurements, regardless of whether manual or automated identifying and monitoring techniques are chosen.

Generally the most important single factor in determining whether manual or automated methods of analysis and level monitoring should be used, is the required frequency interval. In the manual approach costs vary almost directly with the measurement frequency. At monitoring intervals oftener than once a day, or when a continuous watch must be maintained, installation of automatic sensing is more economical, provided reliable automatic sensors are available.

For meaningful control and prevention of pollution the accepted standards and parameters for air and water quality must be known, whether they have been legislated or not. Acceptable levels for each type of impurity from methylmercury to dissolved chlorides or carbon particles must be known by the industrialist or public health official, and this book, with its 1633 references makes a massive attempt to communicate such levels.

As a guideline for the industrial user, data on the toxicity of the various pollutants are included. Such data on the deleterious effects were taken from various government publications. This book is not a treatise of pharmacology or industrial toxicology, but a guide to the accurate evaluation of pollutant levels in a given effluent or ambient substrate.

Whenever applicable, the name of the pollutant is followed by chemical and other information. After this come directions for sampling in air or water or both, as the case may be. Thereafter are given qualitative and quantitative analytical methods suitable for identification and measurement of the degree or level of pollution. This comprehensive information is followed by measurement techniques suitable for repeated or continuous monitoring. Preference is given to methods giving reproducible results in environmental quality control, as recommended by the EPA and other U.S. government agencies. This is followed in each case by many up-to-date references to government publications, patents, and journal articles.

Arrangement is encyclopedic. The book is thus a worthy companion to the author's POLLUTANT REMOVAL HANDBOOK available from the same publisher.

Pollution control is a serious business, and in the design of a monitoring program the specific objectives to be served must be clearly in mind from the very beginning. Otherwise the effort most probably will not result in an efficient and meaningful program. The length of the initial survey, selection of monitoring points, parameter coverage, and sampling frequency all depend on the specific objectives and the timeframe in which the objectives must be achieved. These variables also have a very significant impact on monitoring costs.

Monitoring for the purpose of long-term trend identification, evaluation of standards compliance, and total management of environmental quality must, of course, be a continuing program without end.

It is hoped sincerely that this book, together with its companion volume POLLUTANT REMOVAL HANDBOOK, will do much to establish and sustain such antipollution programs.

A partial and condensed table of contents follows here in which subheadings are indicated below for only the first few entries. The book contains a total of 88 subject entries arranged in an alphabetical and encyclopedic fashion. The subject name refers to the polluting substance, and the text underneath each entry tells how to measure and monitor pollution by said substance:

This book should prove to be very useful as an overall reference volume. The organization is clear, and the text is especially well arranged. The tables and figures are extensive and will be of value to students, engineers, and pollution control authorities.

ISBN 0-8155-0529-9

401 pages